Algebras, Rings and Modules

Lie Algebras and Hopf Algebras

Mathematical
Surveys
and
Monographs
Volume 168

Algebras, Rings and Modules

Lie Algebras and Hopf Algebras

Michiel Hazewinkel
Nadiya Gubareni
V. V. Kirichenko

American Mathematical Society
Providence, Rhode Island

2010 *Mathematics Subject Classification.* Primary 16W30, 05E05, 20C30, 16G20, 17B37, 20G42, 14L05, 81R50, 16W35, 17Bxx; Secondary 05E10, 58B34, 81R60, 81R12, 82C20.

For additional information and updates on this book, visit
www.ams.org/bookpages/surv-168

Library of Congress Cataloging-in-Publication Data

Hazewinkel, Michiel.
Algebras, rings, and modules : Lie algebras and Hopf algebras / M. Hazewinkel, N. Gubareni, V. V. Kirichenko.
p. cm.
Includes bibliographical references and index.
ISBN 978-0-8218-5262-0 (alk. paper)
1. Hopf algebras. 2. Lie algebras. I. Gubareni, Nadezhda Mikhailovna. II. Kirichenko, Vladimir V. III. Title.

QA613.8.H39 2010
512′.55—dc22

2010030618

∞ The paper used in this book is acid-free and falls within the guidelines
established to ensure permanence and durability.
Visit the AMS home page at http://www.ams.org/

10 9 8 7 6 5 4 3 2 1 15 14 13 12 11 10

Contents

Preface

This volume, the third in our series, is about Hopf algebras (mostly) and Lie algebras. It is independent of the first two volumes; though, to be sure, there are some references to them, just as there are references to other books and journals.

The first chapter is devoted to Lie algebras. It is a fairly standard concise treatment of the more established parts of the theory with the exceptions that there is a bit more emphasis on Dynkin diagrams (also pictorially) and that the chapter includes a complete treatment of the correspondence, initiated by Peter Gabriel, between representations of quivers whose underlying graph is a Dynkin diagram and representations of the Lie algebra with that Dynkin diagram, Gabriel, [**4**], [**3**]; Bernstein-Gel'fand-Ponomarev, [**1**]; Dlab-Ringel, [**2**]. The treatment is via the very elegant approach by Coxeter reflection functors of [**1**].

The remaining seven chapters are on Hopf algebras.

The first two of these seven are devoted to the basic theory of coalgebras and Hopf algebras paying special attention to motivation, history, intuition, and provenance. In a way these two chapters are primers[1] on their respective subjects. The remaining five chapters are quite different.

Chapter number four is on the symmetric functions from the Hopf algebra point of view. This Hopf algebra is possibly the richest structured object in mathematics and a most beautiful one. One aspect that receives special attention is the Zelevinsky theorem on *PSH* algebras. The acronym *PSH* stands for 'positive selfadjoint Hopf'. What one is really dealing with here is a graded, connected, Hopf algebra with a distinguished basis that is declared orthonormal, such that each component of the grading is of finite rank, and such that multiplication and comultiplication are adjoint to each other and positive. If then, moreover, there is only one distinguished basis element that is primitive, the Hopf algebra is isomorphic to **Symm**, the Hopf (and more) algebra of symmetric functions with the Schur functions as distinguished basis. Quite surprisingly the second (co)multiplication on **Symm**, which makes each graded summand a ring in its own right and which is distributive over the first one (in the Hopf algebra sense), turns up during the proof of the theorem, the Bernstein morphism. This certainly calls for more investigations.

The enormously rich structure of **Symm** is discussed extensively (various lambda ring structures, Frobenius, and Verschiebung morphims, Adams operations, ... etc. Correspondingly a fair amount of space is given to the big Witt vectors.

---

[1]As a rule a professional mathematician is well advised to stay away from 'primers'; we hope and believe that these are an exception.

Chapter five is on the representations of the symmetric groups; more precisely it is on the direct sum of the RS_n where S_n is the symmetric group on n letters and as an Abelian group RS_n is the Grothendieck group of (virtual) complex representations. The direct sum, with $RS_0 = \mathbf{Z}$ by decree, is given a multiplication and a comultiplication by using induction up from two factor Young subgroups and restriction down to such subgroups. The result is a *PSH* algebra.

That is not at all difficult to see: basically Frobenius reciprocity and the Mackey double coset formula do the job. The distinguished basis is formed by the irreducible representations and there is only one primitive among these, viz the trivial representation of the symmetric group on one letter. It follows that $RS = \bigoplus_{n=0}^{\infty} RS_n$ is a *PSH* algebra and that it is isomorphic to **Symm**.

It also follows that **Symm** itself is *PSH* (with one distinguished primitive). So far that was not clear. At the end of chapter four the situation was that if there were a *PSH* algebra with one distinguished primitive it would be isomorphic to **Symm**, making **Symm** itself also *PSH*.

The stumbling block in proving directly that **Symm** is *PSH* is positivity. There seems to be no direct proof, without going through the representation theory of the symmetric groups, of the fact that the product of two Schur functions is a positive linear combination of (other) Schur functions.

The question of positivity of multiplication and comultiplication, such a vexing matter in the case of **Symm**, becomes a triviality in the case of *RS*. Indeed, the product is describable as follows. Take a representation ρ of S_p and a representation σ of S_q and form the tensor product $\rho \otimes \sigma$ which is a representation of the Young subgroup $S_p \times S_q$ of S_{p+q} and induce this representation up to a representation of all of S_{p+q}. If the two initial representations are real (as opposed to virtual), so is their tensor product and the induced representation of S_{p+q}. Thus multiplication is positive. Comultiplication can be treated in a similar way, or handled via duality.

The isomorphism between **Symm** and *RS* is far more than just an isomorphism of Hopf algebras; it also says things regarding the second (co)multiplications, the various lambda ring structures and plethysms, ... ; see [**7**]. It won't be argued here, but it may well turn out to be the case that among the many Hopf algebras isomorphic to **Symm** the incarnation *RS* is the central one; see loc. cit.

Of course the fact that **Symm** and *RS* are isomorphic (as algebras) is much older than the Zelevinsky theorem; see [**8**], section I.7, for a classical treatment.

Chapter six is about two generalizations of **Symm** that have become important in the last 25 years or so: the Hopf algebra **QSymm** of quasi-symmetric functions and the Hopf algebra **NSymm** of non commutative symmetric functions. They are dual to each other and this duality extends the autoduality of **Symm** via a natural imbedding **Symm** $\subset$ **QSymm** and a natural projection **NSymm** $\longrightarrow$ **Symm**. Both Hopf algebras carry a good deal more structure. The seminal paper which started all this is [**5**], but there, and in a slew of subsequent papers, things were done over a field of characteristic zero and not over the integers as here.

It is somewhat startling to discover how many concepts, proofs, constructions, ... have natural non commutative and quasi analogs; and not rarely these analogs are more elegant than their counterparts in the world of symmetric functions.

Things can be generalized still further to the Hopf algebra *MPR* of permutations of Malvenuto, Poirier, and Reutenauer, [**9**]; [**10**]. This one is twisted autodual and this duality extends the duality between **NSymm** and **QSymm** via a natural imbedding **NSymm** $\longrightarrow$ *MPR* and a natural projection *MPR* $\longrightarrow$ **QSymm**. This is the subject matter of chapter seven. There are still further generalizations, complete with duality, [**6**], but the investigation of these has only just started. Like *MPR* they are Hopf algebras of endomorphisms of Hopf algebras.

Finally, chapter eight contains over fifteen relatively short and shorter outlines, heavily bibliographical, on the roles of Hopf algebras in other parts of mathematics and theoretical physics. Much of this material has not previously appeared in the monographic literature.

In closing, we would like to express our cordial thanks to our friends and colleagues (among them, we especially thank S.A.Ovsienko) who have read portions of preliminary versions of this book and offered corrections, suggestions and other constructive comments improving the entire text. We also give special thanks to A. Katkow and V. V. Sergeichuk who have helped with the preparation of this manuscript.

And last, but not least, we would like to thank a number of anonymous referees; several of whom made some very valuable suggestions (both for corrections and small additions). Needless to say any remaining inaccuracies (and there are always some) are the responsibility of the authors.

References

[1] L. N. Bernstein, I. M. Gel'fand and V. A. Ponomarev, *Coxeter functors and Gabriel's theorem*, Russiam math. Surveys **28** (1973), 17 - 32.

[2] Vlastimil Dlab and Claus Michael Ringel, *Indecomposable representations of graphs and algebras*, Amer. math. Soc., 1976.

[3] Peter Gabriel, *Unzerlegbare Darstellungen I*, Manuscripta mathematica **6** (1972), 71 - 103.

[4] Peter Gabriel, *Indecomposable representations II, Symp. Math. INDAM 11 (Rome. 1971)*, Acad. Press, 1973, 81 - 104.

[5] I. M. Gel'fand, Daniel Krob, Alain Lascoux, Bernard Leclerc, Vladimir S. Retakh and Jean-Yves-Thibon, *Noncommutative symmetric functions*, Adv. Math. **112** (1995), 218-348.

[6] Michiel Hazewinkel, *Hopf algebras of endomorphisms of Hopf algebras*, Arabian J. for Science and Engineering **33**(2C) (Dec. 2008), 239 - 272; arXiv: math.QA/0410364.

[7] Michiel Hazewinkel, *Witt vectors. Part 1*, in: M. Hazewinkel, ed., *Handbook of algebra. Volume 6*, Elsevier, 2009, 319 - 472.

[8] I. G. Macdonald, *Symmetric functions and Hall polynomials*, Clarendon Press, 1995.

[9] C. Malvenuto, and Chr. Reutenauer, *Duality between quasi-symmetric functions and the Solomon descent algebra*, J. of Algebra 177 (1994), 967-982.

[10] Stéphane Poirier and Christophe Reutenauer, *Algèbres de Hopf de tableaux*, Ann. Sci. Math. Québec **19** (1995), 79-90.

Michiel Hazewinkel, Bussum, Nov. 2009
Nadiya Gubareni, Częstochowa, Nov. 2009
Vladimir V. Kirichenko, Kiev, Nov. 2009

CHAPTER 1

Lie algebras and Dynkin diagrams

This chapter contains a short introduction to the theory of Lie algebras and their representations. Lie algebras are the best known class of non-associative algebras. This theory was initiated by the Norwegian mathematician Marius Sophus Lie (1842-1899) in connection with the study of the idea of infinitesimal (symmetry) transformations. The term "Lie algebra" and some more of the main terminology of this theory was introduced by Hermann Weyl (1885-1955). The early more remarkable results in the theory of complex Lie algebras were obtained by E. Cartan, F. Engel, and W. K. J. Killing in 1888-1894.

Although Lie algebras are often studied in their own right, historically they arose by studying Lie groups. A Lie group is a group that is also a manifold for which the group operations (multiplication and taking inverses) are smooth maps. At any point of the Lie group one can form the tangent space. The tangent space at the identity inherits a certain amount of structure making it a Lie algebra, the Lie algebra of the group. A most important aspect of Lie groups is that they turn up as groups of symmetries in particle physics. The correspondence between Lie algebras and Lie groups is used in several ways, including the classification of Lie groups. Any representation of a simple Lie algebra lifts uniquely to a representation of the corresponding compact simple simply connected Lie group, and conversely. Therefore the classification and description of Lie algebras is very important in the theory of Lie groups.

This chapter starts with the basic concepts and notions of the theory of Lie algebras. In section 1.3 solvable and nilpotent Lie algebras are considered. The most interesting theorem in this section is the Engel theorem which connects nilpotentness of a Lie algebra with ad-nilpotentness of its elements.

Section 1.4 is devoted to the study of the main properties of simple and semisimple Lie algebras. Here the important notions of the radical and Killing form of a Lie algebra are introduced and there are a number of theorems concerning the equivalence of various definitions of "semisimple".

The main properties of irreducible representations of finite dimensional Lie algebras are considered in section 1.5. Here the Weyl theorem is proved and also the Ado theorem for finite dimensional Lie algebras. The first theorem states that every finite dimensional representation of such a semisimple Lie algebra is completely reducible. The second theorem states that every finite dimensional Lie algebra L can be viewed as a Lie algebra of square matrices via the commutator bracket. In section 1.6 the Lie theorem is proved which allows one to choose a basis in a vector space V so that the matrices of all representations of given a solvable Lie algebra are upper triangular in this basis.

A detailed study of the representations of the Lie algebra $\mathfrak{sl}(2;k)$ is the subject matter of section 1.8.

A Lie algebra L is a non-associative algebra, but there exists an associative algebra $\mathfrak{U}(L)$ such that L can be embedded in $\mathfrak{U}(L)$, called its universal enveloping algebra. And it is indeed universal, being in a sense the smallest associative algebra containing L. This fundamental result is proved in sections 1.9 and 1.10. There the construction of the universal enveloping algebra of a Lie algebra L and a precise description of this algebra is given according to the Poincaré-Birkhoff-Witt theorem. Free Lie algebras are discussed in section 1.10.

Sections 1.11 - 1.15 are devoted to the classification of simple Lie algebras. This involves what are called **abstract root systems** and the corresponding Dynkin diagrams. The classification of semisimple finite dimensional Lie algebras over an algebraically closed field in terms of these Dynkin diagrams is given in section 1.15. And it is a most wonderful fact that the same diagrams of types A, B, C, D, E, F and G that classify quivers, as considered in [**42**], classify finite reflection groups, compact simple Lie groups and simple Lie algebras, as well.

In sections 1.16 - 1.19 we give a proof of the Gabriel theorem following Bernstein, Gel'fand, Ponomarev using the Weyl group of a quiver, Coxeter functors and the Coxeter transformations. The Gabriel theorem states that there is a bijection between representations of quivers and of Lie algebras of the same type.

We end this chapter with a presentation of some facts about Kac-Moody Lie algebras which can be considered as a generalization of semisimple finite dimensional Lie algebras.

Throughout this chapter, starting with section 1.4, except where otherwise specified, the base field k is considered to be algebraically closed and have characteristic 0, and all Lie algebras are finite dimensional over the field k.

1.1. Lie algebras. Definitions and examples

There are two ways to look at Lie algebras. One of them defines a Lie algebra from the formal algebraic point of view as a vector space with an anticommutative product satisfying a number of conditions. The other one considers Lie algebras from a geometrical point of view as the local picture of Lie groups. In this chapter Lie algebras are treated from the formal algebraic point of view.

Definition. A **Lie algebra** is a vector space L over a field[1] k (typically the real or complex numbers) together with a binary operation $L \times L \to L$, denoted by $(x, y) \mapsto [x, y]$ and called the **Lie bracket** or **commutator** of x and y, which satisfies the following properties:

1. **Bilinearity.**

$$[ax + by, z] = a[x, z] + b[y, z]$$

$$[x, ay + bz] = a[x, y] + b[x, z]$$

for all $a, b \in k$, and $x, y, z \in L$.

2. **Anticommutativity.**

$$[x, x] = 0 \quad \text{for all } x \in L.$$

[1]More generally one considers Lie algebras over any commutative, associative, unital base ring A, especially the integers. Often in such a context the underlying A-module is assumed to be free.

3. **Jacobi identity**.

$$[[x, y], z] + [[z, x], y] + [[y, z], x] = 0$$

for all $x, y, z \in L$.

Note that the first and the second properties together, applied to $[x+y, x+y]$, imply the following property.

2′. **Antisymmetry** (or **anticommutativity**).

$$[x, y] = -[y, x]$$

for all $x, y \in L$.

Conversely, if char$k \neq 2$ then it is easy to verify (putting $x = y$) that property 2′ implies property 2.

Note also that the multiplication in L represented by the Lie bracket is in general not associative, i.e. $[x, [y, z]] \neq [[x, y], z]$ for some $x, y, z \in L$. So Lie algebras are examples of non-associative algebras.

Two elements x, y of a Lie algebra L are said to **commute**, if $[x, y] = 0$. A Lie algebra is called **commutative** (or **Abelian**) if $[x, y] = 0$ for all $x, y \in L$.

A **Lie subalgebra** of a Lie algebra L is a vector subspace $H \subseteq L$ such that $[x, y] \in H$ for all $x, y \in H$. In particular, the vector subspace H is itself a Lie algebra with respect to the operations given in L.

If L is a finite dimensional vector space over a field k the Lie algebra L is called **finite dimensional**, otherwise it is called **infinite dimensional**.

There are several natural sources that provide important examples of Lie algebras.

Examples 1.1.1.
Some examples of Lie algebras are:

1. Three dimensional Euclidean space $\mathbf{R}^3$ with the Lie bracket given by $[u, v] = u \times v$, the cross product (or vector product) of the vectors $u, v \in \mathbf{R}^3$ defined as a vector w satisfying the following requirements:

• the length of w is equal to the product of the lengths of vectors u and v by the sine of the angle θ between them. i.e.

$$|w| = |u| \cdot |v| \sin\theta;$$

• w is orthogonal to both u and v;

• the orientation of the vector triple u, v, w is the same as that of the standard triple of basis vectors.

If u, v have coordinates (u_1, u_2, u_3) and (v_1, v_2, v_3) with respect to an orthogonal basis i, j, k in $\mathbf{R}^3$, then the cross product $u \times v$ can also be represented as the determinant of a matrix:

$$u \times v = \det \begin{pmatrix} i & j & k \\ u_1 & u_2 & u_3 \\ v_1 & v_2 & v_3 \end{pmatrix}.$$

It is easy to show that the cross product is anticommutative, distributive and satisfies the Jacobi identity, so that $\mathbf{R}^3$ together with vector addition and cross product forms a Lie algebra.

2. Any vector space L with the Lie bracket given by $[u,v]=0$ for all $u,v \in L$. This is an Abelian Lie algebra, and all Abelian Lie algebras are of course like this.

3. Let A be an associative algebra with operations of addition $+$ and multiplication. The multiplication symbol $\cdot$ is usually omitted and one writes ab rather that $a \cdot b$. Introduce in A a new operation of multiplication defined by

$$[a,b] = ab - ba,$$

called the **bracket** (or **commutator** (**difference**)) of a and b. Obviously, if A is a commutative algebra the bracket operation is trivial, i.e. $[a,b]=0$ for all $a,b \in A$. If A is not commutative, then it is easy to check, that

1) $[a,a]=0$;

2) $[[a,b],c]+[[b,c],a]+[[c,a],b]=0$

for all $a,b,c \in A$. Therefore with respect to the bracket operation $[\cdot,\cdot]$ A becomes a Lie algebra. So any associative algebra A with associative product ab can be made into a Lie algebra by the bracket $[x,y]$.

4. Let $A = \mathrm{End}_k(V)$ be the endomorphism algebra of a k-vector space V. Define in A a new operation $[x,y]=xy-yx$ for all $x,y \in A$. Then with respect to this operation A becomes a Lie algebra over a field k. In order to distinguish this new algebra structure from the original "composition of endomorphisms" one, it is denoted by $\mathfrak{gl}(V)$ and called the **general linear Lie algebra**. Any subalgebra of $\mathfrak{gl}(V)$ is called a **linear Lie algebra**.

5. Let $A = M_n(k)$ be a set of all $n \times n$ square matrices over a field k. Define $[\mathbf{A},\mathbf{B}] = \mathbf{AB} - \mathbf{BA}$, where $\mathbf{AB}$ is the usual matrix product. With respect to this operation A becomes a Lie algebra which is denoted by $\mathfrak{gl}(n,k)$.

As a vector space, $\mathfrak{gl}(n,k)$ has a basis consisting of the matrix units e_{ij} for $1 \le i,j \le n$. (Here e_{ij} is the $n \times n$ matrix which has a 1 in the ij-th position and all other elements are 0.) It is easy to check that

$$[e_{ij}, e_{kl}] = \delta_{jk}e_{il} - \delta_{il}e_{kj},$$

where δ is the Kronecker delta.

The **Ado theorem** says that any finite dimensional Lie algebra over a field k of characteristic zero is linear, that is, it is isomorphic to a subalgebra of some Lie algebra $\mathfrak{gl}(n,k)$.

6. Let $A = \{\mathbf{X} \in M_n(k) \ : \ \mathrm{Tr}(\mathbf{X}) = 0\}$ be a set of all matrices with trace equal to zero. Then A is a Lie algebra with respect to the bracket operation introduced in the previous example. This algebra is denoted $\mathfrak{sl}(n,k)$. It is a subalgebra of $\mathfrak{gl}(n,k)$ and it is called the **special linear Lie algebra**.

As a vector space, this algebra has a basis consisting of the e_{ij} for $i \neq j$ together with $e_{ii} - e_{i+1,i+1}$ for $1 \le i \le n-1$.

For example, the Lie algebra $\mathfrak{sl}(2;k)$ of 2×2 matrices of trace 0 has a basis given by the three matrices:

$$e=\begin{pmatrix}0&1\\0&0\end{pmatrix};\quad f=\begin{pmatrix}0&0\\1&0\end{pmatrix};\quad h=\begin{pmatrix}1&0\\0&-1\end{pmatrix}$$

with relations $[e,f]=h$, $[h,f]=-2f$, $[h,e]=2e$.

This Lie algebra plays a very important role in physics. But in physics this algebra over the complex numbers is usually considered with a different basis, viz. the basis consisting of the (so-called) Pauli matrices:

$$\sigma_1=\begin{pmatrix}0&1\\1&0\end{pmatrix},\quad \sigma_2=\begin{pmatrix}0&-i\\i&0\end{pmatrix},\quad \sigma_3=\begin{pmatrix}1&0\\0&-1\end{pmatrix}.$$

If we divide each of these matrices by i and write the new matrices:

$$I=\begin{pmatrix}0&-i\\-i&0\end{pmatrix},\quad J=\begin{pmatrix}0&-1\\1&0\end{pmatrix},\quad K=\begin{pmatrix}-i&0\\0&i\end{pmatrix}$$

the following relations result between these matrices:

$$IJ=-JI=K,\quad JK=-KJ=I,\quad KI=-KI=J,\quad I^2=J^2=K^2=-1$$

which is precisely the multiplication table of the quaternions.

7. Let

$$\mathfrak{b}(n,k)=\{\mathbf{X}=(x_{ij})\in M_n(k)\ :\ x_{ij}=0\ \text{ whenever } i>j\}$$

be a set of all upper triangular matrices. This is a Lie algebra with the same Lie bracket operation as in $\mathfrak{gl}(n,k)$.

Another Lie subalgebra of $\mathfrak{gl}(n,k)$ is the algebra

$$\mathfrak{n}(n,k)=\{\mathbf{X}=(x_{ij})\in M_n(k)\ :\ x_{ij}=0\ \text{ whenever } i\geq j\}$$

of all strictly upper triangular matrices with the same Lie bracket operation as in $\mathfrak{gl}(n,k)$.

8. Let $A=\{\mathbf{X}\in M_n(k)\ :\ \mathbf{X}^T=-\mathbf{X}\}$ be a set of all skew symmetric matrices over a field k. Then A is a Lie algebra with respect to the bracket operation. It is denoted $\mathfrak{O}(n,k)$.

9. Let $A=\left\{\mathbf{X}\in M_2(\mathbf{R})\ :\ \mathbf{X}=\begin{pmatrix}*&*\\0&0\end{pmatrix}\right\}$ be the subset of all real 2×2 matrices with the second row 0. Then $A=\mathfrak{aff}(1)$ is the **affine Lie algebra of the line** with a basis given by two matrices $\mathbf{X}=\begin{pmatrix}1&0\\0&0\end{pmatrix}$, $\mathbf{Y}=\begin{pmatrix}0&1\\0&0\end{pmatrix}$, and bracket operation $[\mathbf{X},\mathbf{Y}]=\mathbf{Y}$.

10. Let $\mathbf{R}^4$ be real four dimensional Euclidean space with vectors $v=(x,y,z,t)$. Consider the **Lorentz inner product**

$$\langle v,v\rangle_L=x^2+y^2+z^2-t^2.$$

If $I_{3,1}=\mathrm{diag}(1,1,1,-1)$, then $\langle v,v\rangle_L=v^TI_{3,1}v$. Consider the subset $\mathfrak{l}_{3,1}$ of all real 4×4 matrices M with

$$M^TI_{3,1}+I_{3,1}M=0.$$

Then $\mathfrak{l}_{3,1}$ is a Lie algebra which is called the **Lorentz Lie algebra**.

11. Here is another important example of a Lie algebra that arises in quantum mechanics. The commutator relations between x, y and z as components of angular momentum form a Lie algebra for the three dimensional rotation group, which is isomorphic to SU(2) and SO(3):

$$[L_x, L_y] = i\hbar L_z, \quad [L_y, L_z] = i\hbar L_z, \quad [L_z, L_x] = i\hbar L_y.$$

12. Probably the oldest example of a Lie algebra structure is that given by the Poisson bracket. Let u, v be (real valued) differentiable functions of $2n$ variables $q = (q_1, ..., q_n)$ and $p = (p_1, ..., p_n)$ (and possibly an extra variable t (time)). Then the **Poisson bracket** is

$$\{p, q\} = \sum_{i=1}^{n} \left(\frac{\partial u}{\partial q_i}\frac{\partial v}{\partial p_i} - \frac{\partial u}{\partial p_i}\frac{\partial v}{\partial q_i}\right).$$

The Poisson bracket was introduced in the context of (classical) mechanics by Siméon Denis Poisson in 1809, long before Sophus Lie (1842-1899).

For that matter the Jacobi identity (for the Poisson bracket and certain generalizations) dates from 1862 (Carl Gustav Jacob Jacobi).

See [**78**], [**89**] for some more (free electronically available) information.

Two generalizations of the Poisson bracket play important roles in modern mathematics, mechanics, and physics: the Lie algebra of vector fields on a smooth manifold and the Lie algebra of functions on a symplectic manifold.

13. Let A be an algebra (not necessary associative) over a field k. A k-**derivation** of A is a k-linear mapping $D : A \longrightarrow A$ such that

$$D(ab) = D(a)b + aD(b) \quad \text{for all } a, b \in A.$$

The k-module of derivations carries a Lie algebra structure given by the commutator difference composition

$$[D_1, D_2] = D_1D_2 - D_2D_1.$$

Here $(D_1D_2)(a) = D_1(D_2(a))$.

Derivations are the algebraic version of vector fields in topology, analysis, and differential geometry.

14. Consider a vector space over k with basis L_n, c for $n \in \mathbf{Z}$. Take

$$[L_n, c] = 0, \quad \text{for all } n$$

$$[L_m, L_n] = (m - n)L_{m+n} + \delta_{m+n,0}\frac{m^3 - m}{12}c.$$

This defines a Lie algebra called the **Virasoro algebra** whose representations are of considerable importance in both mathematics and physics.

See [**51**], [**90**] for some more (free electronically accessible) information.

15. Killing the center spanned by c in the Virasoro algebra leaves a (quotient) Lie algebra with the basis L_n, $n \in \mathbf{Z}$, and the Lie bracket

$$[L_m, L_n] = (m - n)L_{m+n}.$$

It can be viewed as the Lie algebra of polynomial vector fields on the circle,

$$L_m = -z^{m+1}\frac{d}{dz}, \quad \text{for } n \in \mathbf{Z}.$$

This Lie algebra is known as the **Witt algebra**[2], named after Ernst Witt (1911-1991). The complex Witt algebra was first defined by E. Cartan (1909), and its analogues over finite fields were studied by E. Witt in the 1930s.

In turn the L_n, $n \geq 0$, span a Lie subalgebra of the Witt algebra. This is the Lie algebra of polynomial vector fields on the line which are zero at the origin.

For some more information, see [**93**], [**91**].

The authors hope that these examples illustrate that Lie algebras tend to turn up anywhere in mathematics and that their significance goes far beyond infinitesimal symmetry transformations.

1.2. Ideals, homomorphisms and representations

Much of the structure theory of groups and rings has suitable analogues for Lie algebras. As usual for algebraic structures, for an arbitrary Lie algebra L there are notions of ideal, quotient Lie algebra, homomorphism and isomorphism.

Definition. A subspace $\mathcal{I}$ of a Lie algebra L is called an **ideal** of L if

$$[x, y] \in \mathcal{I} \quad \text{for all } x \in L \quad \text{and } y \in \mathcal{I}.$$

From property $3'$ of Lie algebra it follows that there is no difference between left and right ideals and any ideal of L is two-sided.

An ideal of a Lie algebra is always a Lie subalgebra. But a Lie subalgebra need not be an ideal. For example, the set of all strictly upper triangular matrices in $\mathfrak{gl}(n, k)$ is a subalgebra of $\mathfrak{gl}(n, k)$, but it is not an ideal for $n \geq 2$.

In the theory of Lie algebras ideals play the role which is played by normal subgroups in group theory and two-sided ideals in ring theory.

Examples 1.2.1.

1. The trivial examples of ideals of a Lie algebra L are 0 (a vector space containing only the zero vector) and the Lie algebra L itself. They are called the **trivial ideals** of L.

2. Let

$$Z(L) = \{z \in L \ : \ [x, z] = 0 \text{ for all } x \in L\}.$$

This is called the **center** of the Lie algebra L. From the Jacobi identity it follows that $Z(L)$ is an ideal of L. Obviously, L is Abelian if and only if $Z(L) = L$.

In example 1.1.1(13) the one dimensional vector space spanned by c is the center of the Virasoro algebra.

3. $\mathfrak{sl}(n, k)$ is an ideal of the Lie algebra $\mathfrak{gl}(n, k)$. This follows from the fact that $\mathrm{Tr}[a, b] = 0$ for all $a, b \in \mathfrak{gl}(n, k)$.

Suppose $\mathcal{I}$ and $\mathcal{J}$ are ideals of a Lie algebra L. Then it is easy to show that $\mathcal{I} + \mathcal{J} = \{x + y \ : \ x \in \mathcal{I}, \ y \in \mathcal{J}\}$ and $\mathcal{I} \cap \mathcal{J}$ are ideals in L.

Proposition 1.2.2. *Let $\mathcal{I}$ and $\mathcal{J}$ be two ideals of a Lie algebra L. Then the Lie product*

$$[\mathcal{I}, \mathcal{J}] = \mathrm{Span}\{[x, y] \ : \ x \in \mathcal{I}, \ y \in \mathcal{J}\}$$

is again an ideal of L. In particular, $[L, L] \subset L$ is an ideal of L.

---

[2]There are several more general Lie algebras of derivations that also go under the name of Witt algebra, see e.g. example 1.2.6 below and [**93**].

Proof. Let $\mathcal{I}$ and $\mathcal{J}$ be two ideals of a Lie algebra L. By definition, $[\mathcal{I},\mathcal{J}]$ is a subspace in L. Because of condition $2'$, $[\mathcal{I},\mathcal{J}] = [\mathcal{J},\mathcal{I}]$. Let $x \in \mathcal{I}$, $y \in \mathcal{J}$ and $a \in L$. Then, by the Jacobi identity,

$$[[x,y],a] + [[y,a],x] + [[a,x],y] = 0.$$

As $[[y,a],x] \in [\mathcal{J},\mathcal{I}] = [\mathcal{I},\mathcal{J}]$ and $[[a,x],y] \in [\mathcal{I},\mathcal{J}]$, it follows that $[[x,y],a] \in [\mathcal{I},\mathcal{J}]$.

A general element $u \in [\mathcal{I},\mathcal{J}]$ has the form $u = \sum c_{ij}[x_i,y_j]$, where $x_i \in \mathcal{I}, y_j \in \mathcal{J}$ and the c_{ij} are scalars. Then for any $v \in L$ there results

$$[u,v] = [\sum c_{ij}[x_i,y_j],v] = \sum c_{ij}[[x_i,y_j],v],$$

where $[[x_i,y_j],v] \in [\mathcal{I},\mathcal{J}]$ as shown above. Hence $[u,v] \in [\mathcal{I},\mathcal{J}]$ and so $[\mathcal{I},\mathcal{J}]$ is an ideal in L. □

For any ideal $\mathcal{I}$ of a Lie algebra L one can construct the quotient Lie algebra $L/\mathcal{I}$ in the usual way. Consider the coset space $L/\mathcal{I} = \{x + \mathcal{I} : x \in L\}$ and define the multiplication by $[x+\mathcal{I}, y+\mathcal{I}] = [x,y] + \mathcal{I}$. This multiplication is well defined. Indeed, suppose $x_1 + \mathcal{I} = x_2 + \mathcal{I}$, then $x_1 - x_2 \in \mathcal{I}$. So

$$[x_1,y] - [x_2,y] = [x_1 - x_2, y] \in \mathcal{I}$$

since $\mathcal{I}$ is an ideal. So

$$[x_1,y] + \mathcal{I} = [x_2,y] + [x_1 - x_2,y] + \mathcal{I} = [x_2,y] + \mathcal{I}.$$

All properties of a Lie algebra now follow easily and so $L/\mathcal{I}$ is a Lie algebra which is called the **quotient Lie algebra** of L by $\mathcal{I}$.

Note that the Witt algebra is the quotient Lie algebra of the Virasoro algebra by the ideal spanned by c.

Definition. Suppose L, H are Lie algebras. A linear map $\varphi : L \to H$ is called a **Lie homomorphism** (Lie algebra morphism) if

$$\varphi([x,y]) = [\varphi(x), \varphi(y)] \quad \text{for all } x, y \in L.$$

The composition of Lie homomorphisms is again a homomorphism, and the Lie algebras over a field k, together with these morphisms, form a category. It is easy to show the following:

Lemma 1.2.3. *Let $\varphi : L \to H$ be a homomorphism of Lie algebras, then*

$$\mathrm{Ker}(\varphi) = \{x \in L \mid \varphi(x) = 0\}$$

is an ideal in L and

$$\mathrm{Im}(\varphi) = \{\varphi(x) \mid x \in L\}$$

is a subalgebra of H.

Definition. A morphism of Lie algebras $\varphi : L \to H$ is called a **monomorphism** if $\mathrm{Ker}(\varphi) = 0$ and φ is called an **epimorphism** if $\mathrm{Im}(\varphi) = H$.

A homomorphism is called an **isomorphism** if it is a monomorphism and an epimorphism simultaneously. An isomorphism of a Lie algebra to itself is an **automorphism**.

There is the usual correspondence between ideals and homomorphisms: any homomorphism φ gives rise to an ideal $\mathrm{Ker}(\varphi)$, and any ideal $\mathcal{I}$ in L is associated with the canonical projection $\pi : L \to L/\mathcal{I}$ which is defined by $\varphi(x) = x + \mathcal{I}$. It is easy to verify the following fundamental theorem of Lie algebras, which is the same as in ring theory:

Theorem 1.2.4.
1. *Suppose $\varphi : L \to H$ is a morphism of Lie algebras. Then $L/\mathrm{Ker}(\varphi) \simeq \mathrm{Im}(\varphi)$.*
2. *If $\mathcal{I}$ is an ideal in L such that $\mathcal{I}$ is contained in $\mathrm{Ker}(\varphi)$, then there is a unique morphism $\psi : L/\mathcal{I} \to H$ such that $\varphi = \psi\pi$, where π is the canonical projection $L \to L/\mathcal{I}$.*
3. *Suppose $\mathcal{I}$ and $\mathcal{J}$ are ideals of a Lie algebra L such that $\mathcal{I} \subset \mathcal{J}$, then $\mathcal{J}/\mathcal{I}$ is an ideal of $L/\mathcal{I}$ and $(L/\mathcal{I})/(\mathcal{J}/\mathcal{I}) \simeq L/\mathcal{J}$.*
4. *If $\mathcal{I}$ and $\mathcal{J}$ are ideals of a Lie algebra L, then $(\mathcal{I}+\mathcal{J})/\mathcal{J} \simeq \mathcal{I}/(\mathcal{I} \cap \mathcal{J})$.*

The Lie algebra

$$[L, L] = \{\sum a[x, y] \ : \ a \in k; \ x, y \in L\}$$

is called the **derived algebra** of a Lie algebra L. By proposition 1.2.2, $[L, L]$ is an ideal in L. Obviously, L is Abelian if and only if $[L, L] = 0$.

For any algebra L the quotient algebra $L/[L, L]$ is Abelian, i.e. a product of any two elements is zero. The following lemma shows that the ideal $[L, L]$ is the minimal ideal with the property that the quotient Lie algebra $L/\mathcal{I}$ is Abelian.

Lemma 1.2.5. *Let $\mathcal{I}$ be an ideal of a Lie algebra L. Then $L/\mathcal{I}$ is Abelian if and only if $\mathcal{I}$ contains the derived algebra $[L, L]$.*

Proof. By definition, a Lie algebra $L/\mathcal{I}$ is Abelian if and only if

$$[x + \mathcal{I}, y + \mathcal{I}] = [x, y] + \mathcal{I} = \mathcal{I},$$

which implies that $[x, y] \in \mathcal{I}$ for all $x, y \in L$. Since $\mathcal{I}$ is an ideal, and so a subspace of L, this means that $[L, L] \subseteq \mathcal{I}$. □

Definition. A Lie algebra L is called **simple** if L has no ideals except 0 and itself, and if moreover $[L, L] \neq 0$.

If L is a simple algebra then $Z(L) = 0$ and $L = [L, L]$.

Definition. Let L be a Lie algebra over a field k, and let V be a vector space over the field k. A Lie homomorphism

$$\varphi : L \to \mathfrak{gl}(V, k)$$

is called a **representation** of L. A representation φ on V is also called an **action** of L on V.

This means that $\varphi([x, y]) = [\varphi(x), \varphi(y)]$. So if $\varphi(x) = S_x \in \mathfrak{gl}(V)$, then $S_{[x,y]} = [S_x, S_y] = S_x \circ S_y - S_y \circ S_x$. If $\varphi(x) = 0$ for all $x \in L$, the representation φ is called **trivial**. If $\mathrm{Ker}(\varphi) = 0$, i.e. φ is a monomorphism, the representation φ is called **faithful**.

A subspace W of a representation space V is called **invariant**, or a **subrepresentation**, if $\varphi(x)(W) \subset W$ for all $x \in L$. The direct sum of two representations φ, ψ in V, W is the vector space $V \oplus W$ with the action $(x, y) \mapsto (\varphi(x), \psi(y))$.

Two representations $\varphi : L \to \mathfrak{gl}(V,k)$ and $\psi : L \to \mathfrak{gl}(W,k)$ are called **equivalent** if there exists a k-vector space isomorphism $\theta : V \to W$ such that $\psi(x) = \theta \circ \varphi(x) \circ \theta^{-1}$ for all $x \in L$.

Definition. A representation φ is called **irreducible** if it has no invariant subspaces except V and 0. A representation φ is called **decomposable** if it is a direct sum of (nontrivial) invariant subspaces.

Example 1.2.6.
Let k be a field and $\operatorname{char} k = p > 0$. Let A be an associative commutative algebra over k which is generated by an element x which satisfies the equality $x^p = 1$. Let $\delta_i : A \to A$ be the derivation defined by $\delta_i(x) = x^i$ for $i = 0, 1, ..., p-1$. More precisely, $\delta_i(1) = 0$, $\delta_i(x^j) = jx^{i+j-1}$ where the exponent is taken modulo p. Then $[\delta_i, \delta_j] = (i-j)\delta_{i+j}$ for $i = 0, 1, ..., p-1$. This is a p-dimensional Lie algebra which is called the **Witt algebra**. See also example 1.1.1(13).

There is a very important representation of a Lie algebra on itself which is called the **adjoint** representation, and denoted by "ad". This is the homomorphism

$$\operatorname{ad} : L \to \mathfrak{gl}(L,k)$$

defined by

$$(\operatorname{ad} x)(y) = [x,y] \quad \text{for all } x, y \in L. \tag{1.2.7}$$

Obviously, ad is a linear transformation. It is needed to show that ad is a Lie homomorphism, i.e. $[\operatorname{ad} x, \operatorname{ad} y](z) = \operatorname{ad}[x,y](z)$. Now, using the Jacobi identity there results

$$\begin{aligned} [\operatorname{ad} x, \operatorname{ad} y](z) &= \operatorname{ad} x \circ (\operatorname{ad} y)(z) - \operatorname{ad} y \circ (\operatorname{ad} x)(z) \\ &= (\operatorname{ad} x)([y,z]) - (\operatorname{ad} y)([x,z]) = [x,[y,z]] - [y,[x,z]] \\ &= [x,[y,z]] + [[x,z],y] = [[x,y],z] = \operatorname{ad}[x,y](z). \end{aligned}$$

Thus the morphism ad is a representation and it is called the **regular** (or **adjoint**) **representation** of L. The adjoint representation plays a fundamental role in the theory of Lie algebras and Lie groups.

Since

$$\operatorname{Ker}(\operatorname{ad}) = \{x \in L \ : \ \operatorname{ad} x = 0\} = \{x \in L \ : [x,y] = 0 \text{ for all } y \in L\},$$

it follows that $\operatorname{Ker}(\operatorname{ad}) = Z(L)$; i.e. the kernel of the regular representation is the center of L. Ideals of L are the same as ad-invariant subspaces. We write $\operatorname{ad} L$ for the **adjoint Lie algebra**, the image of L under ad in $\mathfrak{gl}(L,k)$. Thus $\operatorname{ad} L \simeq L/Z(L)$. If L is a simple Lie algebra, then $Z(L) = 0$ and so ad is a monomorphism, i.e., L is isomorphic to a subalgebra of $\mathfrak{gl}(L,k)$. Thus there results the following statement:

Proposition 1.2.8. *Any simple Lie algebra is isomorphic to a linear Lie algebra.*

In fact a much stronger statement holds, which states that any finite dimensional Lie algebra can be embedded in some $\mathfrak{gl}(V,k)$, where V is finite dimensional vector space. This is the Ado-Iwasawa theorem. A proof for the case when the field k has characteristic zero will be given below, in section 1.4.

If L is a Lie algebra, then a derivation on L is a linear map $\delta : L \to L$ that satisfies the equality

$$\delta([x, y]) = [\delta(x), y] + [x, \delta(y)] \tag{1.2.9}$$

for all $x, y \in L$.

Therefore the adjoint action $\operatorname{ad} x$ is a derivation of L and it is called an **inner derivation**. All other derivations are called **outer**.

1.3. Solvable and nilpotent Lie algebras

From lemma 1.2.5 it follows that the derived algebra $[L, L]$ is the smallest ideal of a Lie algebra L with an Abelian quotient. By proposition 1.2.1 for a Lie algebra L, one can form the following **derived series** of L:

$$L \supseteq L^{(1)} \supseteq L^{(2)} \supseteq \cdots \supseteq L^{(n)} \supseteq \cdots$$

where

$$L^{(0)} = L, \quad L^{(1)} = [L^{(0)}, L^{(0)}], \quad L^{(n+1)} = [L^{(n)}, L^{(n)}] \quad \text{for } n \geq 2.$$

Definition. A Lie algebra L is called **solvable** if $L^{(n)} = 0$ for some $n \geq 1$. The minimal such n is called the (**solvability**) **length** of L.

Remark 1.3.1. Any Lie group gives rise to a Lie algebra (structure on the tangent space at the identity). A connected Lie group is solvable if and only if its Lie algebra is solvable.

Examples 1.3.2.

1. Any Abelian Lie algebra is solvable.
2. A simple Lie algebra is not solvable.
3. The algebra $\mathfrak{b}(n, k)$ of $n \times n$ upper triangular matrices (non zero diagonals are permitted) over a field k is solvable.

Proposition 1.3.3. *Let L be a Lie algebra.*

1. *A subalgebra and a quotient algebra of a solvable algebra L are solvable.*
2. *If $\mathcal{I}$ is a solvable ideal of L and the quotient algebra $L/\mathcal{I}$ is solvable, then L is also solvable.*
3. *If $\mathcal{I}$ and $\mathcal{J}$ are solvable ideals of L, then so is $\mathcal{I} + \mathcal{J}$.*

Proof.

1. If K is a subalgebra of a Lie algebra L then by definition it follows that $K^{(n)} \subset L^{(n)}$. If $\varphi : L \to M$ is an epimorphism of algebras then, by induction on n, it is easy to show that $\varphi(L^{(n)}) = M^{(n)}$.

2. Let $\mathcal{I}^{(m)} = 0$ and $(L/\mathcal{I})^{(n)} = 0$. Applying statement 1 of this proposition to the canonical projection $\pi : L \to L/\mathcal{I}$ one obtains that $\pi(L^{(n)}) = 0$, that is, $L^{(n)} \subset \mathcal{I}$. Then from the equality $(L^{(i)})^{(j)} = L^{(i+j)}$ it follows that $L^{(n+m)} = 0$.

3. Theorem 1.2.4(4) states that $(\mathcal{I}+\mathcal{J})/\mathcal{J} \simeq \mathcal{I}/(\mathcal{I} \cap \mathcal{J})$. Now from statement 1 of this proposition it follows that $\mathcal{I}/(\mathcal{I} \cap \mathcal{J})$ is solvable and so is $(\mathcal{I} + \mathcal{J})/\mathcal{J}$. Now apply statement 2 to the ideals $\mathcal{I} + \mathcal{J}$ and $\mathcal{J}$ to obtain what is required. □

Recall that an associative algebra A is a **nil-algebra** if every element in it is nilpotent. This notion is useless for Lie algebras, because from the definition of a

Lie algebra it follows immediately that any Lie algebra is a nil-algebra. More useful is the notion of nilpotency.

Consider the **lower** (or **descending**) **central series** of a Lie algebra L defined by:

$$L \supseteq L^1 \supseteq L^2 \supseteq \cdots \supseteq L^i \supseteq \cdots$$

where

$$L^0 = L, \quad L^1 = [L, L], \quad L^2 = [L, L^1], \quad L^i = [L, L^{i-1}] \quad \text{for } i \geq 2.$$

Definition. A Lie algebra L is called **nilpotent** if $L^n = 0$ for some n. The minimal such integer n is called the **nilpotency length**.

Examples 1.3.4.
1. Any Abelian Lie algebra is nilpotent.
2. The Lie algebra $\mathfrak{n}(n,k)$ considered in Example 1.1.1(7) is nilpotent.

Note that $L^{(1)} = L^1$, $L^{(2)} = [L^{(1)}, L^{(1)}] \subset [L, L^1] = L^2$ and so on, i.e. $L^{(n)} \subset L^n$ for any n. So there results the following simple statement:

Proposition 1.3.5. *Every nilpotent Lie algebra is solvable.*

Note that the inverse statement is not true. There are solvable algebras which are not nilpotent. For example, the algebra $\mathfrak{b}(n,k)$ of all upper triangular matrices over a field k for $n \geq 2$ is solvable but not nilpotent.

Proposition 1.3.6. *Suppose L is a Lie algebra. Then*
1. *If L is nilpotent then all its subalgebras and quotient algebras are nilpotent.*
2. *If $L/Z(L)$ is a nilpotent algebra then so is L.*
3. *If L is nilpotent and $L \neq 0$, then $Z(L) \neq 0$.*

Proof.
1. If K is a subalgebra of a Lie algebra L then by definition it follows that $K^n \subset L^n$. If $\varphi : L \to M$ is an epimorphism of Lie algebra then, by induction on n, one finds that $\varphi(L^n) = M^n$.
2. Let $L^n \subset Z(L)$. Then $L^{n+1} = [L, L^n] \subset [L, Z(L)] = 0$.
3. This follows from the fact that the last term of the lower central series is contained in $Z(L)$. □

Lemma 1.3.7. *If L is a Lie algebra then for any i and j there is the inclusion* $[L^i, L^j] \subset L^{i+j+1}$.

Proof. This statement is proved by induction on j. Suppose the statement is true for $j-1$. For $i = 0$, by definition, $[L^0, L^j] = [L, L^j] = L^{j+1}$. For $i > 0$, by the Jacobi identity,

$$[L^i, L^j] = [L^i, [L, L^{j-1}]] \subset [L, [L^{j-1}, L^i]] + [L^{j-1}, [L^i, L]].$$

Since, by the induction hypothesis, $[L^{j-1}, L^i] \subset L^{i+j}$, it follows that

$$[L, [L^{j-1}, L^i]] \subset [L, L^{i+j}] = L^{i+j+1}.$$

Since $[L^i, L] = L^{i+1}$, again by the induction hypothesis,

$$[L^{j-1}, [L^i, L]] = [L^{j-1}, L^{i+1}] \subset L^{i+j+1}.$$

Therefore $[L^i, L^j] \subset L^{i+j+1}$. □

Lemma 1.3.8. *Let $\mathcal{I}$ be an ideal of a Lie algebra L. Suppose an element $x \in L$ is a product of n elements from L, such that, moreover, r elements of them belong to $\mathcal{I}$. Then $x \in \mathcal{I}^r$.*

Proof. Let $x = [y, z]$. Suppose $r = p + q$, where p is the number of factors in y and q is the number of factors in z which belong to $\mathcal{I}$. Applying induction on n one can assume that $y \in \mathcal{I}^p$ and $z \in \mathcal{I}^q$. Thus $x \in [\mathcal{I}^p, \mathcal{I}^q]$ and, by lemma 1.3.7, $x \in \mathcal{I}^r$. □

Proposition 1.3.9. *The sum of any two nilpotent ideals of a Lie algebra L is a nilpotent ideal.*

Proof. Suppose $\mathcal{I}$ and $\mathcal{J}$ are nilpotent ideals of a Lie algebra L, and $\mathcal{I}^n = 0$, $\mathcal{J}^m = 0$. Then $(\mathcal{I} + \mathcal{J})^{n+m} = 0$.

Indeed, if $x \in (\mathcal{I} + \mathcal{J})^{n+m}$ then it is a sum of elements from L such that each of them has n factors from $\mathcal{I}$ or m factors from $\mathcal{J}$. From lemma 1.3.8 it follows that $x = 0$. □

Proposition 1.3.10. *Any finite dimensional Lie algebra L has a unique maximal nilpotent ideal R.*

Proof. This is a corollary of the previous proposition. □

Definition. An endomorphism $x \in \operatorname{End}(V)$ is called **nilpotent** if $x^n = 0$ for some n. Let L be a Lie algebra. An element $x \in L$ is called **ad-nilpotent** if the endomorphism $\operatorname{ad} x$ is nilpotent.

Next is one of the main theorems in the theory of nilpotent Lie algebras. It asserts that nilpotence of a finite dimensional Lie algebra and ordinary nilpotence of operators on a vector space are identical concepts. This is the **Engel theorem**. To prove this theorem we consider some elementary facts concerning linear maps and Lie subalgebras of $\mathfrak{gl}(V)$.

Lemma 1.3.11. *If $x \in \mathfrak{gl}(V)$ is a nilpotent endomorphism then the endomorphism $\operatorname{ad} x$ is also nilpotent.*

Proof. Given an element $x \in \mathfrak{gl}(V)$ there are two endomorphisms $\lambda_x, \rho_x \in \operatorname{End} V$ defined by $\lambda_x(y) = xy$ and $\rho_x(y) = yx$. If x is nilpotent, i.e. $x^n = 0$, then λ_x and ρ_x are also nilpotent. Note that λ_x and ρ_x obviously commute. By the binomial theorem, the sum and difference of two commuting nilpotent elements are nilpotent. Since

$$(\operatorname{ad} x)(y) = xy - yx = \lambda_x(y) - \rho_x(y) = (\lambda_x - \rho_x)(y),$$

the endomorphism $\operatorname{ad} x$ is nilpotent. □

Lemma 1.3.12. *Suppose that K is an ideal of a Lie subalgebra L of $\mathfrak{gl}(V, k)$. Let*

$$W = \{v \in V \ : \ t(v) = 0 \text{ for all } t \in K\}.$$

Then W is an L-invariant subspace of V.

Proof. Let $w \in W$ and $y \in L$. It must be shown that $t(yw) = 0$ for all $t \in K$. But since K is an ideal of L there results $ty = yt + [t, y]$, where $[t, y] \in K$. Therefore

$$t(yw) = y(tw) + [t, y](w) = 0.$$

Definition. Let K be a subspace of a Lie algebra L. The set

$$N_L(K) = \{x \in L \ : \ [x, K] \subset K\} \tag{1.3.13}$$

is called the **normalizer** of K in L. The set

$$C_L(K) = \{x \in L \ : \ [x, K] = 0\} \tag{1.3.14}$$

is called the **centralizer** of K in L.

By the Jacobi identity, $N_L(K)$ and $C_L(K)$ are subalgebras in L. In particular, $C_L(L) = Z(L)$. If K is a subalgebra in L, then $K \subset N_L(K)$. If K is an ideal in L, then $N_L(K) = K$.

Theorem 1.3.15. *Let L be a subalgebra of $\mathfrak{gl}(V, k)$, where $V \neq 0$ is a finite dimensional vector space over a field k. If L contains only nilpotent endomorphisms, then there is a nonzero vector $v \in V$ such that $xv = 0$ for all $x \in L$.*

Proof. This will be proved by induction on the dimension $\dim L$. The statement is obvious for $\dim L = 0$ and $\dim L = 1$. Assume $\dim L = n > 1$ and that the statement is true for all cases where $\dim L < n$. Let K be a maximal Lie subalgebra of L. We claim that K is an ideal of L and that $\dim K = \dim L - 1$.

Using ad one can define a Lie homomorphism

$$\varphi : K \longrightarrow \mathfrak{gl}(L/K)$$

in the following way:

$$\varphi(t)(x + K) = (\operatorname{ad} t)(x) + K = [t, x] + K$$

for all $t \in K$ and $x \in L$. This is well-defined since K is a subalgebra of L. Moreover, φ is a Lie morphism, since for $t, s \in K$

$$\begin{aligned}
[\varphi(t), \varphi(s)](x + K) &= \varphi(t)([s, x] + K) - \varphi(s)([t, x] + K) \\
&= ([t, [s, x] + K) - ([s, [t, x]] + K) \\
&= [t, [s, x]] - [s, [t, x] + K = [[t, s], x] + K \\
&= \varphi([t, s])(x + K).
\end{aligned}$$

So $\varphi(K)$ is a subalgebra in $\mathfrak{gl}(L/K)$ and $\dim \varphi(K) < \dim L$. Let $t \in K$. Since t is nilpotent, $\operatorname{ad} t$ is a nilpotent endomorphism on L/K, by lemma 1.3.11. Since $\varphi(t)$ is induced by $\operatorname{ad} t$, $\varphi(t)$ is nilpotent as well. Then by the induction hypothesis there exists a non-zero element $y + K \in L/K$ such that $\varphi(t)(y + K) = (\operatorname{ad} t)(y) + K = [t, y] + K = 0$ for all $t \in K$. Therefore $[t, y] \in K$ for all $y \in L$. Since $y \notin K$, $y \in N_L(K)$ and thus $K \neq N_L(K)$.

Since $N_L(K)$ is a subalgebra in L and K is a maximal proper subalgebra of L, either $N_L(K) = K$ or $N_L(K) = L$. Since $K \neq N_L(K)$, it follows that $N_L(K) = L$ and so K must be an ideal of L.

If $\dim(L/K) > 1$ then the coimage of a one-dimensional subalgebra of L/K is a proper subalgebra which contains K. This contradicts the maximality of K. Therefore $\dim(L/K) = 1$ and $L = K \oplus \operatorname{Span}\{z\}$, where $z \notin K$.

Let

$$W = \{v \in V \ : tv = 0 \ \text{ for all } t \in K\}.$$

Since $\dim K < \dim L$, $W \neq 0$, by the induction hypothesis. Since K is an ideal in L, W is invariant with respect to L, by lemma 1.3.12.

Since $z \notin K$, z is a nilpotent endomorphism on W and it has a nonzero eigenvector $v \in W$ such that $z(v) = 0$. Then any element $x \in L$ can be written in the form $x = t + \alpha z$ for some $t \in K$ and $\alpha \in k$, and so $x(v) = (t + \alpha z)(v) = 0$, as required. □

Theorem 1.3.16 (Engel). *A finite dimensional Lie algebra L is nilpotent if and only if all its elements are* ad-*nilpotent.*

Proof.

1. Necessity. If L is nilpotent then by definition it follows that there is an integer n such that $\operatorname{ad} x_1 \circ \operatorname{ad} x_2 \circ \cdots \circ \operatorname{ad} x_n(y) = 0$ for any $x_1, x_2, ..., x_n, y \in L$. In particular, $(\operatorname{ad} x)^n = 0$ for all $x \in L$. Thus, if a Lie algebra L is nilpotent then every element of it is ad-nilpotent.

2. Sufficiency. This goes by induction on $\dim L$. Let all elements of L be ad-nilpotent. Consider $\operatorname{ad}(L)$ as a subalgebra of $\mathfrak{gl}(L, k)$. Since, by the hypothesis of the theorem, all elements of L are ad-nilpotent, the algebra $\operatorname{ad}(L)$ satisfies the conditions of the previous theorem. So there is a non-zero element $z \in L$ such that $(\operatorname{ad} x)(z) = 0$, for all $x \in L$, i.e., $[L, z] = 0$. Therefore $Z(L) \neq 0$. The elements of $L/Z(L)$ are ad-nilpotent as well. Since the dimension of $L/Z(L)$ is less than $\dim L$, by the induction hypothesis, $L/Z(L)$ is nilpotent. Then, by proposition 1.3.6(2), it follows that L is nilpotent as well. □

1.4. Radical of a Lie algebra. Simple and semisimple Lie algebras

Recall that a Lie algebra is simple if it is not Abelian and has only trivial ideals. As in the theory of associative rings there is the concept of a semisimple Lie algebra which can be formulated in two different ways. The first one is to define a semisimple Lie algebra as a direct sum of simple Lie algebras. Second, one can define a semisimple Lie algebra as a Lie algebra whose radical is 0. As in the theory of associative rings it turns out that in the case of finite dimensional Lie algebras these two concepts are identical. In this chapter various different conditions for a Lie algebra to be semisimple are discussed.

Proposition 1.4.1. *A finite dimensional Lie algebra L has a unique solvable ideal R containing every solvable ideal of L.*

Proof. Let R be a solvable ideal of largest possible dimension. If S is any other solvable ideal of L then $S + R$ is a solvable ideal, by proposition 1.3.3(3). Since $R \subseteq S + R$, $\dim R \leq \dim(S + R)$. But R is a solvable ideal of largest possible dimension. Therefore $\dim R = \dim(S + R)$, hence $S + R = R$. So $S \subset R$. □

Definition. The largest solvable ideal of a finite dimensional Lie algebra is called the **radical** and denoted by $\operatorname{Rad} L$. A non-zero Lie algebra L is called **semisimple** if its radical is equal to zero, that is, L has no nonzero solvable ideals.

Lemma 1.4.2. *If L is any Lie algebra then $L/\operatorname{Rad} L$ is semisimple.*

Proof. Suppose that $\mathcal{J}$ is a solvable ideal of $L/\mathrm{Rad}\,L$. Then there is an ideal $\mathcal{I}$ of L such that $\mathcal{J} = \mathcal{I}/\mathrm{Rad}\,L$. Since $\mathcal{J}$ and $\mathrm{Rad}\,L$ are solvable, $\mathcal{I}$ is solvable as well, by proposition 1.3.3(2). Therefore, by proposition 1.4.1, $\mathcal{I} \subseteq \mathrm{Rad}\,L$, which implies that $\mathcal{I} = \mathrm{Rad}\,L$. Thus $L/\mathrm{Rad}\,L$ has only one solvable ideal, namely 0, and $\mathrm{Rad}\,(L/\mathrm{Rad}\,L) = 0$. □

Proposition 1.4.3. *For any Lie algebra L the following conditions are equivalent:*

1) *L has no nonzero solvable ideals;*
2) *L has no nonzero Abelian ideals.*

Proof.

1) $\Rightarrow$ 2). This follows from the fact that any Abelian ideal is solvable.

2) $\Rightarrow$ 1). Suppose L has no nonzero Abelian ideal and $\mathcal{I}$ is a nonzero solvable ideal of L. Let $\mathcal{I}^{(n)} = 0$ but $\mathcal{I}^{(n-1)} \neq 0$. By proposition 1.2.2, $\mathcal{I}^{(n-1)}$ is an ideal of L and it is Abelian. A contradiction. □

It follows from this proposition that to verify whether L is a semisimple Lie algebra it is sufficient to show that L has no nonzero Abelian ideals.

Definition. Let V be a finite dimensional vector space over a field k. An endomorphism $x \in \mathrm{End}(V)$ is called **diagonalizable** if V has a basis in which x is represented by a diagonal matrix.

An endomorphism $x \in \mathrm{End}(V)$ is called **semisimple**, if it is diagonalizable over an algebraic closure of k.

Note that if k is an algebraically closed field this is equivalent to saying that all roots of its minimal polynomial over the field k are distinct.

Note that any two commuting semisimple endomorphisms can be simultaneously diagonalized, so their sum and difference are semisimple.

Theorem 1.4.4 (Jordan-Chevalley decomposition). *Let V be a finite dimensional vector space over a field k that is algebraically closed or of characteristic zero, $x \in \mathrm{End}_k(V)$. Then there exists a unique decomposition $x = x_s + x_n$, where $x_s, x_n \in \mathrm{End}_k(V)$, which has the following properties:*

1. *x_s is semisimple and x_n is nilpotent; x_s and x_n commute.*
2. *There exist polynomials $p(t), q(t) \in k[t]$ without constant term, such that $x_s = p(x)$ and $x_n = q(x)$. In particular, x_s, x_n commute with any endomorphism commuting with x.*

Proof. The proof that follows is complete only for the algebraically closed case. To deal with the characteristic zero case in general a Galois theoretic argument can be used. In case $\mathrm{char}\,k = p > 0$ and k not algebraically closed the best one can guarantee is a Jordan decomposition over $k^{p^{-\infty}} = \{x \in \overline{k} \ : \ x^{p^n} \in k \text{ for some } n\}$, where $\overline{k}$ is the algebraic closure of k. For details on this see [**13**, section I.4].

Suppose the characteristic polynomial of x is $\prod\limits_{i=1}^{k} (t - a_i)^{m_i}$, where $a_1, a_2, ..., a_k$ are all the distinct eigenvalues of x. Then from linear algebra it is well known that $V = V_1 \oplus V_2 \oplus ... \oplus V_k$, where $V_i = \mathrm{Ker}\,(x - a_i I)^{m_i}$, where I is the identity operator in $\mathrm{End}\,V$, and each V_i is stable under x. Obviously, $(x - a_i I)^{m_i} = 0$ on V_i for all

i. Using the Chinese remainder theorem for the ring $k[t]$ one can find a polynomial $p(t)$ such that $p(t) \equiv a_i (\mathrm{mod}\,(t-a_i)^{m_i})$ for all i. If zero is not an eigenvalue for x add one more congruence $p(t) \equiv 0 (\mathrm{mod}\, t)$. This can be done because t is pairwise prime with all polynomials $(t-a_i)^{m_i}$. Set $q(t) = t - p(t)$. Obviously $p(t)$ and $q(t)$ have no constant term because $p(t) \equiv 0(\mathrm{mod}\, t)$.

Let $x_s = p(x)$ and $x_n = q(x)$. Since $p(x)$ and $q(x)$ are polynomials in x, x_s and x_n commute with each other and with all endomorphisms commuting with x.

Since $p(t) \equiv a_i (\mathrm{mod}\,(t-a_i)^{m_i})$, then $(p(x) - a_i I)_{V_i} = 0$, i.e., $p(x)$ restricted to V_i equals $a_i I$ for all i. This shows that $x_s = p(x)$ acts diagonally on each V_i with a single eigenvalue a_i. This means that x_s is semisimple on V. Since $x_n = x - x_s$, x_n is nilpotent.

Now consider the question of the uniqueness of the decomposition $x = x_s + x_n$. Suppose there are two decompositions: $x = x_s + x_n$ and $x = x'_s + x'_n$. Then $x_s - x'_s = x'_n - x_n$. Because of property 2, proved above, all these endomorphisms commute with each other. Since the difference of semisimple commuting elements is semisimple, $x'_n - x_n$ is semisimple and nilpotent simultaneously. Since the only semisimple and nilpotent endomorphism is zero, $0 = x'_n - x_n = x_s - x'_s$ which means that $x_s = x'_s$ and $x_n = x'_n$. □

The decomposition $x = x_s + x_n$ is called the **Jordan-Chevalley decomposition** (or just the **Jordan decomposition**) of x,; x_s, x_n are called the **semisimple part** and the **nilpotent part** of x, respectively.

Lemma 1.4.5. *Let $x \in \mathrm{End}_k(V)$, where $\mathrm{End}_k(V)$ is given its Lie algebra structure, $\dim_k(V) < \infty$, and let $x = x_s + x_n$ be the Jordan-Chevalley decomposition of x. Then $\mathrm{ad}\, x = \mathrm{ad}\, x_s + \mathrm{ad}\, x_n$ is the Jordan-Chevalley decomposition of $\mathrm{ad}\, x$ in $\mathrm{End}_k(\mathrm{End}_k(V))$.*

Proof. As ad is additive, $\mathrm{ad}\, x = \mathrm{ad}\,(x_s + x_n) = \mathrm{ad}\, x_s + \mathrm{ad}\, x_n$. Since x_n is nilpotent, by lemma 1.3.11, $\mathrm{ad}\, x_n$ is also nilpotent. It remains to show that $\mathrm{ad}\, y$ is semisimple if y is semisimple. Let y be a semisimple element in $\mathrm{End}(V)$. Suppose that $v_1, v_2, ..., v_m$ is a basis of $\overline{V} = V \otimes_k \overline{k}$ in which y has as matrix $\mathrm{diag}(a_1, a_2, ..., a_m)$. Let $\{e_{ij}\}$ be the standard basis in $\mathfrak{gl}(\overline{V}, k)$ with respect to this basis. Then $e_{ij}(v_t) = \delta_{jt}(v_i)$, and $\mathrm{ad}\, y(e_{ij}) = (a_i - a_j)e_{ij}$. So the matrix of $\mathrm{ad}\, y$ is diagonal in this basis, i.e., $\mathrm{ad}\, y$ is semisimple. Thus $\mathrm{ad}\, x_s$ is semisimple. Since x_s and x_n commute, $\mathrm{ad}\, x_s$ and $\mathrm{ad}\, x_n$ also commute, because $[\mathrm{ad}\, x_s, \mathrm{ad}\, x_n] = \mathrm{ad}\,[x_s, x_n]$. Now use theorem 1.4.4 to finish the proof. □

Remark 1.4.6. From now on assume that a Lie algebra L and a vector space V are finite dimensional and they are considered over a field k which is algebraically closed and $\mathrm{char}\, k = 0$, except where otherwise specified. For not a few of the theorems that follow the assumption that k is algebraically closed is not needed, though the proofs are only complete in that case. If so the result is formulated under the assumption $\mathrm{char}\, k = 0$ only and the proof can be completed by an extension of scalars argument (base change argument); see [**46**, section I.8] for some details on this technique. This applies e.g. to theorem 1.4.10, theorem 1.4.14 and theorem 1.4.20.

Lemma 1.4.7. *Let A, B be subspaces in $\mathfrak{gl}(V,k)$, where $A \subset B$ and* $\dim V < \infty$. *Let $M = \{x \in \mathfrak{gl}(V,k) : [x,B] \subseteq A\}$. Suppose that there is an element $x \in M$ such that* $\mathrm{Tr}(xy) = 0$ *for all $y \in M$. Then x is nilpotent.*

Proof. Let $x = x_s + x_n$ be the Jordan-Chevalley decomposition of x. Suppose $v_1, v_2, ..., v_m$ is a basis of V in which x_s is given by the diagonal matrix $\mathrm{diag}(a_1, a_2, ..., a_m)$. Let W be the $\mathbf{Q}$-span of the eigenvalues $a_1, ..., a_m$. Since $\mathrm{char}\, k = 0$, $\mathbf{Q}$ is the prime subfield of k. So $W \subseteq V$. We shall show that $x_s = 0$ or equivalently that $W = 0$. Since V is finite dimensional, it suffices to show that the dual space $W^* = 0$, i.e. any linear function $f : W \to \mathbf{Q}$ is zero.

Let $f \in W^*$ be given. Choose $y \in \mathfrak{gl}(V,k)$ be such that

$$y = \mathrm{diag}(f(a_1), f(a_2), ..., f(a_m))$$

in the fixed basis $v_1, v_2, ..., v_m$ of V. So y is semisimple in $\mathrm{End}_k(V)$, and, by lemma 1.4.5, $\mathrm{ad}\, y$ is semisimple in $\mathrm{End}_k(\mathrm{End}_k(V))$. Let $\{e_{ij}\}$ be the standard basis in $\mathfrak{gl}(V,k)$ with respect to the chosen basis of V. Then from the proof of lemma 1.4.5 $\mathrm{ad}\, x_s(e_{ij}) = (a_i - a_j)e_{ij}$, and $\mathrm{ad}\, y(e_{ij}) = (f(a_i) - f(a_j))e_{ij}$.

By the Lagrange interpolation theorem, there is a polynomial $r(t) \in k[t]$ without constant term such that $r(a_i - a_j) = f(a_i) - f(a_j)$ for each pair $i, j \in \{1, 2, ..., m\}$. Note, if $a_i - a_j = a_k - a_l$ then, by linearity of f, $f(a_i) - f(a_j) = f(a_k) - f(a_l)$. Therefore, $\mathrm{ad}\, y = r(\mathrm{ad}\, x_s)$. By lemma 1.4.5, $\mathrm{ad}\, x_s$ is the semisimple part of $\mathrm{ad}\, x$ and, by theorem 1.4.4, $\mathrm{ad}\, x_s$ may be written in the form of a polynomial without constant term in $\mathrm{ad}\, x$. Therefore $\mathrm{ad}\, y$ by composition is a polynomial without constant term in $\mathrm{ad}\, x$ as well. By hypothesis, $[x,B] \subseteq A$, therefore $[y,B] \subseteq A$, i.e. $y \in M$. By the assumption $\mathrm{Tr}(xy) = 0$, so we obtain the equality $\sum_{i=1}^{m} a_i f(a_i) = 0$. Since $f : W \to \mathbf{Q}$, applying f to this equality gives $\sum_{i=1}^{m} f(a_i)^2 = 0$. As $f(a_i) \in \mathbf{Q}$, $f(a_i) = 0$ for all a_i, $i \in \{1, 2, ..., m\}$. Taking into account that $W = \mathrm{Span}(a_1, a_2, ..., a_m)$, we obtain that $f(u) = 0$ for all $u \in W$, i.e. $f = 0$ and therefore $W^* = 0$. Thus $W = 0$, which means that $x = x_n$ is nilpotent. □

Lemma 1.4.8. *If $x, y, z \in \mathrm{End}_k(V)$, where V is a finite dimensional vector space over a field k, then*

$$\mathrm{Tr}([x,y]z) = \mathrm{Tr}(x[y,z]). \tag{1.4.9}$$

Proof. This follows from the equalities $[x,y]z = xyz - yxz$, $x[y,z] = xyz - xzy$ and the trace property: $\mathrm{Tr}(y(xz)) = \mathrm{Tr}((xz)y)$. □

Theorem 1.4.10. *Let L be a subalgebra of $\mathfrak{gl}(V,k)$, where V is a finite dimensional vector space over a field k of $\mathrm{char}\, k = 0$. Suppose that* $\mathrm{Tr}(xy) = 0$ *for all $x \in [L,L]$, $y \in L$. Then L is solvable.*

Proof. It suffices to show that the derived Lie algebra $[L,L]$ is nilpotent. Then as $L/[L,L]$ is Abelian and $[L,L]$ is nilpotent, both are solvable, by proposition 1.3.5. So, by proposition 1.3.3, it follows that L is solvable.

By lemma 1.3.11 and theorem 1.3.16, to prove that $[L,L]$ is nilpotent is equivalent to prove that each $x \in [L,L]$ is nilpotent. Apply lemma 1.4.7 to the case at hand. Let $A = [L,L]$, $B = L$ and

$$M = \{x \in \mathfrak{gl}(V,k) : [x,L] \subset [L,L]\}.$$

Obviously, $L \subset M$. Suppose, $[x, y]$ is one of the generators of $[L, L]$ and $z \in M$. Then, by (1.4.9),

$$\mathrm{Tr}([x, y]z) = \mathrm{Tr}(x[y, z]) = \mathrm{Tr}([y, z]x) = 0,$$

since $[y, z] \in M$. As $[L, L]$ is a finite dimensional algebra, $\mathrm{Tr}(xy) = 0$ for all $x \in [L, L]$, $y \in L$. Therefore, by lemma 1.4.7, x is nilpotent, and so L is solvable. □

Definition. Let L be a Lie algebra. For any $x, y \in L$ define

$$\kappa(x, y) = \mathrm{Tr}(\mathrm{ad}\, x \circ \mathrm{ad}\, y). \tag{1.4.11}$$

This form is called the **Killing form**[3] of L.

The Killing form is symmetric and bilinear on L. Moreover, it has an "associativity property" in the form that

$$\kappa([x, y]z) = \kappa(x, [y, z]), \tag{1.4.12}$$

which follows from equality (1.4.9). The Killing form is invariant under all automorphisms of L. Indeed, let $\alpha \in \mathrm{Aut}(L)$, then

$$\kappa(\alpha(x), \alpha(y)) = \kappa(x, y) \tag{1.4.13}$$

for all $x, y \in L$. This follows from the symmetry property of Tr, and the relation $\mathrm{ad}\, \alpha(x) = \alpha \circ \mathrm{ad}\, x \circ \alpha^{-1}$.

Using the Killing form there follows almost immediately from theorem 1.4.10 a statement which gives a sufficient condition, in terms of the Killing form, for a Lie algebra L to be solvable. As it will be shown later, in section 1.6, this condition (which is called Cartan's first criterion) is also necessary.

Theorem 1.4.14. *Let L be a finite dimensional algebra over an algebraically closed field k of characteristic zero with the Killing form κ. Suppose $\kappa(x, y) = 0$ for all $x \in L$, $y \in [L, L]$. Then L is solvable.*

Proof. Since $Z(L)$ is solvable, by proposition 1.3.3 it suffices to prove that $L/Z(L)$ is solvable. Consider the adjoint representation $\mathrm{ad} : L \longrightarrow \mathfrak{gl}(L, k)$. Since $\mathrm{Ker}(\mathrm{ad}) = Z(L)$, the proof reduces to the consideration of the subalgebra $\mathrm{ad}\, L \subset \mathfrak{gl}(L, k)$. Now one can apply theorem 1.4.10. □

Let L be a finite dimensional Lie algebra over a field k with a basis $v_1, v_2, \dots, v_n$. Then

$$[v_i, v_j] = \sum_{k=1}^{n} c_{ij}^k v_k, \tag{1.4.15}$$

for certain coefficients $c_{ij}^k \in k$.

Obviously, the (bilinear) equality (1.4.15) uniquely defines the law of multiplication in L. The coefficients c_{ij}^k are called the **structure constants** of L with respect to this basis. Since the Lie bracket is antisymmetric and satisfies the Jacobi identity, the structure constants satisfy the following conditions:

$$c_{ij}^k = -c_{ji}^k \tag{1.4.16}$$

[3]Named after Wilhelm Karl Joseph Killing [1847-1923].

$$\sum_{k=1}^{n}(c_{ij}^k c_{kl}^m + c_{jl}^k c_{ki}^m + c_{li}^k c_{kj}^m) = 0 \tag{1.4.17}$$

for $1 \le i, j, k, l, m \le n$.

Conversely, let L be a finite dimensional algebra L with a basis $v_1, v_2, \dots, v_n$. In order to define a bilinear operation in L, it is enough to define it on the v_i by (1.4.15). In order for L to be a Lie algebra, it is enough to require that the structure constants satisfy (1.4.16) and (1.4.17).

Let L be a finite dimensional Lie algebra with a basis $v_1, v_2, \dots, v_n$ and the Lie bracket defined by (1.4.15). If $x = \sum_{i=1}^{n} x_i v_i \in L$ and $y = \sum_{i=1}^{n} y_i v_i \in L$ then from (1.4.15) it follows that

$$\kappa(x, y) = \sum_{i,j,k,t} c_{it}^k c_{jk}^t x_i y_j. \tag{1.4.18}$$

The tensor $g_{ij} = \sum_{k,t} c_{it}^k c_{jk}^t$ is called the **metric tensor** of the Lie algebra L.

Example 1.4.19.

Let $L = \mathfrak{sl}(2, k)$, and $\operatorname{char} k \neq 2$. Then any element $x \in L$ can be written in the form $x = ae + bh + cf$, where the elements

$$e = \begin{pmatrix} 0 & 1 \\ 0 & 0 \end{pmatrix}; \quad h = \begin{pmatrix} 1 & 0 \\ 0 & -1 \end{pmatrix} \quad f = \begin{pmatrix} 0 & 0 \\ 1 & 0 \end{pmatrix}$$

form a basis of L with relations $[e, f] = h$, $[h, f] = -2f$, $[h, e] = 2e$. Then in the ordered basis $\{e, h, f\}$

$$\operatorname{ad} e = \begin{pmatrix} 0 & -2 & 0 \\ 0 & 0 & 1 \\ 0 & 0 & 0 \end{pmatrix}; \quad \operatorname{ad} h = \begin{pmatrix} 2 & 0 & 0 \\ 0 & 0 & 0 \\ 0 & 0 & -2 \end{pmatrix}; \quad \operatorname{ad} f = \begin{pmatrix} 0 & 0 & 0 \\ -1 & 0 & 0 \\ 0 & 2 & 0 \end{pmatrix}.$$

Therefore the matrix of the Killing form is

$$K = \begin{pmatrix} 0 & 0 & 4 \\ 0 & 8 & 0 \\ 4 & 0 & 0 \end{pmatrix}$$

and $\det K = -128 \neq 0$.

Recall some general notions from the linear algebra. Let $\beta(x, y)$ be a symmetric bilinear form given on a vector space V. Then the subspace

$$\operatorname{Rad} \beta = \{x \in V : \beta(x, y) = 0 \text{ for all } y \in V\}$$

is called the **radical** of β. The bilinear form $\beta(x, y)$ is called **nondegenerate** if $\operatorname{Rad}(\beta) = 0$.

This can be applied to the case of the Killing form κ.

Definition. The set

$$\operatorname{Rad} \kappa = \{x \in L : \kappa(x, y) = 0 \text{ for all } y \in L\}$$

is called the **radical** of κ.

From the associativity of κ it follows that $\operatorname{Rad} \kappa$ is an ideal of L.

The form κ is called **nondegenerate** if $\operatorname{Rad} \kappa = 0$.

There is a simple practical test for checking (non)degeneracy of a form: a form κ is nondegenerate if and only if the matrix of κ in a fixed basis has a nonzero determinant. Example 1.4.19 shows that the Killing form of the Lie algebra $L = \mathfrak{sl}(2, k)$, where char$k \neq 2$, is nondegenerate.

Recall that a Lie algebra L is semisimple if its radical is equal to zero. The following theorem gives another useful test for semisimplicity of a Lie algebra.

Theorem 1.4.20 (Cartan's semisimplicity criterion). *A finite dimensional Lie algebra L over a field k of characteristic zero is semisimple if and only if its Killing form κ is nondegenerate.*

Proof. Assume L is semisimple, i.e., $\operatorname{Rad} L = 0$. Let $S = \operatorname{Rad} \kappa$. By definition, $\operatorname{Tr}(\operatorname{ad} x \circ \operatorname{ad} y) = 0$ for all $x \in S$ and $y \in L$, in particular, for all $y \in [S, S]$. By the Cartan solvability criterion, the algebra $\operatorname{ad}[L, S]$ is solvable, therefore the algebra S is solvable as well. Since S is an ideal in L, $S \subset \operatorname{Rad} L = 0$ and the form κ is nondegenerate.

Conversely, let $S = \operatorname{Rad} \kappa = 0$. Suppose $\mathcal{I}$ is a nonzero Abelian ideal in L. Now, quite generally, $\mathcal{I} \subset S$. Indeed let $x \in \mathcal{I}$ and $y \in L$. This gives a map $\operatorname{ad} x \circ \operatorname{ad} y : L \to \mathcal{I}$ and $(\operatorname{ad} x \circ \operatorname{ad} y)^2 : L \to [\mathcal{I}, \mathcal{I}] = 0$. Thus $(\operatorname{ad} x \circ \operatorname{ad} y)$ is nilpotent, hence $0 = \operatorname{Tr}(\operatorname{ad} x \circ \operatorname{ad} y) = \kappa(x, y)$, i.e. $\mathcal{I} \subset S$. By proposition 1.4.3, L is semisimple. □

Lemma 1.4.21. *Let $\mathcal{I}$ be an ideal in L. Suppose κ is the Killing form of L and $\kappa_{\mathcal{I}}$ is the Killing form of $\mathcal{I}$, considered as a Lie algebra. Then $\kappa_{\mathcal{I}}(x, y) = \kappa(x, y)$ for all $x, y \in \mathcal{I}$.*

Proof. As $\mathcal{I}$ is an ideal in L one can take a basis of $\mathcal{I}$ and extend it to a basis of L. In this basis the matrix of the representation $\operatorname{ad} x$ for any $x \in \mathcal{I}$ has the form

$$\begin{pmatrix} S_x & T_x \\ 0 & 0 \end{pmatrix}$$

where S_x is the matrix of $\operatorname{ad} x$ restricted to $\mathcal{I}$. If $y \in \mathcal{I}$ then the matrix of the representation $\operatorname{ad} x \circ \operatorname{ad} y$ has the form

$$\begin{pmatrix} S_x S_y & S_x T_y \\ 0 & 0 \end{pmatrix}$$

where $S_x S_y$ is the matrix of $\operatorname{ad} x \circ \operatorname{ad} y$ restricted to $\mathcal{I}$. Therefore

$$\kappa(x, y) = \operatorname{Tr}(S_x S_y) = \kappa_{\mathcal{I}}(x, y).$$

□

It is now possible to prove the main structure theorem for semisimple Lie algebras.

Theorem 1.4.22. *Let L be a semisimple finite dimensional Lie algebra. Then there are ideals $L_1, L_2, ..., L_n$ of L, which are simple as Lie algebras, such that $L = L_1 \oplus L_2 \oplus \cdots \oplus L_n$. Moreover, these ideals are unique up to a permutation. Any simple ideal of L is one of L_i entering in this decomposition.*

Proof. Let L be a semisimple Lie algebra and let $\mathcal{I}$ be an ideal in L. Consider the set

$$\mathcal{I}^{\perp} = \{x \in L : \ \kappa(x, y) = 0 \text{ for all } y \in \mathcal{I}\},$$

which is also an ideal by associativity of the Killing form. Since L is semisimple, κ on L is nondegenerate, so $\mathcal{I}^\perp$ is an orthogonal complement to $\mathcal{I}$ and $\dim \mathcal{I} + \dim \mathcal{I}^\perp = \dim L$, i.e. $L = \mathcal{I} + \mathcal{I}^\perp$. The next step is to show that $\mathcal{J} = \mathcal{I} \cap \mathcal{I}^\perp = 0$. Since $\mathcal{J}$ is an ideal, by the previous lemma, its Killing form is equal to the Killing form of L restricted to $\mathcal{J}$, therefore it is equal to zero. Then, by theorem 1.4.14, $\mathcal{J}$ is solvable. Since L is semisimple, $\mathcal{J} = 0$. Thus, $L = \mathcal{I} \oplus \mathcal{I}^\perp$.

Since any ideal of $\mathcal{I}$ is an ideal of L, $\mathcal{I}$ is a semisimple Lie algebra. Analogously, $\mathcal{I}^\perp$ is semisimple. So induction can be applied to the dimension of an algebra. So if $\mathcal{I}$ or $\mathcal{I}^\perp$ is not simple one can repeat this process until a decomposition $L = L_1 \oplus L_2 \oplus \ldots \oplus L_n$ is obtained.

It remains to show that this decomposition is unique. Let $\mathcal{I}$ be any simple ideal in L. Then $\mathcal{J} = [\mathcal{I}, L]$ is an ideal in L and it is contained in $\mathcal{I}$. Since $\mathcal{I}$ is simple, either $\mathcal{J} = \mathcal{I}$ or $\mathcal{J} = 0$. If $\mathcal{J} = 0$ then $\mathcal{I} \subset Z(L)$. Since L is semisimple, $Z(L) = 0$ and hence $\mathcal{I} = 0$. So if $\mathcal{I}$ is a nonzero ideal, then $\mathcal{J} = \mathcal{I}$ is a simple ideal. On the other hand, $\mathcal{J} = [\mathcal{I}, L_1] \oplus [\mathcal{I}, L_2] \oplus \ldots \oplus [\mathcal{I}, L_n]$, therefore there exists a unique index i such that $[\mathcal{I}, L_i] = \mathcal{I}$ and all other summands are equal to zero. As $[\mathcal{I}, L_i]$ is an ideal in L_i and L_i is simple, $L_i = \mathcal{I}$. □

Corollary 1.4.23. *A Lie algebra L is semisimple if and only if it is a direct sum of simple Lie algebras.*

Proof. The direct statement follows from theorem 1.4.22. So there just remains the matter of proving the inverse statement. Suppose $L = L_1 \oplus L_2 \oplus \cdots \oplus L_n$, where the L_i are simple Lie algebras. Let $R = \mathrm{rad} L$. The claim is that $R = 0$. Since for each i L_i is a simple Lie algebra and $[R, L_i] \subseteq R \cap L_i$,

$$[R, L] \subset [R, L_1] \oplus [R, L_2] \oplus \cdots \oplus [R, L_n] = 0$$

This shows that $R \subset Z(L)$. Since $Z(L_i) = 0$ for each simple Lie algebra L_i and $Z(L) = Z(L_1) \oplus Z(L_2) \oplus \cdots \oplus Z(L_n)$ it follows that $R = 0$, that is L is semisimple. □

Corollary 1.4.24. *A semisimple ideal in a finite dimensional Lie algebra L over a field of characteristic zero is a direct summand of L.*

Proof. Let $\mathcal{I}$ be a semisimple ideal in a Lie algebra L. Since $\mathcal{I}^\perp$ is an ideal in L, $\mathcal{J} = \mathcal{I} \cap \mathcal{I}^\perp$ is also an ideal in L, and the Killing form of $\mathcal{J}$ is equal to zero, by lemma 1.4.21. From theorem 1.4.14 it follows that $\mathcal{J} \subset \mathcal{I}$ is a solvable ideal in L, so $\mathcal{J} = 0$, by assumption. Since $\dim \mathcal{I} + \dim \mathcal{I}^\perp = \dim L$, $L = \mathcal{I} \oplus \mathcal{I}^\perp$ is actually a direct sum of ideals. □

Corollary 1.4.25. *If L is a semisimple finite dimensional Lie algebra over a field of characteristic zero then $L = [L, L]$ and all ideals in L and all homomorphic images of L are semisimple. Moreover, each ideal in L is a sum of simple ideals.*

Corollary 1.4.26. *If L is a semisimple Lie algebra and $\mathcal{I}$ is an ideal of L, then L/I is also semisimple.*

1.5. Modules for Lie algebras. Weyl's theorem. Ado's theorem

Let L be a Lie algebra over a field k. Recall that a representation of L is a Lie morphism $\varphi : L \to \mathfrak{gl}(V, k)$, where $V \neq 0$ is a vector space over k. A representation

φ is called **irreducible** if it has no invariant subspaces except V and 0. A representation φ is called **decomposable** if it is a direct sum of nontrivial invariant subspaces. It is called **completely reducible** if it is a direct sum of irreducible representations.

Just as for associative algebras it is often convenient to use together with the language of representations the language of modules. Here is the definition of a module over a Lie algebra.

Definition. Let L be a Lie algebra. A vector space V over a field k, endowed with an action $L \times V \to V$ (denoted by $(x, v) \mapsto x \cdot v$ (or just xv)) is called a **Lie module** for L, or an **L-module**, if the following conditions are satisfied:
1) $(ax + by) \cdot v = a(x \cdot v) + b(y \cdot v)$;
2) $x \cdot (av + bw) = a(x \cdot v) + b(x \cdot w)$;
3) $[x, y] \cdot v = x \cdot (y \cdot v) - y \cdot (x \cdot v)$
for all $x, y \in L; v, w \in V; a, b \in k$.

Note that every Lie algebra L is actually an L-module over itself under the action $(x, y) \mapsto [x, y]$. This one is called the **adjoint L-module** and it corresponds to the adjoint representation of L.

The notions of L-modules and representations of a Lie algebra L are two different languages for describing the same structures. It is easy to show that there is a one-to-one correspondence between L-modules and representations of L. Indeed, let $\varphi : L \to \mathfrak{gl}(V, k)$ be a representation of L. Then V may be viewed as an L-module via the action $x \cdot v = \varphi(x)(v)$. Conversely, for a given L-module V this equation defines a representation $\varphi : L \to \mathfrak{gl}(V, k)$ of L.

A **morphism of L-modules** V and W is a linear map $\varphi : V \to W$ such that

$$\varphi(x \cdot v) = x \cdot \varphi(v) \quad \text{for all } x \in L, v \in V.$$

$\mathrm{Ker}(\varphi)$ is an L-submodule of V, and $\mathrm{Im}(\varphi)$ is an L-submodule of W. All standard morphism theorems hold in this setting.

If φ is a morphism of L-modules, that is an isomorphism of the underlying vector spaces, it is called an **isomorphism of L-modules**. Two modules are **isomorphic** if and only if the corresponding representations are equivalent. A **submodule** U of V is an invariant subspace of V, i.e. $LU \subset U$. Quotient modules are defined in the usual way. Suppose that U is an L-submodule of an L-module V. Then the quotient vector space V/U with

$$x \cdot (v + U) = (x \cdot v) + U \quad \text{for } x \in L, v \in V$$

defines the quotient module V/U. It is easy to check that the action of L is well-defined and satisfies conditions 1), 2), 3) in the definition of an L-module.

An L-module V is called **irreducible**, or **simple** if it is nonzero and it has precisely two submodules: itself and 0. An L-module V is called **decomposable** if there are nonzero submodules U and W of V such that $V = U \oplus V$, otherwise it is called **indecomposable**. An L-module V is called **completely reducible** if it can be written as a direct sum of irreducible submodules.

Just as in the ring theory Schur's lemma holds for irreducible modules over a Lie algebra L. There are different forms for Schur's lemma; here just one of them is given for the case of a algebraically closed field:

Lemma 1.5.1 (Schur's lemma). *Let L be a Lie algebra over a algebraically closed field k and let U be a finite dimensional irreducible L-module. A map $\varphi : U \longrightarrow U$ is an L-module homomorphism if and only if there is a scalar $\alpha \in k$ such that $\varphi = \alpha 1_U$.*

The hypothesis that the field k be algebraically closed is really needed here; otherwise the best one can do is to say that the module endomorphisms form a division algebra.

The remaining part of this section is devoted to prove a fundamental result in the representation theory of Lie algebras, the Weyl theorem, which gives the structure of finite dimensional representations of semisimple algebras.

Here is a generalization of the Killing form. Let L be a semisimple Lie algebra and let $\varphi : L \to \mathfrak{gl}(V, k)$ be a finite dimensional faithful[4] representation of L. Define the corresponding symmetric bilinear form $\beta : L \times L \longrightarrow k$ by

$$\beta(x, y) = \mathrm{Tr}(\varphi(x) \circ \varphi(y)), \quad (x, y \in L) \tag{1.5.2}$$

which is called a **trace form**. This form is associative, i.e.,

$$\beta([x, y], z) = \beta(x, [y, z])$$

and so $R = \mathrm{Rad}\beta$ is an ideal of L. The trace form is nondegenerate, because, by the Cartan solvability criterion, $\varphi(R) \simeq R$ is solvable, and so, L being semisimple, $R = 0$. In particular, if $V = L$ and $\varphi = \mathrm{ad}$ the trace form β coincides with the Killing form.

Let $\varphi : L \to \mathfrak{gl}(V, k)$ be a faithful representation of L with trace form

$$\beta(x, y) = \mathrm{Tr}(\varphi(x) \circ \varphi(y)).$$

Fix a basis $(x_1, x_2, ..., x_n)$ of L. Then there is a uniquely determined dual basis $(y_1, y_2, ..., y_n)$ relative to β such that $\beta(x_i, y_j) = \delta_{ij}$. Write

$$c_\varphi = \sum_{i=1}^{n} \varphi(x_i) \circ \varphi(y_i).$$

It is easy to show that $c_\varphi : V \longrightarrow V$ is an L-module homomorphism and that it commutes with $\varphi(L)$. This L-module endomorphism $c_\varphi \in \mathrm{End}_k(V)$ is called the **Casimir operator**, or **Casimir element**, associated to φ. The trace of c_φ is

$$\mathrm{Tr}(c_\varphi) = \sum_{i=1}^{n} \mathrm{Tr}(\varphi(x_i) \circ \varphi(y_i)) = \sum_{i=1}^{n} \beta(x_i, y_i) = \dim L = n \neq 0.$$

If φ is an irreducible representation, then, by the Schur lemma, c_φ is a scalar multiple of the identity homomorphism, and in this case it is independent of the basis of L which was chosen.

Lemma 1.5.3. *Let $\varphi : L \to \mathfrak{gl}(V, k)$ be a representation of a semisimple Lie algebra L over a field k. Then $\varphi(L) \subset \mathfrak{sl}(V, k)$. In particular, L acts trivially on any one dimensional L-module.*

Proof. Use the fact that $L = [L, L]$ and that $\mathfrak{sl}(V, k)$ is the derived algebra of $\mathfrak{gl}(V, k)$, i.e. $\mathfrak{sl}(V, k) = [\mathfrak{gl}(V, k), \mathfrak{gl}(V, k)]$. □

[4]Faithful means that $\varphi : L \to \mathfrak{gl}(V, k)$ is injective.

Theorem 1.5.4 (H.Weyl). *Let $\varphi : L \to \mathfrak{gl}(V, k)$ be a finite dimensional representation of a semisimple Lie algebra L over a field k. Then φ is completely reducible.*

Proof. If $V = 0$ then the statement is trivial, as $\varphi(L)$ is zero. So assume that $V \neq 0$.

Since there is a one-to-one correspondence between L-modules and representations of L, it is enough to prove that every finite dimensional L-module is a direct sum of irreducible L-modules. It suffices to show that every exact sequence of the form

$$0 \longrightarrow W \longrightarrow V \longrightarrow V/W \longrightarrow 0$$

splits for each submodule W of V.

1. *The special case* $\operatorname{codim} W = 1$. We first prove theorem in the special case when V has an L-submodule W of codimension one. Since L acts trivially on V/W, by lemma 1.5.3, there results an exact sequence

$$0 \longrightarrow W \longrightarrow V \longrightarrow k \longrightarrow 0.$$

Using induction on $\dim W$, things can be reduced to the case where W is an irreducible L-module, as follows. Let W' be a proper nonzero submodule of W. This yields an exact sequence:

$$0 \longrightarrow W/W' \longrightarrow V/W' \longrightarrow M \longrightarrow 0,$$

where $M \cong V/W \cong k$. Since $\dim W/W' < \dim W$, by the induction hypothesis this sequence splits, i.e. there exists a one dimensional L-submodule $\overline{X}$ of V/W' such that

$$V/W' = W/W' \oplus \overline{X}.$$

By the submodule correspondence there exists a submodule $X \subseteq V$ and containing W' such that $\overline{X} = X/W'$. Applying the induction hypothesis to X, we obtain that $X = W' \oplus Y$, where $Y \cong X/W'$ is a one dimensional L-submodule of X. We claim that $V = W \oplus Y$. Consider the canonical projection $\pi : V \to V/W' = W/W' \oplus X/W'$. Then $\pi(W \cap Y) = 0$ which means that $W \cap Y \subseteq W'$. Therefore $W \cap Y \subseteq W' \cap Y = 0$. Since $\dim V = \dim W + \dim Y = \dim W + 1$ and $W \cap Y = 0$, it follows that $V = W \oplus Y$.

Now assume that W is an irreducible submodule of V. It can also be assumed without loss of generality that L acts faithfully on V. Otherwise we can replace L by $L/\operatorname{Ker}\varphi$, which is semisimple, by corollary 1.4.26.

Let $c = c_\varphi$ be the Casimir element of φ. Since c commutes with $\varphi(L)$, c is actually an L-module endomorphism of V; in particular, $c(W) \subset W$ and $\operatorname{Ker}(c)$ is an L-submodule of V. Because L acts trivially on V/W, by lemma 1.5.3, c must do likewise (as a linear combination of products of elements $\varphi(x)$). So c has trace 0 on V/W. On the other hand, by the Schur lemma, c acts as a scalar multiple of the identity homomorphism on the irreducible L-submodule W. This scalar cannot be 0, because that would force $\operatorname{Tr}(c) = 0$. It follows that $\operatorname{Ker}(c)$ is a one dimensional L-submodule of V which intersects W trivially. This is the desired complement to W.

So we have proved the theorem in the special case when $\dim V = \dim W + 1$.

2. *The general case* $\operatorname{codim} W > 1$. Now turn to the general case and reduce it to the previous special case. Suppose that W is any L-submodule of V so that there is an exact sequence:

$$0 \longrightarrow W \longrightarrow V \longrightarrow V/W \longrightarrow 0.$$

Let $M = \operatorname{Hom}_k(V, W)$ be the space of linear maps $V \to W$, viewed as an L-module defined as follows:

$$(x \cdot f)(v) = x \cdot f(v) - f(x \cdot v)$$

for $x \in L$, $v \in V$ and $f \in M$.

Let M_1 be the subspace of M consisting of those maps whose restriction to W is a scalar multiplication. One can check that M_1 is actually an L-submodule. Let $f|_W = \lambda \cdot 1_W$ for some $\lambda \in k$. Then for $x \in L$, $w \in W$,

$$(x \cdot f)(w) = x \cdot f(w) - f(x \cdot w) = a(x \cdot w) - a(x \cdot w) = 0,$$

so that $x \cdot f|_W = 0$. Let M_2 be the subspace of M_1 consisting of those f whose restriction to W is zero. The preceding calculation shows that M_2 is also an L-submodule and that L maps M_1 into M_2. Moreover, $\dim(M_1/M_2) = 1$, because each $f \in M_1$ is determined (modulo M_2) by the scalar $\lambda \in k$. This gives back precisely the situation $0 \to M_2 \to M_1 \to k \to 0$ already treated above.

According to the special case of the theorem, M_1 has a one dimensional submodule complementary to M_2. Let $f : V \to W$ span it, so that after multiplying by a nonzero scalar it can be assumed that $f|_W = 1_W$. To say that L kills f is just to say that

$$0 = (x \cdot f)(v) = x \cdot f(v) - f(x \cdot v),$$

i.e., that f is an L-homomorphism. Therefore $\operatorname{Ker}(f)$ is an L-submodule of V. Since f maps V into W and acts as 1_W on W, it follows that $V = W \oplus \operatorname{Ker}(f)$, as desired. □

The last part of this section is devoted to a proof of the Levi and Ado theorems.

Lemma 1.5.5. *If $\mathcal{I}$ is an ideal of a Lie algebra L, then*

$$\operatorname{rad}(\mathcal{I}) = \operatorname{rad}(L) \cap \mathcal{I}.$$

Proof. Since $\operatorname{rad}(\mathcal{I}) \subset \mathcal{I}$ and any derivation of L maps $\operatorname{rad}(\mathcal{I})$ into itself, $\operatorname{rad}(\mathcal{I})$ is an ideal in L. Since $\operatorname{rad}(\mathcal{I})$ is a solvable ideal, it must be contained in the radical of L. This shows that $\operatorname{rad}(\mathcal{I}) \subseteq \operatorname{rad}(L) \cap \mathcal{I}$. The opposite inclusion follows from the fact that $\operatorname{rad}(L) \cap \mathcal{I}$ is a solvable ideal in $\mathcal{I}$. □

Theorem 1.5.6 (E.E. Levi). *Let L be finite dimensional Lie algebra over a field k with radical* $\operatorname{rad}(L)$. *Then there is a semisimple Lie subalgebra S in L such that*

$$L = S \oplus \operatorname{rad}(L) \tag{1.5.7}$$

Note that a Lie subalgebra $S \subset L$ in (1.5.7) is always semisimple because the natural homomorphism $L \longrightarrow L/\operatorname{rad}(L)$ maps S isomorphically onto the semisimple Lie algebra $L/\operatorname{rad}(L)$. It is easy to see that S is a maximal semisimple subalgebra in L. This semisimple algebra S is called a **Levi subalgebra** (or a **Levi factor**) of L, and the decomposition (1.5.7) is called a **Levi decomposition** of L.

Proof. Let $R = \operatorname{rad}(L)$ be the radical of a Lie algebra L. The theorem will be proved by induction on $\dim R$ which reduces the proof to the main case when R is a minimal Abelian ideal in L.

(a) Suppose that R is not a minimal ideal in L and so there is a nonzero ideal $\mathcal{I}$ of L such that $\mathcal{I} \subsetneqq R$. Then the radical of the Lie algebra $L/\mathcal{I}$ is $R/\mathcal{I}$ and $\dim(R/\mathcal{I}) < \dim R$. Therefore by the induction hypothesis there is a Lie subalgebra $H/\mathcal{I}$ in $L/\mathcal{I}$ such that $L/\mathcal{I} = H/\mathcal{I} \oplus R/\mathcal{I}$, that is $H \cap R = \mathcal{I}$ and $L = H + R$. Since $H/\mathcal{I} \simeq L/R$, $H/\mathcal{I}$ is a semisimple Lie algebra. On the other hand, $\mathcal{I}$ is a solvable ideal in H, so $\mathcal{I}$ is the radical of H. Since $\dim \mathcal{I} < \dim R$, by the induction hypothesis applied to H, there is a Lie subalgebra S in H such that $H = S \oplus \mathcal{I} = S \oplus (H \cap R)$. Therefore $L = H + R = S \oplus (H \cap R) + R = S \oplus R$, since $S \cap R \subset S \cap H \cap R = 0$.

(b) In view of (a) one can assume that $R \neq 0$ and that there is no nonzero ideal of L that is properly contained in R. Since R is a solvable ideal, $[R, R] \neq R$. Therefore $[R, R] = 0$. From lemma 1.5.5 it follows that either $Z(L) = R$ or $Z(L) = 0$. If $Z(L) = R$, then $[L, R] = 0$ and therefore the adjoint representation of L defines a representation of L/R into L. Since L/R is a semisimple Lie algebra, by the Weyl theorem the submodule R has a complement S as required.

(c) Thus one can assume that $R \neq 0$ is an Abelian minimal ideal, $Z(L) = 0$ and $[L, R] = R$. Consider $V = \operatorname{End}(L)$, viewed as a Lie L-module in the usual way, the representation $\sigma : L \to \operatorname{End}(V) = \operatorname{EndEnd}(L)$ being defined via the adjoint representation:

$$\sigma(x)\varphi = [\operatorname{ad} x, \varphi] = \operatorname{ad} x \circ \varphi - \varphi \circ \operatorname{ad} x \tag{1.5.8}$$

for all $x \in L$ and $\varphi \in V$. In other words,

$$(\sigma(x)\varphi)(y) = [x, \varphi(y)] - \varphi([x, y])$$

for all $x, y \in L$ and $\varphi \in V$.

Define three subspaces M, N and P in V as follows:

$$M = \{m \in V \mid m(L) \subset R \text{ and } m|_R = \lambda_m 1_R, \text{ where } \lambda_m \in k\}, \tag{1.5.9}$$

$$N = \{n \in V \mid n(L) \subset R \text{ and } n(R) = 0\} \tag{1.5.10}$$

$$P = \{\operatorname{ad} x \mid x \in R\}$$

It is easy to check that M, N are Lie L-submodules in V, and $N \subset M$. Moreover M/N is a trivial L-module of rank 1, i.e. $M/N \cong k$.

Observe that for any $x \in R$ the operator $m = \operatorname{ad} x$ satisfies (1.5.10), since R is an Abelian ideal in L, i.e. $[R, R] = 0$. Let $x \in L$, $y \in R$ and $\varphi = \operatorname{ad} y \in P$. Then $\sigma(x)\varphi = [\operatorname{ad} x, \varphi] = [\operatorname{ad} x, \operatorname{ad} y] = \operatorname{ad}[x, y] \in P$, since $[x, y] \in R$. Therefore $P = \operatorname{ad} R$ is a Lie submodule in the L-module V, and $P \subset N$.

So there results a chain of Lie L-modules:

$$P \subset N \subset M \subset V. \tag{1.5.11}$$

Since $M/N \cong k$, we have an exact sequence of L-modules:

$$0 \longrightarrow N \xrightarrow{i} M \xrightarrow{\tau} M/N \longrightarrow 0$$

where i is the inclusion, and τ is the map which associates with each $m \in M$ the scalar $\lambda_m \in k$ by which m multiplies elements of R.

Let $x \in L$, $y \in R$ and $m \in M$. Then taking into account that $[L, R] = R$ and conditions (1.5.8), (1.5.9) there results

$$\begin{aligned}(\sigma(x)m)(y) &= [\mathrm{ad}\, x, m](y) = [x, m(y)] - m([x,y]) \\ &= \lambda_m[x,y] - \lambda_m[x,y] = 0\end{aligned}$$

which means that

$$\sigma(x)M \subset N \tag{1.5.12}$$

for all $x \in L$. On the other hand, $[y, m(x)] \in [R, R] = 0$ (since R is Abelian by assumption) and therefore

$$(\sigma(y)m)(x) = [y, m(x)] - m([y,x]) = -m([y,x]) = -\lambda_m[y,x]$$

which shows that

$$\sigma(y)M \subset P \tag{1.5.13}$$

for all $y \in R$.

This means that the map $M/P \xrightarrow{\bar{\tau}} M/N$ is an epimorphism of Lie L/R-modules. Since $\dim M/N = 1$, $\mathrm{Ker}(\bar{\tau})$ is an L/R-submodule in M/P of codimension 1. This means that there is an exact sequence of finite dimensional Lie L/R-modules

$$0 \longrightarrow \mathrm{Ker}(\bar{\tau}) \longrightarrow M/P \xrightarrow{\bar{\tau}} M/N \longrightarrow 0.$$

From theorem 1.5.4 it follows that this sequence splits, since L/R is a semisimple Lie algebra. In other words, there is an element $\psi \in M$ such that $\psi(y) = y$ for each $y \in R$ and $\sigma(x)\psi \in P$ for for each $x \in L$.

Let

$$S = \{x \in L \ : \ \sigma(x)\psi = 0\}.$$

It is easy to check that S is a Lie subalgebra in L. It remains to show that S is the required Lie subalgebra in L, i.e. (i) $S \cap R = 0$, and (ii) $S + R = L$. Suppose $x \in S \cap R$, then $\sigma(x)\psi = -\mathrm{ad}\, x$. So $\mathrm{ad}\,(x) = 0$, i.e. $[x, u] = 0$ for all $u \in L$. This implies $x = 0$, because R is a minimal ideal, and $[L, R] = R$.

For (ii), let $x \in L$. Then $\sigma(x)\psi \in P$, so there is an $y \in R$ such that $\sigma(x)\psi = \mathrm{ad}\, y$. But $\mathrm{ad}\, y = -\sigma(y)\psi$. Therefore $\sigma(x)\psi = -\sigma(y)\psi$, i.e. $\sigma(x+y)\psi = 0$, which means that $x + y \in S$. So $x = (x+y) - y$, where $x + y \in S$ and $-y \in R$. Thus $L = S \oplus R$ as required. □

Remark 1.5.14. The existence of a semisimple subalgebra S as in this theorem was first established by E. E. Levi, see [**61**]. A. I. Mal'tsev proved that any two Levi subalgebras are conjugate, see [**63**].

Theorem 1.5.15 (Ado's theorem). *Let L be a finite dimensional algebra over a field of characteristic 0. Then L has a faithful finite dimensional representation.*

Proof. (a) If $\dim L = 1$ then the Lie algebra L has the faithful representation $\lambda \longrightarrow \begin{pmatrix} 0 & 0 \\ \lambda & 0 \end{pmatrix}$. Therefore any Abelian finite dimensional Lie algebra has a faithful finite dimensional representation given by nilpotent matrices.

(b) If the center $Z(L)$ of a Lie algebra is trivial, then the adjoint representation $\mathrm{ad} : L \longrightarrow \mathfrak{gl}(L)$ is faithful, since $\mathrm{Ker}(\mathrm{ad}) = Z(L)$.

(c) If $Z(L) \neq 0$ construct a representation φ_1 of L which is faithful on $Z(L)$ in the following way. Since $Z(L)$ has a faithful representation, by (a), one can choose such a representation. Then using induction on $\dim L$ and the Levi decomposition (1.5.7) for L, construct an extension of this representation to a representation φ_1 of the entire Lie algebra L.

Then $\varphi = \varphi_1 \oplus \mathrm{ad}$ is a faithful representation of the Lie algebra L. □

Remark 1.5.16. This theorem was proved for the case of an arbitrary base field by K. Iwasawa and Harish-Chandra (see [**45**], [**40**]). Many books on Lie algebras, but by no means all, e.g. [**46**], p.199 and [**14**], §7, discuss the Ado-Iwasawa theorem only in characteristic zero. N. Jacobson, [**46**], p.203ff also treats the characteristic $p > 0$ case.

1.6. Lie's theorem

The first main goal of this section is the prove the important theorem of Sophus Lie which gives an answer to the question: When is there a basis of V in which every element of a Lie subalgebra L of $\mathfrak{gl}(V,k)$ is represented by an upper triangular matrix?

Remark 1.6.1. Everywhere in this section we assume that the base field k is algebraically closed and $\operatorname{char} k = 0$.

Let L be a Lie algebra over a field k, let V be a k-vector space and let $\varphi : L \to \mathfrak{gl}(V,k)$ be a representation. Fix an element $\lambda \in L^*$, where $L^* = \mathrm{Hom}_k(L,k)$ is the dual space.

An element $v \in V$ is called an **eigenvector** for a Lie algebra L if

$$\varphi(x)(v) \in \mathrm{Span}\{v\} \quad \text{for all } x \in L.$$

If the subspace

$$V_\lambda^L = \{v \in V \ : \ \varphi(x)v = \lambda(x)v \ \text{ for } \ \text{ all } \ x \in L\}$$

is nonzero it is called the **weight space** of a Lie algebra L attached to λ, and λ is called a **weight** for L.

Lemma 1.6.2 (Invariance lemma). *Let L be a Lie algebra over a field k, and φ a representation of L on a finite dimensional k-vector space V. If $\mathfrak{h}$ is an ideal of L then for any $\lambda \in \mathfrak{h}^*$ the weight spaces $V_\lambda^{\mathfrak{h}}$ are L-invariant.*

Proof. Assume $V_\lambda^{\mathfrak{h}} \neq 0$. Then the claim is that $\varphi(x)w \in V_\lambda^{\mathfrak{h}}$ for any $w \in V_\lambda^{\mathfrak{h}}$ and $x \in L$, which means that

$$\varphi(y)(\varphi(x)w) = \lambda(y)(\varphi(x)w) \tag{1.6.3}$$

for any $x \in L$, $w \in V_\lambda^{\mathfrak{h}}$ and $y \in \mathfrak{h}$. Taking into account that φ is a Lie morphism and using the equality $ab = ba + [a,b]$ for any $a, b \in L$, (1.6.3) can be rewritten in the form:

$$\varphi(x)\varphi(y)w - \varphi([x,y])w = \lambda(y)(\varphi(x)w). \tag{1.6.4}$$

Since $\mathfrak{h}$ is an ideal of L, $[x,y] \in \mathfrak{h}$. Therefore by the definition of $V_\lambda^{\mathfrak{h}}$ it follows that $\varphi([x,y])w = \lambda([x,y])w$, and $\varphi(y)w = \lambda(y)w$. So (1.6.4) can be written as

$$\lambda(y)(\varphi(x)w) - \lambda([x,y])w = \lambda(y)(\varphi(x)w),$$

which is equivalent to $\lambda([x,y])w = 0$. Therefore it remains to show that $\lambda([x,y]) = 0$ for any $x \in L$ and any $y \in \mathfrak{h}$.

Fix a nonzero element $w \in V_\lambda^{\mathfrak{h}}$ and let $x \in L$. Let n be the maximal integer such that the elements

$$w, \varphi(x)w, \varphi(x)^2 w, \ldots, \varphi(x)^n w$$

are linearly independent in V. Write $W_0 = \mathrm{Span}\{w\}$ and let W_i be the subspace in V which is spanned by the vectors

$$w, \varphi(x)w, \varphi(x)^2 w, \ldots, \varphi(x)^i w,$$

for $i > 0$. This gives an increasing chain of subspaces in V:

$$W_0 \subset W_1 \subset W_2 \subset \cdots$$

Obviously,

$$\varphi(x) W_n \subset W_n = W_{n+i}$$

for any $i > 0$ and $\dim W_n = n+1$.

It is not difficult to show that $\varphi(y) W_m \subset W_m$ for all $m \leq n$ and any $y \in \mathfrak{h}$. This follows by induction using the equality

$$\varphi(y)\varphi(x)w = \varphi(x)\varphi(y)w - \varphi([x,y])w = \lambda(y)\varphi(x)w - \varphi([x,y])w$$

and the fact that $[x,y] \in \mathfrak{h}$.

Now, with respect to the basis $w, \varphi(x)w, \varphi(x)^2 w, ..., \varphi(x)^m w$ of the space W_m the operator $\varphi(y)$, where $y \in \mathfrak{h}$, is upper triangular

$$\begin{pmatrix} \lambda(y) & * & \cdots & * \\ 0 & \lambda(y) & \ddots & \vdots \\ \vdots & \ddots & \ddots & * \\ 0 & \cdots & 0 & \lambda(y) \end{pmatrix}. \tag{1.6.5}$$

This is proved by induction on m. For $m = 1$ the matrix is the 1×1 matrix $(\lambda(y))$. Suppose the triangularity statement holds for all $i < m$. Then

$$\varphi(y)\varphi(x)^m w = \varphi(y)\varphi(x)\varphi(x)^{m-1}w = \varphi(x)\varphi(y)\varphi(x)^{m-1}w - \varphi([x,y])\varphi(x)^{m-1}w$$

Since $[x,y] \in \mathfrak{h}$, the second term $\varphi([x,y])\varphi(x)^{m-1}w \in W_{m-1}$. For the first term apply the induction hypothesis. So

$$\begin{aligned} \varphi(x)\varphi(y)\varphi(x)^{m-1}w &= \varphi(x)(\lambda(y)\varphi(x)^{m-1}w + a_{m-2}\varphi(x)^{m-2}w + \cdots + a_0 w) \\ &= \lambda(y)\varphi(x)^m + a_{m-2}\varphi(x)^{m-1} + \cdots + a_0\varphi(x)w. \end{aligned}$$

This takes care of the induction step and so $\varphi(y)$ has the matrix form (1.6.5). Since $\varphi([x,y]) = [\varphi(x), \varphi(y)]$ on W_n,

$$\mathrm{Tr}(\varphi[x,y]) = \mathrm{Tr}([\varphi(x), \varphi(y)]) = n\lambda([x,y])$$

in W_n. Since $\varphi(x)W_n \subset W_n$, $\varphi(y)W_n \subset W_n$, and $[\varphi(x), \varphi(y)]$ acts as a commutator on W_n it follows that $\mathrm{Tr}([\varphi(x), \varphi(y)]) = 0$. Therefore $n\lambda([x,y]) = 0$, hence $\lambda([x,y]) = 0$, as $\mathrm{char}\, k = 0$.

The lemma is proved. □

Theorem 1.6.6 (Lie's theorem). *Let L be a solvable Lie algebra over a field k of characteristic 0, and $\varphi : L \to \mathfrak{gl}(V,k)$ a representation of L on a finite dimensional vector space V over k. Then there exists a nonzero vector $v \in V$ which is a common eigenvector for all operators $\varphi(x)$, $x \in L$, i.e. $\varphi(x)v = \lambda(x)v$, where $\lambda(x) \in k$.*

Proof. Replacing L by $L/\operatorname{Ker}\varphi$ it can be assumed that the representation is faithful. The theorem will be proved by induction on $n = \dim L$.

If $n = 1$ then $L = \operatorname{Span}\{x\}$ and taking into account that k is an algebraically closed field any eigenvector of $\varphi(x)$ can be used for v.

Suppose the theorem is proved for any Lie algebra of dimension $< n$. As L is a solvable Lie algebra, $[L, L] \subsetneqq L$. So there is a subspace $\mathfrak{h} \subset L$ of codimension 1, which contains $[L, L]$. Then $\mathfrak{h}$ is an ideal in L, because $[\mathfrak{h}, L] \subset [L, L] \subset \mathfrak{h}$. Since $\dim \mathfrak{h} = n - 1$, the induction hypothesis can be applied to yield an eigenvector $v \in V$ such that $\varphi(y)v = \lambda(y)v$ for all $y \in \mathfrak{h}$. This defines a $\lambda \in \mathfrak{h}^*$ such that $V_\lambda^{\mathfrak{h}} \neq 0$. Therefore, by lemma 1.6.1, the vector space $V_\lambda^{\mathfrak{h}}$ is L-invariant. Write $L = \mathfrak{h} + \operatorname{Span}\{x\}$, where $x \in L$. Then $\varphi(x)V_\lambda^{\mathfrak{h}} \subset V_\lambda^{\mathfrak{h}}$. But $V_\lambda^{\mathfrak{h}}$ is a finite dimensional vector space over the algebraically closed field k. Hence the operator $\varphi(x)$ has an eigenvector $v_0 \in V_\lambda^{\mathfrak{h}}$. This is a common eigenvector for the whole algebra L and λ can be extended to a linear function on L such that $\varphi(x)v_0 = \lambda(x)v_0$. The theorem is proved. □

Lie's theorem implies the following corollary.

Corollary 1.6.7. *Let L be a finite dimensional Lie algebra over a field k which is algebraically closed and* $\operatorname{char} k = 0$. *Then*

1. *For any representation φ of a solvable Lie algebra L over a field k on a finite dimensional k-vector space V, there exists a basis of V in which all operators $\varphi(x)$, $x \in L$, are represented by upper triangular matrices.*
2. *A subalgebra $L \subset \mathfrak{gl}(V, k)$, where* $\dim V < \infty$, *is solvable if and only if in some basis the matrices of all operators from L are upper triangular.*
3. *If L is a solvable Lie algebra, then $[L, L]$ is a nilpotent Lie algebra.*

Proof.

1. By Lie's theorem there is a common eigenvector $v \in V$ for $\varphi(L)$. Set $v_1 = v$. Since the subspace $K_1 = \operatorname{Span}\{v_1\}$ is $\varphi(L)$-invariant, one can consider $V_1 = V/K_1$ and the induced quotient map $\varphi(L)$ on V_1. Applying the Lie theorem again there results a common vector v_2' for this quotient. If $v_2 \in V$ is some preimage of it, then $\varphi(L)v_2 \in K_1 + K_2 = V_2$, where $K_2 = \operatorname{Span}\{v_2\}$. Continuing this process there results a chain of subspaces $V_1 \subset V_2 \subset V_3 \subset \cdots$, where $V_i = V_{i-1} + K_i$, $K_i = \operatorname{Span}\{v_i\}$, such that $\varphi(x)v_i \in V_i$ for any $x \in L$. Then $\{v_1, v_2, \ldots\}$ is a basis as desired in which $\varphi(x)$ is represented by an upper triangular matrix for any $x \in L$.
2. If $L \subset \mathfrak{gl}(V, k)$ is solvable and $\dim V < \infty$, condition 1 of this corollary can be applied resulting in a basis with respect to which the matrices of all operators from L are upper triangular.

Conversely, assume that there is a basis of V such that the matrices of all operators from L are upper triangular. Consider the algebra $\mathfrak{h}$ which consists of all upper triangular matrices. This algebra is a solvable Lie algebra. Then L is a Lie subalgebra of $\mathfrak{h}$, and so is solvable as well.

3. By the Engel theorem to prove that $[L, L]$ is nilpotent it suffices to show that the algebra $\operatorname{ad}[L, L]$ is nilpotent. Since $\operatorname{ad} L \subset \mathfrak{gl}(V, k)$ is a solvable subalgebra, statement 1 of this corollary can be applied resulting a basis in which the matrices of all operators from L are upper triangular. But the commutator of two upper

triangular matrices is a strictly upper triangular[5] matrix, and thus a nilpotent matrix. So ad $[L, L]$ is nilpotent, and hence $[L, L]$ is nilpotent as well. □

Using this corollary and theorem 1.4.10 there immediately results the following theorem which gives necessary and sufficient conditions for a Lie algebra to be solvable.

Theorem 1.6.8. *Let L be a subalgebra of $\mathfrak{gl}(V, k)$, where V is a finite dimensional vector space over a field k which is algebraically closed and of* char $k = 0$. *Then L is solvable if and only if* $\mathrm{Tr}(xy) = 0$ *for all* $x \in [L, L]$, $y \in L$.

Proof. The sufficiency part of this theorem is exactly the statement of theorem 1.4.10.

Suppose that L is a solvable Lie algebra. Then, by corollary 1.6.7(2) there is a basis of V such that all the matrices in L are upper triangular, hence all matrices in $[L, L]$ are strictly upper triangular. Then $\mathrm{Tr}(xy) = 0$ for all $x \in [L, L]$, $y \in L$, since xy is strictly upper triangular. □

As several times before the proof given is complete only for the case that the ground field is algebraically closed. As both hypotheses and statement are invariant under extension of scalars the general characteristic zero case follows.

In terms of the Killing form from this theorem and theorem 1.4.14 there immediately follows a test for solvability of a Lie algebra which is known as Cartan's solvability criterion (or Cartan's first criterion):

Theorem 1.6.9 (Cartan's first criterion). *Let L be a finite dimensional algebra with Killing form κ. Then L is solvable if and only if $\kappa(x, y) = 0$ for all $x \in L$, $y \in [L, L]$.*

1.7. The Lie algebra $\mathfrak{sl}(2; k)$. Representation of $\mathfrak{sl}(2; k)$

Consider the special linear algebra $\mathfrak{sl}(n; k) = \{\mathbf{X} \in M_n(k) : \mathrm{Tr}(\mathbf{X}) = 0\}$ which is the set of all matrices with trace equal to zero under the bracket operation. Since $\mathfrak{sl}(n; k)$ is a proper subalgebra of $\mathfrak{gl}(n; k)$, the dimension of $\mathfrak{sl}(n; k)$ is not greater than $n^2 - 1$. On the other hand, there are $n^2 - 1$ linearly independent matrices with zero trace. These are e_{ij} if $i \neq j$ and $h_i = e_{ii} - e_{i+1,i+1}$. These matrices form a basis of $\mathfrak{sl}(n; k)$ and it is called the **standard basis** of $\mathfrak{sl}(n; k)$. So

$$\dim_k \mathfrak{sl}(n; k) = n^2 - 1.$$

Let $n = 2$ and consider $L = \mathfrak{sl}(2; k)$ in the case when k is an algebraically closed field and char $k = 0$. Then as was shown in example 1.1.1(3) $\dim_k L = 3$ and L has the standard basis consisting of 3 matrices:

$$x = \begin{pmatrix} 0 & 1 \\ 0 & 0 \end{pmatrix}, \quad y = \begin{pmatrix} 0 & 0 \\ 1 & 0 \end{pmatrix}, \quad h = \begin{pmatrix} 1 & 0 \\ 0 & -1 \end{pmatrix}.$$

The multiplication in L is completely defined by the equalities:

$$[x, y] = h, \quad [h, x] = 2x, \quad [h, y] = -2y. \tag{1.7.1}$$

[5]Strictly upper triangular means upper triangular with zero diagonal.

Suppose that $\mathcal{I}$ is a proper ideal in L, and $v = ax + by + ch$ is an arbitrary nonzero element from $\mathcal{I}$. Applying two times the transformation $\operatorname{ad} x$ to the element v one finds that $-2bx \in \mathcal{I}$. Applying two times the transformation $\operatorname{ad} y$ to the element v there results that $-2ay \in \mathcal{I}$. If $a \neq 0$ or $b \neq 0$ then $x \in \mathcal{I}$ or $y \in \mathcal{I}$, and from (1.7.1) it follows that $\mathcal{I} = L$. On the other hand, if $a = b = 0$ then h is a non-zero element from $\mathcal{I}$ and therefore $h \in \mathcal{I}$. Again from (1.7.1) it follows that $\mathcal{I} = L$. Thus L is a simple algebra[6].

The first thing to observe is that h acts semisimply on any finite dimensional module V. (At this stage it is only clear that it acts semisimply on the two dimensional space defining $\mathfrak{sl}(2;k)$ and acting on the three dimensional space $\mathfrak{sl}(2;k)$ via the adjoint representation.) To this end consider for each $r \in k$ the eigenspace

$$V_r = \{v \in V : \ h \cdot v = rv\}$$

and let V' be the sum of all these spaces. Note that $V' \neq 0$ as h must have at least one eigenvalue and hence at least one non-zero eigevector in V. It is also a submodule of V by the simple calculation of lemma 1.7.2 below. So if V is irreducible $V' = V$ and V manifests itself as the direct sum of the V_r. Using the Weyl theorem this is true for any finite dimensional representation (complete reducibility).

If r is not an eigenvalue for the endomorphism of V which represents h, $V_r = 0$. Recall that r is called a **weight** of h in V if $V_r \neq 0$ and that then V_r is called a **weight space**. Thus any finite dimensional representation of $\mathfrak{sl}(2;k)$ splits up as a direct sum of weight spaces.

Lemma 1.7.2. *If $v \in V_r$ then*
(i) $x \cdot v \in V_{r+2}$;
(ii) $y \cdot v \in V_{r-2}$;
(iii) $h \cdot v \in V_r$.

Proof. $h \cdot (x \cdot v) = [h,x] \cdot v + x \cdot (h \cdot v) = 2x \cdot v + rx \cdot v = (r+2)x \cdot v$.
The statement (ii) is proved similarly. The statement (iii) follows from the definition. □

From this lemma we immediately obtain that the space V' is a submodule of V. In particular, $V = V'$ if the module V is simple. In this case

$$V = \bigoplus_{r \in k} V_r.$$

Since V is finite dimensional this decomposition must be finite. Thus there must exist some $r \in k$ such that $V_r \neq 0$ and $V_{r+2n} = 0$ for all $n \in \mathbf{N}$. For such an r, any nonzero vector in V_r will be called a **maximal vector** of weight r.

From lemma 1.7.2 it also follows that x, y are nilpotent endomorphisms of V.

Lemma 1.7.3. *Let V be an irreducible L-module. Let $v_0 \in V_r$ be a maximal vector and set $v_{-1} = 0$, $v_i = (1/i!)y^i \cdot v_0$ for $i \geq 0$. Then*
(1) $h \cdot v_i = (r - 2i)v_i$,
(2) $y \cdot v_i = (i+1)v_{i+1}$,

---

[6] Similarly, all $\mathfrak{sl}(n;k)$ are simple, k any field (not necessarily algebraically closed).

(3) $x \cdot v_i = (r - i + 1)v_{i-1}$
for $i \geq 0$.

Proof.
(1) follows from repeated application of lemma 1.7.2.
(2) is just the definition.
(3) The case $i = 0$ is clear from lemma 1.7.2 and the definition of v_0. Let $i \geq 0$, and assume that the statement is true for that i. Then

$$\begin{aligned}(i+1)x \cdot v_{i+1} &= x \cdot y \cdot v_i = [x, y] \cdot v_i + y \cdot x \cdot v_i \\ &= (r - 2i)v_i + (r - i + 1)y \cdot v_{i-1} \\ &= (r - 2i)v_i + i(r - i + 1)v_i = (i + 1)(r - i)v_i.\end{aligned}$$

Then dividing both sides by $i + 1$ yields what is required. □

Thanks to this lemma, the nonzero vectors v_i are all linearly independent. Since $\dim_k V < \infty$, there exists a largest integer m for which $v_m \neq 0$, $v_{m+i} = 0$ for all $i > 0$. From lemma 1.7.2 it follows that the subspace W with basis $(v_0, v_1, v_2, ..., v_m)$ is a nonzero L-submodule in V. Because V is irreducible, $W = V$ and $\dim_k V = r+1$. From lemma 1.7.3(3) for $i = m + 1$ there results $0 = (r - m)v_m$. Since $v_m \neq 0$, it follows that $r = m$. In other words, the weight of a maximal vector is a nonnegative integer. We call it the **highest weight** of V. By lemma 1.7.3(1) the other weights are the m integers $m-2, m-4, \ldots, -m$ and each weight λ occurs with multiplicity one, i.e. $\dim V_\lambda = 1$ if $V_\lambda \neq 0$.

If $\rho_r : \mathfrak{sl}(2; k) \to \mathbf{M}_{r+1}$ is the corresponding matrix representation, then in the ordered basis $(v_0, v_1, v_2, ..., v_m)$

$$\rho_r(x) = \begin{pmatrix} 0 & \mu_1 & 0 & \ldots & 0 \\ 0 & 0 & \mu_2 & \ldots & 0 \\ 0 & 0 & 0 & \ldots & 0 \\ \vdots & \vdots & \vdots & \ddots & \vdots \\ 0 & 0 & 0 & \ldots & \mu_r \\ 0 & 0 & 0 & \ldots & 0 \end{pmatrix},$$

$$\rho_r(y) = \begin{pmatrix} 0 & 0 & 0 & \ldots & 0 & 0 \\ 1 & 0 & 0 & \ldots & 0 & 0 \\ 0 & 1 & 0 & \ldots & 0 & 0 \\ \vdots & \vdots & \vdots & \ddots & \vdots & \vdots \\ 0 & 0 & 0 & \ldots & 1 & 0 \end{pmatrix},$$

$$\rho_r(h) = \begin{pmatrix} r & 0 & 0 & \ldots & 0 \\ 0 & r-2 & 0 & \ldots & 0 \\ 0 & 0 & r-4 & \ldots & 0 \\ \vdots & \vdots & \vdots & \ddots & \vdots \\ 0 & 0 & 0 & \ldots & -r \end{pmatrix},$$

where $\mu_i = (r - i + 1)$.

To summarize:

Theorem 1.7.4. *Let V be an irreducible L-module, where $L = \mathfrak{sl}(2; k)$. Then*

(1) *Relative to* h, V *is a direct sum of weight spaces* V_λ, *where* $\lambda = m,\ m-2, \ldots, -(m-2), -m$, *and* $m+1 = \dim_k V$, $\dim V_\lambda = 1$ *for each* λ.

(2) V *has (up to nonzero scalar multiples) a unique maximal vector, whose weight is* m.

(3) *Any irreducible representation* ρ *of the Lie algebra* $\mathfrak{sl}(2;k)$ *of dimension* $r+1$ *is equivalent to the representation* ρ_r.

Corollary 1.7.5. *Any finite dimensional representation of the Lie algebra* $\mathfrak{sl}(2;k)$ *is equivalent to a representation of the form*

$$\rho_{r_1} \oplus \rho_{r_2} \oplus \cdots \oplus \rho_{r_s}.$$

Proof. This follows from theorems 1.5.4 and 1.7.4, taking into account that the Lie algebra $\mathfrak{sl}(2,k)$ is simple. □

Corollary 1.7.6. *Let* V *be any finite dimensional* L*-module, where* $L = \mathfrak{sl}(2;k)$. *Then the eigenvalues of* h *on* V *are all integers, and each occurs along with its negative (an equal number of times). Moreover, in any decomposition of* V *into a direct sum of irreducible submodules, the number of summands is equal to* $\dim V_0 + \dim V_1$. *Here* $V_0 = \{v \in V\ :\ hv = 0\}$, $V_1 = \{v \in V\ :\ hv = v\}$.

Proof. If $V = 0$ there is nothing to prove. Let $V \neq 0$, then V can be written as a direct sum of irreducible submodules by the Weyl theorem 1.5.4. Since any irreducible submodule is described by theorem 1.7.4, the first assertion of the corollary is obvious.

For the second assertion observe that each irreducible L-submodule has a unique occurrence of either weight 0 or weight 1 but not both. □

As all the weights are integers (and hence exist over the rationals) and $\mathfrak{sl}(2;\bar{k}) = \mathfrak{sl}(2;k) \otimes_k \bar{k}$ these results actually hold over any field of characteristic zero.

1.8. The universal enveloping algebra of A lie algebra

The universal enveloping algebra of a Lie algebra L is an analog of the group algebra of a group. It is an associative algebra which in general is infinite dimensional, but it does allow the use of associative methods. Moreover, it turns out that the ring theoretical properties of this algebra are closely related to the properties of its associated graded algebra. Also an L-module is basically the same thing as a module for the universal enveloping algebra of L.

To understand the basic idea of the construction of a universal enveloping algebra, first note that any associative algebra A over a field k becomes a Lie algebra over k under the commutator difference Lie bracket: $[a,b] = ab - ba$. That is, from an associative product one can construct a Lie bracket by simply taking the commutator (difference) with respect to that associative product. This algebra is denoted A_L.

The construction of the universal enveloping algebra attempts to reverse this process: for a given Lie algebra L over a field k find the "most general" associative k-algebra A with 1 such that the Lie algebra A_L contains L; this algebra A is $\mathfrak{U}(L)$. The important constraint is to preserve the representation theory: representations of a Lie algebra L are in one-to-one correspondence with representations of its associated universal enveloping algebra.

Definition. Let L be a Lie algebra over a field k. A pair $(\mathfrak{U}, h)$, where $\mathfrak{U}$ is an associative algebra with 1 over the field k, and $h : L \to \mathfrak{U}$ is a Lie algebra morphism

$$h([x,y]) = h(x)h(y) - h(y)h(x) \tag{1.8.1}$$

for $x, y \in L$, is an **enveloping algebra** of L.

Definition. An enveloping algebra $(\mathfrak{U}(L), h)$ of a Lie algebra L is called a **universal enveloping algebra** of L if it satisfies the following universal property: for any associative k-algebra A with 1 and any Lie morphism $f : L \to A$ there exists a unique algebra morphism $\varphi : \mathfrak{U}(L) \to A$ such that $\varphi h = f$, i.e. such that the following diagram is commutative.

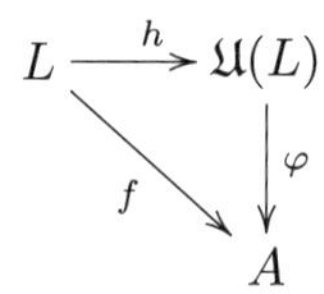

Theorem 1.8.2. *For any Lie algebra L, a universal enveloping algebra exists and is unique up to isomorphisms.*

Proof.

1. Uniqueness. Given another pair $(\mathfrak{U}', h')$ satisfying the same hypotheses, there result morphisms $\varphi : \mathfrak{U}(L) \to \mathfrak{U}'$, $\psi : \mathfrak{U}' \to \mathfrak{U}(L)$. By definition, there is a unique dotted map making the following diagram

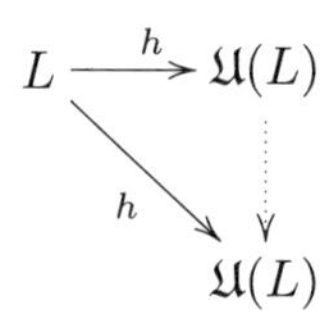

commute. But $1_{\mathfrak{U}(L)}$ and $\psi\varphi$ both do the trick, so $\psi\varphi = 1_{\mathfrak{U}(L)}$. Similarly, $\varphi\psi = 1_{\mathfrak{U}'(L)}$.

2. Existence. Let $\mathfrak{T}(L)$ be the tensor algebra on L (see section 2.1, [**42**]). Recall that $\mathfrak{T}(L) = \sum_{n=0}^{\infty} T^n L$, where $T^0 L = k$, $T^n L = L \otimes \cdots \otimes L$ (n copies) for $n > 0$. Let $\mathcal{J}$ be the two-sided ideal in $\mathfrak{T}(L)$ generated by all elements of the form $x \otimes y - y \otimes x - [x, y]$, where $x, y \in L$. Define

$$\mathfrak{U}(L) = \mathfrak{T}(L)/\mathcal{J},$$

and let $\pi : \mathfrak{T}(L) \to \mathfrak{U}(L)$ be the canonical quotient morphism. Note that $\mathcal{J} \subset \bigoplus_{i>0} T^i L$, so π maps $T^0 L = k$ isomorphically into $\mathfrak{U}(L)$. The claim is that $(\mathfrak{U}(L), h)$ is a universal enveloping algebra of L, where $h : L \to \mathfrak{U}(L)$ is the composite of $L \simeq T^1 L$ with the restriction of π to $T^1 L$. Indeed, let $f : L \to A$ be as in the definition. The universal property of $\mathfrak{T}(L)$ yields an algebra homomorphism $\varphi' : \mathfrak{T}(L) \to A$ which extends f and sends 1 to 1. The special property (1.8.1) of f forces all $x \otimes y - y \otimes x - [x, y]$ to lie in $\mathrm{Ker}\,(\varphi')$, so φ' induces a homomorphism $\varphi : \mathfrak{U}(L) \to A$ such that $\varphi h = f$. The uniqueness of φ is evident, since 1 and $\mathrm{Im}(h)$ together generate $\mathfrak{U}(L)$. $\square$

Let L be a Lie algebra with k-basis $\{x_i : i \in I\}$ and $[\ ,\ ]$ defined by $[x_i, x_j] = \sum_k c_{ij}^k x_k$. Then $\mathfrak{U}(L)$ is the quotient of the free (associative) algebra on the set $\{x_i\}$ by the ideal generated by the elements $x_i x_j - x_j x_i - [x_i, x_j]$. So $\mathfrak{U}(L)$ may be described as the associative k-algebra generated by the elements $\{x_i : i \in I\}$ with the relations $x_i x_j - x_j x_i = \sum_k c_{ij}^k x_k$.[7]

Examples 1.8.3.

1. Let L be an Abelian Lie algebra. Then the ideal $\mathcal{J}$ above is generated by all $x \otimes y - y \otimes x$. This means that $\mathfrak{U}(L)$ coincides with the symmetric algebra $\mathfrak{S}(L)$, considered in section 2.1, [**42**]. In particular, if L is a 1-dimensional Abelian Lie algebra then $\mathfrak{U}(L)$ is the polynomial algebra $k[x]$. In the general case, if L is an n-dimensional Abelian Lie algebra, $\mathfrak{U}(L)$ is a polynomial algebra in n variables.

2. Let L be the Lie algebra with a basis $\{x, y, z\}$ over a field k and bracket products as follows:

$$[x, y] = z; \quad [y, z] = x; \quad [z, x] = y.$$

Then the associative k-algebra $k[x, y, z]$ with relations

$$xy - yx = z; \quad yz - zy = x; \quad zx - xz = y$$

is the universal enveloping algebra $\mathfrak{U}(L)$ of L.

3. The Lie algebra $\mathfrak{sl}(2; k)$ of example 1.1.1(6) has the basis

$$e = \begin{pmatrix} 0 & 1 \\ 0 & 0 \end{pmatrix}; \quad f = \begin{pmatrix} 0 & 0 \\ 1 & 0 \end{pmatrix}; \quad h = \begin{pmatrix} 1 & 0 \\ 0 & -1 \end{pmatrix}$$

with relations

$$[e, f] = h, \quad [h, f] = -2f, \quad [h, e] = 2e.$$

Thus $\mathfrak{U}(\mathfrak{sl}(2; k))$ is the associative algebra generated by e, f, g with relations:

$$he - eh = 2e, \quad hf - fh = -2f, \quad ef - fe = h.$$

4. Let $\mathfrak{H}$ be the Heisenberg Lie algebra of upper triangular matrices of the form

$$\begin{pmatrix} 0 & * & * \\ 0 & 0 & * \\ 0 & 0 & 0 \end{pmatrix}$$

which has the basis

$$e = \begin{pmatrix} 0 & 0 & 0 \\ 0 & 0 & 1 \\ 0 & 0 & 0 \end{pmatrix}; \quad f = \begin{pmatrix} 0 & 1 & 0 \\ 0 & 0 & 0 \\ 0 & 0 & 0 \end{pmatrix}; \quad h = \begin{pmatrix} 0 & 0 & 1 \\ 0 & 0 & 0 \\ 0 & 0 & 0 \end{pmatrix}$$

with relations $[f, e] = h$, $[f, h] = [e, h] = 0$. Then $\mathfrak{U}(\mathfrak{H})$ is the associative algebra generated by elements e, f, g with relations:

$$fe - ef = h, \quad fh - hf = 0, \quad eh - he = 0.$$

The Weyl algebra W_1 of differential operators with polynomial coefficients, realized as an algebra of operators on the ring of polynomials $k[x]$, is generated by x (multiplication by x) and $\frac{d}{dx}$ (differentiation), and these operators satisfy $\frac{d}{dx}x - x\frac{d}{dx} = 1$.

---

[7] In some books this statement is considered as the definition of the universal enveloping algebra $\mathfrak{U}(L)$.

So the Weyl algebra W_1 is the quotient of the universal enveloping algebra of the Heisenberg algebra by the ideal generated by $h - 1$.

Remark 1.8.4. If M is a left (right) L-module, then the action of L on M gives a representation of L, i.e. a morphism of Lie algebras from L to the associative algebra $\mathrm{End}_k(M)$, which by the universal property of $\mathfrak{U}(L)$ extends to an action of $\mathfrak{U}(L)$ on M. This makes M into a $\mathfrak{U}(L)$-module. For example, if $m \in M$, $x, y \in L$ consider the monomial xy in $\mathfrak{U}(L)$. Then $(xy)m = x(y(m))$ (and if M is a right module $m(xy) = ((m)x)y$).

Conversely, if M is a $\mathfrak{U}(L)$-module then M is also an L-module by restriction.

Thus, there is a bijective correspondence between L-modules and $\mathfrak{U}(L)$-modules. Moreover, it is easy to show that M is a simple as an L-module if and only if M is simple as a $\mathfrak{U}(L)$-module. Therefore the theory of representations of Lie algebras can be considered as a part of the theory of representations of associative algebras.

Remark 1.8.5. Let L be a Lie algebra and $\mathfrak{U}(L)$ its universal enveloping algebra. Consider the morphism defined by $x \in L \subset \mathfrak{U}(L)$ goes to $1 \otimes x + x \otimes 1 \in \mathfrak{U}(L) \otimes \mathfrak{U}(L)$. This is a Lie algebra morphism and so there is a unique morphism of associative algebras $\mathfrak{U}(L) \to \mathfrak{U}(L) \otimes \mathfrak{U}(L)$ associated to it. This turns $\mathfrak{U}(L)$ into a Hopf algebra, see chapter 3.

Thus the universality property of $\mathfrak{U}(L)$ induces extra structure. This frequently happens with universal and adjoint constructions, see [**41**].

Remark 1.8.6. The situation $A \mapsto A_L$, the functor that associates the Lie algebra A_L to the associative unital algebra A and the functor $L \mapsto \mathfrak{U}(L)$ that associates the universal enveloping algebra of a Lie algebra to L is an example of an adjoint functor situation. This means that there is a functorial bijection

$$\mathbf{Lie}(L, A_L) \simeq \mathbf{Ass}(\mathfrak{U}(L), A).$$

Taking $A = \mathfrak{U}(L)$ in the functorial equation above there results a morphism $L \longrightarrow \mathfrak{U}(L)_L$ corresponding to the identity. This is the morphism h of the definition of a universal enveloping algebra above.

See e.g. [**62**] for a good deal of material on adjoint functors and their importance and pervasiveness in mathematics.

The functor $A \mapsto A_L$ is a kind of forgetful functor: it forgets parts of the structure of A. So $\mathfrak{U}(L)$ is to be thought of as the free associative algebra on the Lie algebra L. In category theory many of these adjoint functor problems are called 'universal problems.

1.9. Poincaré-Birkhoff-Witt theorem

This section is about the fundamental result known as the Poincaré-Birkhoff-Witt theorem, which gives a precise description of the universal enveloping algebra $\mathfrak{U}(L)$. The most important consequence of this theorem is that every Lie algebra L is naturally (i.e. functorially) embedded in $\mathfrak{U}(L)$ by $h : L \to \mathfrak{U}(L)$. The proof given here follows N. Jacobson [**46**] and P. Garrett [**36**].

Let L be a Lie algebra. Consider the tensor algebra $\mathfrak{T}(L)$ of L. If $\{x_i\}_{i \in I}$ is a basis of the Lie algebra L, where we fix a total ordering in I, then the tensor monomials $x_{i(1)} \otimes \cdots \otimes x_{i(n)}$ of degree n form a basis in $T^n L$ for any $n \geq 1$. Let $\mathcal{J}$ be the two-sided ideal in $\mathfrak{T}(L)$ generated by all elements of the form $x \otimes y - y \otimes x - [x, y]$, where $x, y \in L$. Then $\mathfrak{U}(L) = \mathfrak{T}(L)/\mathcal{J}$ is the universal enveloping algebra of $\mathfrak{T}(L)$. Let

$$\pi : \mathfrak{T}(L) \to \mathfrak{U}(L) = \mathfrak{T}(L)/\mathcal{J}$$

be the canonical projection. Then the images $\pi(x_{i(1)} \otimes \cdots \otimes x_{i(m)})$ in $\mathfrak{U}(L)$ of the tensor monomials $x_{i(1)} \otimes \cdots \otimes x_{i(n)}$ span the universal enveloping algebra over a field k.

Now let I be a totally ordered index set and let $\{u_i : i \in I\}$ be a basis of V. Such a basis is called (totally) ordered. To each monomial $u_\Psi = u_{k(1)} \cdots u_{k(n)}$ assign the sequence $\Psi = \{k(1), (k(2), \ldots, k(n)\}$. Such a sequence Ψ is called **nondecreasing** if

$$k(1) \leq k(2) \leq \cdots \leq k(n).$$

The number n is called the **length** of the sequence Ψ and it is also called the **length** of the corresponding monomial. This is written $l(\Psi) = n$. The monomial $u_\Psi = u_{k(1)} \cdots u_{k(n)}$ is called **standard** (or **ordered**) if the corresponding sequence Ψ is nondecreasing. A pair of numbers (i, j) forms an **inversion** in a sequence Ψ if $i < j$ but $k(i) > k(j)$. The number of inversions in the sequence Ψ is called the **defect** of the monomial u_Ψ. One of the main goals of this section is to prove that the standard monomials form a basis for the universal enveloping algebra $\mathfrak{U}(L)$.

Lemma 1.9.1. *There exists a linear map* $\sigma : \mathfrak{T}(L) \to \mathfrak{T}(L)$ *such that*

$$\sigma(1) = 1, \quad \sigma(x_{i(1)} \otimes x_{i(2)} \otimes \ldots \otimes x_{i(n)}) = x_{i(1)} \otimes x_{i(2)} \otimes \cdots \otimes x_{i(n)} \tag{1.9.2}$$

if $i(1) \leq i(2) \leq \ldots \leq i(n)$, *and*

$$\begin{aligned} &\sigma(x_{i(1)} \otimes \cdots \otimes x_{i(t)} \otimes x_{i(t+1)} \otimes \cdots \otimes x_{i(n)}) \\ &= \sigma(x_{i(1)} \otimes \cdots \otimes x_{i(t+1)} \otimes x_{i(t)} \otimes \cdots \otimes x_{i(n)}) \\ &+ \sigma(x_{i(1)} \otimes \cdots \otimes [x_{i(t)}, x_{i(t+1)}] \otimes \cdots \otimes x_{i(n)}). \end{aligned} \tag{1.9.3}$$

whenever $i(t) > i(t+1)$.

Proof. This lemma is proved by induction on the degree n and defect m. Let $\{x_i : i \in I\}$ be an ordered basis of L over a field k. First, define σ to be identity on $T^0L = k$ and $T^1L = L$, i.e. $\sigma(1) = 1$ and $\sigma(x_i) = x_i$ for all $x_i \in L$. Now fix n. Assume that σ has been defined for all tensor monomials of degree $\leq n-1$ in $\mathfrak{T}(L)$ and that is has properties (1.9.2) and (1.9.3). It needs to be shown that σ can be extended to a map with properties (1.9.2) and (1.9.3) on all tensor monomials of degree $\leq n$ in $\mathfrak{T}(L)$.

For monomials of degree n with $i(t) > i(t+1)$ set

$$\begin{aligned} \sigma(x_{i(1)} \otimes \cdots \otimes x_{i(n)}) &= \sigma(x_{i(1)} \otimes \cdots \otimes x_{i(t+1)} \otimes x_{i(t)} \otimes \cdots \otimes x_{i(n)}) \\ &+ \sigma(x_{i(1)} \otimes \cdots \otimes [x_{i(t)}, x_{i(t+1)}] \otimes \cdots \otimes x_{i(n)}). \end{aligned} \tag{1.9.4}$$

The first term on the right side of (1.9.4) is of smaller defect and the other one is of lower length (degree). Thus induction can be used on the degree of tensor monomials, and for each fixed degree induction on defect.

If a monomial $x_{i(1)} \otimes \cdots \otimes x_{i(n)}$ has a defect 0 or 1 properties (1.9.2) and (1.9.3) follow immediately. Therefore it remains only to check that this map is well-defined for monomials with defect ≥ 2. Suppose there are two pairs: $i(k)$, $i(k+1)$ with $i(k) > i(k+1)$ and $i(l)$, $i(l+1)$ with $i(l) > i(l+1)$. There are only two essential cases which must be checked: $l > k+1$ and $l = k+1$. Consider each case separately.

1) $l > k+1$. Without loss of generality one can simplify the notations by taking $n = 4$ and write

$$u = x_{i(k)}, \quad v = x_{i(k+1)}, \quad w = x_{i(l)}, \quad z = x_{i(l+1)},$$

since all other factors in the tensor monomials are left untouched. Then first applying the inductive definition (1.9.4) to the pair u, v results in:

$$\sigma(u \otimes v \otimes w \otimes z) = \sigma(v \otimes u \otimes z \otimes w) + \sigma([u,v] \otimes z \otimes w) =$$

$$= \sigma(v \otimes u \otimes z \otimes w) + \sigma(v \otimes u \otimes [w,z]) + \sigma([u,v] \otimes z \otimes w) + \sigma([u,v] \otimes [w,z].$$

On the other hand applying (1.9.4) first to the pair w, z there results

$$\sigma(u \otimes v \otimes w \otimes z) = \sigma(u \otimes v \otimes z \otimes w) + \sigma(u \otimes v \otimes [w,z]) =$$

$$= \sigma(v \otimes u \otimes z \otimes w) + \sigma([u,v] \otimes z \otimes w) + \sigma(v \otimes u \otimes [w,z]) + \sigma([u,v] \otimes [w,z]).$$

It is clear that this is the same expression as the one obtained before.

2) $l = k+1$. Thus $i(l+1) < i(k+1) < i(k)$. This time one can consider that $n = 3$ and write

$$u = x_{i(k)}, \quad v = x_{i(k+1)}, \quad w = x_{i(l+1)}.$$

Applying sequentially the inductive definition (1.9.4) to the pairs u, v; u, w and v, w in that order there results:

$$\begin{aligned}\sigma(u \otimes v \otimes w) &= \sigma(v \otimes u \otimes w) + \sigma([u,v] \otimes w)\\ &= \sigma(v \otimes w \otimes u) + \sigma(v \otimes [u,w]) + \sigma([u,v] \otimes w)\\ &= \sigma(w \otimes v \otimes u) + \sigma([v,w] \otimes u) + \sigma(v \otimes [u,w]) + \sigma([u,v] \otimes w).\end{aligned}$$

Now first applying the inductive definition (1.9.4) to interchanging pair v, w there results

$$\begin{aligned}\sigma(u \otimes v \otimes w) &= \sigma(u \otimes w \otimes v) + \sigma(x \otimes [v,w])\\ &= \sigma(w \otimes u \otimes v) + \sigma([u,w] \otimes v) + \sigma(u \otimes [v,w])\\ &= \sigma(w \otimes v \otimes u) + \sigma(w \otimes [u,v]) + \sigma([u,w] \otimes v) + \sigma(u \otimes [v,w]).\end{aligned}$$

So it remains to show that

$$\begin{aligned}F = \sigma([v,w] \otimes u) + \sigma(v \otimes [u,w]) + \sigma([u,v] \otimes w)\\ -\sigma(w \otimes [u,v]) - \sigma([u,w] \otimes v) - \sigma(u \otimes [v,w]) = 0.\end{aligned} \tag{1.9.5}$$

But all summands in (1.9.5) are tensor monomials of degree $\leq n-1$ and so the inductive definition (1.9.4) can be applied to (1.9.4), which gives

$$\begin{aligned}F =& \sigma(u \otimes [v,w]) + \sigma([[v,w],u])) + \sigma([u,w] \otimes v)\\ &+ \sigma([v,[u,w]]) + \sigma(w \otimes [u,v]) + \sigma([[u,v],w])\\ &- \sigma(w \otimes [u,v]) - \sigma([u,w] \otimes v) - \sigma(u \otimes [v,w])\\ &= \sigma([[v,w],u]) + \sigma([v,[u,w]]) + \sigma([[u,v],w]) = 0,\end{aligned}$$

by the Jacobi identity, the property of antisymmetry and linearity of σ. Thus σ is well-defined for monomials of degree less than or equal to n, as well. Therefore σ can be extended linearly to all monomials of degree $\leq n$ in $\mathfrak{T}(L)$. In this way, σ is defined for all monomials from $\mathfrak{T}(L)$ and satisfies properties (1.9.2), (1.9.3).

The lemma is proved. $\square$

Theorem 1.9.6 (Poincaré-Birkhoff-Witt). *The standard monomials form a basis for the universal enveloping algebra $\mathfrak{U}(L) = \mathfrak{T}(L)/\mathcal{J}$ of a Lie algebra L over a field k. In particular, the Lie algebra L itself injects into the enveloping algebra $\mathfrak{U}(L)$. Thus, any Lie algebra over a field is isomorphic to a Lie subalgebra of an associative algebra.*

Proof. Let L be a Lie algebra over a field k with universal enveloping algebra $\mathfrak{U}(L) = \mathfrak{T}(L)/\mathcal{J}$, where $\mathcal{J}$ is the two-sided ideal in $\mathfrak{T}(L)$ generated by all elements of the form $x \otimes y - y \otimes x - [x, y]$ for $x, y \in L$. Let $\pi : \mathfrak{T}(L) \to \mathfrak{U}(L)$ be the canonical projection and let $h : L \to \mathfrak{U}(L)$ be the restriction of π to L. Let $\{x_i\}_{i \in I}$ be a basis of L, and write $y_i = h(x_i)$. The task at hand is to show that the elements $h(x_{i(1)}) \cdots h(x_{i(m)}) = \pi(x_{i(1)} \otimes \cdots \otimes x_{i(m)})$, $m \in \mathbf{Z}^+$, $i(1) \le i(2) \le \cdots \le i(m)$, together with 1, form a basis of $\mathfrak{U}(L)$, i.e. they are linearly independent over k and span $\mathfrak{U}(L)$.

1. The first step is to prove that this set of elements spans $\mathfrak{U}(L)$. This statement is proved by induction on the length and the defect of monomials. Obviously, the set of all unordered (= not necessarily standard) monomials $y_\Psi = y_{k(1)} \cdots y_{k(n)}$ spans $\mathfrak{U}(L)$ as a vector space.

Given a monomial $y_\Psi = y_{k(1)} \cdots y_{k(n)}$ with an inversion $k(t) > k(t+1)$ replace it by

$$y_{k(1)} \cdots y_{k(t+1)} y_{k(t)} \cdots y_{k(n)} + y_{k(1)} \cdots [y_{k(t)}, y_{k(t+1)}] \cdots y_{k(n)} \tag{1.9.7}$$

using that $x_{k(t)} \otimes x_{k(t+1)} - x_{k(t+1)} \otimes x_{k(t)} - [x_{k(t)}, x_{k(t+1)}] \in \mathcal{J}$.

The second summand of (1.9.7) is a linear combinations of tensor monomials of smaller degree and the first summand is of smaller defect. So for each fixed degree the inductive assumption on the number of defects can be applied.

2. Second, prove that the set of elements

$$y_{i(1)} \cdots y_{i(m)}, \quad m \in \mathbf{Z}^+, \quad i(1) \le i(2) \le \cdots \le i(m),$$

together with 1, is linearly independent in $\mathfrak{U}(L)$.

From lemma 1.9.1 it follows that there exists a linear map

$$\sigma : \mathfrak{T}(L) \to \mathfrak{T}(L)$$

with properties (1.9.2) and (1.9.3), which implies $\sigma(\mathcal{J}) = 0$, i.e. $\mathrm{Ker}(\pi) = \mathcal{J} \subseteq \mathrm{Ker}(\sigma)$. Then there is a morphism (of vector spaces) $\varphi : \mathfrak{U}(L) \to \mathfrak{T}(L)$:

$$\begin{array}{ccc} \mathfrak{T}(L) & \xrightarrow{\pi} & \mathfrak{U}(L) \\ \downarrow \sigma & \swarrow \varphi & \\ \mathfrak{T}(L) & & \end{array}$$

such that $\varphi\pi = \sigma$. Because σ acts as the identity on any linear combinations of standard monomials, by lemma 1.9.1 it follows that the images of the standard monomials in $\mathfrak{U}(L)$ are a linearly independent set.

If $h : L \to \mathfrak{U}(L)$ is the restriction of π to L, then φh is the identity map. Therefore h is injective.

This completes the proof. □

Remark 1.9.8. A basis of $\mathfrak{U}(L)$ consisting of standard monomials as constructed in theorem 1.9.6 is often called a **PBW basis**.

Corollary 1.9.9. *Let H be a sub Lie algebra of L, and extend an ordered basis $\{h_j : j \in J\}$ of H to an ordered basis $\{h_j : j \in J\} \cap \{x_i : i \in I\}$ of L. Then the homomorphism $\mathfrak{U}(H) \to \mathfrak{U}(L)$ induced by the injection $H \to L \to \mathfrak{U}(L)$ is itself injective, and $\mathfrak{U}(L)$ is a free $\mathfrak{U}(H)$-module with free basis consisting of all $x_{i(1)} \cdots x_{i(m)}$, $i(1) \leq i(2) \leq \cdots \leq i(m)$, together with 1.*

Let L be a trivial Lie algebra, i.e. a vector space V made into a Lie algebra by declaring all brackets to be zero. Then the ideal $\mathcal{J}$ in $\mathfrak{T}(L)$ generating the universal enveloping algebra $\mathfrak{U}(L) = \mathfrak{T}(L)/\mathcal{J}$ is generated by elements $x \otimes y - y \otimes x$, $x, y \in L$, and $\mathfrak{T}(L)/\mathcal{J}$ is just the symmetric algebra $\mathfrak{S}(L)$. So the universal enveloping algebra of the trivial Lie algebra is the symmetric algebra.

Consider now the general case when L is an arbitrary Lie algebra with its tensor algebra $\mathfrak{T}(L)$ and universal enveloping algebra $\mathfrak{U}(L) = \mathfrak{T}(L)/\mathcal{J}$. If $\mathcal{I}$ is the two sided ideal in $\mathfrak{T}(L)$ generated by all $x \otimes y - y \otimes x$, $x, y \in L$, then $\mathfrak{S}(L) = \mathfrak{T}(L)/\mathcal{I}$ is the symmetric algebra of L and $\mathfrak{S}(V) = \bigoplus_{i=0}^{\infty} S^i L$ (see section 2.1, [**42**]).

Write $\mathfrak{T} = \mathfrak{T}(L)$, $\mathfrak{S} = \mathfrak{S}(L)$, and $\mathfrak{U} = \mathfrak{U}(L)$. Also let $T^m = T^m L$, $S^m = S^m L$. Define a filtration on $\mathfrak{T}$ by $T_m = T^0 \oplus T^1 \oplus \cdots \oplus T^m$, and let $U_m = \pi(T_m)$, $U_{-1} = 0$, where π is the canonical algebra morphism from $\mathfrak{T}$ to $\mathfrak{U}$. Then $U_m U_n \subset U_{m+n}$ and $U_m \subset U_{m+1}$. Define $G^m = U_m / U_{m-1}$ and $\mathfrak{G} = \mathfrak{G}(L) = \bigoplus_{m=0}^{\infty} G^m$ with a multiplication $G^m \times G^n \to G^{m+n}$ induced by the multiplication in $\mathfrak{U}$ and extended to a bilinear map $\mathfrak{G} \times \mathfrak{G} \to \mathfrak{G}$, making $\mathfrak{G}$ a graded associative algebra with 1.

Since π maps T^m into U_m, the composite linear map $\varphi_m : T^m \to U_m \to G^m = U_m \setminus U_{m-1}$ makes sense. It is surjective, because $\pi(T_m \setminus T_{m-1}) = U_m \setminus U_{m-1}$. Therefore the maps φ_m yield a linear map $\varphi : \mathfrak{T} \to \mathfrak{S}$, which is surjective.

Lemma 1.9.10. *The map $\varphi : \mathfrak{T}(L) \to \mathfrak{G}(L)$ is an algebra morphism. Moreover, $\varphi(\mathcal{I}) = 0$, so φ induces an algebra morphism $w : \mathfrak{S}(L) = \mathfrak{T}(L)/\mathcal{I} \to \mathfrak{G}(L)$ which is surjective.*

Proof. Let $x \in T^m$, $y \in T^p$. By definition of the product in $\mathfrak{G}$, $\varphi(xy) = \varphi(x)\varphi(y)$, so it follows that φ is a homomorphism. Let $x \otimes y - y \otimes x$ $(x, y \in L)$ be a generator of $\mathcal{I}$. Then $\pi(x \otimes y - y \otimes x) \in U_2$, by definition. On the other hand, since $x \otimes y - y \otimes x - [x, y] \in \mathcal{J}$, $\pi(x \otimes y - y \otimes x) = \pi([x, y]) \in U_1$, whence $\varphi(x \otimes y - y \otimes x) \in U_1/U_1 = 0$. It follows that $\mathcal{I} \subset \operatorname{Ker}\varphi$. So $\mathfrak{G}(L)$ is commutative graded associative algebra. Hence, by the the universal property of the symmetric algebra, there is a unique algebra homomorphism

$$w : \mathfrak{S}(L) = \mathfrak{T}(L)/\mathcal{I} \to \mathfrak{G}(L)$$

extending the linear map $L \to \mathfrak{G}(L)$, $x \mapsto \varphi(x)$. Since the $\varphi(x)$ generate $\mathfrak{G}(L)$ as an algebra, this map is surjective.

In fact it follows from the Poincaré-Birkhoff-Witt theorem that the algebras $\mathfrak{S}$ and $\mathfrak{G}$ are isomorphic.

Theorem 1.9.11 (Poincaré-Birkhoff-Witt). *Let L be a Lie algebra over a field k with universal enveloping algebra $\mathfrak{U} = \mathfrak{U}(L)$ and symmetric algebra $\mathfrak{S} = \mathfrak{S}(L)$, and let $\mathfrak{G}$ be the graded algebra associated with the filtration defined above on $\mathfrak{U}$. Then the homomorphism $w : \mathfrak{S} \to \mathfrak{G}$, constructed in lemma* 1.9.10, *is an isomorphism of algebras.*

Proof. Let

$$x_M = x_{i(1)} \cdots x_{i(m)} = \pi(x_{i(1)} \otimes \cdots \otimes x_{i(m)}), \quad m \in \mathbf{Z}^+,$$

be an ordered monomial of length $l(M) = m$. The elements $w(x_M)$ are the images under w of the monomial basis of $S^m(L)$. By lemma 1.9.10, w is surjective. So it is enough to prove that w is injective. Suppose $w(y) = 0$, where $y \in \mathfrak{G}$. There results a relation of the form

$$\sum_{l(M)=n} c_M x_M = \sum_{l(M)<n} c_M x_M$$

with some non-zero coefficients on the left hand side. But any non-trivial relation between the x_M can be written in the above form by moving the terms of the highest length to one side. So $y = 0$, i.e. w is an isomorphism. □

1.10. Free Lie algebras

Let L be a Lie algebra over a field k and $X = \{v_i : i \in I\}$ a set of elements from L. Let L_X be the smallest subalgebra in L which contains the set X. If $L = L_X$ one says that L is generated by the set X. In this case one also says that the elements v_i $(i \in I)$ generate L.

Definition. A **free Lie algebra** on a set X is a pair $(\mathfrak{F}(X), \iota)$, consisting of a Lie algebra $\mathfrak{F}(X)$ and map

$$\iota : X \longrightarrow \mathfrak{F}(X)$$

with the following universal property:

Given any mapping φ of X into a Lie algebra M, there exists a unique Lie algebra homomorphism

$$\psi : \mathfrak{F}(X) \to M$$

extending φ, i.e. such that the diagram

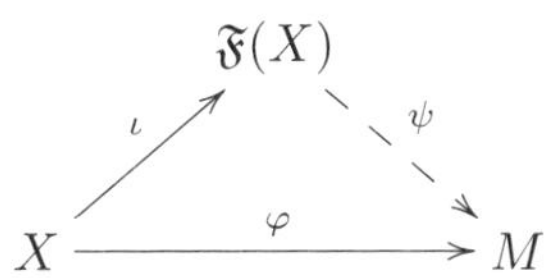

(1.10.1)

commutes.

The Poincaré-Birkhoff-Witt theorem makes it possible to establish the existence and uniqueness of a free Lie algebra generated by a set X.

Proposition 1.10.2. *Given a nonempty set X, there is a unique (up to isomorphism) free Lie algebra $\mathfrak{F}(X)$ over a field k on X. The image of X in $\mathfrak{F}(X)$ generates $\mathfrak{F}(X)$.*

Proof. First the existence of such an algebra is proved. Let V be the k-vector space with X as basis. Form the tensor algebra $\mathfrak{T}(V)$ and view it as a Lie algebra via the bracket operation. Let $\mathfrak{F}(X)$ be the Lie subalgebra of $\mathfrak{T}(V)$ generated by X. Given any map $\varphi : X \to M$, extend φ first to a linear map $V \to M$, and let $i_M : M \longrightarrow \mathfrak{U}(M)$ be the canonical injection, where $\mathfrak{U}(M)$ is the universal Lie algebra of M. Then, as $\mathfrak{T}(V)$ is the free associative algebra on V there is a canonical associative algebra morphism $\beta : \mathfrak{T}(V) \to \mathfrak{U}(M)$ such that the diagram

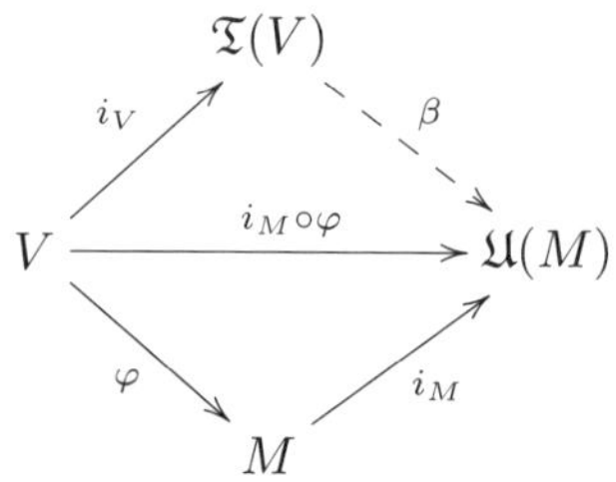

(1.10.3)

commutes.

Then for any $x \in X$, the commutativity of this diagram implies that

$$\beta(i_V(x)) = i_M(\varphi(x)). \tag{1.10.4}$$

Therefore considering (1.10.4) as a homomorphism of Lie algebras via the bracket operation there results that $\beta(\mathfrak{F}(X)) \subseteq \mathrm{Im}(i_M)$. Since by the Poincaré-Birkhoff-Witt theorem the map $i_M : M \longrightarrow \mathfrak{U}(M)$ is injective, the map $\psi = i_M^{-1} \circ \beta$ is the desired Lie algebra morphism $\psi : \mathfrak{F}(X) \to M$ making (1.10.1) commute.

It remains to show that the morphism ψ defined above is unique. First observe that $\psi(i_V(x)) = \varphi(x)$ for all $x \in X$ and that the elements $i_V(x)$ generate the Lie algebra $\mathfrak{F}(X)$. Since ψ is a Lie algebra homomorphism, ψ is completely determined on all elements of $\mathfrak{F}(X)$, which shows the uniqueness of ψ. Thus, $\mathfrak{F}(X)$ is a free Lie algebra generated by X.

The uniqueness (up to a unique isomorphism) of such an algebra $\mathfrak{F}(X)$ is proved in the same way as the uniqueness of the tensor algebra (see proposition 2.1.2, [**42**]). □

The construction $X \mapsto \mathfrak{F}(X)$ obviously extends to a functor from the category of sets to the category of Lie algebras and so the cardinality of X completely determines the free Lie algebra $\mathfrak{F}(X)$ and is called its **rank**.

By the Poincare-Birkhoff-Witt theorem the symmetric algebra and the free Lie algebra are of the same size (meaning that if the both sides are graded by giving elements of X degree 1 then they are isomorphic as graded vector spaces). This can be used to describe the dimension of the (vector space) summand of the free Lie algebra of any given degree.

Note that if $L = \mathfrak{F}(X)$ is free on a set X, then a vector space V can be given an L-module structure simply by assigning to each $x \in X$ an element of the Lie algebra $\mathfrak{gl}(V)$ and extending canonically.

Finally, if $L = \mathfrak{F}(X)$ is free on X, and if R is the ideal of L generated by elements f_j (j running over some index set), L/R is said to be the Lie algebra with generators x_i and relations $f_j = 0$, where the x_i are the images in L/R of the elements of X.

The following result is known as Shirshov's theorem or the Shirshov-Witt theorem. It was independently proved in 1953 by A. I. Shirshov (1921-1981) and in 1956 by Ernst Witt (see [**76**], [**95**]).

Theorem 1.10.5. *A subalgebra M of a free Lie algebra over a field k is itself a free Lie algebra.*

The main step in the proof of this theorem by A.I.Shirshov is the remarkable result that if a set of homogeneous elements in a free Lie algebra has the property that no element of it is contained in the subalgebra generated by the other elements, then this subset is a free generating set for the subalgebra it generates. For proofs of this freeness theorem see the original publications of A. I. Shirshov [**76**], E. Witt [**95**] and also [**71**].

1.11. Examples of simple Lie algebras

In the theory of Lie algebras the simple algebras are very important. As in many parts of mathematics the "simples" play the special role of "elementary building blocks". In this case for Lie algebras. And, on the other hand, the Lie algebra of any compact Lie group is a direct sum of simple Lie algebras and one-dimensional Lie algebras with zero Lie bracket. The classification of simple Lie algebras over an algebraically closed field of characteristic zero was done by W. Killing and E. Cartan. Simple Lie algebras over the reals were classified by E. Cartan. For information in the case of characteristic see e.g. [**81**]. A main fact is that there are several additional series of finite dimensional simple Lie algebras that are analogues of infinite dimensional Lie algebras of differential operators in characteristic zero. In characteristics 2, 3, 5 there is still more and the classification job is not yet finished.

In the case of an algebraically closed field of characteristic zero all simple finite dimensional Lie algebras have been explicitly listed. A remarkable result is that over such a field all simple finite dimensional Lie algebras are divided into two essential types: "classical" Lie algebras and "exceptional" Lie algebras. There are four series of Lie algebras of classical type: A_n, B_n, C_n, D_n and further there are five exceptional simple Lie algebras G_2, F_4, E_6, E_7, E_8.

This section contains a brief description of all these finite dimensional simple Lie algebras over an algebraically closed field k of characteristic zero.

1. $A_n = \mathfrak{sl}(n+1)$

As in example 1.1.1(5) the algebra $A_n = \mathfrak{sl}(n+1;k)$ is the subalgebra of the algebra of all $(n+1)\times(n+1)$-matrices over the field k with trace equal to zero and with the usual bracket $[\mathbf{X},\mathbf{Y}] = \mathbf{XY} - \mathbf{YX}$. So

$$A_n = \{\mathbf{X} \in M_{n+1}(k) \ : \mathrm{Tr}(\mathbf{X}) = 0\}, \quad n \geq 1$$

The dimension of A_n over k is $(n+1)^2 - 1$.

2. $B_n = \mathfrak{so}(2n+1)$

This is the subalgebra of the algebra of all $(2n+1)\times(2n+1)$-matrices with the following property

$$B_n = \{\mathbf{X} \in M_{2n+1}(k) \ : \mathbf{XB} + \mathbf{B}^T\mathbf{X} = 0\}, \quad n \geq 2,$$

where

$$\mathbf{B} = \begin{pmatrix} 1 & 0 & 0 \\ 0 & 0 & \mathbf{E}_n \\ 0 & \mathbf{E}_n & 0 \end{pmatrix}$$

and $\mathbf{E}_n \in M_n(k)$ is the identity matrix. The bracket (Lie product) is $[\mathbf{X},\mathbf{Y}] = \mathbf{XY} - \mathbf{YX}$.

The dimension of B_n over k is $n(2n+1)$.

If $k = \mathbf{C}$ is the field of complex numbers, the Lie algebra $\mathfrak{so}(2n+1;\mathbf{C}) = \mathfrak{so}(2n+1)$ consists of the $(2n+1) \times (2n+1)$ skew symmetric complex matrices with trace equal to zero.

3. $C_n = \mathfrak{sp}(2n)$

This is the subalgebra of the algebra of all $2n \times 2n$-matrices with the following property

$$C_n = \{\mathbf{X} \in M_{2n}(k) \ : \mathbf{XC} + \mathbf{C}^T\mathbf{X} = 0\}, \quad n \geq 3,$$

where

$$\mathbf{C} = \begin{pmatrix} 0 & \mathbf{E}_n \\ -\mathbf{E}_n & 0 \end{pmatrix},$$

and $\mathbf{E}_n \in M_n(k)$ is the identity matrix.

The dimension of C_n over k is $n(2n+1)$.

4. $D_n = \mathfrak{so}(2n)$

This is the subalgebra of the algebra of all $2n \times 2n$-matrices with the following property

$$D_n = \{\mathbf{X} \in M_{2n}(k) \ : \mathbf{XD} + \mathbf{D}^T\mathbf{X} = 0\}, \quad n \geq 4,$$

where

$$\mathbf{D} = \begin{pmatrix} 0 & \mathbf{E}_n \\ \mathbf{E}_n & 0 \end{pmatrix},$$

and $\mathbf{E}_n \in M_n(k)$ is the identity matrix.

The dimension of D_n over k is $n(2n-1)$.

If $k = \mathbf{C}$ is the field of complex numbers, the algebra $\mathfrak{so}(2n;C) = \mathfrak{so}(2n)$ consists of $2n \times 2n$ skew symmetric complex matrices with with trace equal to zero.

The finite dimensional Lie algebras A_n, B_n, C_n and D_n, considered above, are called the "classical" simple Lie algebras. Besides these there are five so-called "exceptional" simple Lie algebras, which are described now.

5. G_2

The algebra G_2 is the algebra of derivations of the Cayley-Dickson algebra (octonions). This is a 14-dimensional Lie algebra.

6. F_4

The algebra F_4 is the algebra of derivations of the unique exceptional Jordan algebra. This exceptional Jordan algebra can be represented as the algebra of Hermitian matrices of order three over the Cayley-Dickson algebra (octonions). It is a 52-dimensional Lie algebra.

7. E_6

The algebra E_6 is the linear envelope of the derivations and the multiplications by elements of the simple exceptional Jordan algebra. This is a 78-dimensional Lie algebra.

8. E_7 and E_8

These are a 133-dimensional and 248-dimensional Lie algebra respectively. The algebras E_7 and E_8 are also connected with the Cayley-Dickson algebra.

For details on the exceptional Lie algebras and realizations (models) of them, see e.g. [**47**].

1.12. Abstract root systems and the Weyl group

The classification of semisimple Lie algebras, as it will be given in section 1.15, is reduced to the classification of certain objects, called root systems. The main properties of these root systems may be understood by studying abstract root systems and that is what will be done in this section. In the next section it will be shown in what way root systems gives rise to (somethings called) Dynkin diagrams.

Let V be a fixed Euclidean space, i.e. a finite dimensional vector space over the field of the real numbers $\mathbf{R}$ endowed with a positive definite symmetric bilinear form (α, β) for $\alpha, \beta \in V$, which is often called the standard inner product. Let $P_\alpha = \{\beta \in V : (\beta, \alpha) = 0\}$ be the hyperplane defined by α. Define a linear mapping $s_\alpha : V \longrightarrow V$ by the formula:

$$s_\alpha(\beta) = \beta - \frac{2(\beta, \alpha)}{(\alpha, \alpha)}\alpha \tag{1.12.1}$$

which is called the **reflection** in α. This is a linear transformation of V which fixes all points in the hyperplane P_α and sends any vector orthogonal to this hyperplane into its negative; indeed:

$$s_\alpha(\alpha) = \alpha - \frac{2(\alpha, \alpha)}{(\alpha, \alpha)}\alpha = -\alpha.$$

Of course, nonzero vectors proportional to α yield the same reflection. Moreover,

$$s_\alpha^2(\beta) = s_\alpha(\beta) - s_\alpha(\frac{2(\beta, \alpha)}{(\alpha, \alpha)}\alpha) = \beta - \frac{2(\beta, \alpha)}{(\alpha, \alpha)}\alpha - \frac{2(\beta, \alpha)}{(\alpha, \alpha)}(-\alpha) = \beta$$

i.e. s_α^2 is the identity transformation of V.

It is also easy to check that s_α preserves the inner product, that is

$$(s_\alpha(x), s_\alpha(y)) = (x, y) \quad \text{for all } x, y \in V$$

Definition. Let V be a finite dimensional Euclidean space over a field k with the standard Euclidean inner product denoted by $(\ , \ )$. An **abstract root system** is a pair (V, Φ), where Φ is a finite set of non-zero vectors in V (called **roots**), such that the following properties hold:

1. The root system $\Phi \neq 0$ spans V, i.e. $V = \text{Span}\{\Phi\}$.

2. The only scalar multiples of a root $\alpha \in \Phi$ that belong to Φ are α itself and $-\alpha$, that is, if $\alpha \in \Phi$ then $n\alpha \in \Phi$ if and only if $n = 1$ or $n = -1$.

3. For any two roots $\alpha, \beta \in \Phi$

$$s_\alpha(\beta) = \beta - 2\frac{(\beta, \alpha)}{(\alpha, \alpha)}\alpha \in \Phi. \tag{1.12.2}$$

4. (Integrality condition). For any two roots $\alpha, \beta \in \Phi$

$$\langle \beta, \alpha \rangle = 2\frac{(\beta, \alpha)}{(\alpha, \alpha)} \in \mathbf{Z}. \tag{1.12.3}$$

In what follows we shall often call an abstract root system simply a **root system** [8].

In view of property 3, the integrality condition is equivalent to stating that β and its reflection $s_\alpha(\beta)$ differ by an integer multiple of α.

Note that the product

$$\langle\, , \rangle : \Phi \times \Phi \to \mathbf{Z}$$

defined by (1.12.3) is not an inner product. It is not necessarily symmetric and it is linear only in the first argument.

The **rank** of an abstract root system Φ is the dimension of V.

There are strong geometric constraints on how a root system can look. E.g. the angle between two unequal roots can only be one of 30°, 45°, 60°, 90°, 120°, 135°, 150°, 180°. This follows from the following statement.

Lemma 1.12.4 (Finiteness lemma). *For any roots* $\alpha, \beta \in \Phi$ with $\beta \neq \pm\alpha$

$$\langle \alpha, \beta \rangle \langle \beta, \alpha \rangle \in \{0, 1, 2, 3\}.$$

Proof. Since the inner product is a positive bilinear form,

$$(\alpha + t\beta, \alpha + t\beta) \geq 0$$

for all $t \in \mathbf{R}$. This means that the discriminant of the polynomial

$$t^2(\alpha, \alpha) + 2t(\alpha, \beta) + (\beta, \beta)$$

should be non-positive, i.e.

$$(\alpha, \beta)^2 - (\alpha, \alpha)(\beta, \beta) \leq 0.$$

Thus

$$\frac{(\alpha, \beta)^2}{(\alpha, \alpha)(\beta, \beta)} \leq 1. \tag{1.12.5}$$

(Note that (1.12.5) is the well-known Schwartz inequality.)

Taking into account the integrality condition (1.12.3), it follows that

$$\langle \alpha, \beta \rangle \langle \beta, \alpha \rangle = \frac{4(\alpha, \beta)^2}{(\alpha, \alpha)(\beta, \beta)} \leq 4$$

and so $\langle \alpha, \beta \rangle \langle \beta, \alpha \rangle$ is a non-negative integer ≤ 4. If it were be equal to 4, then the equal sign in (1.12.5) would hold making β a multiple of α which by axiom 2 would imply $\beta = \pm\alpha$. □

For the Euclidean space V the standard inner product can be written in the geometrical form $(\alpha, \beta) = \|\alpha\| \cdot \|\beta\| \cos\theta$, where θ is the angle between the vectors α and β, $\|x\|$ is the Euclidean norm of a vector $x \in V$. The finiteness lemma describes all possibilities for the angle θ and the coefficient $\dfrac{\|\beta\|}{\|\alpha\|}$ for $\beta \neq \pm\alpha$ and $\|\beta\| \geq \|\alpha\|$ as represented in the following table.

[8] In some of the literature condition 2 is omitted. Then what is called "root system" here is referred to as a reduced root system.

$\langle\alpha,\beta\rangle$	$\langle\beta,\alpha\rangle$	θ	$\dfrac{\|\beta\|}{\|\alpha\|}$
0	0	$\pi/2$	undetermined
1	1	$\pi/3$	1
-1	-1	$2\pi/3$	1
1	2	$\pi/4$	$\sqrt{2}$
-1	-2	$3\pi/4$	$\sqrt{2}$
1	3	$\pi/6$	$\sqrt{3}$
-1	-3	$5\pi/6$	$\sqrt{3}$

This table also yields the following useful statement.

Lemma 1.12.6. *Let α and β be roots, $\|\beta\| \geq \|\alpha\|$ and $\beta \neq \pm\alpha$.*

1. *If $(\alpha,\beta) > 0$ (i.e. the angle between α and β is strictly acute), then $\alpha - \beta$ is a root.*

2. *If $(\alpha,\beta) < 0$ (i.e. the angle between α and β is strictly obtuse), then $\alpha + \beta$ is a root.*

Proof. By (1.12.2), $s_\beta(\alpha) = \alpha - \langle\alpha,\beta\rangle\beta$ is a root. From the table above it follows that if $(\alpha,\beta) > 0$ then $\langle\alpha,\beta\rangle = 1$, which implies that $\alpha - \beta$ is a root. And if the angle between α and β is strictly acute, then $\langle\alpha,\beta\rangle = -1$, and so $\alpha + \beta$ is a root. □

Definition. If α and β are roots that are not scalar multiplies of each other the set $\{\beta + i\alpha : \ i \in \mathbf{Z}\} \cap \Phi$ is called the α-**string** of roots containing β.

Lemma 1.12.7 ('string property'). *The α-string of roots containing β is of the form $\{\beta + i\alpha : \ -p \leq i \leq q\}$ where $p, q \geq 0$ and*

$$p - q = \langle\beta,\alpha\rangle \leq 4. \tag{1.12.8}$$

Proof. Let $p, q \in \mathbf{Z}^+$ be the largest integers for which $\beta - p\alpha \in \Phi$ and $\beta + q\alpha \in \Phi$. Suppose there exists an integer $i \in [-p, q]$ such that $\beta + i\alpha \notin \Phi$. Then there exist integers $s, t \in [-p, q]$, $s < t$ such that $\beta + (s+1)\alpha \notin \Phi$, $\beta + s\alpha \in \Phi$, $\beta + (t-1)\alpha \notin \Phi$, $\beta + t\alpha \in \Phi$. From lemma 1.12.6 it follows that $(\alpha, \beta + s\alpha) \geq 0$ and $(\alpha, \beta + t\alpha) \leq 0$ simultaneously, which is impossible, since $s < t$ and $(\alpha,\alpha) > 0$. Therefore all vectors of the form $\{\beta + i\alpha : \ -p \leq i \leq q\}$ are roots. In particular, $s_\alpha(\beta + q\alpha) = \beta - p\alpha$. Since $s_\alpha(\beta + q\alpha) = \beta - \langle\beta,\alpha\rangle\alpha - q\alpha$, there results that $p - q = \langle\beta,\alpha\rangle \leq 4$. □

Example 1.12.9.

Let $\dim V = 1$, then there is only one root system of rank 1 consisting of two nonzero vectors $\{\alpha, -\alpha\}$. This root system is called A_1:

$$A_1 : \qquad -\alpha \longleftarrow \bullet \longrightarrow \alpha$$

This is the root system of $\mathfrak{sl}(2;\mathbf{C})$.

Two abstract root systems may be combined by regarding the Euclidean spaces they span as mutually orthogonal subspaces of a common Euclidean space. An abstract root system which does not arise from such a combination is said to be **irreducible**. More strongly, an abstract root system Φ is called **irreducible** (or **indecomposable**) if it cannot be decomposed into a union $\Phi_1 \cup \Phi_2$ of two disjoint orthogonal nonempty subsets Φ_1, Φ_2 of Φ. Any root system can be presented as a union of a number of irreducible root systems.

Example 1.12.10.
Let $\dim V = 2$, then up to scalars there are four different root systems of rank 2: $A_1 \times A_1$, A_2, B_2 and G_2.

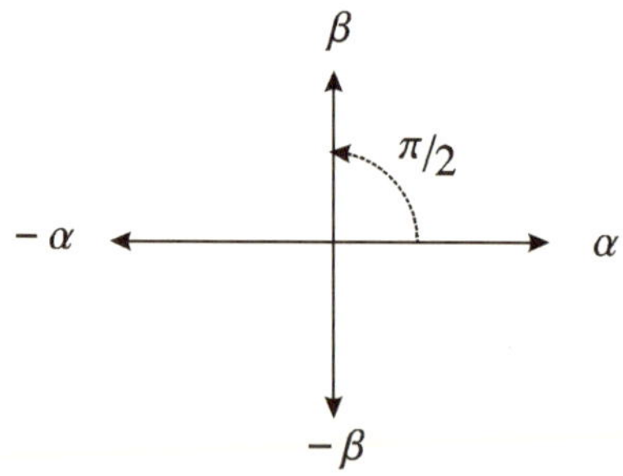

$A_1 \times A_1$

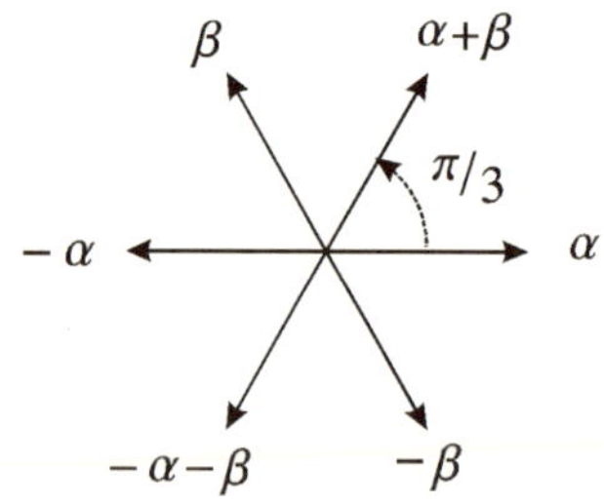

A_2

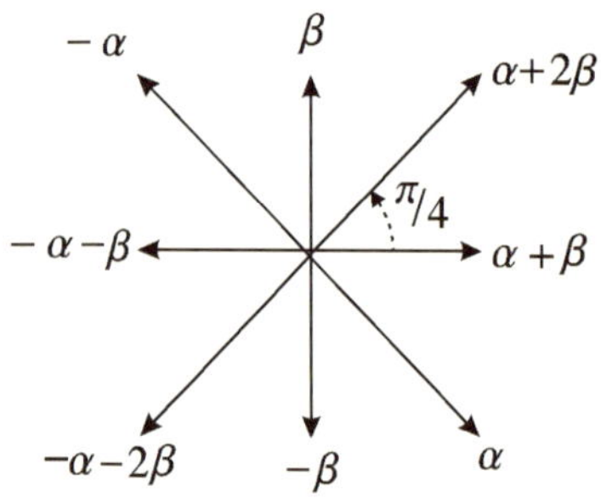

B_2

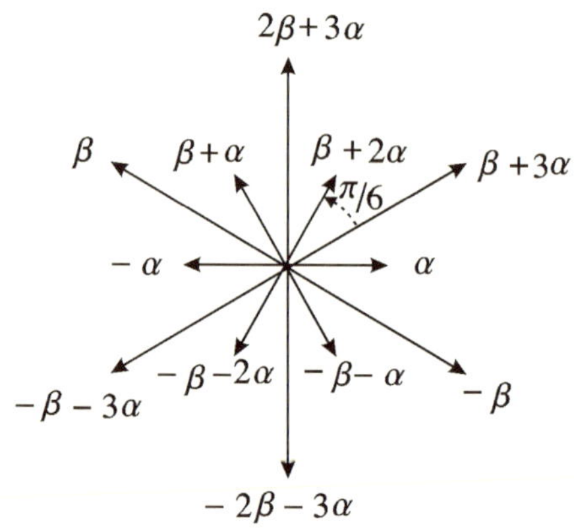

G_2

The first one $A_1 \times A_1$ is the root system of $\mathfrak{sl}(2;\mathbf{C}) \times \mathfrak{sl}(2;\mathbf{C}) \cong \mathfrak{so}(4;\mathbf{C})$. It consists of four vectors $\{-\beta, -\alpha, \beta, \alpha\}$, such that α and β are orthogonal, and any ratio $\|\alpha\| : \|\beta\|$ is permissible. $A_1 \times A_1$ is the direct sum of two root systems A_1, and so it is an example of a reducible system. The other three root systems are irreducible systems.

A_2 is the root system of $\mathfrak{sl}(3;\mathbf{C})$. It consists of six vectors of the same norm, and the angle between adjacent vectors is equal to $\pi/3$.

B_2 is the root system of $\mathfrak{so}(5, \mathbf{C}) \cong \mathfrak{sp}(4; \mathbf{C})$. It consists of eight vectors of two different lengths with ratio $\|\beta\| : \|\alpha\| = \sqrt{2}$ and the angle between adjacent vectors is equal to $\pi/4$.

The root system G_2 consists of twelve vectors of two different lengths with ratio $\|\beta\| : \|\alpha\| = \sqrt{3}$ and the angle between adjacent vectors is equal to $\pi/6$.

Example 1.12.11.
Besides the direct sums of A_1 with one of those of rank 2, up to scalars there are only three irreducible root systems of rank 3: A_3, B_3 and C_3.

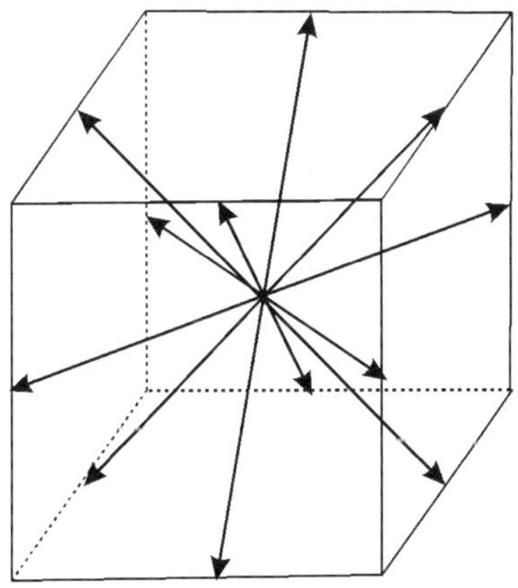

A_3

A_3 is the root system of $\mathfrak{sl}(4; \mathbf{C}) \cong \mathfrak{so}(6; \mathbf{C})$. It consists of 12 vectors of the same norm. Take a cube with center at the origin, edges parallel to the coordinate axes. The root vectors of A_3 are the twelve vectors from the origin to the midpoints of the 12 edges.

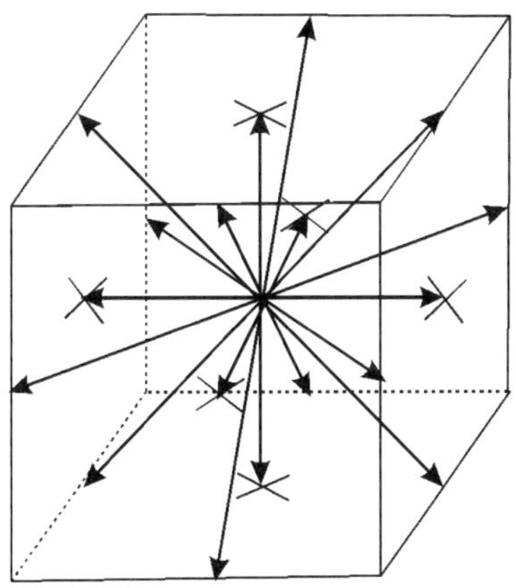

B_3

B_3 is the root system of $\mathfrak{so}(7; \mathbf{C})$. It consists of 18 vectors. Take a cube with center at the origin, edges parallel to the coordinate axes of length 2. So the eight vertices are at positions $(\pm 1, \pm 1, \pm 1, \pm 1)$. The root vectors of B_3 are the six vectors of length 1 from the origin to the midpoints of the six faces, and the twelve vectors of length $\sqrt{2}$ from the origin to the midpoints of the 12 edges.

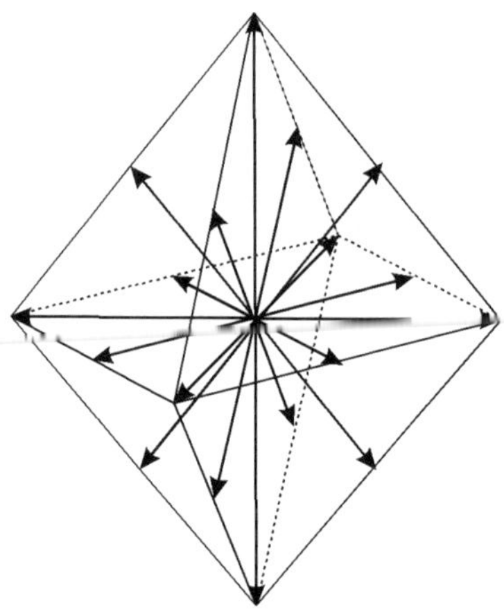

C_3

C_3 is the root system of $\mathfrak{sp}(6;\mathbf{C})$. It consists also of 18 vectors. Take the regular octahedron with vertices at $(\pm 2,0,0)$, $(0,\pm 2,0)$, $(0,0,\pm 2)$. The root vectors of C_3 are the six vectors of length 2 from origin to these six vertices, and the twelve vectors of length $\sqrt{2}$ from the origin to the midpoints of the 12 edges.

It is easy to prove the following statement:

Proposition 1.12.12. *An abstract root system (V,Φ) is decomposable if and only if $V = V_1 \oplus V_2$, where $V_1 \perp V_2$ and $\Phi = \Phi_1 \cup \Phi_2$ where $\Phi_1 = V_1 \cap \Phi_1$ and $\Phi_2 = V_2 \cap \Phi_2$.*

Two irreducible abstract root systems (V_1,Φ_1) and (V_2,Φ_2) are **isomorphic** if there is a vector space isomorphism $\varphi : V_1 \to V_2$ such that $\varphi(\Phi_1) = \Phi_2$ and which preserves distance up to a (constant) scalar factor, i.e. $(\varphi(a),\varphi(b)) = d(a,b)$ for all $a,b \in V_1$ and a constant $d \in \mathbf{R}^+$.

Let (V,Φ) be an abstract root system, and let $f : V \to \mathbf{R}$ be a linear function such that $f(\alpha) \neq 0$ for all $\alpha \in \Phi$. The set

$$\Phi^+ = \{\alpha \in \Phi \ : \ f(\alpha) > 0\}$$

of Φ is called the subset of **positive roots** of Φ. Analogously, the set

$$\Phi^- = \{\alpha \in \Phi \ : \ f(\alpha) < 0\}$$

of Φ is called the subset of **negative roots** of Φ. It is easy to show that the subset Φ^+ satisfies the following properties:

1) for each root $\alpha \in \Phi$ exactly one of the roots α, $-\alpha$ is contained in Φ^+;
2) for any $\alpha,\beta \in \Phi^+$ such that $\alpha+\beta$ is a root, $\alpha+\beta \in \Phi^+$.

Note that $\Phi^- = -\Phi^+$ and $\Phi = \Phi^+ \cup \Phi^-$.

A positive root is called **simple** if it cannot be represented as a sum of two positive roots. The set of all simple roots is denoted $\Delta \subset \Phi^+$.

The set Δ is called **indecomposable** if it cannot be decomposed into a union of orthogonal non-empty subsets $\Delta = \Delta_1 \cup \Delta_2$; i.e. such that $(\Delta_1, \Delta_2) = 0$.

A root θ with maximal value of $f(\theta)$ is called a **highest root** (or a **maximal root**). The following theorem shows, in particular, that for each choice of positive roots there exists a unique set of simple roots so that the positive roots are exactly those roots that can be expressed as a combination of simple roots with non-negative coefficients.

Proposition 1.12.13.

1. *If $\alpha, \beta \in \Delta$, $\alpha \neq \beta$, then $\alpha - \beta \notin \Phi$ and $(\alpha, \beta) \leq 0$. If θ is a highest root, then $(\theta, \alpha) \geq 0$ and $\theta + \alpha \notin \Phi$ for any $\alpha \in \Phi^+$.*
2. *Any positive root is a linear combination of simple roots with nonnegative integer coefficients. Any negative root is a linear combination of simple roots with nonpositive integer coefficients.*
3. *If $\alpha \in \Phi^+ \backslash \Delta$, then $\alpha - \gamma$ is a root for some $\gamma \in \Delta$. Moreover, $\alpha - \gamma \in \Phi^+$.*
4. *The simple roots are linearly independent.*

Proof.

1. Let $\alpha, \beta \in \Delta$, $\alpha \neq \beta$. Suppose $\gamma = \alpha - \beta \in \Phi$. Then either $\gamma \in \Phi^+$ or $-\gamma \in \Phi^+$. If $\gamma \in \Phi^+$, then $\alpha = \beta + \gamma \in \Phi^+$ which contradicts the simplicity of α. Analogously if $-\gamma \in \Phi^+$, then $\beta = \alpha + \gamma \in \Phi^+$. Another contradiction.

If $\alpha, \beta \in \Delta$, then by the previous argument $\alpha - \beta$ is not a root and so from (1.12.1) it follows that $p = 0$. And thus $p - q \leq 0$, that is, $(\alpha, \beta) \leq 0$.

Since θ is a highest root and $f(\theta + \alpha) = f(\theta) + f(\alpha) > f(\theta)$, there results that $\theta + \alpha \notin \Phi$. Then from lemma 1.12.7 it follows that $q = 0$. So $p \geq 0$ and $(\theta, \alpha) \geq 0$.

2. This is proved by induction on the value of $f(\alpha)$. Since Δ is a finite set we can order the finite set $S = \{f(\alpha_i) \mid \alpha_i \in \Delta\}$. If α is a simple root then everything is done. If $\alpha \in \Phi^+$ and α is not simple then $\alpha = \beta + \gamma$, where β and γ are in Φ^+. Then $f(\alpha) > f(\beta) > 0$ and $f(\alpha) > f(\gamma) > 0$. Applying induction on the ordered set S, it follows that β and γ are linear combinations of simple roots with nonnegative integer coefficients, and hence so is α. Taking into account that $\Phi^- = -\Phi^+$ we obtain the second statement.

3. Suppose the opposite, i.e. $\alpha - \gamma$ is not a root for any $\gamma \in \Delta$. Then, by the string property, $(\alpha, \gamma) \leq 0$ for any $\gamma \in \Delta$. Hence $(\alpha, \alpha) = (\alpha, \sum\limits_{\alpha_i \in \Delta} x_i \alpha_i) \leq 0$, since all $x_i \geq 0$, by statement 2. But $(\alpha, \alpha) > 0$ as $\alpha \in \Phi^+ \backslash \Delta$. This contradiction shows that there exists a $\gamma \in \Delta$ such that $\alpha - \gamma \in \Phi \cup \{0\}$. Since $\alpha \notin \Delta$ and $\gamma \in \Delta$, there results that $\alpha - \gamma \in \Phi$.

To see that $\alpha - \gamma$ is a positive root let $\alpha - \gamma = \beta < 0$, then $\gamma = \alpha + (-\beta)$, which contradicts the simplicity of γ.

4. Suppose that the simple roots $\alpha_1, \alpha_2, ..., \alpha_n \in \Delta$ are not linearly independent. Collecting all the terms with positive coefficients to one side we obtain an element:

$$\varepsilon = \sum_{\alpha_i \in \Delta'} x_i \alpha_i = \sum_{\alpha_j \in \Delta''} y_j \alpha_j, \tag{1.12.14}$$

where all the x_i, y_j are positive numbers. Then

$$(\varepsilon, \varepsilon) = (\sum_{\alpha_i \in \Delta'} x_i \alpha_i, \sum_{\alpha_i \in \Delta'} x_i \alpha_i) = (\sum_{\alpha_i \in \Delta'} x_i \alpha_i, \sum_{\alpha_j \in \Delta''} y_j \alpha_j).$$

The left hand side of this equality is the inner product of a vector with itself and hence nonnegative. The right hand side is a sum of nonpositive terms by statement 1 because Δ' and Δ'' are disjoint. Hence there is a contradiction unless $\varepsilon = 0$. On the other hand $f(\varepsilon) = f(\sum\limits_i x_i\alpha_i) > 0$, since $\alpha_i \in \Phi^+$ and at least one of the x_i is strictly positive. So we must have $x_i = 0$ for all i. And similarly $y_j = 0$ for all j. □

Proposition 1.12.15. *A highest root (with respect to a chosen functional f) θ in an indecomposable root system Φ is unique. It is a linear combination of simple roots with positive coefficients. $(\theta, \alpha) \geq 0$ for all $\alpha \in \Phi^+$ and $(\theta, \alpha) > 0$ for at least one $\alpha \in \Delta$.*

Proof. Let Δ be a set of simple roots of Φ. Suppose that θ is a highest root of an indecomposable root system Φ. Obviously, $\theta \neq 0$. By proposition 1.12.13(2), $\theta = \sum\limits_{\alpha_i \in \Delta} x_i\alpha_i$, where α_i are simple roots and all integers $x_i \geq 0$. We shall show that all the $x_i \geq 1$. Suppose that this is not the case, i.e. there exists i such that $x_i = 0$. Let $\alpha_j \in \Delta$ such that $x_j = 0$. Then $\theta + \alpha_j \notin \Phi$, by proposition 1.12.13(1), and $\theta - \alpha_j \notin \Phi$, by proposition 1.12.13(2). Therefore, by the string property, $(\theta, \alpha_j) = 0$. On the other hand, $(\theta, \alpha_j) = \sum\limits_{\alpha_i \in \Delta} x_i(\alpha_i, \alpha_j)$. Since $(\alpha_i, \alpha_j) \leq 0$, by proposition 1.12.13(1), the only possibility for $(\theta, \alpha_j) = 0$ is $(\alpha_i, \alpha_j) = 0$ for all α_i such that $x_i \neq 0$ and all α_j such that $x_j = 0$. Then, $(\beta, \gamma) = 0$ for every $\beta \in V_1 = \mathrm{Span}\{\alpha_i \in \Delta \; : \; x_i \neq 0\}$ and every $\gamma \in V_2 = \mathrm{Span}\{\alpha_j \in \Delta \; : \; x_j = 0\}$. Since $V_1 \oplus V_2 = \mathrm{Span}(\Phi)$, $V_1 \perp V_2$, and both V_1 and V_2 are nonempty, the root system Φ is decomposable, by proposition 1.12.12. This contradiction shows that all integers $x_i \geq 1$.

Suppose that there exist two highest roots θ_1 and θ_2 in an indecomposable root system Φ. Then $\theta_1 + \theta_2$ and $\theta_1 - \theta_2$ are not roots, since $f(\theta_1 + \theta_2) > f(\theta_1)$, and $f(\theta_1 - \theta_2) = 0$. Therefore, by the string property, $(\theta_1, \theta_2) = 0$. As shown above, $\theta_1 = \sum\limits_{\alpha_i \in \Delta} x_i\alpha_i$ with $x_i \geq 1$ and $\theta_2 = \sum\limits_{\alpha_i \in \Delta} y_i\alpha_i$ with $y_i \geq 1$. By theorem 1.12.13(1), $(\theta_1, \alpha_i) \geq 0$ for all $\alpha_i \in \Delta$. By theorem 1.12.13(3), there exists a simple root α_k such that $\theta_1 - \alpha_k \in \Phi^+$. Since θ_1 is a highest root, $\theta_1 + \alpha_k \notin \Phi^+$. Therefore, by the string property, $(\theta_1, \alpha_k) > 0$. Since $\theta_2 = \sum\limits_{\alpha_i \in \Delta} y_i\alpha_i$ with $y_i \geq 1$, there results, that $(\theta_1, \theta_2) \neq 0$. This contradiction shows that there exists only one highest root in an indecomposable root system. □

Definition. Let Φ be a root system in the Euclidean vector space V. A subset $S \subset V$ is called a **base** of the root system Φ if

(1) S is a vector space basis of V;

(2) every root $\beta \in \Phi$ can be written in the form $\beta = \sum\limits_{\alpha \in S} x_\alpha \alpha$, where the $x_\alpha \in \mathbf{Z}$, are all simultaneously non-positive or non-negative.

From proposition 1.12.13 there immediately results the following statement.

Corollary 1.12.16. *The set of all simple roots $\Delta \subset \Phi$ is a base of Φ.*

Examples 1.12.17. In all examples given below ε_i is the vector from a vector space $\mathbf{R}^n$ with i-th entry 1 and all other entries zero. Therefore $(\varepsilon_i, \varepsilon_j) = \delta_{ij}$.

These examples can be used to give a description of the roots for all examples of root systems considered in example 1.12.10 and example 1.12.11.

A_1. Write $\alpha = \varepsilon$, then the root system $\Phi = \{-\alpha, \alpha\}$. Take $f \in V^*$ given by $f(\varepsilon) = 1$, then

$$\Phi^+ = \Delta = \{\alpha\},$$

and the highest root $\theta = \alpha$.

A_2. Here

$$\Phi = \{\varepsilon_i - \varepsilon_j \ : \ i \neq j; \ i, j = 1, 2, 3\}.$$

There are 6 such roots, and all of them have the same length. Take $f \in V^*$ given by $f(\varepsilon_1) = 3$, $f(\varepsilon_2) = 2$ and $f(\varepsilon_3) = 1$. Then

$$\Phi^+ = \{\varepsilon_1 - \varepsilon_2, \ \varepsilon_1 - \varepsilon_3, \ \varepsilon_2 - \varepsilon_3\}.$$

Write $\alpha = \varepsilon_1 - \varepsilon_2$, $\beta = \varepsilon_2 - \varepsilon_3$, which are simple roots. Then $\Delta = \{\alpha, \beta\}$, $\Phi = \{\pm\alpha, \ \pm\beta, \ \pm(\alpha + \beta)\}$,

$$\Phi^+ = \{\alpha, \beta, \alpha + \beta\}.$$

The highest root is $\theta = \varepsilon_1 - \varepsilon_3 = \alpha + \beta$.

B_2. Here $\Phi = \{\pm\varepsilon_1 \pm \varepsilon_2; \ \pm\varepsilon_1; \ \pm\varepsilon_2\}$. Take $f \in V^*$ given by $f(\varepsilon_1) = 2$, and $f(\varepsilon_2) = 1$. Then

$$\Phi^+ = \{\varepsilon_1 + \varepsilon_2, \ \varepsilon_1 - \varepsilon_2, \ \varepsilon_1, \ \varepsilon_2\}.$$

Write $\alpha = \varepsilon_1 - \varepsilon_2$, $\beta = \varepsilon_2$, which have different lengths and the angle between them is equal to $3\pi/4$. They are all simple roots. Then $\Delta = \{\alpha, \beta\}$, and

$$\Phi^+ == \{\alpha, \beta, \ \alpha + \beta, \ \alpha + 2\beta\}.$$

The highest root is $\theta = \varepsilon_1 + \varepsilon_2 = \alpha + 2\beta$.

G_2. Here $\Phi = \{\pm(\varepsilon_1 - \varepsilon_2), \ \pm(\varepsilon_1 - \varepsilon_3), \ \pm(\varepsilon_2 - \varepsilon_3), \ \pm(\varepsilon_1 + \varepsilon_2 - 2\varepsilon_3), \ \pm(\varepsilon_1 + \varepsilon_3 - 2\varepsilon_2), \ \pm(\varepsilon_2 + \varepsilon_3 - 2\varepsilon_1)\}$. This gives 12 roots as were shown in example 1.12.10.

Take $f \in V^*$ given by $f(\varepsilon_1) = 2$, $f(\varepsilon_2) = 1$ and $f(\varepsilon_3) = 4$. Write $\alpha = \varepsilon_1 - \varepsilon_2$, $\beta = \varepsilon_2 + \varepsilon_3 - 2\varepsilon_1$, which have different lengths and the angle between them is equal to $5\pi/6$. They are simple roots. Then $\Phi = \{\pm\alpha, \ \pm\beta, \ \pm(\alpha+\beta), \ \pm(2\alpha+\beta), \ \pm(3\alpha+\beta), \ \pm(3\alpha+\beta)\}$,

$$\Phi^+ = \{\alpha, \beta, \ \alpha + \beta, \ 2\alpha + \beta, \ 3\alpha + \beta, \ 3\alpha + 2\beta\},$$

and $\Delta = \{\alpha, \beta\}$. The highest root is $\theta = 3\alpha + 2\beta$.

A_3. Here $\Phi = \{\pm(\varepsilon_i - \varepsilon_j) \ : \ i, j = 1, 2, 3, 4\}; \ i \neq j$. Take $f \in V^*$ given by $f(\varepsilon_1) = 4$, $f(\varepsilon_2) = 3$, $f(\varepsilon_3) = 2$, and $f(\varepsilon_4) = 1$. Then $\Phi^+ = \{\varepsilon_i - \varepsilon_j \ : i < j; \ i, j = 1, 2, 3, 4\}$. The number of roots is 12, and all of them have the same length. Write $\alpha_1 = \varepsilon_1 - \varepsilon_2$, $\alpha_2 = \varepsilon_2 - \varepsilon_3$, $\alpha_3 = \varepsilon_3 - \varepsilon_4$. Then

$$\Phi^+ = \{\alpha_1, \ \alpha_2, \ \alpha_3, \ \alpha_1 + \alpha_2, \ \alpha_1 + \alpha_2 + \alpha_3, \ \alpha_2 + \alpha_3\}.$$

The highest root is $\theta = \varepsilon_1 - \varepsilon_4 = \alpha_1 + \alpha_2 + \alpha_3$.

B_3. Here $\Phi = \{\pm\varepsilon_i \pm \varepsilon_j \ (i \neq j); \ \pm\varepsilon_i; \ i, j = 1, 2, 3\}$. Take $f \in V^*$ given by $f(\varepsilon_1) = 3$, $f(\varepsilon_2) = 2$ and $f(\varepsilon_3) = 1$. Then $\Phi^+ = \{\varepsilon_i + \varepsilon_j \ (i \neq j); \ \varepsilon_i - \varepsilon_j \ (i < j); \ \varepsilon_i; \ i, j = 1, 2, 3\} = \{\varepsilon_1 - \varepsilon_2, \ \varepsilon_1 - \varepsilon_3, \ \varepsilon_2 - \varepsilon_3, \ \varepsilon_1 + \varepsilon_2, \ \varepsilon_1 + \varepsilon_3, \ \varepsilon_2 + \varepsilon_3, \ \varepsilon_1, \varepsilon_2, \ \varepsilon_3\}$. Write $\alpha_1 = \varepsilon_1 - \varepsilon_2$, $\alpha_2 = \varepsilon_2 - \varepsilon_3$, $\alpha_3 = \varepsilon_3$, which are simple roots. Then $\Phi^+ =$

$\{\alpha_1,\ \alpha_2,\ \alpha_3,\ \alpha_1+\alpha_2,\ \alpha_2+\alpha_3,\ \alpha_1+\alpha_2+\alpha_3,\ \alpha_1+\alpha_2+2\alpha_3,\ \alpha_2+2\alpha_3,\ \alpha_1+2\alpha_2+2\alpha_3\}$. The highest root is $\theta = \varepsilon_1 + \varepsilon_2 = \alpha_1 + 2\alpha_2 + 2\alpha_3$.

C_3. Here $\Phi = \{\pm\varepsilon_i \pm \varepsilon_j\ (i \neq j);\ \pm 2\varepsilon_i;\ i,j = 1,2,3\}$. Take $f \in V^*$ given by $f(\varepsilon_1) = 3$, $f(\varepsilon_2) = 2$ and $f(\varepsilon_3) = 1$. Write $\alpha_1 = \varepsilon_1 - \varepsilon_2$, $\alpha_2 = \varepsilon_2 - \varepsilon_3$, $\alpha_3 = 2\varepsilon_3$, which are simple roots. Then $\Phi^+ = \{\alpha_1,\ \alpha_2,\ \alpha_3,\ \alpha_1+\alpha_2,\ \alpha_2+\alpha_3,\ \alpha_1+\alpha_2,\ 2\alpha_2+\alpha_3,\ \alpha_1+\alpha_2+\alpha_3,\ \alpha_1+2\alpha_2+\alpha_3,\ 2\alpha_1+2\alpha_2+\alpha_3$. The highest root is $\theta = 2\varepsilon_1 = 2\alpha_1 + 2\alpha_2 + \alpha_3$.

Let Φ be a root system in a vector space V. For each root $\alpha \in \Phi$ there is a reflection s_α defined by (1.12.1). As was shown above s_α^2 is the identity mapping on V. So every reflection is an invertible linear map on V and one can consider the group of all reflections s_α which is a subgroup in $GL(V)$.

Definition. The subgroup of $GL(V)$ generated by all reflections s_α defined by (1.12.1) for each $\alpha \in \Phi$, is called the **Weyl group** of Φ, and denoted by $W(\Phi)$.

Proposition 1.12.18. *The Weyl group $W(\Phi)$ of any root system Φ is finite.*

Proof. By condition 4 of the definition of a root system, the elements of the group $W(\Phi)$ permute Φ. Therefore there is a group homomorphism $\tau : W(\Phi) \longrightarrow S(\Phi)$, where $S(\Phi)$ is the group of all permutations of Φ. It needs to be shown that the homomorphism τ is injective. Suppose $g \in W(\Phi)$ and $g \in \operatorname{Ker}\tau$. Then, by definition, $g(\alpha) = \alpha$ for all $\alpha \in \Phi$. But Φ spans the vector space V, so $g(\beta) = \beta$ for all $\beta \in V$, which implies that g must be the identity map. Thus τ is an injective homomorphism. Since Φ is a finite set, the group of permutations $S(\Phi)$ is also finite and so is $W(\Phi)$. □

Proposition 1.12.19. *The Weyl group $W(\Phi)$ permutes the roots of Φ, i.e. if $\alpha \in \Phi$ and $w \in W$ then $w(\alpha) \in \Phi$.*

Proof. It is sufficient to prove that $s_\alpha(\beta) \in \Phi$ for all $\alpha, \beta \in \Phi$, since the reflections s_α generate W. If $\beta = \pm\alpha$ this is obvious. Suppose that $\beta \neq \pm\alpha$. Consider the α-string of roots containing β:

$$-p\alpha + \beta, \ldots, \beta, \ldots, q\alpha + \beta. \tag{1.12.20}$$

Then from lemma 1.12.7 we obtain that

$$s_\alpha(\beta) = \beta - 2\frac{(\beta,\alpha)}{(\alpha,\alpha)}\alpha = \beta - 2(p-q)\alpha$$

which is one of the roots in (1.12.20). Thus $s_\alpha(\beta) \in \Phi$. □

Lemma 1.12.21. *Let $\alpha \in \Delta$. If $\beta \in \Phi^+$ and $\beta \neq \alpha$ then $s_\alpha(\beta) \in \Phi^+$.*

Proof. By proposition 1.12.13(3), β can be expressed in the form $\beta = \sum\limits_{\alpha_i \in \Delta} x_i\alpha_i$ where x_i are nonnegative integers. Since $\beta \neq \alpha$, there is some $x_i \neq 0$ with $\alpha_i \neq \alpha$. Since $s_\alpha(\beta) \in \Phi$, by proposition 1.12.19, we can can express $s_\alpha(\beta)$ as a linear combination of the elements of Δ. The coefficient of α_i in $s_\alpha(\beta)$ remains x_i. Since $x_i > 0$, and $s_\alpha(\beta) \in \Phi$, from proposition 1.12.13(2) it follows that $s_\alpha(\beta) \in \Phi^+$. □

Theorem 1.12.22. *Let Φ_1^+ and Φ_2^+ be two sets of positive roots in Φ. Then there exists $w \in W(\Phi)$ such that $w(\Phi_1^+) = \Phi_2^+$.*

Proof. Denote $m = |\Phi_1^+ \cap \Phi_2^-|$. We prove this theorem by induction on m. If $m = 0$ then $\Phi_1^+ = \Phi_2^+$ and so $w = 1$ has the required property.

Suppose $m \neq 0$. Let Δ_1 be a set of simple roots in Φ_1. Since $\Delta_1 \subset \Phi_2^+$ implies $\Phi_1^+ \subset \Phi_2^+$, which contradicts $m > 0$, it follows that there is $\alpha \in \Delta_1 \cap \Phi_2^-$. Consider a set $s_\alpha(\Phi_1^+)$, which a set of positive roots in Φ. By lemma 1.12.21, $s_\alpha(\Phi_1^+)$ contains all roots in Φ_1^+ except α, together with $-\alpha$. Thus, $|s_\alpha(\Phi_1^+) \cap \Phi_2^+| = m - 1$. By induction hypothesis there exists $w \in W$ such that $w(s_\alpha(\Phi_1^+)) = \Phi_2^+$. Let $w_1 = ws_\alpha$. Then $w_1(\Phi_1^+) = \Phi_2^+$, as required. □

Corollary 1.12.23. *Let Δ_1 and Δ_2 be two sets of simple roots in Φ. Then there exists $w \in W(\Phi)$ such that $w(\Delta_1) = \Delta_2$.*

Proof. Let Φ_1^+ and Φ_2^+ be two sets of positive roots in Φ containing Δ_1 and Δ_2 respectively. Then, by theorem 1.12.21, there exists $w \in W(\Phi)$ such that $w(\Phi_1^+) = \Phi_2^+$. Then $w(\Delta_1)$ is a set of simple roots contained in Φ_2^+, so $w(\Delta_1) = \Delta_2$. □

Remark 1.12.24. Since $s_\alpha^2 = 1$ for any $\alpha \in \Phi$, the Weyl group $W(\Phi)$ is a particular example of a Coxeter group $W(\Gamma)$, as considered in section 1.14 below. This means that it has a special presentation in which each generator x_i is of order 2, and the relations other than $x_i^2 = 1$ are of the form $(x_ix_j)^{m_{ij}} = 1$. The generators are the reflections given by simple roots, and m_{ij} is 2,3,4 or 6 depending on whether the roots i and j make an angle of 90, 120, 135 or 150 degrees, i.e. whether in the corresponding Dynkin diagram, see below, they are unconnected, connected by a single edge, connected by a double edge, or connected by a triple edge. The **length** of a Weyl group element is the length of the shortest word representing that element in terms of these standard generators.

All information about the Weyl group of an abstract root system Φ can be obtained from a special matrix, which is called the Cartan matrix and which is discussed in the next section.

1.13. Cartan matrices and Dynkin diagrams

This section describes in what way abstract root systems give rise to Dynkin diagrams.

Let (V, Φ) be an abstract root system and $\Delta = \{\alpha_1, \alpha_2, ..., \alpha_n\}$ a set of simple roots. The Weyl group of an abstract root system Φ permutes the roots. Define a matrix called the Cartan matrix in the following way:

Definition. Let (V, Φ) be an abstract root system and let $\Delta = \{\alpha_1, \alpha_2, ..., \alpha_n\}$, where $n = \dim V$, be a set of simple roots. Fix an order on the set of simple roots. Then the matrix $C = (c_{ij})$ defined by

$$c_{ij} = 2\frac{(\alpha_i, \alpha_j)}{(\alpha_i, \alpha_i)}, \tag{1.13.1}$$

$i, j = 1, ..., n$, is called the **Cartan matrix** of (V, Φ).

The refection s_{α_i} takes an another simple root α_j into

$$\alpha_j - \frac{2(\alpha_j, \alpha_i)}{(\alpha_i, \alpha_i)}\alpha_i,$$

and so the i-th row of the Cartan matrix describes precisely where the group element s_{α_i} takes the other simple roots α_j.

Example 1.13.2.

1. Let the root system be A_3 with simple roots $\alpha_1, \alpha_2, \alpha_3$ considered in example 1.12.17. Then the Cartan matrix with respect to this ordered base is

$$C = \begin{pmatrix} 2 & -1 & 0 \\ -1 & 2 & -1 \\ 0 & -1 & 2 \end{pmatrix}.$$

2. For the root system B_3 with ordered simple roots $\alpha_1, \alpha_2, \alpha_3$ considered in example 1.12.17 the Cartan matrix is

$$C = \begin{pmatrix} 2 & -1 & 0 \\ -1 & 2 & -1 \\ 0 & -2 & 2 \end{pmatrix}.$$

The following theorem gives the main properties of such a Cartan matrix C.

Theorem 1.13.3. *If $C = (c_{ij})$ is a Cartan matrix, then*

1) $c_{ii} = 2$ *for all* $i = 1, ..., n$.

2) c_{ij} *is a non-positive integer if* $i \neq j$. *If* $c_{ij} \neq 0$ *then* $c_{ji} \neq 0$.

3) *All principal minors of C are positive. In particular,* $\det C > 0$. *Therefore, by Sylvester's criterion, C is positive definite.*

Proof.

1. This is trivial.

2. This property follows from lemma 1.12.6(1).

3. Note that C can be written as the product of two matrices $A \cdot B$, where $A = \operatorname{diag}(a_{11}, ..., a_{nn})$ is a diagonal matrix with $a_{ii} = \dfrac{2}{(\alpha_i, \alpha_i)} > 0$ and $B = (b_{ij})$, where $b_{ij} = (\alpha_i, \alpha_j)$. Since B is a symmetric positive definite matrix, all principal minors of B are positive, by Sylvester's criterion. Therefore all principal minors of C are also positive, and C is a positive definite matrix. □

Since the scalar product of two different simple roots is non-positive, the off-diagonal elements of the Cartan matrix can be only $0, -1, -2$ and -3. From the Schwarz inequality it follows that $(\alpha_i, \alpha_j)^2 \leq (\alpha_i, \alpha_i)(\alpha_j, \alpha_j)$, where the inequality is strict unless α_i and α_j are proportional. This cannot happen for $i \neq j$ since the simple roots are linearly independent. Thus there results that

$$c_{ij}c_{ji} < 4, \tag{1.13.4}$$

for $i \neq j$. Hence it follows that if $c_{ij} = -2$ or -3 then $c_{ji} = -1$.

For any Cartan matrix $C = (c_{ij})$, $i, j = 1, ..., n$ with respect to the ordered simple roots $\alpha_1, \alpha_2, \ldots, \alpha_n$, there is a corresponding graph, the **Dynkin diagram** $D(C)$, which is obtained in the following way. $D(C)$ is a graph with n vertices labelled by the simple roots $\alpha_1, \alpha_2, \ldots, \alpha_n$ (or $1, 2, \ldots, n$) and the vertices labelled by two simple roots α_i and α_j are connected by $c_{ij}c_{ji}$ edges. If $c_{ij}c_{ji} > 1$ and $|c_{ij}| < |c_{ji}|$ then the corresponding edge is marked with an arrow in the direction from i towards j.

Example 1.13.5.
The Dynkin diagrams for the following Cartan matrices

$$\begin{pmatrix} 2 & 0 \\ 0 & 2 \end{pmatrix}, \begin{pmatrix} 2 & -1 \\ -1 & 2 \end{pmatrix}, \begin{pmatrix} 2 & -1 \\ -2 & 2 \end{pmatrix}, \begin{pmatrix} 2 & -1 \\ -3 & 2 \end{pmatrix}$$

are:

Example 1.13.6.
1. The Dynkin diagram of the root system A_3 with the Cartan matrix C considered in example 1.13.2(1) is

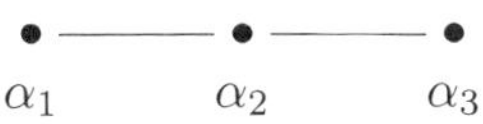

2. The Dynkin diagram of the root system B_3 with the Cartan matrix C considered in example 1.13.2(2) is

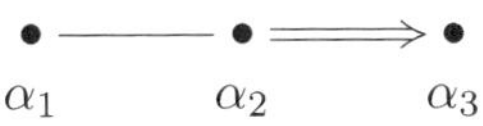

Obviously, $D(C)$ is a connected graph if and only if the set of simple roots Δ is indecomposable.

Although a given root system has more than one possible set of a simple roots, the Weyl group acts transitively on such choices, by corollary 1.12.23. Consequently, the Dynkin diagram is independent of the choice of simple roots; it is determined by the root system itself. Conversely, given two root systems with the same Dynkin diagram, one can match up roots, starting with the roots in the base, and show that the systems are in fact the same.

Thus the problem of classifying abstract root systems reduces to the problem of classifying all possible Dynkin diagrams. The problem of classifying irreducible abstract root systems reduces to the problem of classifying connected Dynkin diagrams. Dynkin diagrams encode the inner product on V in terms of the basis Δ, and the condition that this inner product must be positive definite turns out to be all that is needed to get the desired classification.

Theorem 1.13.7. *The Dynkin diagrams of the indecomposable Cartan matrices are as follows:*

A_n :

$n \geq 1$

B_n :
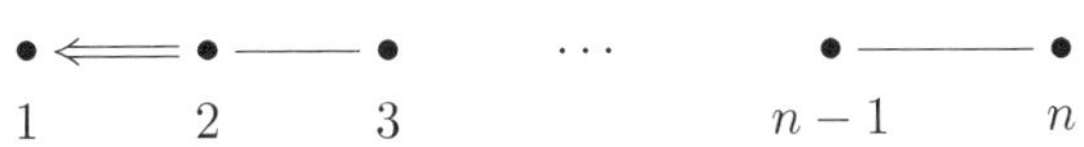

$n \geq 2$

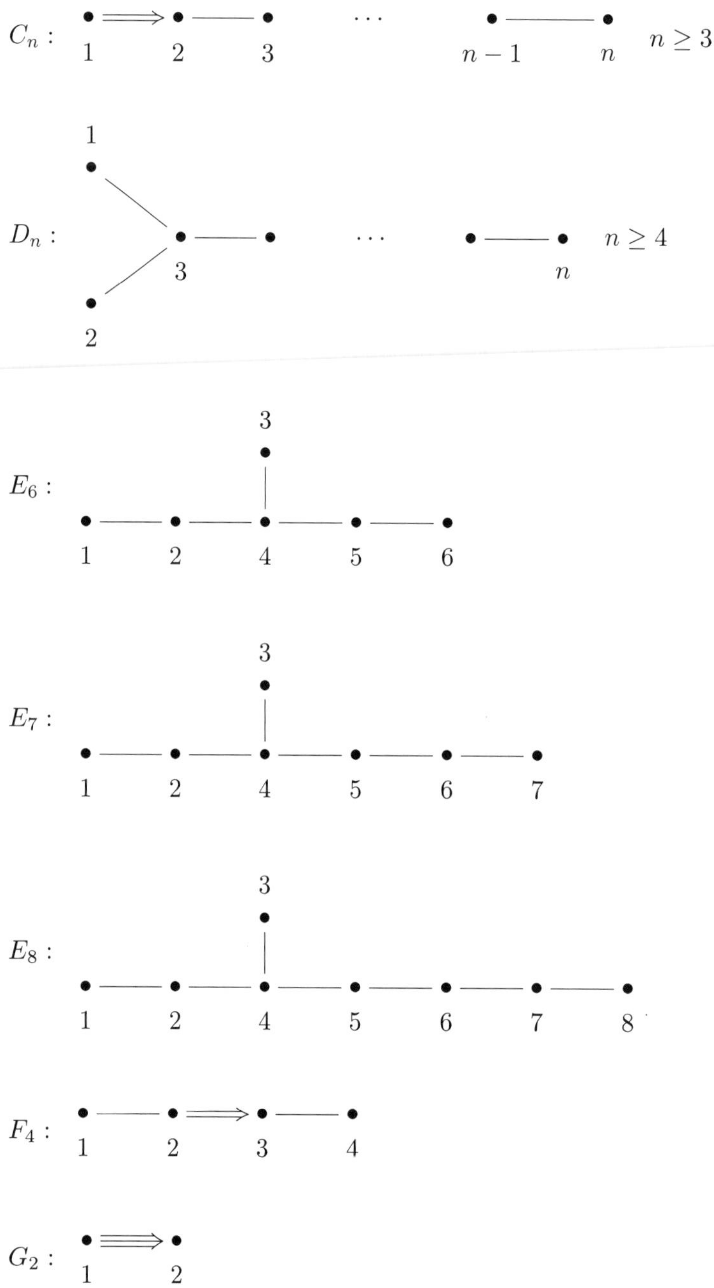

Figure 1.13.1.

Remark 1.13.8. Note the restriction on n in these diagrams. This is done to avoid "double listing". For $n = 1$ the diagram B_1 is identical with A_1. So one takes $n \geq 2$ for series B_n. C_2 is the same diagram as B_2, so one requires $n \geq 3$ for C_n. The diagram D_3 is identical with A_3. The diagram D_2 contains two vertices and no edge, so it is a decomposable diagram which is in fact $A_1 \oplus A_1$. The diagram D_1 can be interpreted as the empty diagram. So $n \geq 4$ for the class D_n.

Irreducible root systems are named according to their corresponding connected Dynkin diagrams. There are four infinite families (A_n, B_n, C_n and D_n, called the **classical root systems**) and five exceptional cases (E_6, E_7, E_8, F_4 and G_2, called the **exceptional root systems**). In every case, the subscript indicates the rank of the root system. The following table lists some other properties.

$\mathbf{\Phi}$	$\|\mathbf{\Phi}\|$	$\|\mathbf{\Phi}^<\|$	$\mathbf{I}$	$\|\mathbf{W}\|$
A_n $(n \geq 1)$	$n(n+1)$		$n+1$	$(n+1)!$
B_n $(n \geq 2)$	$2n^2$	$2n$	2	$2^n n!$
C_n $(n \geq 3)$	$2n^2$	$2n(n-1)$	2	$2^n n!$
D_n $(n \geq 4)$	$2n(n-1)$		4	$2^{n-1} n!$
E_6	72		3	51840
E_7	126		2	2903040
E_8	240		1	696729600
F_4	48	24	1	1152
G_2	12	6	1	12

Table 1.13.1.

Here $|\mathbf{\Phi}^<|$ denotes the number of short roots (if all roots have the same length they are taken to be long by definition), $\mathbf{I}$ denotes the determinant of the Cartan matrix, and $|\mathbf{W}|$ denotes the order of the Weyl group, i.e. the number of symmetries of the root system.

1.14. Coxeter groups and Coxeter diagrams

In the classification theory of semisimple Lie algebras the Weyl groups and Dynkin diagrams play a significant role. They are particular cases of more general symmetry groups, known as Coxeter groups, and the Coxeter diagrams associated with these groups.

A **Coxeter group**, named after H.S.M.Coxeter (1907-2003), is an abstract group that admits a formal description in terms of mirror symmetries. Indeed, the finite Coxeter groups are precisely the finite Euclidean reflection groups; the symmetry groups of regular polyhedra are examples. However not all Coxeter groups are finite, and not all can be described in terms of symmetries and Euclidean reflections.

Coxeter groups find applications in many areas of mathematics. Examples of finite Coxeter groups include the symmetry groups of regular polytopes, and the Weyl groups of simple Lie algebras. Examples of infinite Coxeter groups include the triangular groups corresponding to regular tesselations of the Euclidean plane and the hyperbolic plane, and the Weyl groups of infinite dimensional Kac-Moody algebras.

Definition. A **Coxeter group** W is a group which can be presented as follows:

$$\{x_1, x_2, ..., x_n \ : \ (x_i x_j)^{m_{ij}} = 1\}$$

where $m_{ii} = 1$ and $m_{ij} = m_{ji} \geq 2$ for $i \neq j$. The condition $m_{ij} = \infty$ means no relation of the form $(x_i x_j)^{m_{ij}}$ is imposed.

From this definition it immediately follows that:

1) the relation $m_{ii} = 1$ means that $x_i^2 = 1$ for all i; the generators are involutions.

2) If $m_{ij} = 2$ then the generators x_i and x_j commute.

Some examples of the finite Coxeter groups and their properties are given in the following table:

Type	**Rank**	**Order**	**Polytope**
A_n	n	$(n+1)!$	n-simplex
$B_n = C_n$	n	$2^n n!$	n-cube/n-cross-polytope
D_n	n	$2^{n-1}n!$	n-uniform prisms
G_2	2	12	6-gon
$I_2(n)$	2	$2n$	n-gon
H_3	3	120	icosahedron/dodecahedron
F_4	4	1152	24-cell
H_4	4	14400	120-cell/600-cell
E_6	6	51840	E_6 polytope
E_7	7	2903040	E_7 polytope
E_8	8	696729600	E_8 polytope

Table 1.14.1

For example, A_2 is the group of symmetries of the equilateral triangle. More generally, A_n is the group of symmetries of an n-dimensional simplex, which is just the group of permutations of $n+1$ vertices.

B_n is the group of symmetries of a hypercube or hyperoctahedron in dimension n.

$I_2(n)$ $(n \geq 3)$ is the group of symmetries of a regular n-sided polygon centered at the origin. Note that $I_2(6) = G_2$.

H_3 is the group of symmetries of the dodecahedron or icosahedron.

Definition. A **Coxeter diagram** Γ (or **Coxeter graph**) is a finite undirected graph with a set of vertices S whose edges are labelled with integers ≥ 3 or with the symbol ∞. If $i, j \in S$ let m_{ij} denote the label on the edge joining $i \neq j$. The label $m_{ij} = 3$ is often omitted when depicting Coxeter diagrams. By convention, if two vertices $i \neq j$ are not connected by an edge, $m_{ij} = 2$. Also, by convention, $m_{ii} = 1$ for all $i \in S$.

To each Coxeter diagram Γ with set of vertices S in an obvious way there corresponds a Coxeter group, which will be denoted by $W(\Gamma)$.

Definition. The **Coxeter matrix** (of a Coxeter group) is the $n \times n$ symmetric matrix with entries m_{ij}, where $m_{ii} = 1$ and $m_{ij} \geq 2$ for $i \neq j$.

To each Coxeter group G corresponds a unique Coxeter matrix M_G. It turns out that the inverse statement is also true. I.e. to each symmetric matrix with positive integer and ∞ entries and with 1's on the diagonal corresponds a Coxeter group (see [**15**], chapter V, §4).

The Coxeter matrix $M = (m_{ij})$ can be conveniently encoded by a Coxeter diagram Γ_M, as per the following rules:

1) The vertices of the graph are labeled by the generator subscripts.
2) Two vertices i and j are connected by an edge if and only if $m_{ij} \geq 3$.
3) An edge is labeled with the value of m_{ij} whenever it is 4 or greater.

In particular, two generators commute if and only if they are not connected by an edge. Furthermore, if a Coxeter diagram has two or more connected components, the associated group is the direct product of the groups associated to the individual components.

Examples 1.14.1.

1. If the Coxeter diagram has the form • • then the corresponding Coxeter group is $\mathbf{Z}_2 \times \mathbf{Z}_2$.

2. If the Coxeter diagram has the form

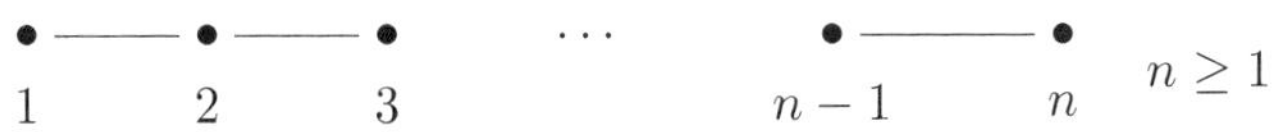

then the corresponding Coxeter group is the permutation group S_{n+1}, the symmetric group on $n+1$ vertices.

3. If the Coxeter diagram has the form $\bullet \overset{n}{\text{——}} \bullet$ $(n \geq 4)$, then the corresponding Coxeter group is the dihedral group D_{2n} of order $2n$, which is the symmetry group of the regular n-gon (see example 1.2.2, [**42**]). This group has the presentation $\{x, y \ : x^2 = y^2 = (xy)^n = 1\}$ and it is isomorphic to the semidirect product of the cyclic groups $\mathbf{Z}_n$ and $\mathbf{Z}_2$ with $\mathbf{Z}_2$ acting on $\mathbf{Z}_n$ by inversion: $\mathbf{Z}_n \rtimes_\varphi \mathbf{Z}_2$, where $\varphi(0)$ is the identity and $\varphi(1)$ is inversion.

Example 1.14.2.

The symmetry groups of all the regular polytopes are finite Coxeter groups. The dihedral groups, which are the symmetry groups of the regular polygons, form the series $I_2(p)$. The symmetry group of the n-cube is the same as that of the n-cross-polytope, namely BC_n. The symmetry group of the regular dodecahedron and the regular icosahedron is H_3. In dimension 4 there are 3 special regular polytopes, the 24-cell, the 120-cell, and the 600-cell. For a lot of information on these three beautiful four dimensional objects see e.g. the corresponding articles in the Wikipedia; e.g. http://en.wikipedia.org/wiki/120-cell. The first has symmetry group F_4, while the other two have symmetry group H_4. The Coxeter groups of types D_n, E_6, E_7 and E_8 are the symmetry groups of certain semiregular polytopes.

The corresponding Coxeter diagrams of these irreducible finite Coxeter groups are given by the following pictures:

A_n :

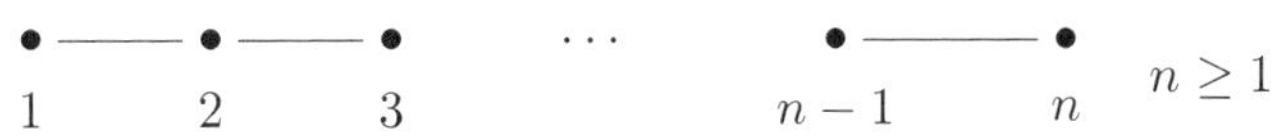

$B_n = C_n$:

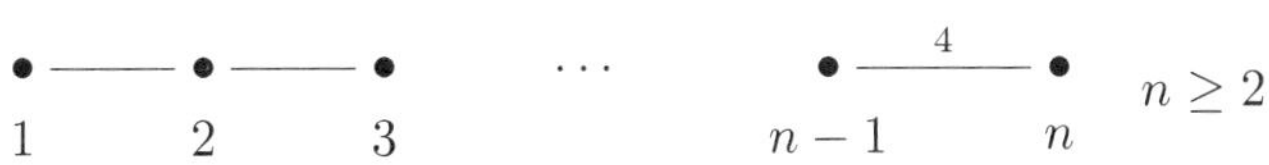

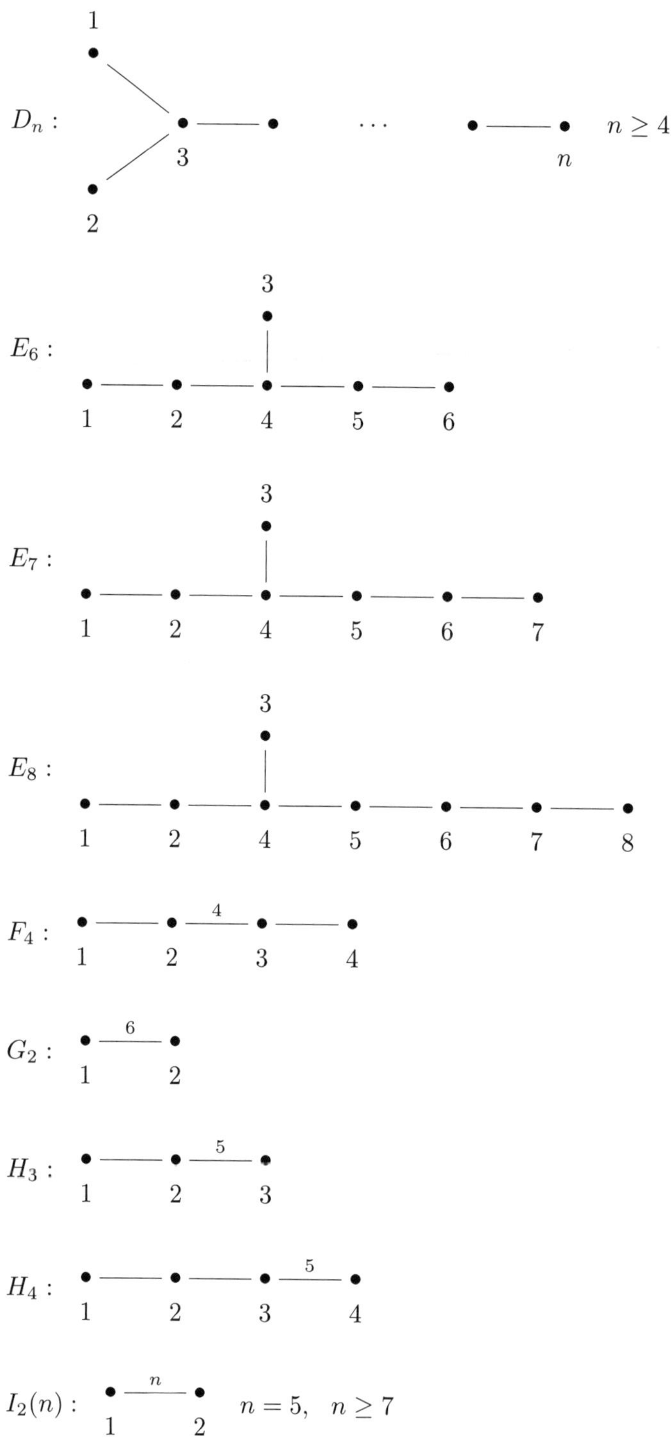

D_n :
1
2
3
n
$n \geq 4$
E_6 :
1 2 3 4 5 6
E_7 :
1 2 3 4 5 6 7
E_8 :
1 2 3 4 5 6 7 8
F_4 :
4
1 2 3 4
G_2 :
6
1 2
H_3 :
5
1 2 3
H_4 :
5
1 2 3 4
$I_2(n)$:
n
1 2
$n = 5, \quad n \geq 7$

Figure 1.14.1.

Remark 1.14.3. Instead of $\bullet \overset{4}{\text{———}} \bullet$ in a Coxeter diagram one also writes $\bullet = \bullet$ and instead of $\bullet \overset{6}{\text{———}} \bullet$ one also uses $\bullet \equiv \bullet$. Thus

is an alternative version of F_4.

The following theorem gives the classification of all finite Coxeter groups in terms of Coxeter diagrams (see [**15**], chapter VI, §4, theorem 1, theorem 2).

Theorem 1.14.4. *Let Γ be a connected Coxeter diagram. Then the corresponding irreducible Coxeter group $W(\Gamma)$ is finite if and only if Γ is one of the Coxeter diagrams represented in the list of the previous figure* 1.14.1.

Example 1.14.5.
The affine Weyl groups form a second important series of Coxeter groups. These are not finite themselves, but each contains a normal Abelian subgroup such that the corresponding quotient group is finite. In each case, the quotient group is itself a Weyl group, and the Coxeter diagram is obtained from the Coxeter diagram of the corresponding finite Weyl group by adding one additional vertex and one or two additional edges.

A list of Coxeter diagrams of irreducible affine Weil groups is as follows:

$\widetilde{A}_1$:

$\widetilde{A}_n$:

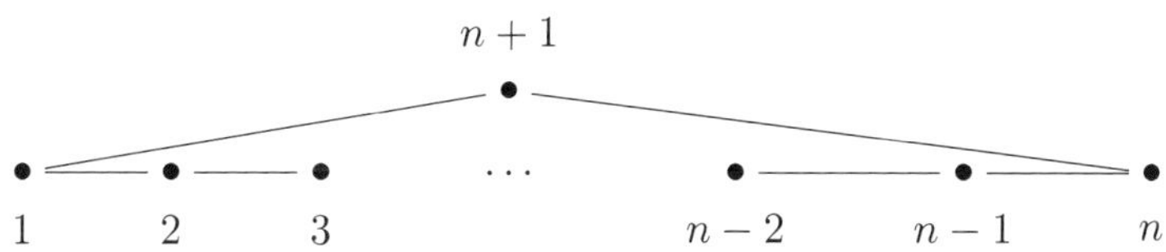

$n \geq 2$

$\widetilde{B}_2$:

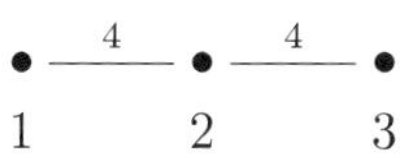

$\widetilde{B}_n$:

$n \geq 3$

$\widetilde{C}_n$:

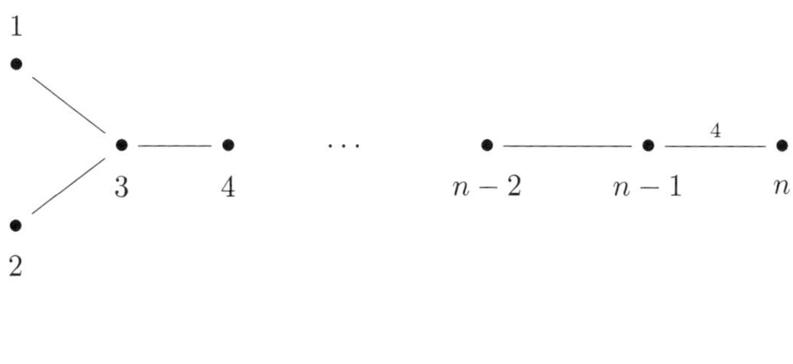

$n \geq 3$

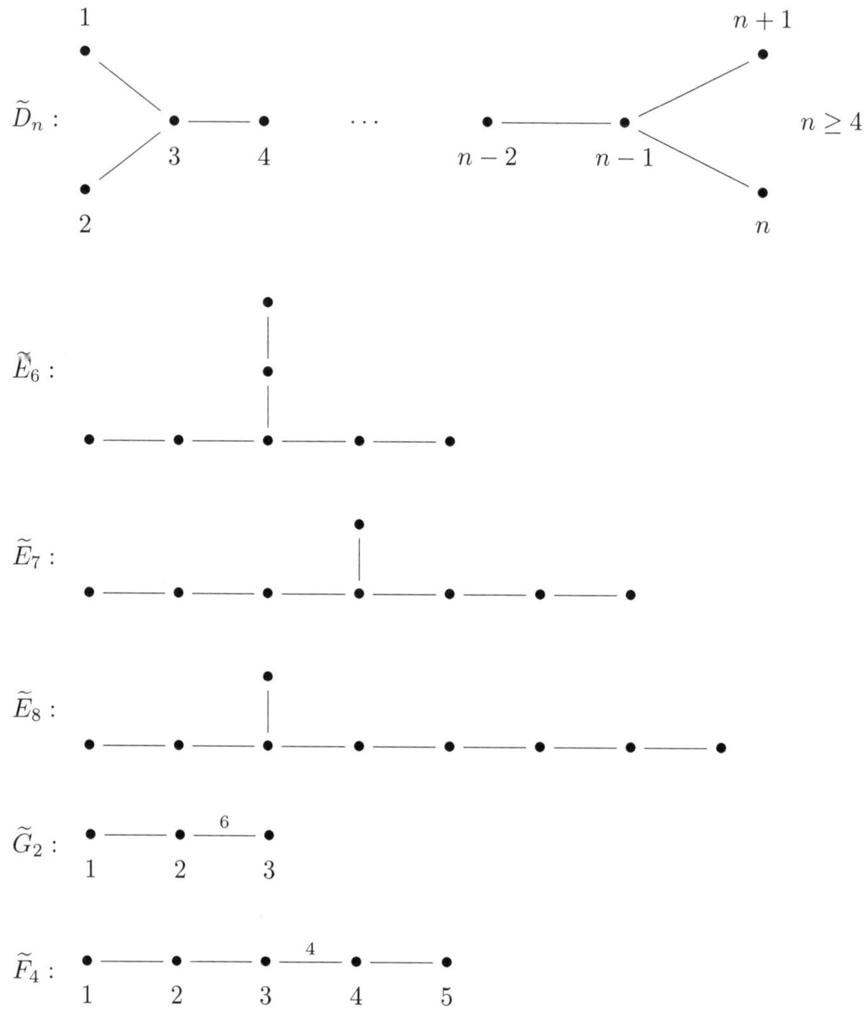

Figure 1.14.2.

Let Γ be a Coxeter diagram with a finite vertex set S, corresponding to a matrix $M = (m_{ij})$. Let V be the vector space $V = \mathbf{R}^{(S)}$, and let $(e_i)_{i \in S}$ be the canonical basis in V. Then the bilinear form B_Γ (or B_M) on V defined by

$$B_\Gamma(e_i, e_j) = -\cos(\frac{\pi}{m_{ij}}) \tag{1.14.6}$$

is called the **bilinear form of the Coxeter diagram** Γ associated with the matrix M. Since M is a symmetric matrix, B_Γ is a symmetric bilinear form. It is easy to see that

$$B_\Gamma(e_i, e_i) = 1 \quad \text{and } B_\Gamma(e_i, e_j) \leq 0 \quad \text{if } i \neq j. \tag{1.14.7}$$

The following theorems give the connection between finite Coxeter diagrams and the corresponding bilinear forms (see [**15**], chapter V, §4, theorem 2 and chapter VI, §4, theorem 4).

Theorem 1.14.8. *The Coxeter group $W(\Gamma)$ corresponding to a Coxeter diagram Γ is finite if and only if the corresponding symmetric bilinear form B_Γ is positive definite and (hence) non-degenerate.*

Theorem 1.14.9. *Let Γ be a connected Coxeter diagram with finite vertex set S. The corresponding quadratic form q_Γ is nonnegative definite and degenerate if and only if the Coxeter diagram Γ is one of the form presented in figure 1.14.2, i.e. corresponds to one of the affine Weyl groups.*

Let Γ, S, V be as above. For each $i \in S$ define the mapping: $\sigma_i : V \longrightarrow V$ by

$$\sigma_i(x) = x - 2B_\Gamma(e_i, x)e_i$$

which is called a **reflection** on V.

Note that from (1.14.7) it follows that each reflection σ_i preserves the bilinear form B_M:

$$\begin{aligned} B_\Gamma(\sigma_i(x), \sigma_i(y)) &= B_\Gamma(x - 2B_\Gamma(e_i, x)e_i,\ y - 2B_\Gamma(e_i, y)e_i) \\ &= B_\Gamma(x, y) - 2B_\Gamma(x, e_i)B_\Gamma(e_i, y) - 2B_\Gamma(e_i, y)B_\Gamma(x, e_i) \\ &\quad + 4B_\Gamma(x, e_i)B_\Gamma(e_i, y)(e_i, e_i) = B_\Gamma(x, y), \end{aligned}$$

and σ_i^2 is the identity mapping for each i:

$$\begin{aligned} \sigma_i^2(x) &= \sigma_i(x - 2B_\Gamma(e_i, x)e_i) \\ &= x - 2B_\Gamma(e_i, x)e_i - 2B(e_i, x - 2B_\Gamma(e_i, x)e_i)e_i \\ &= x - 2B_\Gamma(e_i, x)e_i - 2B_\Gamma(e_i, x)e_i + 4B_\Gamma(e_i, B(e_i, x)e_i)e_i = x. \end{aligned}$$

1.15. Root systems of semisimple Lie algebras

In this section it is shown how one can construct a root system for a semisimple Lie algebra over an algebraically closed field k of characteristic 0 (for example the field of complex numbers $\mathbf{C}$).

Throughout in this section, except when it is explicitly stated otherwise, L is a semisimple Lie algebra over an algebraically closed field k of characteristic 0.

To start, here is the very useful notion of a Cartan subalgebra.

Definition. An element $x \in L$ is called **semisimple** if ad x is semisimple, i.e. ad x is diagonalizable. A **Cartan subalgebra** $\mathfrak{h}$ of a Lie algebra L is a maximal Abelian subalgebra with the additional property that all its elements are semisimple in L. If L is a finite dimensional Lie algebra, then such a subalgebra exists by the Jordan-Chevalley decomposition (theorem 1.4.4).

Examples 1.15.1.

1. Let $L = \mathfrak{gl}(n, \mathbf{C})$, then a Cartan subalgebra $\mathfrak{h}$ of L is the Lie algebra of all diagonal matrices over $\mathbf{C}$.

2. Let $L = \mathfrak{sl}(n, \mathbf{C})$, then a Cartan subalgebra $\mathfrak{h}$ of L is the Lie algebra of all diagonal matrices over $\mathbf{C}$ with trace equal to zero. The dimension of $\mathfrak{h}$ is $n - 1$.

Remark 1.15.2. In the general case of an arbitrary Lie algebra a Cartan subalgebra is introduced in another way. A **Cartan subalgebra** of a Lie algebra L is a nilpotent subalgebra which is equal to its normalizer in L. If L is a semisimple

Lie algebra it can be proved that this definition coincides with definition given above (see, for example, [**75**]).

Let $\mathfrak{h}$ be a Cartan subalgebra of a Lie algebra L and let $\mathfrak{h}^* = \mathrm{Hom}(\mathfrak{h}, k)$ be the linear space dual to $\mathfrak{h}$. If $h \in \mathfrak{h}$, then $\mathrm{ad}\, h$ is a linear operator on the vector space L which is semisimple, i.e. diagonal in some basis by the definition of $\mathfrak{h}$. Since $\mathfrak{h}$ is Abelian, these operators commutate with each other. Therefore by a well known result of linear algebra there is a basis with respect to which all these operators are diagonal. So there is a decomposition $L = \bigoplus_{\lambda} L_\lambda$, where $\lambda \in \mathfrak{h}^*$ and

$$L_\lambda = \{x \in L \ : \ [h, x] = \lambda(h)x \text{ for all } h \in \mathfrak{h}\}.$$

This decomposition is a grading of Lie algebras, i.e. $[L_\lambda, L_\mu] \subset L_{\lambda+\mu}$, which follows from the Jacobi identity. Indeed, if $y \in L_\lambda$, $z \in L_\mu$,

$$\begin{aligned} \mathrm{ad}\, h([y,z]) &= [(\mathrm{ad}\, h(y), z] + [y, (\mathrm{ad}\, h(z)] \\ &= [\lambda(h)y, z] + [y, \lambda(\mu)z] = (\lambda(h) + \mu(h))[y, z]. \end{aligned}$$

From this property it follows immediately that $\mathrm{ad}\, x$ is nilpotent for any $x \in L_\alpha$ with $\alpha \neq 0$.

A further important property is that if $\alpha, \beta \in \mathfrak{h}^*$, and $\alpha + \beta \neq 0$, then the subspace L_α is orthogonal to L_β with respect to the Killing form κ on L.

Since $\alpha+\beta \neq 0$, there is an element $h \in \mathfrak{h}$ such that $(\alpha+\beta)(h) \neq 0$. Let $x \in L_\alpha$ and $y \in L_\beta$, then $\kappa([h,x], y) = -\kappa([x,h], y) = -\kappa(x, [h,y])$, since the Killing form is associative. Therefore $\alpha(h)\kappa(x,y) = -\beta(h)\kappa(x,y)$, i.e. $(\alpha+\beta)(h)\kappa(x,y) = 0$.

So the following statement has been proved.

Proposition 1.15.3.
1. *If* $\alpha, \beta \in \mathfrak{h}^*$, *then* $[L_\alpha, L_\beta] \subset L_{\alpha+\beta}$.
2. *If* $\alpha, \beta \in \mathfrak{h}^*$, *and* $\alpha + \beta \neq 0$, *then* $\kappa(L_\alpha, L_\beta) = 0$.

Note that $L_0 = C_L(\mathfrak{h})$, the centralizer of $\mathfrak{h}$ in L. Since $\mathfrak{h}$ is an Abelian subalgebra, $\mathfrak{h} \subseteq L_0$. If L is semisimple, it can be proved that $\mathfrak{h} = L_0$ (see [**43**]). From this last fact the following result can be extracted.

Proposition 1.15.4. *The restriction of the Killing form to a Cartan subalgebra* $\mathfrak{h}$ *is non-degenerate.*

Proof. Since L is a finite dimensional semisimple Lie algebra over a field K of characteristic 0, the Killing form κ is non-degenerate on L, by theorem 1.4.20. Moreover $L_0 = \mathfrak{h}$ and L_0 is orthogonal to L_α for each nonzero $\alpha \in \mathfrak{h}^*$. Therefore if $x \in L_0$ is orthogonal to L_0 it would be the case that $\kappa(x, L) = 0$, which implies $x = 0$. □

The set of all nonzero elements $\lambda \in \mathfrak{h}^*$ for which $L_\lambda \neq 0$ is denoted by Φ. An element $\lambda \in \Phi$ is called a **root** of L with respect to $\mathfrak{h}$ and the vector space L_λ is called a **root subspace** of L with respect to $\mathfrak{h}$. There are a finite number of roots for any finite dimensional Lie algebra L. Therefore there results a **root space decomposition** (or **Cartan decomposition**) of L:

$$L = \mathfrak{h} \oplus \bigoplus_{\lambda \in \Phi} L_\lambda. \tag{1.15.5}$$

Example 1.15.6.

Let L be the Lie algebra $\mathfrak{sl}(n, \mathbf{C})$ with the Cartan subalgebra $\mathfrak{h}$ of diagonal matrices. Let $y = \operatorname{diag}(\lambda_1, ..., \lambda_n) \in \mathfrak{h}$ and set $\omega_i(y) = \lambda_i$. Then $\omega_i - \omega_j : \mathfrak{h} \to \mathbf{C}$ is a root if $i \neq j$. A nonzero element of $L_{\omega_i - \omega_j}$ is E_{ij}, the matrix with a 1 at location (i, j) and 0 elsewhere.

In more detail, the root space of $\omega_i - \omega_j$ is the one-dimensional vector space kE_{ij} and the root space decomposition is

$$\mathfrak{sl}(n) = \mathfrak{h} \oplus \bigoplus_{i \neq j} kE_{ij}.$$

Using $[E_{ij}, E_{rs}] = \delta_{jr}E_{is} - \delta_{is}E_{rj}$ it is tedious but not difficult to calculate the entries of the Killing form with respect to the basis $E_{11} - E_{ii}$, $i = 2, ..., n$; E_{ij}, E_{ji}, $i < j$. The result is the block diagonal matrix

$$\left(\begin{array}{c|c|c|c}
\begin{matrix}4n & & \\ & \ddots & \\ & & 4n\end{matrix} & & & \\ \hline
 & \begin{matrix}0 & 2n \\ 2n & 0\end{matrix} & & \\ \hline
 & & \ddots & \\ \hline
 & & & \begin{matrix}0 & 2n \\ 2n & 0\end{matrix}
\end{array}\right),$$

consisting of $(n-1)$ blocks $(4n)$ and $\frac{1}{2}n(n-1)$ blocks $\begin{pmatrix} 0 & 2n \\ 2n & 0 \end{pmatrix}$. Thus indeed the root spaces of a root α and a root β such that $\alpha + \beta \neq 0$ are orthogonal to each other, and the Killing form is non-degenerate on $\mathfrak{sl}(n)$ and on $\mathfrak{h}$.

Example 1.15.7.

Let $L = \mathfrak{sl}(n, \mathbf{C})$ be again the Lie algebra with the Cartan subalgebra $\mathfrak{h}$ of diagonal matrices. Then

$$V = \mathfrak{h}^* = \Big\{\sum_{i=1}^{n} \xi_i \omega_i \ : \ \sum_{i=1}^{n} \xi_i = 0\Big\}.$$

The reflection s_α associated to $\omega_i - \omega_j$ interchanges ω_i and ω_j and leaves all other ω_k invariant. Hence $(\omega_i - \omega_j)^* \in \mathfrak{h}$ is $\operatorname{diag}(0, ..., 0, 1, 0, ..., 0, -1, 0, ..., 0)$ with 1 at the i-th place and -1 at the j-th place. A base (or a set of simple roots) is $\alpha_1 = \omega_1 - \omega_2$, $\alpha_2 = \omega_2 - \omega_3$, $\ldots$, $\alpha_{n-1} = \omega_{n-1} - \omega_n$.

In proposition 1.15.4 it was proved that the restriction of κ to $\mathfrak{h}$ is non-degenerate. This makes it possible to associate to every linear operator $\alpha \in \mathfrak{h}^*$ the unique element $t_\alpha \in \mathfrak{h}$ such that

$$\alpha(h) = \kappa(t_\alpha, h)$$

for all $h \in \mathfrak{h}$. In particular, Φ corresponds to the subset $\{t_\alpha \in \mathfrak{h} \ : \ \alpha \in \Phi\} \subset \mathfrak{h}$.

Proposition 1.15.8. *Let L be a semisimple Lie algebra with a Cartan subalgebra $\mathfrak{h}$ and a root set Φ. Then*

1. Φ *spans* $\mathfrak{h}^*$.
2. *If* $\alpha \in \Phi$, *then* $-\alpha \in \Phi$.
3. *If* $x \in L_\alpha$, $y \in L_{-\alpha}$, $\alpha \in \Phi$ *then* $[x, y] = \kappa(x, y)t_\alpha$.
4. *If* $\alpha \in \Phi$, *then* $[L_\alpha, L_{-\alpha}] \subset \mathfrak{h}$ *is a one dimensional subspace with basis* t_α.
5. $\alpha(t_\alpha) = \kappa(t_\alpha, t_\alpha) \neq 0$ *for all* $\alpha \in \Phi$.
6. *If* $\alpha \in \Phi$ *then for any* $0 \neq x_\alpha \in L_\alpha$ *there exists an element* $y_\alpha \in L_{-\alpha}$ *such that the elements* $x_\alpha, y_\alpha, h_\alpha = [x_\alpha, y_\alpha]$ *span a three dimensional simple subalgebra of* L *which is isomorphic to* $\mathfrak{sl}(2, k)$ *via*

$$x_\alpha \mapsto \begin{pmatrix} 0 & 1 \\ 0 & 0 \end{pmatrix}, \quad y_\alpha \mapsto \begin{pmatrix} 0 & 0 \\ 1 & 0 \end{pmatrix}, \quad h_\alpha \mapsto \begin{pmatrix} 1 & 0 \\ 0 & -1 \end{pmatrix}.$$

7. $h_\alpha = \dfrac{2t_\alpha}{\kappa(t_\alpha, t_\alpha)}$; $h_\alpha = h_{-\alpha}$.

Proof.

1. If Φ does not span $\mathfrak{h}^*$ then there is an element $h \in \mathfrak{h}$ such that $\alpha(h) = 0$ for all $\alpha \in \Phi$. This implies that $[h, L_\alpha] = 0$ for all $\alpha \in \Phi$, so $[h, L] = 0$, which is impossible.

2. Suppose $-\alpha \notin \Phi$, then $L_{-\alpha} = 0$. So $\kappa(L_\alpha, L_{-\alpha}) = 0$; also by proposition 1.15.3 $\kappa(L_\alpha, L_\beta) = 0$ for all $\beta \neq -\alpha$. Therefore $\kappa(L_\alpha, L) = 0$, which contradicts the non-degeneracy of the Killing form.

3. Let $h \in \mathfrak{h}$. Then

$$\kappa(h, [x, y]) = \kappa([h, x], y) = \alpha(h)\kappa(x, y) = \kappa(t_\alpha, h)\kappa(x, y) = \kappa(\kappa(x, y)t_\alpha, h)$$

which implies that h is orthogonal to $[x, y] - \kappa(x, y)t_\alpha$. Then by proposition 1.15.4 $[x, y] - \kappa(x, y)t_\alpha = 0$.

4. This follows from the previous property taking into account the fact that $[L_\alpha, L_{-\alpha}] \neq 0$ and $[L_\alpha, L_{-\alpha}] \subset L_0 = \mathfrak{h}$, by proposition 1.15.3(1).

5. Suppose $\alpha(t_\alpha) = \kappa(t_\alpha, t_\alpha) = 0$ for all $\alpha \in \Phi$. Then choosing $x \in L_\alpha$, $y \in L_{-\alpha}$ with $\kappa(x, y) = 1$ we obtain

$$[x, y] = t_\alpha, \quad [t_\alpha, x] = [t_\alpha, y] = 0.$$

So the subspace of L spanned by the elements x, y, t_α forms a three dimensional algebra U which is solvable. Acting as ad on L there results that $\operatorname{ad} t_\alpha$ is nilpotent, by corollary 1.6.7. Since t_α is also semisimple, by the definition of $\mathfrak{h}$, we obtain that $\operatorname{ad} t_\alpha = 0$, i.e. $t_\alpha \in Z(L) = 0$, which contradicts the choice of t_α.

6. Let $0 \neq x_\alpha \in L_\alpha$. Since $\kappa(t_\alpha, t_\alpha) \neq 0$ and $\kappa(x_\alpha, L_{-\alpha}) \neq 0$, there is an $y_\alpha \in L_{-\alpha}$ such that $\kappa(x_\alpha, y_\alpha) = \dfrac{2}{\kappa(t_\alpha, t_\alpha)}$. Set

$$h_\alpha = \frac{2t_\alpha}{\kappa(t_\alpha, t_\alpha)}, \tag{1.15.9}$$

then $[x_\alpha, y_\alpha] = h_\alpha$, by property 3. The element h_α is called the **coroot** of α. Moreover, $[h_\alpha, x_\alpha] = \dfrac{2}{\alpha(t_\alpha)}[t_\alpha, x_\alpha] = \dfrac{2\alpha(t_\alpha)}{\alpha(t_\alpha)}x_\alpha = 2x_\alpha$ and similarly, $[h_\alpha, y_\alpha] = -2y_\alpha$. Thus

$$[x_\alpha, y_\alpha] = h_\alpha, \quad [h_\alpha, x_\alpha] = 2x_\alpha, \quad [h_\alpha, y_\alpha] = -2y_\alpha \tag{1.15.10}$$

and so the elements $x_\alpha, y_\alpha, h_\alpha$ span a three dimensional subalgebra of L with the same multiplication table as the simple Lie algebra $\mathfrak{sl}(2, k)$ (see example 1.1.1(6)).

7. Since t_α is defined by the relation $\kappa(t_\alpha, h) = \alpha(h)$ for all $h \in \mathfrak{h}$, this shows that $t_\alpha = t_{-\alpha}$, and so $h_\alpha = h_{-\alpha}$. □

Proposition 1.15.11.
1. $\dim L_\alpha = 1$ *for any* $\alpha \in \Phi$.
2. *If* $\alpha \in \Phi$, *then* $n\alpha \in \Phi$ *if and only if* $n = 1$ *or* $n = -1$.
3. *If* $\alpha, \beta \in \Phi$, *then* $\beta(h_\alpha) \in \mathbf{Z}$ *and* $\beta - \beta(h_\alpha)\alpha \in \Phi$.
4. *If* $\alpha, \beta, \alpha + \beta \in \Phi$, *then* $[L_\alpha, L_\beta] = L_{\alpha+\beta}$.
5. *Let* $\alpha, \beta \in \Phi$ *and* $\beta \neq \pm\alpha$. *Let* p, q *be the maximal integers such that* $\beta - p\alpha$ *and* $\beta + q\alpha$ *are roots. Then* $\beta + i\alpha \in \Phi$ *for* $-p \leq i \leq q$ *and* $p - q = \beta(h_\alpha)$.
6. *The root spaces* L_α, *where* $\alpha \in \Phi$, *generate* L *as a Lie algebra.*

Proof. Let $\alpha \in \Phi$ and let the elements $x_\alpha, y_\alpha, h_\alpha$ be given by (1.15.10). Consider the subspace W_α of L spanned by y_α, h_α and all $L_{t\alpha}$ for $t = 1, 2, 3, \ldots$

$$W_\alpha = \{h_\alpha\} \oplus \{y_\alpha\} \oplus L_\alpha \oplus L_{2\alpha} \oplus L_{3\alpha} \oplus \cdots$$

Then, from the definition of a Cartan subalgebra and propositions 1.15.3, 1.15.8, it follows that W_α is invariant under the operators $x_\alpha, y_\alpha, h_\alpha$. Since $\operatorname{ad} h_\alpha = \operatorname{ad}[x_\alpha, y_\alpha] = [\operatorname{ad} x_\alpha, \operatorname{ad} y_\alpha]$, the trace of $\operatorname{ad} h_\alpha$ on W_α is equal to 0. From (1.15.9) it follows that $\alpha(h_\alpha) = 2$. Therefore calculating the trace of $\operatorname{ad} h_\alpha$ on W_α explicitly this trace is $2(-1 + n_\alpha + n_{2\alpha} + n_{3\alpha} + \cdots)$, where $n_{t\alpha} = \dim L_{t\alpha}$. Hence $n_\alpha = 1$ and $n_{t\alpha} = 0$ for $t > 1$. In particular, $\dim L_\alpha = 1$.

From proposition 1.15.8(6) it follows that for any $\alpha \in \Phi$ the elements $x_\alpha, y_\alpha, h_\alpha$ given by (1.15.10) span a Lie subalgebra S_α isomorphic to the simple Lie algebra $\mathfrak{sl}(2; k)$ whose representations were described in section 1.7. Suppose, $\alpha, \beta \in \Phi$. From lemma 1.7.3 it follows that all eigenvalues of the operator h_β are integers, therefore

$$\beta(h_\alpha) = \kappa(t_\beta, h_\alpha) = \frac{2\kappa(t_\beta, t_\alpha)}{\kappa(t_\alpha, t_\alpha)} \in \mathbf{Z}. \tag{1.15.12}$$

These numbers are called the **Cartan numbers**.

Suppose, $\alpha, \beta \in \Phi$, and $\alpha \neq \pm\beta$. Consider $M = \sum\limits_{i \in \mathbf{Z}, \beta + i\alpha \in \Phi} L_{\beta+i\alpha}$. By the conditions 1 and 2 proved above, all subspaces $L_{\beta+i\alpha}$ are one-dimensional and $\beta + i\alpha \neq 0$. Therefore M is an S_α-module of L with one dimensional weight spaces for distinct integral weights $\beta(h_\alpha) + 2i$. Obviously, not both 0 and 1 can occur as weights of this form. Therefore M is irreducible, by corollary 1.7.6. Let q (resp. p) be the largest integer for which $\beta + q\alpha$ (resp. $\beta - p\alpha$) is a root, then the highest (resp. lowest) weight must be $\beta(h_\alpha) + 2q$ (resp. $\beta(h_\alpha) - 2p$). Moreover the weights on M form an arithmetic progression with difference 2, by theorem 1.7.4. This implies that the roots $\beta + i\alpha$ form a string

$$\beta - p\alpha, \beta - (p-1)\alpha, \ldots, \beta, \ldots, \beta + q\alpha,$$

which is called the **α-string through** β. Note also that $(\beta - p\alpha)(h_\alpha) = -(\beta + q\alpha)(h_\alpha)$, or $\beta(h_\alpha) = p - q$.

Suppose $\alpha, \beta, \alpha + \beta \in \Phi$, then $\operatorname{ad} L_\alpha$ maps L_β onto $L_{\alpha+\beta}$. Since each of these subspaces is one dimensional, $[L_\alpha, L_\beta] = L_{\alpha+\beta}$.

The theorem is proved. □

It will now be shown that the roots of a semisimple Lie algebra over the algebraically closed field k can be taken to lie in a real Euclidean vector space.

Since by proposition 1.15.4 the restriction of the Killing form to the Cartan subalgebra $\mathfrak{h}$ is non-degenerate, one can transfer the bilinear form κ from $\mathfrak{h}$ to $\mathfrak{h}^*$ by defining

$$(\gamma, \delta) = \kappa(t_\gamma, t_\delta) \tag{1.15.13}$$

for all $\gamma, \delta \in \mathfrak{h}^*$. So

$$\beta(h_\alpha) = \kappa(t_\beta, h_\alpha) = \frac{2\kappa(t_\beta, t_\alpha)}{\kappa(t_\alpha, t_\alpha)} = \frac{2(\beta, \alpha)}{(\alpha, \alpha)} \in \mathbf{Z}. \tag{1.15.14}$$

Since by proposition 1.15.8(1) Φ spans $\mathfrak{h}^*$, one can choose a basis of $\mathfrak{h}^*$ consisting of roots $\{\alpha_1, ..., \alpha_n\}$. Any root $\beta \in \Phi$ can be written uniquely as a linear combination

$$\beta = c_1\alpha_1 + c_2\alpha_2 + \cdots + c_n\alpha_n,$$

where the $c_i \in k$. The claim is that all $c_i \in \mathbf{Q}$, i.e. are rational numbers. For each $j = 1, ..., n$, $(\beta, \alpha_j) = \sum\limits_{i=1}^{n} c_i(\alpha_i, \alpha_j)$, and multiplying both sides of the j-th equality by $2/(\alpha_j, \alpha_j)$ gives

$$\frac{2(\beta, \alpha_j)}{(\alpha_j, \alpha_j)} = \sum_{i=1}^{n} \frac{2(\alpha_i, \alpha_j)}{(\alpha_j, \alpha_j)} c_i. \tag{1.15.15}$$

Now consider (1.15.15) as a system of n linear algebraic equations with respect to unknowns c_i, $i = 1, ..., n$, with coefficients that are integral because of (1.15.14).

Since $(\alpha_1, ..., \alpha_n)$ is a basis of $\mathfrak{h}^*$ and the form $(\ ,\ )$ is non-degenerate, the matrix $((\alpha_i, \alpha_j))$ is nonsingular. Therefore the matrix of coefficients of the system (1.15.14) is also nonsingular, and so the system possesses a unique solution over the field $\mathbf{Q}$.

Note, that the matrix $((\alpha_i, \alpha_j))$ of system (1.15.15) is exactly the same as the Cartan matrix for the abstract root system $\Delta = \{\alpha_1, ..., \alpha_n\}$ given by (1.13.1).

Denote by E the real vector space of $\mathfrak{h}^*$ spanned by all roots from Φ. Then the form $(\ ,\ )$ restricts to a real scalar product on E. Let $\gamma, \delta \in \mathfrak{h}^*$, then

$$(\gamma, \delta) = \kappa(t_\gamma, t_\delta) = \sum_{\alpha \in \Phi} \alpha(t_\gamma)\alpha(t_\delta) = \sum_{\alpha \in \Phi} (\alpha, \gamma)(\alpha, \delta).$$

In particular, $(\gamma, \gamma) = \sum\limits_{\alpha \in \Phi} (\alpha, \gamma)^2$ is a sum of squares of rational numbers, so it is positive if $\gamma \neq 0$. Therefore the form on E is positive definite, and this form gives E the structure of a Euclidean space. So the following statement has been proved.

Proposition 1.15.16. *The restriction of the Killing form to E is real and positive definite. E is a real finite dimensional Euclidean vector space.*

Summarizing the main properties of root spaces from propositions 1.15.8, 1.15.11, 1.15.16 gives the following theorem.

Theorem 1.15.17. *Let L be a semisimple Lie algebra with a Cartan subalgebra $\mathfrak{h}$ and a root set Φ. Suppose that $\alpha, \beta \in \Phi$ and $\beta \neq \pm\alpha$. Then*

1. *The set $\Phi \neq 0$ spans a real Euclidean space E and does not contain zero.*
2. *If $\alpha \in \Phi$, then $n\alpha \in \Phi$ if and only if $n = 1$ or $n = -1$.*
3. $\beta(h_\alpha) = \dfrac{2(\beta, \alpha)}{(\alpha, \alpha)} \in \mathbf{Z}$ *and* $\beta - \beta(h_\alpha)\alpha \in \Phi$.

4. *Let p, q be the maximal integers such that $\beta - p\alpha$ and $\beta + q\alpha$ are roots. Then $\beta + i\alpha \in \Phi$ for $-p \leq i \leq q$ and $p - q = 2\dfrac{(\alpha, \beta)}{(\alpha, \alpha)}$.*

Comparing this theorem with the definition of an abstract root system as introduced in section 1.12 it follows that the root set Φ of a semisimple Lie algebra L over an algebraically closed field k of characteristic 0 is exactly an abstract root system.

Thus, to each semisimple Lie algebra over an algebraically closed field of characteristic 0 there corresponds a root system, which is easily classified by Dynkin diagrams. If L is a simple Lie algebra, then the corresponding root system is irreducible. It turns out that the inverse statement is also true. I.e. to each irreducible root system there corresponds a simple Lie algebra. In other words there is a bijection between the set of simple Lie algebras and the set of abstract irreducible root systems, which are described by the Dynkin theorem. This famous statement is known as the **Cartan-Killing classification theorem**:

Theorem 1.15.18. *For any simple finite dimensional Lie algebra L over an algebraically closed field of characteristic 0, the set of all its roots is an irreducible root system. Two simple finite dimensional Lie algebras are isomorphic if and only if their root systems are isomorphic. To any irreducible root system corresponds a simple finite dimensional Lie algebra.*

For more details on this classification theorem see e.g. [**43**, section 8.5], or, especially, [**92**, section 3.7].

From this theorem and theorem 1.13.7 there immediately follows the main classification theorem for simple Lie algebras over an algebraically closed field of characteristic 0.

Theorem 1.15.19. *A finite dimensional Lie algebra over an algebraically closed field of characteristic 0 is simple if and only if the Dynkin diagram of its root system is one of the following list $A_n, B_n, C_n, D_n, E_6, E_7, E_8, G_2, F_4$ as depicted in figure* 1.13.1.

Example 1.15.20.
Let $L = \mathfrak{sl}(n, \mathbf{C})$ with Cartan subalgebra $\mathfrak{h}$. We find $\langle \alpha_i^*, \alpha_j \rangle = 0$ if $i < j - 1$ or $i > j + 1$, $\langle \alpha_i^*, \alpha_i \rangle = 2$, $\langle \alpha_{i-1}^*, \alpha_i \rangle = \langle \alpha_{i+1}^*, \alpha_i \rangle = -1$. It follows that the Dynkin diagram of $\mathfrak{sl}(n, \mathbf{C})$ is A_{n-1}. The Weyl group of $L = \mathfrak{sl}(n, \mathbf{C})$ is S_n.

1.16. The Weyl group of a quiver

In the previous sections the connection between the simple Lie algebras and the Dynkin diagrams has been described. This connection used root systems, the Weyl groups, the Cartan matrices and the corresponding Dynkin diagrams. It turns out that a similar construction can be used to obtain a connection between quivers of finite type and Dynkin diagrams. This is a famous theorem proved by Bernstein, Gel'fand and Ponomarev in 1973 (see [**8**]), which can be considered a generalization of the Gabriel theorem to the case of an arbitrary field (see [**34**]). The next sections of this chapter will be devoted to a proof of this theorem. This proof uses the Weyl group of a quiver and Coxeter functors.

In section 2.6, [**42**] finite acyclic quivers (i.e. without loops and oriented cycles) and corresponding integral quadratic forms were studied. The main goal of this section is to define the Weyl group of a quiver. This group is defined as the Weyl group of the corresponding integral quadratic form. The section continues with the study of its main properties.

To start with, here is definition of the Weyl group for an integral quadratic form $q(x)$ on $\mathbf{Z}^n$ of the following form:

$$q(x) = \sum_{i=1}^{n} x_i^2 + \sum_{i<j} a_{ij}x_i x_j, \tag{1.16.1}$$

where $x = (x_1, x_2, \dots, x_n) \in \mathbf{Z}^n$, $a_{ij} \in \mathbf{Z}$.

Suppose that $\mathfrak{B}(\ ,\ )$ is the corresponding bilinear symmetric form on $\mathbf{Z}^n$, defined by

$$\mathfrak{B}(\alpha, \beta) = q(\alpha + \beta) - q(\alpha) - q(\beta) \tag{1.16.2}$$

for any $\alpha, \beta \in \mathbf{Z}^n$. It is easy to see that

$$q(\alpha) = \frac{1}{2}\mathfrak{B}(\alpha, \alpha) \tag{1.16.3}$$

for any $\alpha \in \mathbf{Z}^n$. Let $\{e_1, e_2, \dots, e_n\}$ be the canonical basis of $\mathbf{Z}^n$, where e_i is the vector with 1 in the i-th position and zeroes elsewhere. Then

$$\mathfrak{B}(e_i, e_j) = \begin{cases} 2, & \text{if } i = j \\ a_{ij}, & \text{otherwise} \end{cases} \tag{1.16.4}$$

where it is assumed that $a_{ij} = a_{ji}$ for $i > j$.

Definition. The mapping $s_i : \mathbf{Z}^n \longrightarrow \mathbf{Z}^n$ given by the formula:

$$s_i(x) = x - \frac{2\mathfrak{B}(x, e_i)}{\mathfrak{B}(e_i, e_i)} e_i = x - \mathfrak{B}(x, e_i)e_i \tag{1.16.5}$$

is called the **simple reflection** at i.

Note that the action of s_i on a vector x changes only its i-th coordinate.

Lemma 1.16.6. *Let s_i be a reflection of a quadratic form q with associated bilinear symmetric form $\mathfrak{B}$. Then*

1. *s_i preserves the bilinear form $\mathfrak{B}$, i.e. $\mathfrak{B}(s_i(x), s_i(y)) = \mathfrak{B}(x, y)$.*
2. *$s_i^2(x) = x$. In particular, s_i is an automorphism of $\mathbf{Z}^n$.*
3. *$s_i(e_i) = -e_i$.*

Proof.

1. Since $\mathfrak{B}(e_i, e_i) = 2$, there results

$$\mathfrak{B}(s_i(x), s_i(y)) = \mathfrak{B}(x - \mathfrak{B}(x, e_i)e_i, y - \mathfrak{B}(y, e_i)e_i) = \mathfrak{B}(x, y) -$$
$$\mathfrak{B}(x, e_i)\mathfrak{B}(e_i, y) - \mathfrak{B}(y, e_i)\mathfrak{B}(x, e_i) + \mathfrak{B}(x, e_i)\mathfrak{B}(y, e_i)\mathfrak{B}(e_i, e_i) = \mathfrak{B}(x, y)$$

which shows that any simple reflection preserves the bilinear form.

2. $s_i^2(x) = x - \mathfrak{B}(x, e_i)e_i - \mathfrak{B}(x, e_i)e_i + 2\mathfrak{B}(x, e_i)e_i = x$, so s_i is an automorphism of $\mathbf{Z}^n$.

3. $s_i(e_i) = e_i - \mathfrak{B}(e_i, e_i)e_i = e_i - 2e_i = -e_i$. □

Definition. The group generated by all simple reflections of a given quadratic form q is called the **Weyl group** of q and denoted by $W(q)$.

Recall that a **quiver** $Q = (VQ, AQ, s, e)$ is a directed graph which consists of (nonempty) sets VQ, AQ and two mappings $s, e : AQ \to VQ$. The elements of VQ are called **vertices** (or **points**), and those of AQ are called **arrows**. In this chapter only finite quivers are considered, that is the quivers with finite (nonempty) sets VQ and AQ. Without loss of generality the set of vertices VQ will be considered to be $\{1, 2, ..., n\}$. Each arrow $\sigma \in AQ$ starts at the vertex $s(\sigma)$ and ends at the vertex $e(\sigma)$. The vertex $s(\sigma)$ is called the **start** (or **initial**, or **source**) **vertex** and the vertex $e(\sigma)$ is called the **end** (or **terminal**, or **target**) **vertex** of σ.

In what follows all quivers considered in this chapter will be finite connected and acyclic, i.e. without loops or oriented cycles.

For such a quiver Q there is a (nonsymmetrical) bilinear form defined by

$$< \alpha, \beta > = \sum_{i=1}^{n} \alpha_i \beta_i - \sum_{\sigma \in AQ} \alpha_{s(\sigma)} \beta_{e(\sigma)}, \tag{1.16.7}$$

where $\alpha, \beta \in \mathbf{Z}^n$. The corresponding quadratic form is

$$q(\alpha) = < \alpha, \alpha > = \sum_{i=1}^{n} \alpha_i^2 - \sum_{\sigma \in AQ} \alpha_{s(\sigma)} \alpha_{e(\sigma)}. \tag{1.16.8}$$

This is an integral quadratic form on $\mathbf{Z}^n$. Introduce the corresponding **symmetric bilinear form** by

$$(\alpha, \beta) = < \alpha, \beta > + < \beta, \alpha > . \tag{1.16.9}$$

It is easy to see that

$$q(\alpha) = \frac{1}{2}(\alpha, \alpha) \tag{1.16.10}$$

for any $\alpha \in \mathbf{Z}^n$ and

$$(e_i, e_j) = \begin{cases} 2, & \text{if } i = j \\ -t_{ij}, & \text{otherwise} \end{cases}$$

where $\{e_1, e_2, \ldots, e_n\}$ is the canonical basis of $\mathbf{Z}^n$, t_{ij} is the number of arrows from the vertex i to the vertex j.

Definition. The Weyl group of the quadratic form (1.16.8) is called the **Weyl group of the quiver** Q and denoted by $W(Q)$.

Let Γ be a finite connected acyclic undirected graph with a set of vertices $\Gamma_0 = \{1, 2, \ldots, n\}$ and consider the set of natural numbers $\{t_{ij}\}$, where $t_{ij} = t_{ji}$ is the number of edges between the vertex i and the vertex j. For such a graph its corresponding quadratic form is defined by

$$q(x) = \sum_{i=1}^{n} x_i^2 - \sum_{i<j} t_{ij} x_i x_j. \tag{1.16.11}$$

The corresponding bilinear symmetric form is hence given by

$$(e_i, e_j) = \begin{cases} 2, & \text{if } i = j \\ -t_{ij}, & \text{otherwise} \end{cases} \tag{1.16.12}$$

Remark 1.16.13. Let Q be a quiver and let $\Gamma = \overline{Q}$ be its underlying graph (obtained from Q by deleting the orientation of the arrows). Then the quadratic forms and the corresponding bilinear symmetric forms of Q and Γ are the same.

Therefore the Weyl groups of Q and Γ are the same, as well. Thus, the quadratic form, the bilinear symmetric form and the Weyl group of the quiver do not depend on the orientation.

Recall that a quadratic form q is **positive definite** if $q(\alpha) > 0$ for all $0 \neq \alpha \in \mathbf{Z}^n$. A quadratic form q is **positive semi-definite** (or **nonnegative definite**) if $q(\alpha) \geq 0$ for all $\alpha \in \mathbf{Z}^n$.

Then there is the following main statement (see [**42**], theorem 2.5.2).

Theorem 1.16.14. *Suppose Γ is a connected simply laced graph (i.e. a graph without loops and multiple edges) and let q be its quadratic form.*

(1) *q is positive definite if and only if Γ is one of the simple Dynkin diagrams A_n, D_n, E_6, E_7 or E_8 presented in figure* 1.13.1.

(2) *if q is not positive definite then it is positive semi-definite if and only if Γ is a one of the extended Dynkin diagram $\tilde{A}_n$, $\tilde{D}_n$, $\tilde{E}_6$, $\tilde{E}_7$ or $\tilde{E}_8$; moreover,* $\mathrm{rad}(q) = \mathbf{Z}\delta$, *where δ is the vector with entries as given by the numbers in the graphs presented in figure* 1.16.1.

(3) *if q is neither positive definite nor positive semi-definite, then there is a vector $\alpha \geq 0$ with $q(\alpha) < 0$ and $(\alpha, e_i) \leq 0$ for all i.*

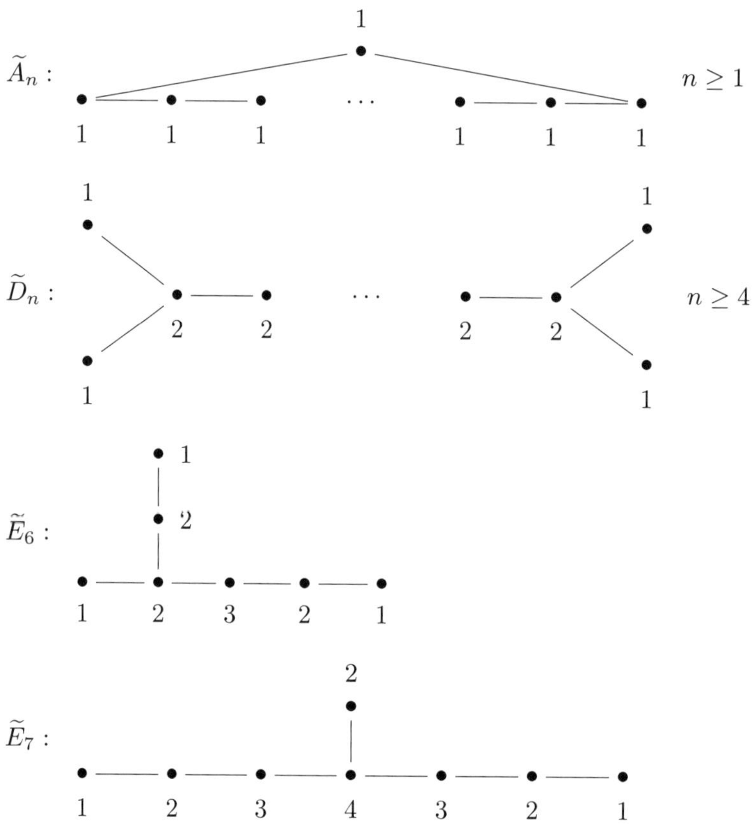

$\widetilde{E}_8$:

Figure 1.16.1.

From now on, if not stated otherwise, by a graph Γ we shall always mean a simple Dynkin diagram. The set

$$\Phi = \{\alpha \in \mathbf{Z}^n \,:\, q(\alpha) = 1\} \tag{1.16.15}$$

is the **set of roots** of the quadratic form q of Γ.

Note that each e_i is a root. These roots are called the **simple roots** of Γ. By lemma 1.16.1, the reflection s_i maps e_i to $-e_i$. Each root $\alpha \in \Phi$ can be written in the form $\sum\limits_i k_i e_i$. A root α is called **positive** if α is not a zero vector and all $k_i \geq 0$, and it is called **negative** if α is not a zero vector and $k_i \leq 0$ for all i. Denote by Φ^+ the set of all positive roots and by Φ^- the set of all negative roots of a simple Dynkin diagram Γ. The following lemma will be needed. It follows immediately from lemma 2.5.3 in [**42**].

Lemma 1.16.16. *Let Γ be a simple Dynkin diagram. Then the set of roots Φ is finite and $\Phi = \Phi^+ \cup \Phi^-$.*

Lemma 1.16.17. *Let Γ be a simple Dynkin diagram and Φ be the set of all roots of the quadratic form q of Γ. Then*

1. *The reflection s_i maps any root to a root.*
2. *The set $\Phi^+ \backslash \{e_i\}$ is invariant under the reflection s_i.*

Proof.

1. Let $\alpha \in \Phi$, i.e. $q(\alpha) = 1$. Since $(\alpha, \alpha) = 2q(\alpha) = 2$ and s_i preserves the bilinear symmetric form, by lemma 1.16.6, there results that $q(s_i(\alpha)) = \frac{1}{2}(s_i(\alpha), s_i(\alpha)) = \frac{1}{2}(\alpha, \alpha) = 1$, i.e. $s_i(\alpha) \in \Phi$.

2. Let $\alpha \in \Phi^+ \backslash \{e_i\}$. Since, by theorem 1.16.14, the quadratic form q is positive definite

$$0 \leq q(\alpha \pm e_i) = \frac{1}{2}(\alpha \pm e_i, \alpha \pm e_i) = q(\alpha) \pm (\alpha, e_i) + 1 = 2 \pm (\alpha, e_i),$$

and hence

$$-2 \leq (\alpha, e_i) \leq 2.$$

If $(\alpha, e_i) = 2$ then $q(\alpha - e_i) = 0$ and so $\alpha = e_i$, which is a contradiction. If $(\alpha, e_i) \leq 0$, then $s_i(\alpha) = \alpha - (\alpha, e_i)e_i \geq \alpha$ and so $s_i(\alpha) \in \Phi^+ \backslash \{e_i\}$. Suppose that $(\alpha, e_i) = 1$, i.e. the i-th coordinate of α is strictly positive. Then $s_i(\alpha) = \alpha - e_i$. Since α is not a multiple of e_i, it must involve some other e_j. Since the j-th coordinates of $s_i(\alpha)$ and α are the same, $s_i(\alpha)$ is also a positive root in this case. □

For any $\alpha = (\alpha_1, \alpha_2, \ldots, \alpha_n) \in \Phi$ define its **height** as $h(\alpha) = \sum\limits_{i=1}^{n} \alpha_i$.

Proposition 1.16.18. *Let Γ be a simple Dynkin diagram and $W(\Gamma)$ be the Weyl group of Γ. Then*

1. *The Weyl group $W(\Gamma)$ is finite.*
2. *Every root α of Γ is of the form $\alpha = w(e_i)$, where $w \in W(\Gamma)$ and e_i is a simple root.*

Proof.
1. By lemma 1.16.17 the elements of $W(\Gamma)$ permute the set of roots Φ. Since the simple roots e_i span $\mathbf{Z}^n$, and the action of $W(\Gamma)$ on Δ is faithful

$$W(\Gamma) \hookrightarrow \mathrm{Sym}(\Phi),$$

$W(\Gamma)$ is a subgroup of the group of permutations of Φ. Since Φ is a finite set by lemma 1.16.17, $W(\Gamma)$ is also finite.

2. Let $\alpha \in \Phi$, then $-\alpha \in \Phi$ as well. Since $\Phi = \Phi^+ \cup \Phi^-$, first assume that $\alpha \in \Phi^+$. If $h(\alpha) = 1$ then α is a simple root. Suppose $h(\alpha) > 1$. Since $(\alpha, \alpha) = 2q(\alpha) = 2 = \sum\limits_{i=1}^{n} \alpha_i(\alpha, e_i)$ and all $\alpha_i \geq 0$, there is an i so that $(\alpha, e_i) > 0$. Then $s_i(\alpha) = \alpha - (\alpha, e_i)e_i$ is another positive root of strictly smaller height, by lemma 1.16.17. Continuing by induction on the height of vectors there results a proof. Analogously, $-\alpha = ws_i(e_i)$, and that proves the statement for negative roots. □

1.17. Reflection functors

Reflection functors (also called partial Coxeter functors) were first introduced by I.N.Bernstein, I.M.Gel'fand, V.A.Ponomarev in [**8**] as a main tool for a new proof and better understanding of the Gabriel theorem.

Definition. Let Q be a quiver. A vertex $i \in VQ$ is called a **sink** if there is no arrow $\sigma \in AQ$ such that $s(\sigma) = i$.

Example 1.17.1.

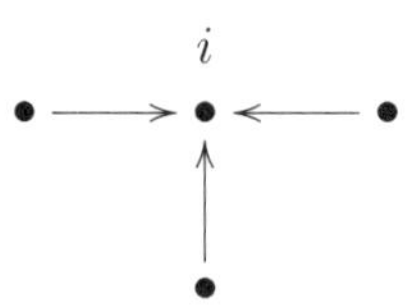

A vertex $i \in VQ$ is called a **source** if there is no arrow $\sigma \in AQ$ such that $e(\sigma) = i$.

Example 1.17.2.

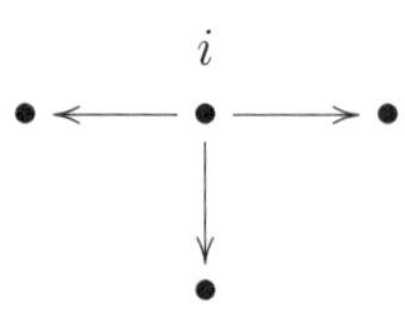

Definition. Let Q be a quiver and $i \in VQ$. Then $s_iQ = \overline{Q_i}$ is the quiver obtained from Q by reversing the directions of all arrows incident with the vertex i.

Note that $s_i^2 Q = Q$.

Example 1.17.3.
If Q is the quiver

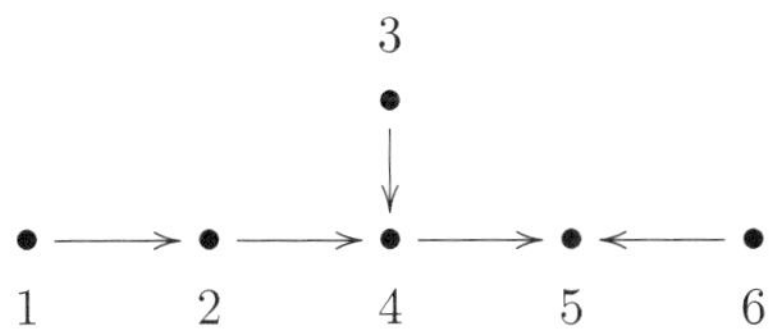

then $s_4Q = \overline{Q_4}$ is the quiver

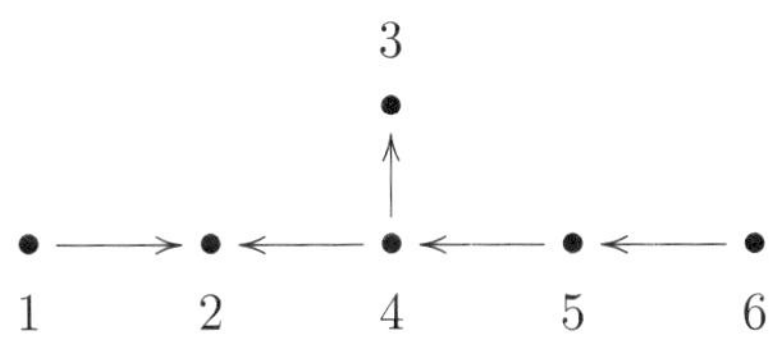

and s_2s_4Q is the quiver

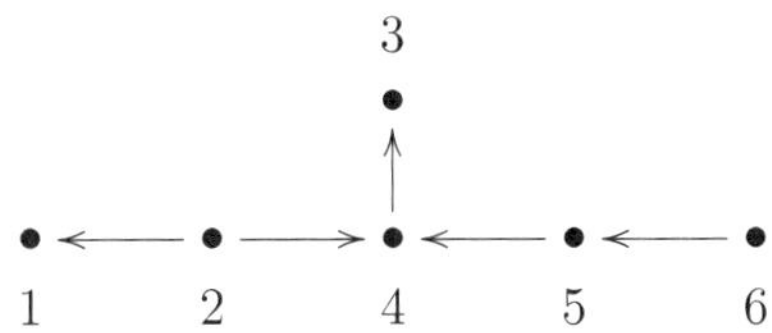

Let $Q = (VQ, AQ, s, e)$ be a finite quiver and let K be a field. A **representation** $V = (V_x, V_\sigma)$ of Q over K is a family of vector spaces V_x $(x \in VQ)$ together with a family of linear mappings $V_\sigma : V_{s(\sigma)} \to V_{e(\sigma)}$ $(\sigma \in AQ)$. In this section we shall always assume that all quivers are finite, and all representations are finite dimensional, that is, for every such representation V all vector spaces V_i are finite dimensional over K. All (finite dimensional) representations of a quiver Q over the field K form a category denoted by $\mathrm{Rep}_K(Q)$, in which the objects are representations V and morphisms $f = (f_x) : V \to W$ are defined as a family of linear mappings $f_x : V_x \to W_x$, $x \in VQ$, such that for each $\sigma \in AQ$ the diagram

$$\begin{array}{ccc} V_{s(\sigma)} & \xrightarrow{f_{s(\sigma)}} & W_{s(\sigma)} \\ {\scriptstyle V_\sigma}\downarrow & & \downarrow{\scriptstyle W_\sigma} \\ V_{e(\sigma)} & \xrightarrow{f_{e(\sigma)}} & W_{e(\sigma)} \end{array}$$

commutes.

In this category in an obvious way one can define the zero representation, subrepresentations and quotient representations, direct sum of representations, indecomposable and decomposable representations (see section 2.4, [**42**]). Thus, $\mathrm{Rep}_K(Q)$ is an additive category. Every object of this category is isomorphic to a finite direct sum of indecomposable representations. Since $\mathrm{Rep}_K(Q)$ is a Krull-Schmidt category [9] (see corollary 2.4.2, [**42**]), this decomposition is unique up to isomorphism and permutation of summands. So the problem of classifying finite

[9]Recall that an additive category $\mathfrak{C}$ is said to be Krull-Schmidt category provided the endomorphism ring $\mathrm{End}(X)$ of any indecomposable object X of $\mathfrak{C}$ is a local ring. In a Krull-Schmidt

dimensional representations of a finite quiver Q over a field K reduces to classifying the indecomposable representations of Q.

Definition. Let Q be a quiver and $i \in VQ$ be a sink of Q. Introduce the **left reflection functor**

$$F_i^+ : \operatorname{Rep}(Q) \longrightarrow \operatorname{Rep}(\overline{Q_i})$$

as follows.

Let V be a representation of Q, then

$$(F_i^+V)_k = \begin{cases} V_k & \text{if } k \neq i \\ \operatorname{Ker} \varphi_i & \text{if } k = i \end{cases}$$

where

$$\varphi_i : \bigoplus_{\substack{\sigma \in AQ \\ e(\sigma)=i}} V_{s(\sigma)} \to V_i$$

is a canonical map defined by $(x_{e(\sigma)})_\sigma \mapsto \sum_\sigma V_\sigma(x_{e(\sigma)})$. Further $(F_i^+V)_\sigma = V_\sigma$ for all $\sigma \in AQ$ with $e(\sigma) \neq i$, otherwise $(F_i^+V)_\sigma = \pi_\sigma \circ \iota_i$, the composition of the natural embedding ι_i of $(F_i^+V)_i$ into $\bigoplus_{\substack{\sigma \in AQ \\ e(\sigma)=i}} V_{s(\sigma)}$ and the projection π_σ of this sum onto the term $V_{s(\sigma)}$ for each $\sigma \in AQ$ with $e(\sigma) = i$:

$$(F_i^+V)_\sigma : (F_i^+V)_i \xrightarrow{\iota_i} \bigoplus_{\substack{\sigma \in AQ \\ e(\sigma)=i}} V_{s(\sigma)} \xrightarrow{\pi_\sigma} V_{s(\sigma)}$$

On the morphisms the functor F_i^+ is defined in a natural way. Let V, W be two representations of Q and let $f : V \to W$ be a morphism in $\operatorname{Rep}(Q)$. Then $F_i^+(f) = g : F_i^+V \to F_i^+W$ is defined as follows: $g_k = f_k$ for all $k \neq i$ and g_i is the unique homomorphism such that the following diagram

$$\begin{array}{ccccccc}
0 \longrightarrow & (F_i^+V)_i & \xrightarrow{\iota_i} & \bigoplus_{\substack{\sigma \in AQ \\ e(\sigma)=i}} V_{s(\sigma)} & \xrightarrow{\varphi_i} & V_i \\
& \Big\downarrow g_i & & \Big\downarrow \oplus f_{s(\sigma)} & & \Big\downarrow f_i \\
0 \longrightarrow & (F_i^+W)_i & \xrightarrow{\iota_i'} & \bigoplus_{\substack{\sigma \in AQ \\ e(\sigma)=i}} W_{s(\sigma)} & \xrightarrow{\varphi_i'} & W_i
\end{array}$$

commutes.

The definition of the functor F_i^- is dual to this. Let $i \in VQ$ be a source of Q. The **right reflection functor**

$$F_i^- : \operatorname{Rep}(Q) \to \operatorname{Rep}(\overline{Q_i})$$

is defined as follows.

Let V be a representation of Q, then

$$(F_i^-V)_k = \begin{cases} V_k & \text{if } k \neq i \\ \operatorname{Coker}(\psi_i) & \text{if } k = i \end{cases}$$

category any object is a finite direct sum of indecomposable objects, and such a decomposition is unique up to isomorphism and permutation of summands.

where ψ_i is the canonical map:

$$\psi_i : V_i \longrightarrow \bigoplus_{\substack{\sigma\in AQ\\ s(\sigma)=i}} V_{e(\sigma)}.$$

Further $(F_i^- V)_\sigma = V_\sigma$ for all $\sigma \in AQ$ with $s(\sigma) \neq i$, otherwise $(F_i^- V)_\sigma = \pi_i \circ \iota_\sigma$, the composition of the natural embedding ι_σ of $V_{e(\sigma)}$ into $\bigoplus_{\substack{\sigma\in AQ\\ s(\sigma)=i}} V_{e(\sigma)}$ and the natural projection π_i of this sum onto the term $(F_i^- V)_i$ for each σ such that $s(\sigma) = i$.

On the morphisms the functor F_i^- is defined in a natural way. Let V, W be two representations of Q and let $f : V \to W$ be a morphism in $\mathrm{Rep}(Q)$. Then $F_i^-(f) = g : F_i^-(V) \to F_i^-(W)$ is defined as follows: $g_k = f_k$ for all $k \neq i$ and g_i is a unique homomorphism such that the following diagram

$$\begin{array}{ccccccc}
V_i & \xrightarrow{\psi_i} & \bigoplus_{\substack{\sigma\in AQ\\ s(\sigma)=i}} V_{e(\sigma)} & \xrightarrow{\pi_i} & (F_i^- V)_i & \longrightarrow & 0 \\
\Big\downarrow f_i & & \Big\downarrow (f_{e(\sigma)}) & & \Big\downarrow g_i & & \\
W_i & \xrightarrow{\psi_i'} & \bigoplus_{\substack{\sigma\in AQ\\ s(\sigma)=i}} W_{e(\sigma)} & \xrightarrow{\pi_i'} & (F_i^- W)_i & \longrightarrow & 0
\end{array}$$

commutes.

Let V be a representation of a quiver Q with $VQ = \{1, 2, \dots, n\}$. In this case the **dimension vector** of V is the vector $\underline{\dim}\, V = (\dim V_i)_{i\in VQ} \in \mathbf{N}^n$.

Denote by $L_i = V = (V_1, V_2, \dots, V_n)$ the simple indecomposable representation corresponding to the vertex i such that $\dim V_i = 1$ and $\dim V_j = 0$ for $j \neq i$. Thus $\underline{\dim}\, L_i = e_i = (0, \dots, 1, \dots 0)$ which is the vector with 1 in the i-th entry and 0 otherwise.

Proposition 1.17.4. *Let Q be a quiver, and let V be an indecomposable representation of Q.*

1. *If $i \in VQ$ is a sink, then either $V = L_i$ or*

$$\varphi_i : \bigoplus_{\substack{\sigma\in AQ\\ e(\sigma)=i}} V_{s(\sigma)} \longrightarrow V_i$$

is surjective.

2. *If $i \in VQ$ is a source, then either $V = L_i$ or*

$$\psi_i : V_i \longrightarrow \bigoplus_{\substack{\sigma\in AQ\\ s(\sigma)=i}} V_{e(\sigma)}$$

is injective.

Proof.

1. Let W be a complement of $\mathrm{Im}\, \varphi_i$ in V_i. Then

$$V = V' \oplus V''$$

where

$$V'_k = \begin{cases} W & \text{if } k = i \\ 0 & \text{otherwise} \end{cases}$$

and $V''_i = \operatorname{Im} \varphi_i$. Since V is indecomposable, one of these summands has to be zero. If the first summand is zero, then φ_i has to be surjective. If the second summand is zero, then the first has to be of the desired form, because V is supposed to be indecomposable.

2. The proof is similar. This time the kernel of ψ_i splits off. □

Proposition 1.17.5. *Let Q be a quiver, and let V be a representation of Q.*
1. *If*

$$\varphi_i : \bigoplus_{\substack{\sigma \in AQ \\ e(\sigma)=i}} V_{s(\sigma)} \longrightarrow V_i$$

is surjective, then

$$F_i^- F_i^+ V \cong V$$

.

2. *If*

$$\psi_i : V_i \longrightarrow \bigoplus_{\substack{\sigma \in AQ \\ s(\sigma)=i}} V_{e(\sigma)}$$

is injective, then

$$F_i^+ F_i^- V \cong V$$

.

Proof.
1. Note that if i is a sink of Q then it is a source in $s_i Q$, so the functor $F_i^- F_i^+ : \operatorname{Rep}(Q) \longrightarrow \operatorname{Rep}(Q)$ is defined. Define a natural transformation $\Phi_i : F_i^- F_i^+ \longrightarrow \mathrm{id}$ of functors from $\operatorname{Rep}(Q)$ to $\operatorname{Rep}(Q)$ as follows. For any representation V of Q consider the diagram

$$\begin{array}{ccccccccc} 0 \longrightarrow & (F_i^+ V)_i & \xrightarrow{\iota_i} & \bigoplus_{\substack{\sigma \in AQ \\ e(\sigma)=i}} V_{s(\sigma)} & \xrightarrow{\varphi_i} & V_i & & \\ & \| & & \| & & & & \\ & (F_i^+ V)_i & \xrightarrow{\nu_i} & \bigoplus_{\substack{\sigma \in AQ \\ s(\sigma)=i}} (F_i^+ V)_{e(\sigma)} & \xrightarrow{\pi_i} & (F_i^- F_i^+ V)_i & \longrightarrow 0 \end{array}$$

with exact rows. Since, by definition, $\nu_i = \sum_\sigma (\pi_\sigma \iota_i) = (\sum_\sigma \pi_\sigma) \circ \iota_i = \iota_i$, the first square of this diagram commutes. By definition, $(F_i^- F_i^+ V)_i$ is the cokernel of the kernel of ν_i. Therefore, by the homomorphism theorem, there is an injective homomorphism $\tau_i : F_i^- F_i^+ V_i \longrightarrow V_i$ making this diagram commutative:

$$\begin{array}{ccccccccc}
0 & \longrightarrow & F_i^+ V_i & \xrightarrow{\iota_i} & \bigoplus\limits_{\substack{\sigma\in AQ\\ e(\sigma)=i}} V_{s(\sigma)} & \xrightarrow{\varphi_i} & V_i & & \\
 & & \Vert & & \Vert & & \uparrow{\scriptstyle \tau_i} & & \\
 & & F_i^+ V_i & \xrightarrow{\nu_i} & \bigoplus\limits_{\substack{\sigma\in AQ\\ s(\sigma)=i}} (F_i^+ V)_{e(\sigma)} & \xrightarrow{\pi_i} & (F_i^- F_i^+ V)_i & \longrightarrow & 0
\end{array}$$

Define $\Phi_i(V) : F_i^- F_i^+ V \longrightarrow V$ as follows:

$$(\Phi_i(V))_k : V_k \longrightarrow V_k$$

is the identity if $k \neq i$ and $(\Phi_i(V))_i = \tau_i$. Since φ_i is surjective, τ_i is also surjective, therefore τ_i is an isomorphism. Thus, $F_i^- F_i^+ V \cong V$.

2. In a similar way define a natural transformation $\Psi_i : \mathrm{id} \longrightarrow F_i^+ F_i^-$ of functors from $\mathrm{Rep}(Q)$ to $\mathrm{Rep}(Q)$. For any representation V of Q consider the diagram

$$\begin{array}{ccccccccc}
 & & V_i & \xrightarrow{\psi_i} & \bigoplus\limits_{\substack{\sigma\in AQ\\ s(\sigma)=i}} V_{e(\sigma)} & \xrightarrow{\pi_i} & (F_i^- V)_i & \longrightarrow & 0 \\
 & & & & \Vert & & \Vert & & \\
0 & \longrightarrow & (F_i^+ F_i^- V)_i & \xrightarrow{\iota_i} & \bigoplus\limits_{\substack{\sigma\in AQ\\ e(\sigma)=i}} (F_i^- V)_{s(\sigma)} & \xrightarrow{\eta_i} & (F_i^- V)_i & &
\end{array}$$

with exact rows. Since $\eta_i = \bigoplus_\sigma (F_i^- V)_\sigma = \bigoplus_\sigma (\pi_i \iota_\sigma) = \pi_i \circ \bigoplus_\sigma \iota_\sigma = \pi_i$, the second square of this diagram commutes. By definition, $(F_i^+ F_i^- V)_i$ is the kernel of the cokernel of η_i. Therefore, by the homomorphism theorem, there is a surjective homomorphism $\mu_i : V_i \longrightarrow (F_i^+ F_i^- V)_i$ making this diagram commutative:

$$\begin{array}{ccccccccc}
 & & V_i & \xrightarrow{\psi_i} & \bigoplus\limits_{\substack{\sigma\in AQ\\ s(\sigma)=i}} V_{e(\sigma)} & \xrightarrow{\pi_i} & (F_i^- V)_i & \longrightarrow & 0 \\
 & & \downarrow{\scriptstyle \mu_i} & & \Vert & & \Vert & & \\
0 & \longrightarrow & (F_i^+ F_i^- V)_i & \xrightarrow{\iota_i} & \bigoplus\limits_{\substack{\sigma\in AQ\\ e(\sigma)=i}} (F_i^- V)_{s(\sigma)} & \xrightarrow{\eta_i} & (F_i^- V)_i & &
\end{array}$$

Define $\Psi_i(V) : V \longrightarrow F_i^+ F_i^- V$ as:

$$(\Psi_i(V))_k : V_k \longrightarrow V_k$$

is the identity if $k \neq i$ and $(\Psi_i(V))_i = \mu_i$. Since ψ_i is injective, μ_i is also injective, therefore μ_i is an isomorphism. Thus, $F_i^+ F_i^- V \cong V$. $\square$

Proposition 1.17.6. *Let Q be a quiver, and let V be an indecomposable representation of Q. Then F_i^+V and F_i^-V (whenever defined) are either indecomposable or 0.*

Proof. The proposition will be proved for F_i^+V; the case F_i^-V is proved similarly. By proposition 1.17.4 it follows that either

$$\varphi_i : \bigoplus_{\substack{\sigma\in AQ\\ e(\sigma)=i}} V_{s(\sigma)} \longrightarrow V_i$$

is surjective or $V = L_i$. In the last case $F_i^+V = 0$. So it can be assumed that φ is surjective. Assume that F_i^+V is decomposable

$$F_i^+V = X \oplus Y$$

with $X, Y \neq 0$. Since, by proposition 1.17.5, $F_i^-F_i^+V \cong V$ is indecomposable, we must have $F_i^-X = 0$, or $F_i^-Y = 0$. Suppose that $F_i^-Y = 0$. Then $F_i^+F_i^-Y = 0$. Since Φ_i is an isomorphism, there results that $Y = 0$. This shows that F_i^+V is indecomposable. □

Proposition 1.17.7. *Let Q be a quiver and V a representation of Q.*
1. *Let i be a sink and let*

$$\varphi_i : \bigoplus_{\substack{\sigma\in AQ\\ e(\sigma)=i}} V_{s(\sigma)} \longrightarrow V_i$$

be surjective. Then

$$\underline{\dim}(F_i^+V) = s_i(\underline{\dim}(V))$$

2. *Let i be a source and let*

$$\psi_i : V_i \longrightarrow \bigoplus_{\substack{\sigma\in AQ\\ s(\sigma)=i}} V_{e(\sigma)}$$

be injective. Then

$$\underline{\dim}(F_i^-V) = s_i(\underline{\dim}(V))$$

Proof. Only the first statement will be proved, the second one is handled similarly. Let $i \in Q$ be a sink and let

$$\varphi_i : \bigoplus_{\substack{\sigma\in AQ\\ e(\sigma)=i}} V_{s(\sigma)} \longrightarrow V_i$$

be surjective. Then from the exact sequence

$$0 \longrightarrow \operatorname{Ker}\varphi_i \longrightarrow \bigoplus_{\substack{\sigma\in AQ\\ e(\sigma)=i}} V_{s(\sigma)} \xrightarrow{\varphi_i} V_i \longrightarrow 0$$

it follows that

$$\dim \operatorname{Ker}\varphi_i = \sum_{\substack{\sigma\in AQ\\ e(\sigma)=i}} \dim V_{s(\sigma)} - \dim V_i$$

Therefore

$$(\underline{\dim}(F_i^+V) - \underline{\dim}(V))_i = \sum_{\substack{\sigma\in AQ\\ e(\sigma)=i}} \dim V_{s(\sigma)} - 2\dim V_i = -(\underline{\dim}(V), e_i)$$

and

$$(\underline{\dim}(F_i^+V) - \underline{\dim}(V))_j = 0, \quad j \neq i$$

This implies

$$\underline{\dim}(F_i^+V) - \underline{\dim}(V) = -(\underline{\dim}(V), e_i)e_i \iff$$

$$\underline{\dim}(F_i^+V) = \underline{\dim}(V) - (\underline{\dim}(V), e_i)e_i = s_i(\underline{\dim}(V))$$

□

From propositions 1.17.4 -1.17.7 the Bernstein-Gel'fand-Ponomarev theorem follows:

Theorem 1.17.8 (Bernstein-Gel'fand-Ponomarev)

1. *Let i be a sink of a quiver Q, and let V be an indecomposable representation of Q. Then there are only two cases:*

(a) *$V = L_i$ and $F_i^+V = 0$.*

(b) *F_i^+V is nonzero and indecomposable, $F_i^-F_i^+V \cong V$ and $\underline{\dim}(F_i^+V) = s_i(\underline{\dim}V)$.*

2. *Let i be a source of a quiver Q, and let V be an indecomposable representation of Q. Then there are only two cases:*

(a) *$V = L_i$ and $F_i^-V = 0$.*

(b) *F_i^-V is nonzero and indecomposable, $F_i^+F_i^-V \cong V$ and $\underline{\dim}(F_i^-V) = s_i(\underline{\dim}V)$.*

1.18. Coxeter functors and Coxeter transformations

In the proof of the Gabriel theorem for an arbitrary field following Bernstein, Gel'fand, Ponomarev (see [**8**]) a most important role is played by Coxeter functors and Coxeter elements (or Coxeter transformations).

This section describes the Coxeter functors and the Coxeter elements and their main properties.

Definition. An ordered sequence $a_1, a_2, \ldots, a_n$ of the vertices of Q is called $(+)$-**admissible** (or **sink-admissible**, or an **admissible sequence of sinks**) in Q if

1) the vertex a_1 is a sink in the quiver Q;

2) the vertex a_i is a sink in the quiver $s_{a_{i-1}}s_{a_{i-2}}\cdots s_{a_1}Q$ for all $2 \leq i \leq n$.

Dually, an ordered sequence $a_1, a_2, \ldots, a_{n-1}, a_n$ of the vertices of Q is called $(-)$-**admissible** (or **source-admissible**) in Q if

1) the vertex a_1 is a source in the quiver Q;

2) the vertex a_i is a source in the quiver $s_{a_{i-1}}s_{a_{i-2}}\cdots s_{a_1}Q$ for all $2 \leq i \leq n$.

It is easy to see that if $a_1, a_2, \ldots, a_n$ is a sink-admissible sequence in Q then $a_n, a_{n-1}, \ldots, a_1$ is a source-admissible sequence in Q.

Since for a sink-admissible (or source-admissible) sequence $a_1, a_2, \ldots, a_n$ every arrow in $s_{a_n}s_{a_{n-1}}\cdots s_{a_1}Q$ is reversed exactly twice, there results

$$s_{a_n}s_{a_{n-1}}\cdots s_{a_1}Q = Q.$$

Note that a sink-admissible (source-admissible) sequence does not always exist. But if the quiver is a simple Dynkin diagram these sequences always exist. More generally there is the following statement.

Lemma 1.18.1. *An admissible sequence of sinks (sources) always exists in a quiver Q if and only if Q has no oriented cycles.*

Proof. The proof of this lemma easily follows from proposition 11.3.8 and corollary 11.3.10, [**42**].

Definition. Let Q be a quiver whose underlying graph Γ is a simple Dynkin diagram. In this case Q is simply called a **Dynkin quiver**.

Assume that $1, 2, \ldots, n$ is an admissible sequence of sinks in Q. The functors

$$C^+ = F_n^+ F_{n-1}^+ \ldots F_1^+ : \mathrm{Rep}(Q) \to \mathrm{Rep}(Q) \tag{1.18.2}$$

and

$$C^- = F_1^- F_2^- \ldots F_n^- : \mathrm{Rep}(Q) \to \mathrm{Rep}(Q) \tag{1.18.3}$$

are called the **Coxeter functors** associated with an admissible ordering $\{1, 2, \ldots, \}$ of the vertices of Q.

The **Coxeter element** c of $W(Q)$ corresponding to this admissible ordering is defined as

$$c = s_n s_{n-1} \cdots s_1. \tag{1.18.4}$$

It is easy to see that

$$c^{-1} = s_1 s_2 \ldots s_n. \tag{1.18.5}$$

Lemma 1.18.6. *The Coxeter functor C^+ (resp. C^-) is well defined, i.e. it is independent of the choice of a sink-admissible sequence.*

Proof. The statement will only be proved for the functor C^+. For the functor C^- the proof is similar.

Note that if i and j are two different sinks of a quiver Q then they are not joined by an arrow. In this case the two functors F_i^+ and F_j^+ commute, i.e. $F_i^+ F_j^+ = F_j^+ F_i^+$, since each of these functors does only do something to vertices which are joined by an arrow with the corresponding sink.

Suppose that there are two sink-admissible sequences

$$a_1, a_2, \ldots, a_n \quad \text{and } b_1, b_2, \ldots, b_n$$

of the vertices of a quiver Q. Assume that $a_1 = b_j$. Then the vertices $b_1, b_2, \ldots, b_{j-1}$ are not joined with a_1 by an arrow. Therefore

$$F_{b_n}^+ \cdots F_{b_j}^+ \cdots F_{b_1}^+ = F_{b_n}^+ \cdots F_{b_{j+1}}^+ F_{b_{j-1}}^+ \cdots F_{b_1}^+ F_{a_1}^+.$$

Applying a similar argument for the vertex a_2, then for vertex a_3 and so on, there results that $F_{b_n}^+ \cdots F_{b_1}^+ = F_{a_n}^+ \cdots F_{a_1}^+$. □

Applying induction to theorem 1.17.8 there is the following corollary.

Theorem 1.18.7. *Let Q be a Dynkin quiver, and V a representation of Q. Then either $C^- C^+(V) \cong V$ or $C^+(V) = 0$. In the first case $\underline{\dim}(C^+ V) = c(\underline{\dim} V)$.*

Lemma 1.18.8. *Let Q be a Dynkin quiver and $c = s_n s_{n-1} \cdots s_1 \in W(Q)$ a Coxeter element. Then c has no nonzero fixed points in $\mathbf{Z}^n$.*

Proof. Suppose c has a fixed point $\alpha \in \mathbf{Z}^n$, i.e.

$$\alpha = c(\alpha) = s_n s_{n-1} \cdots s_1(\alpha). \tag{1.18.9}$$

Since $s_n^2 = 1$, there results

$$s_n(\alpha) = s_{n-1}s_{n-2}\dots s_1(\alpha). \tag{1.18.10}$$

Taking into account that s_i changes only i-th coordinate of the vector α, from (1.18.10) it follows that

$$s_n(\alpha) = \alpha$$

and

$$s_{n-1}s_{n-2}\dots s_1(\alpha) = \alpha. \tag{1.18.11}$$

Continuing this process

$$\alpha = s_{n-1}(\alpha) = s_{n-2}(\alpha) = \dots = s_1(\alpha).$$

Therefore $(\alpha, e_i) = 0$ for each $i = 1, 2, \dots, n$, which implies that $q(\alpha) = \frac{1}{2}(\alpha, \alpha) = 0$. Since, by theorem 1.16.14, the quadratic form q is positive definite, there results that $\alpha = 0$. □

Lemma 1.18.12. *Let Q be a Dynkin quiver with set of roots Φ. Then for any $\alpha \in \Phi$ there is an $m \in \mathbf{N}$ such that $c^m\alpha \not> 0$.*

Proof. Since $W(Q)$ is a finite group, by proposition 1.16.18, and $c \in W(Q)$, there is an $M \in \mathbf{N}$ such that $c^M = 1$. But then the vector $\beta = (1 + c + c^2 + \dots + c^{M-1})\alpha$ is a fixed point for the operator c. By lemma 1.18.8 it follows that $\beta = 0$. Therefore the assumption that $c^m\alpha > 0$ for all k implies $\beta > 0$, which is a contradiction. □

1.19. The Gabriel theorem

In 1972 P. Gabriel, [**34**], gave a full description of quivers of finite representation type over an algebraically closed field (see also theorem 2.6.1, [**42**]). He also proved that there is a bijection between the isomorphism classes of indecomposable representations of a quiver Q and the set of positive roots of the quadratic form corresponding to this quiver.

Another proof of this theorem for an arbitrary field using Coxeter functors and the Weyl group has been given by Bernstein, Gelfand, Ponomarev (see [**8**]).

In this section the second part of the Gabriel theorem will be proved in the general case, i.e. for an arbitrary field k, following Bernstein, Gel'fand, Ponomarev.

Theorem 1.19.1 (Gabriel, Bernstein, Gelfand, Ponomarev). *Let Q be a Dynkin quiver, and let K be a field. Then Q is of finite representation type over the field K, and the map $V \mapsto \underline{\dim}(V)$ is a one-to-one correspondence between the finite dimensional indecomposable representations of Q and the positive roots.*

Proof. Let Q be a Dynkin quiver, and let V be an indecomposable representation of Q with a dimension vector $\underline{\dim}V = \alpha$. Let $\{1, 2, \dots, n\}$ be a sink-admissible sequence of the vertices of Q, which exists by corollary 11.3.10, [**42**]. Let $C^+ = F_n^+F_{n-1}^+\cdots F_1^+$ be the corresponding Coxeter functor, and let $c = s_ns_{n-1}\cdots s_1$ be the Coxeter element. Then for every k the dimension vector of $(C^+)^kV$ is $c^k\alpha$ if for every $i = 1, 2, \dots, k$, $c^i(\alpha) > 0$. By theorem 1.18.7 $(C^+)^kV$ is either indecomposable (with positive dimension) or zero. By lemma 1.18.12 there exists a number m such that $c^m\alpha \not> 0$. For this m it must be the case that $(C^+)^m V = 0$. Choose the smallest m such that $(C^+)^m V = 0$. Then there exists some i such that:

$$F^+_{i-1} \cdots F^+_1 (C^+)^{m-1} V \neq 0, \quad F^+_i F^+_{i-1} \cdots F^+_1 (C^+)^{m-1} V = 0$$

which means, by theorem 1.17.8, that $F^+_{i-1} \cdots F^+_1 (C^+)^{m-1} V = L_i$. Then

$$V \cong (C^-)^{m-1} F^-_1 F^-_2 \ldots F^{-1}_{i-1} L_i,$$

which shows that the dimension vector α of V is of the form:

$$\alpha = c^{-(m-1)} s_1 s_2 \ldots s_{i-1}(e_i).$$

Since α is a positive vector and s_k preserves the bilinear symmetric form for each k by lemma 1.16.6, $(\alpha, \alpha) = (e_i, e_i) = 2$, so it is a positive root.

The representation V must be the unique indecomposable representation of Q with dimension vector $\underline{\dim} V = \alpha$. Suppose W is another indecomposable representation of Q with dimension vector $\underline{\dim} W = \alpha$. Then one can choose the same sequence of reflection functors to send W to L_i and, since the computation on the dimension vector α is the same it follows that

$$V \cong (C^-)^{m-1} F^-_1 F^-_2 \ldots F^{-1}_{i-1} L_i \cong W$$

Conversely, suppose α is a positive root. By lemma 1.18.12, there exists a number m such that $c^{m+1}\alpha \not> 0$. Let $\beta = c^m \alpha$. Choose the shortest expression of the form $s_i s_{i-1} s_{i-2} \ldots s_1 \beta$ which is not a positive root. Then $\gamma = s_{i-1} s_{i-2} \ldots s_1 \beta$ is a positive root which becomes a negative root by applying s_i. This means, by proposition 1.16.18, that $\gamma = e_i$. So the representation $V = (C^-)^m F^-_1 F^-_2 \cdots F^-_{i-1} L_i$ is an indecomposable representation of dimension α. □

1.20. Generalized Cartan matrices and Kac-Moody Lie algebras

The previous sections have been devoted to finite dimensional Lie algebras. This section contains some facts about a new class of Lie algebras which was discovered in 1967 by V. G. Kac ([**48**]). They are infinitely dimensional Lie algebras. These Lie algebras can be defined similarly to finite-dimensional ones using generalized Cartan matrices. Simultaneously and independently, the same class of algebras was discovered by R. Moody ([**64**]) in his study of root systems. These algebras are known today as **Kac-Moody algebras**. They are mostly infinite dimensional Lie algebras and form a natural generalization of finite dimensional semisimple Lie algebras.

The isomorphism classes of simple complex finite-dimensional Lie algebras can be parametrized by the Cartan matrices. From theorem 1.15.18 it follows that every simple Lie algebra over an algebraically closed field of characteristic 0 can be seen as coming from an irreducible root system. Concretely, the simple Lie algebra involved can be described in terms of the Chevalley generators and the Serre relations using the Cartan matrix corresponding to this irreducible root system in the following way.

Let $C = (c_{ij})$ be the Cartan matrix of a root system with base $\alpha_1, \alpha_2, \ldots, \alpha_n$. Suppose e_i, f_i, h_i is the standard basis of the Lie algebra $\mathfrak{sl}(\alpha_i) \cong \mathfrak{sl}(2, k)$ for $i = 1, 2, \ldots, n$, where k is an algebraically closed field of characteristic 0. Then the elements e_i, f_i, h_i for $i = 1, 2, \ldots, n$ are linearly independent and satisfy the following relations

S1. $[h_i, h_j] = 0$ for all i, j.
S2. $[e_i, f_i] = h_i$ for each i and $[e_i, f_j] = 0$ if $i \neq j$.
S3. $[h_i, e_j] = c_{ji} e_j$ and $[h_i, f_j] = -c_{ji} f_j$ for all i, j.
S4. $(\mathrm{ad}\, e_i)^{1-c_{ji}}(e_j) = 0$ and $(\mathrm{ad}\, f_i)^{1-c_{ji}}(f_j) = 0$ if $i \neq j$

where $\mathrm{ad} : \mathfrak{g} \longrightarrow \mathrm{End}(\mathfrak{g})$, $\mathrm{ad}(x)(y) = [x, y]$ is the adjoint representation of $\mathfrak{g}$.

The elements e_i, f_i, h_i for $i = 1, 2, \ldots, n$ are called the **Chevalley generators**. Following J.-P. Serre the relations S1-S3 are called the commutation relations or the **Weyl relations**. The relations S4 are also often called the **Serre relations**. The following theorem proved by J.-P. Serre in [**75**] states that these relations completely determine a finite dimensional complex semisimple Lie algebra.

Theorem 1.20.1. (Serre's theorem). *Let C be the matrix of a root system. Let $\mathfrak{g}$ be the Lie algebra over an algebraically closed field k of characteristic 0 which is generated by elements e_i, f_i, h_i for $i = 1, 2, \ldots, n$ subject to the defining relations S1-S4. Then $\mathfrak{g}$ is finite dimensional and semisimple with Cartan subalgebra $\mathfrak{h}$ spanned by $h_1, h_2, \ldots, h_n$, and its root system has the Cartan matrix C.*

Any semisimple finite dimensional Lie algebra $\mathfrak{g}$ has an associated Cartan matrix $C = (c_{ij})$, $i, j = 1, ..., n$ constructed via the root system. This matrix has the properties given in theorem 1.13.3. These properties can be used for the general definition of a Cartan matrix.

Definition. A square matrix $C = (c_{ij})$ is called an (**abstract**) **Cartan matrix** if the following conditions hold:

(K1) C is an indecomposable matrix.
(K2) c_{ij} is a non-positive integer, moreover, if $c_{ij} = 0$ then $c_{ji} = 0$.
(K3) $c_{ii} = 2$ for all i.
(K4) C is a positive definite matrix.

A matrix $C = (c_{ij})$ is said to be **symmetrizable** if there exists a diagonal matrix $D = \mathrm{diag}(d_1, d_2, \ldots, d_n)$ with positive integers such that DC is a symmetric matrix. Note that any Cartan matrix C whose diagram is a simply laced diagram (even with cycles) is always symmetrizable because it is symmetric.

Serre's theorem states that any finite dimensional complex semisimple Lie algebra $\mathfrak{g}(C)$ can be constructed starting with a Cartan matrix C associated to a root system using the Chevalley generators and the relations S1-S4 (see [**75**]).

Thus, finite dimensional complex simple Lie algebras are in one-to-one correspondence with Cartan matrices defined as above. Instead of taking the Cartan matrix associated to a root system it is possible to start with a more general matrix, slightly relaxing the defining conditions on a Cartan matrix, and in the same way as in the Serre theorem, using the Serre relations, one obtains a wider class of Lie algebras called Kac-Moody algebras. These algebras are mostly infinite dimensional and provide a very natural generalization of finite dimensional simple Lie algebras.

Definition. A **generalized Cartan matrix** is a square matrix $C = (c_{ij})$ with integer entries satisfying conditions (K1)-(K3).

With a generalized Cartan matrix one can associate a diagram in the same way as for an ordinary Cartan matrix. Let $C = (c_{ij})$, $i, j = 1, 2, \dots, n$, be a generalized Cartan matrix. The vertices of the diagram correspond to numbers $1, 2, \dots, n$, and the i-th vertex is connected with j-th vertex by $c_{ij}c_{ji}$ edges. Unfortunately, in the general case, this diagram characterizes neither the matrix nor the corresponding Lie algebra. Nevertheless it may be useful because this diagram is connected if and only if the matrix is indecomposable.

R. V. Moody showed in [**64**] that any generalized Cartan matrix whose diagram contains no cycles is symmetrizable. In this case the diagram of a generalized Cartan matrix is a valued graph which characterizes this matrix.

There is a natural and rather obvious bijection between valued graphs and symmetrizable generalized Cartan matrices. To a valued graph with edge labels (d_{ij}, d_{ji}) associate the generalized Cartan matrix $C = c_{ij}$ with $c_{ii} = 2$, $c_{ij} = -d_{ij}$ if i and j are connected and $c_{ij} = 0$ if i and j are not connected. This matrix is symmetrizable because, by the definition of the notion of a valued graph, there are natural numbers f_i such that $d_{ij}f_j = d_{ji}f_i$. On the other hand, if $C = (c_{ij})$ is a symmetrizable generalized Cartan matrices, then the corresponding diagram $\Gamma(C) = (\Gamma_0, \Gamma_1)$ of C has a set of vertices $\Gamma_0 = \{1, 2, \dots, n\}$ and a $\{i, j\} \in \Gamma_1$ if and only if $c_{ij} \neq 0$. A valuation on $\Gamma(C)$ is given by a set of numbers d_{ij} with $d_{ii} = 0$ and $d_{ij} = c_{ij}$ for $i \neq j$. Since C is a symmetrizable matrix there are positive integers $f_1, f_2, \dots, f_n$ such that $f_i c_{ij} = f_j c_{ji}$ which shows that the diagram $\Gamma(C)$ thus obtained is a valued graph.

In this correspondence connectedness of the valued graph is the same as indecomposability of the Cartan matrix.

The associated Kac-Moody algebra $\mathfrak{g}(C)$ is the Lie algebra over the complex field $\mathbf{C}$ given by the following:

1. A generalized $n \times n$ Cartan matrix $C = (c_{ij})$ of rank r.
2. A vector space $\mathfrak{h}$ over $\mathbf{C}$ of dimension $2n - r$.
3. A set of n linear independent elements $h_i \in \mathfrak{h}$ and a set of n linear independent elements h_i^* of the dual space, such that $h^*(h) = c_{ij}$. (The h_i are known as **coroots** while the h_i^* are known as **roots**.)

Definition. The **Kac-Moody algebra** associated to a generalized Cartan matrix C is a Lie algebra $\mathfrak{g}(C)$ defined by $3n$ generators e_i, f_i, h_i subject to the relations S1-S4 given above.

A Kac-Moody algebra $\mathfrak{g}(C)$ is a finite dimensional simple Lie algebra if and only if C is a positive definite matrix. Otherwise there result infinite dimensional Lie algebras. Therefore Kac-Moody algebras are infinite dimensional analogues of finite dimensional semisimple Lie algebras.

Under the assumption that a generalized Cartan matrix C is indecomposable and symmetrizable, the matrix C cannot be negative definite or negative semidefinite.

Kac-Moody algebras $\mathfrak{g}(C)$ associated to symmetrizable indecomposable generalized Cartan matrices C are divided into three classes:

1) finite dimensional simple Lie algebras.

2) infinite dimensional Kac-Moody algebras of affine type, or affine Lie algebras.

3) Kac-Moody algebras of indefinite type.

The symmetrizable indecomposable generalized Cartan matrices of finite and affine type have been completely classified. Affine Lie algebras correspond to generalized Cartan matrices satisfying conditions (K1)-(K3) which are symmetrizable and positive semidefinite. It follows that the rank of such $n \times n$-matrices C is $n-1$.

V. G. Kac ([**50**]) classified all generalized Cartan matrices by associating with every such matrix a graph, known as an **extended Coxeter-Dynkin diagram**. Depending on the properties of the generalized Cartan matrices involved affine Lie algebras are divided into two groups: non-twisted and twisted. The diagrams corresponding to non-twisted affine Lie algebras are $\widetilde{A}_n$, $\widetilde{C}_n$, $\widetilde{B}_n$, $\widetilde{D}_n$, $\widetilde{E}_6$, $\widetilde{E}_7$, $\widetilde{E}_8$, $\widetilde{F}_4$ and $\widetilde{G}_2$ as given in figure 1.14.2.

V. G. Kac and R. V. Moody in their work generalized the notions of root systems and corresponding Lie algebras to Cartan matrices of arbitrary quivers (see [**49**], [**64**]). V. G. Kac in 1980 generalized the Gabriel theorem to the case of arbitrary quivers (see [**49**]). In particular, he proved the following theorem.

Theorem 1.20.2. (V.G.Kac, [**49**]) *For an arbitrary quiver Q, the set of dimension vectors of indecomposable representations of Q does not depend on the orientation of the arrows in Q. The dimension vectors of indecomposable representations correspond to the positive roots of the corresponding root system.*

In the theory of Kac-Moody algebras one distinguishes between real roots and imaginary roots. In the theorem above, real roots correspond to dimension vectors for which there is exactly one indecomposable representation, while imaginary roots correspond to dimension vectors for which there are families of indecomposable representations. If a positive root α is real, then $q(\alpha) = 1$. If it is imaginary, then $q(\alpha) \leq 0$.

Theorem 1.20.3. (V.G.Kac, [**49**]) *Let Q be a quiver, and $\alpha > 0$ a dimension vector. Then there is an indecomposable representation of Q of dimension α if and only if α is a root. If α is a real root, there is a unique indecomposable representation of dimension α. If α is an imaginary root, there are infinitely many indecomposable representations of dimension α.*

Further generalization of Kac-Moody Lie algebras are super Lie algebras and Borcherds algebras. The latter are connected to monstrous moonshine and vertex operator algebras (see [**32**], [**52**], [**35**], [**11**], [**12**], [**7**]).

1.21. Historical notes

Lie algebras appeared in mathematics at the end of the nineteen century in connection with the study of Lie groups, and in implicit form somewhat earlier in mechanics (see [**60**]). Note however that the Lie algebra structure given by the Poisson bracket dates from 1809. This one is not necessarily connected to infinitesimal symmetry transformations, but instead is closely tied up with the Hamiltonian formalism in mechanics. The common prerequisite for such a concept to arise was the concept of an 'infinitesimal transformation', which goes back at least to the time of the origin of infinitesimal calculus. The fact that integrals of class C^2 of the Hamilton equation are closed with respect to the Poisson brackets, which satisfy structure theory, representations, classification and applications of certain infinite Lie algebras known as Kac-Moody algebras, e.g. [**50**], [**87**], [**88**], modular Lie algebras (Lie algebras over fields of characteristic $p > 0$), e.g. [**74**], [**81**], infinite dimensional Lie algebras in general, [**3**], free Lie algebras, [**71**], vertex operators, Borcherds algebras and monstrous moonshine, [**32**], [**52**], [**35**], [**11**], [**12**], super Lie algebras, [**7**], graded infinite dimensional Lie algebras and their classification, [**58**], etc. etc.

References

[1] I. D. Ado, *Note on the representation of finite continuous groups by means of linear substitutions*, Izv. Fiz.-Mat. Obsch. (Kazan') **7** (1935), 1–43 (Russian).

[2] I. D. Ado, *The representation of Lie algebras by matrices*, Transl. Amer. Math. Soc. **9**:1 (1962), 308–327 Uspekhi Mat. Nauk **2** (1947), 159–173.

[3] Ralph K. Amado, Ian Stewart, *Infinite dimensional Lie algebras*, Noordhoff, 1974.

[4] I. Assem, D. Simson, A. Skowroński, *Elements of the representation theory of associative algebras. Vol.1 Techniques of representations theory*, Cambridge Univ. Press, 2006.

[5] M. Auslander, M. I. Platzeck, I. Reiten, *Coxeter functors without diagrams*, Trans. Amer. Math. Soc. **250** (1979), 1–46.

[6] M. Auslander, I. Reiten, S. Smalø, *Representation theory of Artin algebras*, Cambridge University Press, 1995.

[7] Yu. A. Bahturin, A. A. Mikhalev, V. M. Petrogradsky, M. V. Zaičev, *Infinite dimensional Lie super algebras*, W. de Gruyter, 1992.

[8] L. N. Bernstein, I. M. Gel'fand, V. A . Ponomarev, *Coxeter functors and Gabriel's theorem* , Uspekhi Math. Nauk, **28**:2(170) (1973), 19–33 (Russian); English translation: Russian Math. Survey, **28**:2 (1973), 17–32.

[9] G. Birkhoff, *Representability of Lie algebras and Lie groups by matrices*, Ann. Math. **38**:2 (1937), 526–532.

[10] G. Birkhoff, *Representation of Jordan and Lie algebras*, Trans. Amer. Math. Soc. **65** (1949), 116–136.

[11] R. E. Borcherds, *Monstrous moonshine and monstrous Lie super algebras*, Inv. Math. **109** (1992), 405–444.

[12] R. E. Borcherds, *What is moonshine*, Proc. Int. Cong. Math., Berlin, 1998, Vol.**1**, Birkhäuser, 1998, 607–615.

[13] A.Borel, *Linear algebraic groups*, 2-nd ed., Springer, 1991.

[14] N. Bourbaki, *Groups et algèbres de Lie. Chap.1-3*, Paris, Hermann, 1960.

[15] N. Bourbaki, *Groups et algèbres de Lie. Chap.4-6*, Paris, Hermann, 1968.

[16] N. Bourbaki, *Groups et algèbres de Lie. Chap.7-8*, Paris, Hermann, 1975.

[17] Jon Brundan, *Representation theory*, `http://darkwing.uoregon/edu/~brundan`

[18] Elie J. Cartan, *Sur la structure des groupes de transformations finis et continues*, Œuvres complètes, partie I, Gauthier-Villars (CNRS), 1952, 137–184.

[19] H. S. M. Coxeter, *Discrete groups generated by reflexions*, Ann. Math. **35**:2 (1934), 588–621.

[20] H. S. M. Coxeter, *The complete enumeration of finite groups of the form* $R_i^2 = (R_iR_j)^{k_{ij}} = 1$, J. London Math. Soc. **10** (1935), 21–25.

[21] Pierre-Paul Crivel, *Une histoire du théorème de Poincaré-Birkhoff-Witt*, Expo. Math. **22** (2004), 145–184.

[22] H. Derksen, *Quiver representations and linear algebra*, `http://www.math.lsa.umich.edu/~hderksen/math711.w01/quivers`

[23] H. Derksen, J. Weyman, A. Zelevinsky, *Quivers with potentials and their representations and mutations*, `http://arxiv.org/abs/0704.0649`, 2007.

[24] V. Dlab, C. M. Ringel, *Indecomposable representations of graphs and algebras*, Memoirs Amer. Math. Soc., vol. **173**, 1976.

[25] Yu. A. Drozd, *Coxeter transformatiuons and representations of partially ordered sets*, Funktsional. Anal. i Priložen. **8**:3 (1974) 34–42 (Russian); English translation: Functional Anal. Appl. **8** (1974) 219–225.

[26] Yu. A. Drozd, *Representations of bisected posets and reflection functors*, In Algebras and Modules. II. CMS Conference proceedings **24**, Amer. Math. Soc., 1998, 153–165.
[27] E. B. Dynkin, *Classification of the simple Lie groups*, Mat. Sbornik **18**:60 (1946), 347–352.
[28] E. B. Dynkin, *The structure of semi-simple Lie algebras*, Uspekhi Mat. Nauk **2** (1947), 59–127.
[29] E. B. Dynkin, *Semi-simple subalgebras of semi-simple Lie algebras*, Mat. Sbornik **30** (1952) 349–462; English transl.: Amer. Math. Soc. Trans. Ser. **6**:2 (1957), 111–244.
[30] E. B. Dynkin, *Maximal subgroups of the classical groups*, Trudy Moskov. Mat. Obshcsestva **1** (1952), 39-166; English transl.: Amer. Math. Soc. Trans. Ser. **6**:2 (1957), 245–378.
[31] K. Erdmann, M. H. Wildon, *Introduction to Lie algebras*, Springer, 2006.
[32] Igor Frenkel, James Lepowsky, Arne Meurman, *Vertex operator algebras and the monster*, Acad. Press, 1988.
[33] W. Fulton, J. Harris, *Representation theory: a first course*, Springer-Verlag, New York, 1971.
[34] P. Gabriel, *Unzerlegbare Darstellungen*, Manuscripts Math. **6** (1972), 71–103.
[35] Terry Gannon, *Moonshine beyond the monster*, Cambridge Univ. Press, 2006.
[36] Paul Garrett, *Poincaré-Birkhoff-Witt theorem*, preprint Dept. Math., Univ. of Minnesota, 2005; see `http/www.math.umn.edu/garrett`.
[37] I. M. Gel'fand, V. A. Ponomarev, *Free modular lattices and their representations*, Collection of articles dedicated to the memory of I.G.Petrovski (1901-1973), IV. Uspekhi Mat. Nauk **29**:6(180) (1974), 3–58; English translation: Russian Math. Surveys, **29**:6 (1974), 1–56.
[38] I. M. Gel'fand, V. A. Ponomarev, *Lattices, representations and their related algebras I*, Uspekhi Mat. Nauk **31**:5(191) (1976), 71–88; English translation: Russian Math. Surveys **31**:5 (1976), 67–85.
[39] W. A. de Graaf, *Constructing faithful matrix representations of Lie algebras*, Proceedings of the 1997 Internat. Symp. on Symbolic and Algebraic Computation (ISSAC'97, 1997, Kihei HI, USA), ACM Press, New York, 54–59.
[40] Harish-Chandra, *Faithful representations of Lie algebras*, Ann. Math. **50** (1949), 68–76.
[41] Michiel Hazewinkel, *Niceness theorems*, to appear in Proc. WWCA conference, Waterloo, Canada, May 2008 and in Proc. Abel Symposium Tromsø, Norway, June 2008; arXiv: 0810.5691 v1
[42] Michiel Hazewinkel, Nadiya Gubareni, V. V. Kirichenko, *Algebras, rings and modules*, vol.II, Springer, 2007.
[43] J. E. Humphreys, *Introduction to Lie algebras and representation theory*, Springer-Verlag, New York, 1978.
[44] J. E. Humphreys, *Reflection groups and Coxeter groups*, Cambridge studies in advanced mathematics, **29**, 1990.
[45] K. Iwasawa, *On the representation of Lie algebras*, Japanese Journal of Mathematics, **19** (1948), 405–426.
[46] N. Jacobson, *Lie algebras*, Interscience tracts in pure and applied Mathematics, **10**, Interscience, New York, 1962; Dover reprint, 1979.
[47] N. Jacobson, *Exceptional Lie algebras*, Marcel Dekker, 1971.
[48] Victor G. Kac, *Simple irreducible graded Lie algebras of finite growth*, Izvestia Akad. Nauk, USSR **32** (1968), 1271–1311.
[49] Victor G. Kac, *Infinite root systems, representations of graphs and invariant theory*, Invent. Math. **56**:1 (1980), 57–92.
[50] Victor G. Kac, *Infinite dimensional Lie algebras*, Cambridge Univ. Press, 2-nd ed., 1985.
[51] Victor G. Kac, *Virasoro algebra*, In: M.Hazewinkel (ed.), Encyclopaedia of Mathematics, **9** (1993), 420–422, KAP, `http://eom.springer.de/V/vo96710.htm`
[52] Victor G. Kac, *Vertex algebras for beginners*, Amer. Math. Soc., 1997.
[53] Victor G. Kac et al., *Course notes for MIT. 18.7 45: Introduction to Lie algebras.* `http://www.math.mit.edu/~lesha/7451ec`
[54] W. Killing, *Die Zusammensetzung der stetigen endlichen Transformationsgruppen*, I,II,II,IV *Math. Ann.*, **31** (1988), 252–290; ibid. **33** (1989) , 1–48; ibid. **34** (1989), 57–122; ibid. **36** (1990), 161–189.
[55] M. Kleiner, H. R. Tyler, *Admissible sequences and the preprojective component of a quiver*, Advances in Math. **192** (2005), 376–402.

CHAPTER 2

Coalgebras: motivation, definitions, and examples

The following seven chapters are intended to give the reader a first glimpse of the world of Hopf algebras and quantum groups. The time is long past that just this topic, this small part of algebra, could be covered adequately in one full sized monograph. So this treatment is necessarily very selective. The aim is to make readers aware of this wonderful world and to try to convey some of the beauty and elegance of the ideas, constructions, results, applications, and proofs. By now this is in the main a highly noncommutative world. Indeed most of the examples treated have an underlying algebra or coalgebra that is free, respectively cofree and thus maximally non commutative, respectively maximally non co commutative. For the permutations example (chapter 7) both apply. It is a section of noncommutative algebra that tends to illustrate well that the noncommutative versions of things are often more elegant and understandable than the commutative ones, once one has found the *right* noncommutative versions. In the commutative world things are often obscured, or even messed up, by all kinds of counting coefficients.

Many of the objects discussed below, such as coalgebras, bialgebras, Hopf algebras,... form categories. And many of the constructions are functorial. This point is not stressed. We are assuming that the reader is sufficiently acquainted with the categorical point of view, feeling, and philosophy, to recognize when things are functorial or not.[1]

We shall always work with modules, algebras, coalgebras, bialgebras, Lie algebras, Hopf algebras, and even more structure-rich objects, over a fixed Noetherian commutative integral domain k with unit element. The examples: k is a field and $k = \mathbf{Z}$, the ring of integers, are the most important. Unadorned tensor products are always tensor products over k. The category of k-modules is denoted $\mathbf{Mod}_k$ and correspondingly the morphisms between two $\mathbf{Mod}_k$ objects are denoted $\mathbf{Mod}_k(M, N)$.

The main examples of (graded) Hopf algebras treated in this book are: **Symm**, the Hopf algebra of symmetric functions, RS, the Hopf algebra of representations of all the symmetric groups (which is isomorphic to **Symm**), **NSymm**, the Hopf algebra of noncommutative symmetric functions, **QSymm**, the Hopf algebra of quasisymmetric functions, and MPR, the Hopf algebra of permutations.

These can with right be called combinatorial Hopf algebras in the loose sense that the (deeper) study of them involves quite a bit of combinatorics. There are more such (graded) 'combinatorial' Hopf algebras or bialgebras. An important one is the bialgebra of Young tableau, or coplactic bialgebra studied in Dieter Blessenohl and Manfred Schocker [**5**].

---

[1]The time is also long past that one could refer (with some justice) to category theory as 'general nonsense'. However, the more sophisticated and powerful techniques of category theory sofar do not play a significant role in the theory of Hopf algebras and quantum groups.

Further 'combinatorial Hopf algebras' are the peak algebras whose study was initiated by Nyman [**24**] and Stembridge [**32**]. Still more Hopf algebras with a lot of combinatorics are various Hopf algebras of trees (and Feynman diagrams), some of which make a brief appearance in section 8.8.

In addition there are "combinatorial Hopf algebras" based on (families of) posets [**15**], graphs [**29**], [**23**], geometric lattices [**17**] and matroids [**27**], [**28**], [4], [**10**], [**11**], [**12**].

There is also a technical meaning of the phrase 'combinatorial Hopf algebra' introduced in [**2**]. A combinatorial Hopf algebra in this technical sense is a graded Hopf algebra H together with an algebra morphism to the base ring k, called a character or a multiplicative linear functional. Many of the "combinatorial Hopf algebras" mentioned in the papers quoted above are also 'combinatorial' in this technical sense.

As has already been indicated chapters 2-8 by no means constitute a treatise on all of Hopf algebra theory. For instance, two large parts, the theory of the Hopf algebras attached to affine algebraic groups, [**3**], chapter 4 and to formal groups, [**30**] are not discussed in this volume, except for a brief section, 3.11, on formal groups.

2.1. Coalgebras and 'addition formulae'

It is customary to motivate and define coalgebras as in some sense the duals of algebras. As a rule by writing the axioms that define an algebra in diagram form and than reversing all arrows.

However, the roots of coalgebras go further back in time, to before the concept of an algebra was formulated. They go back to the so called addition formulae of special functions. The simplest of these are the trigonometric addition formulae

$$\begin{aligned} \sin(\alpha+\beta) &= \sin(\alpha)\cos(\beta)+\cos(\alpha)\sin(\beta) \\ \cos(\alpha+\beta) &= \cos(\alpha)\cos(\beta)-\sin(\alpha)\sin(\beta). \end{aligned} \tag{2.1.1}$$

There are many, many more addition formulae. Indeed, they could fill a whole book easily. It was a great discovery ([**21, 33**] [2], see also [**34, 35, 36, 37**]), that special functions and all these (classical) [3] addition formulae come from representations of suitable groups. Let us see how this comes about.[4] Let G be a group, or more generally a monoid (semigroup with unit element e), and let ρ be a finite dimensional representation of G in a vector space V. Choose a basis of V and let

$$(\rho_{i,j}(g))_{i,j=1}^{n} \tag{2.1.2}$$

be the matrix of $\rho(g)$ with respect to the chosen basis. The $\rho_{i,j}$, span a finite dimensional space, C_ρ , of functions on G. Let $u_1, u_2, \dots, u_m$ be a basis of this space of functions. Because ρ is a representation

$$(\rho_{i,j}(gh)) = \sum_{t=1}^{n} \rho_{i,t}(g)\rho_{t,j}(h) \tag{2.1.3}$$

[2] Willard Miller Jr. and N. Ya. Vilenkin made the discovery independently.

[3] The same relation holds for q-special functions versus quantum groups, see loc. cit.

[4] The same point of view is e.g. taken in [**26**]. However, there the unfortunate assumption is made that the matrix entry functions are independent which need not be the case. Independence holds of course for instance for the case of an irreducible representation of a compact group (because of the orthogonality relations in that case).

for all $g, h \in G$. It follows that for every $u \in C_\rho$. there is an element

$$\mu(u) = \sum_{i,j=1}^{n} a_{i,j} u_i \otimes u_j \in C_\rho \otimes C_\rho \tag{2.1.4}$$

such that

$$u(gh) = \sum_{i,j=1}^{n} a_{i,j} u_i(g) u_j(h), \ \forall g, h \in G. \tag{2.1.5}$$

Moreover, this element $\mu(u) \in C_\rho \otimes C_\rho$ is unique. Indeed if $\sum b_{ij} u_i \otimes u_j$ also did the job we would have $\sum_j \left(\sum_i (a_{i,j} - b_{i,j}) u_i(g)\right) u_j(h) = 0$ for all $g, h \in G$, which, because of the independence of the u_j (as functions of h), implies $\sum_i (a_{i,j} - b_{i,j}) u_i(g) = 0$ for all j, $g \in G$. And then independence of the u_j (as functions of g) gives $a_{i,j} = b_{i,j}$ for all i, j. Thus there is a unique well defined morphism of vector spaces

$$\mu : C_\rho \longrightarrow C_\rho \otimes C_\rho \tag{2.1.6}$$

defined by the representation ρ.

Note that (2.1.5) is an addition law (written multiplicatively). The addition law of the sin and cos functions comes from the representation of the real line (or the circle)

$$\alpha \mapsto \begin{pmatrix} \cos(\alpha) & \sin(\alpha) \\ -\sin(\alpha) & \cos(\alpha) \end{pmatrix}.$$

The so-called comultiplication μ of (2.1.6) is coassociative, meaning that

$$(\mathrm{id} \otimes \mu)\mu = (\mu \otimes \mathrm{id})\mu : C_\rho \longrightarrow C_\rho \otimes C_\rho \otimes C_\rho \tag{2.1.7}$$

or in diagram form

$$\begin{array}{ccc} C_\rho & \xrightarrow{\ \mu\ } & C_\rho \otimes C_\rho \\ {\scriptstyle \mu}\big\downarrow & & \big\downarrow{\scriptstyle (\mathrm{id}\otimes\mu)\mu} \\ C_\rho \otimes C_\rho & \xrightarrow[(\mu\otimes\mathrm{id})\mu]{} & C_\rho \otimes C_\rho \otimes C_\rho \end{array} \tag{2.1.8}$$

This comes from the fact that the multiplication of G is associative (and the representation property).

There is also a special morphism

$$\varepsilon : C_\rho \longrightarrow k, \quad u \mapsto u(e) \tag{2.1.9}$$

where e is the identity element of the monoid G. By (2.1.5), putting in resp. $h = e$ or $g = e$ the morphism satisfies

$$u(g) = \sum a_{i,j} u_i(g) \varepsilon(u_j), \quad u(h) = \sum a_{i,j} u_i(g) u_j(h) \tag{2.1.10}$$

or in terms of the comultiplication μ.

$$(\mathrm{id} \otimes \varepsilon)\mu = \mathrm{id}, \quad (\varepsilon \otimes \mathrm{id})\mu = \mathrm{id} \tag{2.1.11}$$

which in diagram terms looks like

$$\begin{array}{ccc} C_\rho & = & C_\rho \\ {\scriptstyle\mu}\downarrow & & \downarrow{\scriptstyle\cong} \\ C_\rho \otimes C_\rho & \xrightarrow[(\mathrm{id}\otimes\varepsilon)]{} & C_\rho \otimes k \end{array} \qquad \text{and} \qquad \begin{array}{ccc} C_\rho & = & C_\rho \\ {\scriptstyle\mu}\downarrow & & \downarrow{\scriptstyle\cong} \\ C_\rho \otimes C_\rho & \xrightarrow[(\varepsilon\otimes\mathrm{id})]{} & k \otimes C_\rho \end{array} \tag{2.1.12}$$

where the two right hand vertical arrows of the two diagrams are the canonical identifications.

A **coalgebra** C over a base ring k (commutative with unit element) is now a k-module C together with a comultiplication morphism μ, and a counit morphism ε

$$\mu : C \longrightarrow C \otimes C, \quad \varepsilon : C \longrightarrow k \tag{2.1.13}$$

such that the diagrams (2.1.8) and (2.1.12) are commutative (with C_ρ replaced by C).

2.2. Coalgebras and decompositions

An **algebra** over k is a k-module A with a bilinear map $A \times A \longrightarrow A$. That is, there is a recipe for taking two elements of A and composing them to get a third new element made up of these two. At this stage we are not concerned with any other properties this composition law might have (such as associativity, Lie, alternative, ...).

Of course a bilinear map $m : A \times A \longrightarrow A$ is the same thing as a k-module morphism

$$\mu : A \otimes A \longrightarrow A. \tag{2.2.1}$$

A good example is the word-concatenation-algebra. Let $\mathcal{X}$ be an alphabet and let $\mathcal{X}^*$ be the monoid (under concatenation) of all words over $\mathcal{X}$ including the empty word. Let $k\mathcal{X}^*$ be the free module over k with basis $\mathcal{X}^*$. Define a multiplication on $k\mathcal{X}^*$ by assigning to two words $v, w \in \mathcal{X}^*$ their concatenation (denoted by $*$)

$$m(v, w) = v * w \tag{2.2.2}$$

and extending bilinearly. Thus if $v = [4, 3, 5, 1]$ and $w = [7, 1, 1]$ are two words over the natural numbers $\mathbf{N} = \{1, 2, \ldots\}$ their concatenation is $v * w = [4, 3, 5, 1, 7, 1, 1]$ [5] This multiplication is associative and the empty word, [ ], serves as a two sided unit element. This algebra is of course also the monoid algebra on $\mathcal{X}^*$ over k and the free associative algebra on $\mathcal{X}$ over k.

Of course not all algebras have this very nice kind of monomial structure where there is a basis such that multiplying (composing) two basis elements gives either zero [6] or another basis element. But quite a few of the more important ones in fact do. These include the monoid and group algebras, the matrix algebras $M^{n \times n}(k)$ of square n by n matrices with entries in k and even the shuffle algebras Shuffle($\mathcal{X}$),

[5] We shall often work with words over the natural numbers in this and the following chapters. Whence, in order to avoid confusion, the square bracket notation for words. It is unlikely that something like [**13, 9, 14, 3**] (in context) will be confused with a bibliographic reference, or that there will be confusion with the integral part of a real number or a Lie bracket (commutator difference product) $[x, y]$. There is no way in which typography can keep up with the needs of mathematical notation.

[6] This is to be thought of as that the two objects cannot be composed.

see below, over a field of characteristic zero, even though this last one at first sight does not look at all like that.

Sort of dually, a **coalgebra** over k is a k-module complete with a decomposition map

$$\mu : C \longrightarrow C \otimes C. \tag{2.2.3}$$

As in the algebra case above, think of the k-module C as a free module over k with as basis some kind of objects (such as words). Then μ of such a basis object gives all possible ways of decomposing that object into two other basis objects. A good example is the **word-cut-coalgebra** structure on $k\mathcal{X}^*$ which is given by

$$\mu[x_1, x_2, \dots, x_m] = \sum_{i=0}^{m} [x_1, \dots, x_i] \otimes [x_{i+1}, \dots, x_m]. \tag{2.2.4}$$

For example

$$\begin{aligned} \mu[4,3,5,1] = [\,] \otimes [4,3,5,1] + [4] \otimes [3,5,1] + [4,3] \otimes [5,1] \\ + [4,3,5] \otimes [1] + [4,3,5,1] \otimes [\,]. \end{aligned} \tag{2.2.5}$$

Coassociativity says that the decomposition rule μ, satisfies

$$(\mathrm{id} \otimes \mu)\mu = (\mu \otimes \mathrm{id})\mu : C \longrightarrow C \otimes C \otimes C \tag{2.2.6}$$

which in the present context can be thought of as follows. If one wants to break up an object into three pieces, than first break it up into two pieces (in all possible ways) and than break up the left halves into two pieces in all possible ways. Or, after the first step, break the right halves into two pieces in all possible ways. Coassociativity now says that it does not matter which of the two procedures is followed.

The word-cut-coalgebra just defined is coassociative.

There is also a counit morphism given by

$$\begin{aligned} &\varepsilon([\,]) = 1 \\ &\varepsilon(w) = 0 \quad \text{ if } w \text{ has length } \geq 1 \end{aligned} \tag{2.2.7}$$

which satisfies $(\mathrm{id} \otimes \varepsilon)\mu = \mathrm{id}$, $(\varepsilon \otimes \mathrm{id})\mu = id$.

2.3. Dualizing the idea of an algebra

Now let's do what is customary. I.e., write out the definition of an associative algebra with unit over k in terms of diagrams; and then reverse all arrows to get the (formal) definition of a coassociative coalgebra with counit over k [7].

An **associative algebra** over k with unit is a k-module A equipped with two k-module morphisms

$$m : A \otimes A \longrightarrow A, \quad e : k \longrightarrow A \tag{2.3.1}$$

[7] There also exist 'co'-versions of other notions of algebras, for instance co-Lie-algebras, see **[20]**.

such that the following three diagrams are commutative [8]

$$\begin{CD} A\otimes A\otimes A @>{m\otimes \mathrm{id}}>> A\otimes A \\ @V{\mathrm{id}\otimes m}VV @VV{m}V \\ A\otimes A @>>{m}> A \end{CD} \tag{2.3.2}$$

$$\begin{CD} A\otimes k @>{\mathrm{id}\otimes e}>> A\otimes A \\ @V{\cong}VV @VV{m}V \\ A @>>{=}> A \end{CD} \qquad \begin{CD} k\otimes A @>{e\otimes \mathrm{id}}>> A\otimes A \\ @V{\cong}VV @VV{m}V \\ A @>>{=}> A \end{CD} \tag{2.3.3}$$

Dually, a **coassociative coalgebra** over k with counit is a k-module C equipped with two k-module morphisms

$$\mu : C \longrightarrow C\otimes C, \qquad \varepsilon : C \longrightarrow k \tag{2.3.4}$$

such that the following three diagrams are commutative

$$\begin{CD} C @>>{\mu}> C\otimes C \\ @V{\mu}VV @VV{\mu\otimes \mathrm{id}}V \\ C\otimes C @>>{\mathrm{id}\otimes\mu}> C\otimes C\otimes C \end{CD} \tag{2.3.5}$$

$$\begin{CD} C @= C \\ @V{\mu}VV @VV{\cong}V \\ C\otimes C @>>{\mathrm{id}\otimes\varepsilon}> C\otimes k \end{CD} \qquad \begin{CD} C @= A\otimes C \\ @V{\mu}VV @VV{\cong}V \\ C\otimes C @>>{m}> k\otimes C \end{CD} \tag{2.3.6}$$

Let

$$\mathrm{tw} : \ M\otimes N \longrightarrow N\otimes M, \quad x\otimes y \mapsto y\otimes x \tag{2.3.7}$$

be the twist (switch, flip) isomorphism of tensor products of k-modules.

A k-algebra is **commutative** if the following diagram commutes

$$\begin{CD} A\otimes A @>>{\mathrm{tw}}> A\otimes A \\ @V{m}VV @VV{m}V \\ A @= A \end{CD} \tag{2.3.8}$$

and, dually, a coalgebra is **cocommutative** if the corresponding reversed arrow diagram commutes:

$$\begin{CD} C @= C \\ @V{\mu}VV @VV{\mu}V \\ C\otimes C @>>{\mathrm{tw}}> C\otimes C \end{CD} \tag{2.3.9}$$

---

[8] Very strictly speaking $id\otimes m$ is a morphism from $A\otimes(A\otimes A)$ to $A\otimes A$. So we are silently using the canonical identifications $A\otimes(A\otimes A)\cong A\otimes A\otimes A\cong(A\otimes A)\otimes A$.

2.4. Some examples of coalgebras

The word-cut-coalgebra of section 2.2 above is one example of a coalgebra. Other examples are the coalgebras constructed in section 2.1 for every finite dimensional representation of an (associative) monoid. Some more examples follow

Example 2.4.1. The (standard) matrix coalgebra. Let $M_{\text{coalg}}^{n\times n}$ be the free module over k of rank n^2 with basis $E_{i,j}, i,j = 1,\dots,n$. Define a comultiplication morphism and a counit morphism by

$$\mu(E_{i,j}) = \sum_{s=1}^{n} E_{i,s} \otimes E_{s,j}, \qquad \varepsilon(E_{i,j}) = \delta_{i,j} = \begin{cases} 1, & \text{if } i = j \\ 0, & \text{if } i \neq j. \end{cases} \tag{2.4.2}$$

Example 2.4.3. The coalgebra of functions on a finite monoid (or group), the convolution coalgebra. Let G be a finite monoid and consider the k-module $\mathrm{Func}(G, k)$ of all k-valued functions on G. A basis for $\mathrm{Func}(G, k)$ is $\{\delta_g : g \in G\}$, $\delta_g(h) = 1$ if $g = h$ and $\delta_g(h) = 0$ if $g \neq h$, $g, h \in G$. The comultiplication and counit are given by

$$\mu(\delta_g) = \sum_{hh'=g} \delta_h \otimes \delta_{h'}, \qquad \varepsilon(f) = f(e),\ f \in \mathrm{Func}(G, k) \tag{2.4.4}$$

so that ε is evaluation at the identity element of G.

Example 2.4.5. The sin-cos coalgebra. Let $C_{\sin-\cos}$ be the rank two free module over k with basis $\{c, s\}$. Define a comultiplication and counit by

$$\begin{aligned} &\mu(s) = s \otimes c + c \otimes s, \quad \mu(c) = c \otimes c - s \otimes s \\ &\varepsilon(s) = 0,\ \varepsilon(c) = 1. \end{aligned} \tag{2.4.6}$$

Then $C_{\sin-\cos}$ is a cocommutative coalgebra. This is of course the coalgebra of the representation

$$\alpha \mapsto \begin{pmatrix} \cos(\alpha) & \sin(\alpha) \\ -\sin(\alpha) & \cos(\alpha) \end{pmatrix} \tag{2.4.7}$$

of the circle group (or the real line), see section 2.1. Note that though the representation involved is only defined over the reals, the corresponding coalgebra is defined over the integers **Z**.

Example 2.4.8. The graded cofree coalgebra over a rank one module. Let $\mathrm{CoF}(k)$ be the free module over k: with basis $\{Z_0, Z_1, Z_2, \dots\}$. Define

$$\mu(Z_n) = \sum_{i=0}^{n} Z_i \otimes Z_{n-i}, \quad \varepsilon(Z_n) = 0, \quad \text{if } n \geq 1, \quad \varepsilon(Z_0) = 1. \tag{2.4.9}$$

Then $\mathrm{CoF}(k)$ is a cocommutative k-coalgebra with counit. Just what the phrase ‘graded cofree coalgebra’ might mean will be explained later.

This coalgebra is also called the **divided power coalgebra**. The reason is that if we define, say over **Q**,

$$d_n(x) = \frac{x^n}{n!} \tag{2.4.10}$$

then, by the binomial formula,

$$d_n(x+y) = \sum_{i+j=n} d_i(x) d_j(y). \tag{2.4.11}$$

Now consider the triangular representation of the monoid over **Q** (or **R**, or **C**, it does not matter which) given by

$$\rho(x) = \begin{pmatrix} 1 & d_1(x) & d_2(x) & \dots & d_n(x) \\ 0 & 1 & d_1(x) & \ddots & \vdots \\ 0 & 0 & 1 & \ddots & d_2(x) \\ \vdots & \vdots & \ddots & \ddots & d_1(x) \\ 0 & 0 & \dots & 0 & 1 \end{pmatrix} \tag{2.4.12}$$

Then the subcoalgebra of CoF(k) spanned by the $Z_0, Z_1, \dots, Z_n$ is precisely the coalgebra, C_ρ, of this representation, see section 2.1, and 2.4.11 is the corresponding addition formula.

Example 2.4.13. The coalgebra of representative functions on a monoid. In this example k is a field. Consider a monoid G and consider the k-module of functions Func(G, k), just like in subsection 2.4.3 above, except that this time the monoid need not be finite.

Call an element of Func(G, k), i.e. a function f on G, representative if there exist a finite number of functions g_i, h_i, such that for all x, $y \in G$

$$f(xy) = \sum_i g_i(x)h_i(y). \tag{2.4.14}$$

The terminology comes from the observation that the matrix entries of a finite dimensional representation of G are representative. [9]

For any function f define its left translate with respect to an element $y \in G$ by $L_y f(x) = f(yx)$ and its right translate by $R_y f(x) = f(xy)$. Let Lf be the subspace spanned by all left translates and Rf the space spanned by all the right translates of f.

Lemma 2.4.15. *The following properties are equivalent:*
(i) *The function f is representative.*
(ii) *The space Lf is finite dimensional.*
(iii) *The space Rf is finite dimensional.*
Moreover if these properties hold than the g_i, and h_i, in (2.4.14) *can be chosen to be representative.*

Proof. If f is representative than by (2.4.14) every left translate is a linear combination of the finite number of functions h_i, and every right translate is a linear combination of the finite number of functions g_i. This proves (i)$\mapsto$ (ii), (iii). Now suppose (ii) holds and let $h_1, \dots h_m$ be a basis of the space Lf. Then every left translate, f_x, of f is a linear combination (with coefficients depending on x) of the $h_1, \dots h_m$, implying (2.4.14). Similarly (iii) implies (i).

It remains to prove the final statement of the lemma. Again let $h_1, \dots h_m$ be a basis of the space Lf. A left translate of a left translate of f is a left translate of

[9] Depending on k and the group G the representative functions from Func(G, k) can be, indeed sometimes have to be, pretty weird. For instance take $k =$**Q**, $G = \mathrm{GL}_n(\mathbf{C})$, where **C** is the field of complex numbers and GL_n stands for the general linear group. Let $f(x)$ be the $(1,1)$ entry of the $n \times n$ matrix x if that matrix has all its entries in **Q** and zero otherwise. Then f is representative.

f. Thus, $Lh_i \subset Lf$, the h_i are representative and there are for each i finitely many functions $b_{ij}(y)$ such that $h_j(yz) = \sum\limits_i b_{j,i}h_i(z)$. Then

$$f(xyz) = \sum_i g_i(xy)h_i(z) = \sum_j g_j(x)h_j(yz) = \sum_{ij} g_j(x)b_{ij}(y)h_i(z)$$

and because the h_i, are independent it follows that

$$g_i(xy) = \sum_j g_j(x)b_{j,i}(y).$$

Construction 2.4.16. Define $\mathrm{Func}(G, k)_{\mathrm{repr}}$ as the k-module of representative functions on G. Define a comultiplication by the formula

$$\mu(f) = \sum_i g_i \otimes h_i \Leftrightarrow f(xy) = \sum_i g_i(x)h_i(y). \tag{2.4.17}$$

This is well defined because the canonical morphism

$$\mathrm{Func}(G,\, k) \otimes \mathrm{Func}(G,\, k) \longrightarrow \mathrm{Func}(G \times G,\, k)$$

is injective. Also define

$$\varepsilon(f) = f(e) \tag{2.4.18}$$

where e is the identity element of G. Then using lemma 2.4.15, one sees that $(\mathrm{Func}(G,\, k)_{\mathrm{repr}},\, \mu,\, \varepsilon)$ is a coalgebra. Note that by construction it is the span of the functions coming from finite dimensional representations of G (as in section 2.1). This fits well with the main theorem of coalgebras below in section 2.6).

2.4.19. Example of representative functions. The d_n of (H 1.4.10) are representative.

Remark 2.4.20. Consider the coalgebra of functions on a finite monoid as defined in example 2.4.3. Note that

$$\delta_g(xy) = \sum_{hh'=g} \delta_h(x)\delta_{h'}(y). \tag{2.4.21}$$

Thus if G is finite all elements of $\mathrm{Func}(G,\, k)$ are representative, and, moreover the definitions used for the comultiplication in the two cases, (2.4.4), (2.4.17), agree in this case. Note, however, that (2.4.4) is not useable in the infinite case in general.

Example 2.4.22. The binomial coalgebra. Consider the module of polynomials in one variable, $k[X]$ over k, with basis $1,\, X,\, X^2, \ldots$. Consider the comultiplication and counit

$$\begin{aligned} &X^n \mapsto \sum_{i=0}^{n} \binom{n}{i} X^i \otimes X^{n-i} = (1 \otimes X + X \otimes 1)^n \\ &\varepsilon(X^n) = 0 \text{ for } n \geq 1,\ \varepsilon(1) = 1. \end{aligned} \tag{2.4.23}$$

These define a coalgebra as is easily verified. This is the underlying coalgebra of the affine line Hopf algebra, see example 3.4.20 below.

Example 2.4.24. Set coalgebra. Let S be any set and consider the free module kS over k with basis S. Define a comultiplication and counit by

$$\mu(s) = s \otimes s, \quad \varepsilon(s) = 1,\, \forall s \in S. \tag{2.4.25}$$

Then kS becomes a coalgebra.

2.5. Sub coalgebras and quotient coalgebras

The following definition seems natural. A **sub coalgebra** of a coalgebra C is a submodule $C' \subset C$ such that $\mu(C') \subset C' \otimes C'$. If k is a field there is no problem. But if k is not a field there is a tricky point here. Namely, the natural map $C' \otimes C' \xrightarrow{i^{\otimes 2}} C \otimes C$ is not automatically injective. And if it is not μ is not well-defined on C'. So in the case of coalgebras over a ring we must add the condition that the tensor square of the inclusion is injective, or, better, define a sub algebra as a coalgebra (C', μ') together with an injective morphism of coalgebras $C' \to C$.

A well-known example of a module inclusion of which the tensor square is not injective is the following. Take $C = \mathbf{Q}[X, Y]$, the ring of commutative polynomials in two variables over the rationals (or any other field). Consider $C' = (X, Y)$ the (twosided) ideal generated by X and Y. The element $X \otimes Y - Y \otimes X$ of $C' \otimes C'$ is nonzero, but $i^{\otimes 2}(X \otimes Y - Y \otimes X)$ is zero in $C \otimes C$.

So add the injectivity of the tensor square. Then

$$(C', \mu|_{C'}, \varepsilon|_{C'}) \tag{2.5.1}$$

is a coalgebra.

A **quotient coalgebra** of a coalgebra C is a quotient module $C'' = C/I$ where I is a submodule such that

$$\mu(I) \subset \mathrm{Im}(I \otimes C + C \otimes I) \tag{2.5.2}$$

where on the right hand side of (2.5.2) is meant the sum of the images of $I \otimes C$ and $C \otimes I$ in $C \otimes C$.Then μ induces a morphism $\mu'' : C/I \longrightarrow C/I \otimes C/I$ and if also $I \subset \mathrm{Ker}(\varepsilon)$, with induced morphism $\varepsilon'' : C/I :\longrightarrow k$,

$$(C'', \mu_{C''}, \varepsilon_{C''}) \tag{2.5.3}$$

is a coalgebra. A submodule I such that (2.5.2) holds is called a **coideal**.

2.6. The main theorem of coalgebras

In this section k is a field. Coalgebras over a field have a very important finiteness property, viz that every element of a coalgebra over a field is contained a a finite dimensional subcoalgebra. This is called the main theorem of coalgebras, or also, the fundamental theorem of coalgebras. Ultimately this property comes from the fact that an element of a tensor product is a *finite* sum.

Theorem 2.6.1. (Main theorem of coalgebras over fields.) *Let C be a coalgebra with counit over a field k, and let $c \in C$ be an element of C. Then there is a finite dimensional subcoalgebra $C' \subset C$ that contains c.*

Notation 2.6.2. It is useful to have a notation for the iterates of the comultiplication. Set

$$\begin{aligned} \mu_0 &= \varepsilon, \quad \mu_1 = \mathrm{id}, \quad \mu_2 = \mu : C \longrightarrow C^{\otimes 2} \\ \mu_3 &= (\mathrm{id} \otimes \mu)\mu = (\mu \otimes \mathrm{id})\mu : C \longrightarrow C^{\otimes 3}, \ldots, \\ \mu_n &= (\mathrm{id} \otimes \mu_{n-1})\mu = (\mu \otimes \mu_{n-2})\mu = \cdots = (\mu_{n-2} \otimes \mu_2)\mu \\ &= (\mu_{n-1} \otimes \mathrm{id})\mu : C \longrightarrow C^{\otimes n}. \end{aligned} \tag{2.6.3}$$

Proof of theorem 2.6.1. [10] Let $c \in C$ and write

$$\mu_3(c) = \sum_{i,j} x_i \otimes y_{i,j} \otimes z_j. \tag{2.6.4}$$

It is easy to show that one can take the $\{x_i\}$ and $\{z_j\}$ to be independent sets of elements. Indeed, take a basis $\{u_i, : i \in I\}$. Then any element in $C \otimes C \otimes C$ can be written as a unique finite sum

$$\sum_{i_1,i_2,i_3} a_{i_1,i_2,i_3} u_{i_1} \otimes u_{i_2} \otimes u_{i_3} = \sum_{i_1,i_3} u_{i_1} \otimes \left(\sum_{i_2} a_{i_1,i_2,i_3} u_{i_2} \right) \otimes u_{i_3}.$$

Let C' be the subspace of C that is generated by the $y_{i,j}$ in (2.6.4). Since $c = (\varepsilon \otimes \mathrm{id} \otimes \varepsilon)\mu_3(c)$, $c \in C'$. Now

$$\mu_4(c) = (\mu \otimes \mathrm{id} \otimes \mathrm{id})\mu_3(c) = (\mathrm{id} \otimes \mu \otimes \mathrm{id})\mu_3(c)$$

and because the z_j are independent it follows that

$$\sum_i x_i \otimes \mu(y_{i,j}) = \sum \mu(x_i) \otimes y_{i,j} \in C \otimes C \otimes C'$$

and because the x_i are independent it follows that

$$\mu(y_{i,j}) \in C \otimes C'. \tag{2.6.5}$$

Similarly $\mu(y_{i,j}) \in C' \otimes C$ and an application of the lemma below concludes the proof.

Lemma 2.6.6. *Let V and W be vector spaces over a field k, and let $V' \subset V$ and $W' \subset W$ be the subspaces. Then*

$$V' \otimes W \cap V \otimes W' = V' \otimes W'. \tag{2.6.7}$$

The proof of the lemma is simple and left to the reader. (Take bases of V' and W' and complete these to bases of V and W.) Note, however, that it is here that the proof breaks down for, say, coalgebras over Noetherian integral domains whose underlying module is free. As a matter of fact the main theorem of coalgebras is not generally true for coalgebras over rings such as e.g. the ring of integers **Z**, [3].

2.7. Cofree coalgebras

Let M be a k-module. The tensor algebra on M is constructed as follows. Let $T^iM = M^{\otimes i}$ be the i-th tensor power of the module M. Let

$$\psi_{i,j} : T^iM \otimes T^jM \longrightarrow T^{i+j}M \tag{2.7.1}$$

be the canonical isomorphism. Define

$$TM = \bigoplus_i T^iM, \qquad m : TM \otimes TM \longrightarrow TM,$$

$$x \otimes y \in T^iM \otimes T^iM \mapsto \psi_{i,j}(x \otimes y). \tag{2.7.2}$$

[10] This proof comes from [**13**], p. 25. We do not know where this elegant proof originated. The theorem is due to Sweedler, [**31**]; the proof there is long and laborious and makes heavy use of duality, but has the advantage that it can be generalized to the case of reflexive modules over a Noetherian integral domain, for instance to the case of coalgebras over the integers whose underlying module is a free Abelian group of countable rank, see [**16, 9**].

Then with $1 \in k = T^0M \subset TM$ as unit element, TM is the free algebra on M over k. This means the following

Definition 2.7.3. The **free k-algebra** on a k-module M is a k-algebra FM, together with a k-module morphism $M \xrightarrow{i} FM$ such that for any k-algebra A and any k-module morphism $M \xrightarrow{\varphi} A$ there is a unique k-algebra morphism $\widetilde{\varphi}: FM \longrightarrow A$ such that $\widetilde{\varphi}i = \varphi$.

It follows immediately that FM is unique up to isomorphism (if it exists), and it is a simple straightforward matter to prove that the tensor algebra TM just constructed does the job. The morphism i is the inclusion $M = T^1M \subset TM$. Given $\varphi : M \longrightarrow A$, the k-algebra morphism $\widetilde{\varphi} : FM \longrightarrow A$ is given by $\widetilde{\varphi}(x_1 \otimes \cdots \otimes x_n) = m_n(\varphi(x_1), \ldots, \varphi(x_n)) = \varphi(x_1) \ldots \varphi(x_n)$.

The question arises whether the dual object: a cofree coalgebra over a k-module M exists. By definition this would be a k-coalgebra with the following universality property.

Definition 2.7.4. The **cofree k-coalgebra** over a k-module M is a coalgebra CM together with a k-module morphism $p : CM \longrightarrow M$ such that for any k-coalgebra C and morphism of k-modules $\varphi : C \longrightarrow M$ there is a unique morphism of k-coalgebras $\widehat{\varphi} : C \longrightarrow CM$ such that $p\widehat{\varphi} = \varphi$.

The definition guarantees uniqueness. Existence is another matter. As it turns out the cofree k-coalgebra over a module does exist for the case of a module M over a Noetherian integral domain k [**16**]. Even the description, let alone the proofs, involves a lot of technicalities. To avoid these consider the case of a free module of finite rank over a Noetherian integral domain k.

First consider the following example of a coalgebra.

Example 2.7.5. The tensor k-coalgebra over a k-module M, $\mathrm{CoF}(M)$. The underlying module of the tensor k-coalgebra $\mathrm{CoF}(M) = (TM, \mu, \varepsilon)$ is the same as that of the tensor algebra. The comultiplication is given by

$$\mu(x_1 \otimes x_2 \otimes \cdots \otimes x_n) = \sum_{i+j=n} \varphi_{i,j}^{-1}(x_1 \otimes x_2 \otimes \cdots \otimes x_n) \in \bigoplus_{i=0}^{n} T^iM \otimes T^{n-i}M. \quad (2.7.6)$$

and the counit is the canonical projection on the 0 th summand $T^0M = k$. Note that for $M = k$, we get back $\mathrm{CoF}(k)$ of example 2.4.8.

However, this tensor k-coalgebra is not the cofree k-coalgebra over M. For instance consider the following multiplicative k-coalgebra.

Example 2.7.7. The coalgebra of the multiplicative group. The underlying module is the k-module of polynomials in one variable $k[X]$ with basis $1 = X^0, X, X^2, X^3, \ldots$. The comultiplication and counit are given by

$$\mu(X^n) = X^n \otimes X^n, \quad \varepsilon(X^n) = 1. \quad (2.7.8)$$

Now consider the following morphism of k-modules

$$\varphi : k[X] \longrightarrow k = M, X \mapsto 1, \; X^n \mapsto 0 \text{ for } n \neq 1. \quad (2.7.9)$$

Then there is no morphism of k-coalgebras $\varphi : k[X] \longrightarrow \mathrm{CoF}(k)$ such that $p\widehat{\varphi} = \varphi : k[X] \longrightarrow k$ simply because there is no element $x \in \mathrm{CoF}(k)$ such that $\mu(x) = x \otimes x$ [11] (as is easily verified).

As it turns out $\mathrm{CoF}(k)$ is not large enough and what is needed is a certain completion, the recursive or representative completion. This completion will now be described in the case that M is a free module of finite rank over a Noetherian integral domain k.

Let $X_1, X_2, \ldots, X_m$ be a basis of M. Then the noncommutative monomials of degree n are a basis for $T^n M$, and TM is the k-module of noncommutative polynomials in the $X_1, X_2, \ldots, X_m$. Consider

$$\widehat{T}M = \prod_{i=0}^{\infty} T^i M \tag{2.7.10}$$

the k-module of noncommutative power series in the indeterminates $X_1, X_2, \ldots, X_m$. This module can be identified with the module of functions $\mathrm{Func}(X^*, k)$ of k-valued functions on the free monoid of all words in the alphabet $X = \{X_1, X_2, \ldots, X_m\}$. Now let

$$\widehat{T}M_{\mathrm{repr}} \subset \widehat{T}M = \mathrm{Func}(X^*, k) \tag{2.7.11}$$

be the submodule of representative functions (in the sense of example 2.4.13) [12]. It turns out that (2.7.11) is a subcoalgebra of $\widehat{T}M$ and that together with the induced canonical projection

$$\hat{T}M_{\mathrm{repr}} \subset \hat{T}M \mapsto T^1 M = M \tag{2.7.12}$$

it is the cofree k-coalgebra over M.

If C is a k-coalgebra and $\varphi : C \longrightarrow M$ is a morphism of k-modules, then the corresponding morphism $\widehat{\varphi} : C \longrightarrow \widehat{T}M_{\mathrm{repr}}$ is given by

$$\widehat{\varphi}(c) = (\varepsilon(c), \varphi(c), \varphi^{\otimes 2}\mu(c), \ldots, \varphi^{\otimes n}\mu_n(c), \ldots). \tag{2.7.13}$$

In case k is a field the theorem that $\widehat{T}M$ is the cofree coalgebra over M is due to [**6**], the Block-Leroux theorem. For the general case where k is not necessarily a field, see [**16**].

In case $M = k$, i.e. $m = 1$, the representative power series in the variable X_1 are precisely the recursive or rational power series; that is those power series that can be written in the form

$$\sum_{i=0}^{\infty} r_i X_1^i = (a_0 + a_1 X_1 + \cdots + a_t X_1^t)^{-1}(b_0 + b_1 X_1 + \cdots + b_s X_1^s). \tag{2.7.14}$$

For finitely many variables this still holds with recursiveness replaced by Schützenberger recursiveness, or, equivalently, Schützenberger rationality of noncommutative power series [13], see [**16**] for details. For infinitely many variables there are appropriate generalizations, loc. cit.

---

[11] Such an element of a coalgebra is called group-like.

[12] In example 2.4.13 the underlying ring k was supposed to be a field. However, the same definition applies more generally.

[13] Even for commutative power series in more than one variable these notions of recursiveness are more general than one might naively guess.

2.8. Algebra - coalgebra duality

As seen the formal definition of a coalgebra is dual to that of an algebra in the sense that all arrows get reversed. Taking dual modules also reverses arrows. Thus one wonders whether taking dual modules, $M \mapsto M^* = \mathrm{Hom}_k(M, k)$, would take coalgebras into algebras, and vice versa. To some extent that is the case.

Construction 2.8.1. The dual algebra of a coalgebra. Let C be a k-coalgebra, and consider the dual module $C^* = \mathbf{Mod}_k(M, k)$. There is a natural morphism

$$C^* \otimes C^* \xrightarrow{\psi} (C \otimes C)^*, \ \psi(f \otimes g)(c \otimes d) = f(c)g(d). \tag{2.8.2}$$

The dual of the comultiplication $\mu : C \longrightarrow C \otimes C$ is a morphism $\mu^* : (C \otimes C)^* \longrightarrow C^*$ and composing this with (2.8.2) gives a k-module morphism

$$m : C^* \otimes C^* \longrightarrow C^*. \tag{2.8.3}$$

which is associative because μ is coassociative [14]. Moreover the dual of $\varepsilon : C \longrightarrow K$ is a morphism of k-modules

$$e = \varepsilon^* : k \cong k^* \longrightarrow C^* \tag{2.8.4}$$

and it is a straightforward exercise to show that $(A = C^*, \ m = \mu^*\psi, \ e = \varepsilon^*)$ is an associative algebra with unit.

Example 2.8.5. The dual algebra of the matrix coalgebra. Consider the matrix coalgebra $M^{n\times n}_{\text{coalg}}$ of example 2.4.1. The underlying module is free with basis $E_{i,j}, \ i,j = 1, \dots, n$. So the dual module is also free with dual basis, say $F_{i,j}$,

$$\langle F_{i,j}, \, E_{r,s}\rangle = \delta_{i,r}\delta_{j,s} \tag{2.8.6}$$

(Kronecker delta). Then by definition the multiplication on $(M^{n\times n}_{\text{coalg}})^*$ is determined by

$$\begin{aligned} \langle F_{i,j}F_{u,v}, \, E_{r,s}\rangle &= \langle F_{i,j} \otimes F_{u,v}, \, \mu(E_{r,s})\rangle \\ &= \langle F_{i,j} \otimes F_{u,v}, \, \sum_t (E_{r,t} \otimes E_{t,s})\rangle \\ &= \delta_{i,r}\delta_{v,s}\delta_{j,u} = \delta_{j,u}\langle F_{i,v}, E_{r,s}\rangle \end{aligned} \tag{2.8.7}$$

and we see that

$$F_{i,j}F_{u,v} = \delta_{j,u}F_{i,v} \tag{2.8.8}$$

so that if we interpret the $F_{i,j}$ as the elementary matrices the usual matrix multiplication is obtained. Further the unit element of $M^{n\times n}_{\text{coalg}}$ is determined by

$$\langle e(1), \, E_{u,v}\rangle = \langle 1, \, \varepsilon(E_{u,v})\rangle = \delta_{u,v} \tag{2.8.9}$$

so that $e(1) = \sum_n F_{u,u}$, the usual matrix unit. Thus the dual algebra of the matrix coalgebra $M^{n\times n}_{\text{coalg}}$ is the familiar algebra $M^{n\times n}_{\text{alg}}(k)$ of $n \times n$ matrices over k.

Example 2.8.10. The dual algebra of the coalgebra of functions on a finite monoid. Now consider the coalgebra of functions on a finite monoid as in example 2.4.3. The underlying module is free with basis $\delta_g, \ g \in G$. Thus the dual module is also free with basis $h, h \in G$ and, as the notation suggests

$$\langle h, \, \delta_g\rangle = \begin{cases} 1, & \text{if } h = g \\ 0, & \text{if } h \neq g. \end{cases} \tag{2.8.11}$$

[14]And μ is compatible with the various natural iterated tensor product identifications.

The multiplication on $\mathrm{Func}(G, k)^*$ is determined by

$$\langle m(h_1 \otimes h_2), \delta_g\rangle = \langle h_1 \otimes h_2, \mu(\delta_g)\rangle = \langle h_1 \otimes h_2, \sum_{hh'=g} \delta_h \otimes \delta_{h'}\rangle \tag{2.8.12}$$

and it follows that $m(h_1 \otimes h_2) = h_1h_2$, so that $\mathrm{Func}(G, k)^* = kG$ the monoid (group) algebra of G over k.

Example 2.8.13. The dual algebra of the divided power coalgebra over a field of characteristic zero. Consider the divided power coalgebra of 2.4.8. Recall that as a module over k it is free with basis d_n, $n = 0, 1, 2, \ldots$.
Thus the dual module is the module of all formal sums

$$\sum_{n=0}^{\infty} r_n h_n,\ r_n \in k,\ \langle \sum_{n=0}^{\infty} r_n h_n, \sum_{j=0}^{m} b_j d_j\rangle = \sum_{j=0}^{m} b_j d_j \tag{2.8.14}$$

and the multiplication on the dual module is given by

$$(\sum_{n=0}^{\infty} r_n h_n)(\sum_{n=0}^{\infty} s_n h_n) = \sum_{n=0}^{\infty}\Big(\sum_{i=0}^{n} \binom{n}{k} r_i s_{n-i}\Big) h_n. \tag{2.8.15}$$

Now suppose that k is a field of characteristic zero (or more generally a $\mathbf{Q}$-algebra). It then readily follows that

$$\sum_{n=0}^{\infty} r_n h_n \mapsto \sum_{n=0}^{\infty} r_n \frac{x^n}{n!} \tag{2.8.16}$$

gives an isomorphism between the dual algebra of the divided power coalgebra and the power series algebra $k[[x]]$.

Of course if k is a field of characteristic $p > 0$ things are completely different.

Construction 2.8.17. The zero dual coalgebra of an algebra.[15] To get an coalgebra from an algebra is a completely different story (compared to obtaining an algebra from a coalgebra). The trouble is that the natural morphism of modules (2.8.2), i.e.

$$A^* \otimes A^* \xrightarrow{\psi} (A \otimes A)^*,\ \psi(f \otimes g)(c \otimes d) = f(c)g(d) \tag{2.8.18}$$

runs the wrong way for the dual of a multiplication to define a comultiplication and that it is not an isomorphism unless M is free of finite rank over k.

To avoid additional complications from now in this section 2.8 k is a field. [16]

Let A be an algebra over a field k. The zero dual k-module A^0 of A is the module of all functionals on A that have a finite codimensional ideal in their kernel. I.e., writing $A^* = \mathbf{Mod}_k(A, k)$ as usual

$$A^0 = \{f \in A^* : \exists I \subset A \text{ such that } \dim_k(A/I) < \infty \text{ and } I \subset \mathrm{Ker}(f)\} \tag{2.8.19}$$

where I is supposed to be a (two-sided) ideal of the algebra A. Thus f factors through a finite dimensional quotient algebra of A. Another way of expressing (2.8.19) is to consider the left and right translates of functionals on A. For

[15] The zero dual is sometimes also called the **finite dual**.

[16] This is very possibly not the optimal choice for duality questions. See the remarks on reflexive modules in 2.11.

each $a \in A$ and $f \in A^*$ define the left translate by a of f and the right translate by a of f by, respectively

$$L_a f(x) = f(ax), \quad R_a f(x) = f(xa) \tag{2.8.20}$$

and let Lf and Rf be the k-modules spanned by the $L_a f$ and $R_a f$ respectively. Then

Lemma 2.8.21. *Let A be an algebra over a field k, and let $f \in A^*$ be a functional on A. The following are equivalent*

(i) *Lf is finite dimensional*

(ii) *Rf is finite dimensional*

(iii) *$f \in A^0$*

Proof. Suppose that (iii) holds and let f be a finite codimensional ideal contained in $\operatorname{Ker}(f)$. Then for all $a \in I$, $x \in A$, $L_a f(x) = f(ax) = 0$ so that Lf is a quotient module of A/I and hence finite dimensional. Similarly $(iii) \Longrightarrow (ii)$.

Now assume (i). Let $J = \{a \in A : L_a(f) = 0\}$. Then J is a right ideal of A. Indeed, if $a \in J$, then for all $r, x \in A$, $L_{ar} f(x) = f(arx) = L_a f(rx) = 0$. The algebra A acts on the right on the finite dimensional right A-module A/J. This module is finite dimensional because of (i). Let I be the kernel of the algebra morphism $A \longrightarrow \operatorname{End}(A/J)$. Then $J \subset \operatorname{Ker}(f)$. Indeed let $b \in I$. Then $tb \in J$ for all $t \in A$. Hence $L_{tb} f = 0$ and so $f(tbx) = 0$ for all $t, x \in A$. Taking $t = x = 1$ gives $f(b) = 0$, proving (iii) because $\operatorname{Ker}(f)$ is of finite codimension. Similarly $(ii) \Longrightarrow (iii)$.

Note the (family) likeness of this lemma with the notion of 'representative function' on an infinite group as discussed in 2.4.13 above. Both are really manifestations of the same idea, which also involves recursiveness as in the construction of cofree coalgebras. See also lemma 2.8.26 and example 2.8.27 below.

Lemma 2.8.22. *Let A be an algebra over a field k, and let $f \in A^0$. Let $m^* : A^* \longrightarrow (A \otimes A)^*$ be the dual of the multiplication morphism of A. Then*

$$m^*(f) \in A^0 \otimes A^0 \subset (A \otimes A)^*.$$

Proof. Pick a basis $g_1, \dots, g_m$ for $Lf \subset A^*$. Now

$$m^*(f)(x \otimes y) = f(xy) = L_x f(y) = \sum_i h_i(x) g_i(y) \tag{2.8.23}$$

so that

$$m^*(f) = \sum_i h_i \otimes g_i$$

for certain $h_i \in A^*$. Because left translates of left translates of f are left translates of f, $Lg_i \subset Lf$, so that the g_i, are in A^0. Choose $a_1, \dots, a_m \in A$ such that $g_i(a_j) = \delta_{i,j}$. Then, using (2.8.23)

$$R_{a_j} f(z) = f(za_j) = \sum_i h_i(z) g_i(a_j) = h_j(z)$$

showing that the h_j are all right translates of f, and because of lemma 1.8.22 and because right translates of right translates are right translates it follows that the h_j are all in A^0 proving the lemma.

Theorem 2.8.24. (Existence of the zero dual coalgebra of an algebra). *Let A be an algebra over a field k. Then the dual of the multiplication induces a comultiplication $\mu : A^0 \longrightarrow A^0 \otimes A^0$ which together with the the morphism $\varepsilon : A^0 \longrightarrow k$, that is the dual of the unit map $e : k \longrightarrow A$, makes A^0 into a coassociative coalgebra with counit.*

Example 2.8.25. The coalgebra dual of a monoid algebra (group algebra). Let G be an associative monoid and let kG be the monoid algebra of finite k-linear combinations $\sum a_g g$ of elements of G.

Lemma 2.8.26. *A function f on G with values in k is representative if and only if the induced function $\overline{f} : kG \longrightarrow k$ is in kG^0.*

Proof. If f is representative the span of the left translates of f is finite dimensional. That immediately says that $L\overline{f}$ is finite dimensional. The other way is even more direct. [17]

It follows that the coalgebra dual of kG is the coalgebra of representative functions on G: $kG^0 = \mathrm{Func}(G,\, k)_{\mathrm{repr}}$.

Example 2.8.27. The zero dual of a free associative algebra. Let $X = \{X_i : i \in I\}$ be a finite or infinite set of indeterminates and let $k\langle X\rangle$ be the free associative algebra in these indeterminates. Let M be the free module over k with basis X. Then the zero dual of $k\langle X\rangle$ is the cofree k-coalgebra over $M : k\langle X\rangle^0 = \widehat{T}M_{\mathrm{repr}}$. As of course, morally speaking, should be the case. For k a field this is due to [**6**]. It still holds for k a Noetherian integral domain, [**16**]. See loc. cit. for proofs.

Construction 2.8.28. The double dual morphisms. Let A be a k-algebra. Define

$$\varphi_A : A \longrightarrow (A^0)^*,\ a \mapsto \varphi_A(a) \tag{2.8.29}$$

as the restriction to A^0 of $A^* \longrightarrow k,\ f \ \mapsto f(a)$.

Let C be a k-coalgebra. Define

$$\psi_C : C \longrightarrow (C^*)^0,\ c \mapsto \psi(c),\ \psi(c)(f) = f(c). \tag{2.8.30}$$

Lemma 2.8.31. *For all $c \in C$, $\psi_C(c) \in (C^*)^0$.*

Proof. Let $f, g \in C^*$, let $c \in C$ and $\mu(c) = \sum_i^\infty c_{i,1} \otimes c_{i,2}$. Then

$$\begin{aligned}(L_g\psi_C(c))(f) &= \psi_C(c)(gf) = \langle gf,\, c\rangle = \langle g \otimes f,\, \mu(c)\rangle \\ &= \sum_i g(c_{i,1})f(c_{i,2}) = \sum_i \psi_C(c_{i,1})(g)\psi_C(c_{i,2})(f)\end{aligned}$$

so that

$$L_g\psi_C(c) = \sum_i \psi_C(c_{i,1})(g)\psi_C(c_{i,2})$$

is for each g a linear combination of the finite number of functionals $\psi_C(c_{i,2})$. This proves the lemma.

[17] The attentive reader will have noticed in any case that there is a great similarity between the arguments used in example 2.4.13 and the arguments leading up to theorem 2.8.24.

In case A is a finite dimensional algebra over a field, $A^0 = A^*$ and the double-dual morphism φ_A is an isomorphism. And if C is a finite dimensional coalgebra over a field, $(C^*)^0 = (C^*)^*$ and the double dual morphism ϕ_C is an isomorphism. Thus we find:

Theorem 2.8.32. (Algebra-coalgebra duality for finite dimensional modules over a field). *Let A be a finite dimensional algebra over a field k and let C be a finite dimensional coalgebra over the field k. Then there is an isomorphism of algebras, respectively coalgebras*

$$\varphi_A : A \xrightarrow{\cong} (A^0)^*,\ \psi_C : C \xrightarrow{\cong} (C^*)^0. \tag{2.8.33}$$

In general, i.e., for not necessarily finite dimensional objects, this is far from being the case. For instance if A is an infinite dimensional field extension of k, e.g. $A = k(X)$, $A^0 = 0$.

However, there is still an adjointness results between the functors $A \mapsto A^0$ and $C \mapsto C^*$.

Define

$$\begin{aligned} \alpha &: \mathbf{Coalg}_k(C, A^0) \longrightarrow \mathbf{Alg}_k(A, C^*),\ f \mapsto f^*\varphi_A \\ \beta &: \mathbf{Alg}_k(A, C^*) \longrightarrow \mathbf{Coalg}_k(C, A^0),\ g \mapsto g^0\psi_C. \end{aligned} \tag{2.8.34}$$

Theorem 2.8.35. (Adjointness for the coalgebra-algebra duality functors). *The α and β of* (2.8.34) *are mutually inverse bijections.*

The proof amounts to no more than a careful writing out of all the definitions.

2.8.36. Coideals and sub coalgebras. From now on in this section 2.8 k is a field unless something else is explicitly specified.

Let (C, μ, ε) be a coalgebra over k. A subspace $J \subset C$ is called

- a **left coideal** if $\mu(J) \subset C \otimes J$
- a **right coideal** if $\mu(J) \subset J \otimes C$
- a (two-sided) **coideal** if $\mu(J) \subset C \otimes J + J \otimes C$ and $\varepsilon(J) = 0$
- a **sub coalgebra** if $\mu(J) \subset J \otimes J$

Of course there are also suitable notions in general for coalgebras over a k that is not necessarily a field. But then some extra care must be taken, because it is then not guaranteed, given a submodule J, that $J \otimes J \longrightarrow C \otimes C$ is injective so that one cannot talk about the comultiplication on J induced by that of C.

2.8.37. Coalgebra-algebra duality and quotients. Let C be a coalgebra over a field k and let C^* be its dual algebra. Further let $\langle f, c\rangle = f(c)$ be the duality pairing

$$C^* \otimes C \longrightarrow k.$$

Then by definition of the multiplication on C^*

$$\langle f_1 f_2, c\rangle = \langle f_1 \otimes f_2, \mu(c)\rangle = \sum \langle f_1, c_{1,i}\rangle\langle f_2, c_{2,i}\rangle$$

if

$$\mu(c) = \sum_i c_{1,i} \otimes c_{2,i}. \tag{2.8.38}$$

Given a subspace $J \subset C$ define $J^\perp \subset C^*$ to be the subspace

$$J^\perp = \{f \in C^* : \langle f, J\rangle = 0\}. \tag{2.8.39}$$

Sort of dually for a subspace $I \subset C^*$ define

$$I^\perp = \{c \in C : \langle I, c\rangle = 0\}. \tag{2.8.40}$$

The situation is not entirely symmetric because C^* is the full algebraic dual of C but C is much smaller than the algebraic dual of C^* unless C is finite dimensional. This reflects itself for instance in the observation that

- for $J \subset C$ $\;J = J^{\perp\perp}$, but for $I \subset C^*$ only $I \subset I^{\perp\perp}$ is guaranteed

and the latter inclusion may well be strict as the following example shows.

Example 2.8.41. Let

$$\begin{aligned} \mathrm{CoF}(k) &= kX_0 \oplus kX_1 \oplus kX_2 \oplus \cdots \\ \mu(X_m) &= \sum_{i+j=m} X_i \otimes X_j, \quad \varepsilon(X_m) = \delta_{0,m} \end{aligned} \tag{2.8.42}$$

be the graded cofree coalgebra over k. Then its dual is the algebra of power series in one indeterminate, $k[[Y]]$, and the duality pairing is given by

$$\langle Y^m, X_n\rangle = \delta_{m,n}. \tag{2.8.43}$$

Now for I take the subalgebra of polynomials $k[Y] \subset k[[Y]]$. Then $I^\perp = 0$ and so $I^{\perp\perp} = k[[Y]] \supsetneqq k[Y]$.

A little exploratory calculation shows that the orthogonal of a coideal has a tendency to be a sub algebra, that the orthogonal of an sub coalgebra is an ideal etc. These yield twelve possible implications as follows.

Let $J \subset C$ be a subspace of the coalgebra C and let $J^\perp$ be its orthogonal in C^*. Then there are the following six potential implications.

1. J is a sub algebra of C $\quad\overset{?}{\Longleftrightarrow}\quad$ $J^\perp$ is an ideal in C^*
2. J is a coideal in C $\quad\overset{?}{\Longleftrightarrow}\quad$ $J^\perp$ is a sub algebra of C^*
3. J is a left (right) coideal in C $\quad\overset{?}{\Longleftrightarrow}\quad$ $J^\perp$ is a left (right) ideal in C^*

and sort of dually, let $I \subset C^*$ be a subspace of the algebra C^* and let $I^\perp$ be its orthogonal in C. Then there are the following six more potential implications to consider.

4. I is an ideal in C^* $\quad\overset{?}{\Longleftrightarrow}\quad$ $I^\perp$ is a sub coalgebra of C
5. I is a sub algebra of C^* $\quad\overset{?}{\Longleftrightarrow}\quad$ $I^\perp$ is a coideal in C
6. I is a left (right) ideal in C^* $\quad\overset{?}{\Longleftrightarrow}\quad$ $I^\perp$ is a left (right) coideal in C

Theorem 2.8.44. *Nine of the twelve implications listed above hold; the three exceptions are* $4 \Leftarrow$, $5 \Leftarrow$, $6 \Leftarrow$.

In all these three cases the difficulty lies in the fact that for $I \subset C^*$ it is not necessary true that $I = I^{\perp\perp}$.

It is somewhat curious that several of these implications are not discussed at all in the standard books [**13**], [**31**], or the book [**1**]. The potential implications $4 \Leftarrow$ and $6 \Leftarrow$ are not mentioned in any way in these three books and for potential implication $5 \Leftarrow$ there is only a short remark in the Abe book that it need not always hold; nothing else, in particular no counter example. Such an example is given below.

Proofs. The proofs are very similar (for the first group of six implications; and in a slightly different way for the three of the second group that hold), so only some of them will be given. Proofs of all nine that hold can be found in [**31**] and the Abe book and all nine except $2 \Leftarrow$ and $3 \Leftarrow$ are also proved in [**13**].

Take a basis $\{x_u : u \in U\}$ of $J \subset C$ and complement this with $\{y_v : v \in V\}$ to a basis $\{x_u : u \in U\} \cup \{y_v : v \in V\}$ of all of C.

Proof of $1 \Rightarrow$. Let f be an element of $J^\perp$ so that f is zero on the x_u; let g be an arbitrary element of C^*. Now because J is a sub coalgebra $\mu(x)$ is of the form

$$\mu(x) = \sum a_{u,v} x_u \otimes x_v$$

for any $x \in J$, so that

$$\langle fg, x\rangle = \langle f \otimes g, \mu(x)\rangle = \sum a_{u,v}\langle f, x_u\rangle\langle g, x_v\rangle = 0$$

proving that fg is in $J^\perp$. Similarly one shows that $gf \in J^\perp$. And so $J^\perp$ is an ideal.

Proof of $1 \Leftarrow$. Let $x \in C$. Then $\mu(x)$, written out with respect to the chosen basis looks like

$$\begin{aligned} \mu(x) &= \sum a^{x,x}_{u,u'} x_u \otimes x_{u'} + \sum a^{x,y}_{u,v} x_u \otimes y_v \\ &\quad + \sum a^{y,x}_{v,u} y_v \otimes x_u + \sum a^{y,y}_{v,v'} y_v \otimes y_{v'}. \end{aligned} \tag{2.8.45}$$

Now among the elements of C^* are the δ_{x_u} and δ_{y_v} defined by

$$\delta_{x_u}(x_{u'}) = \delta_{u,u'}, \quad \delta_{x_u}(y_v) = 0, \quad \delta_{y_v}(y_{v'}) = \delta_{v,v'}, \quad \delta_{y_v}(x_u) = 0$$

and of these the δ_{y_v} are in $J^\perp$. By the hypothesis the products $\delta_{y_v}\delta_{y'_v}$, $\delta_{y_v}\delta_{x_u}$, $\delta_{x_u}\delta_{y_v}$ are all in $J^\perp$. So for instance

$$0 = \langle \delta_{x_u}\delta_{y_v}, x\rangle = \langle \delta_{x_u} \otimes \delta_{y_v}, \mu(x)\rangle = a^{x,y}_{u,v}$$

so that all the $a^{x,y}_{u,v}$ are zero. Similarly the $a^{y,y}_{v,v'}$ and the $a^{y,x}_{v,u}$ are zero, so that $\mu(x)$ is of the form

$$\mu(x) = \sum a^{x,x}_{u,u} x_u \otimes x_{u'}$$

proving what is desired.

Proof of $2 \Rightarrow$. Let f, g be in $J^\perp$ so that they are zero on the x_u and let $x \in J$. Because J is an ideal $\mu(x)$ is of the form

$$\mu(x) = \sum a^{x,x}_{u,u'} x_u \otimes x_{u'} + \sum a^{x,y}_{u,v} x_u \otimes y_v + \sum a^{y,x}_{v,u} y_v \otimes x_u$$

and as f and g are both zero on the x_u it follows that $\langle fg, x\rangle = \langle f \otimes g, \mu(x)\rangle = 0$. Further, as $\varepsilon(J) = 0$ it follows that ε, which is the unit of C^*, satisfies $\langle 1, J\rangle = 0$ by the definition of the pairing $\langle , \rangle$. Thus $1 \in J^\perp$.

For the second group of six implications take a basis $\{x_u : u \in U\}$ of $J^\perp \subset C$ and complement this with $\{y_v : v \in V\}$ to a basis $\{x_u : u \in U\} \cup \{y_v : v \in V\}$ for all C.

Proof of 4 $\Rightarrow$. Let $x \in I^\perp$ and write $\mu(x)$ as in (2.8.45) above and rewrite this as

$$\mu(x) = \sum_u c_u \otimes x_u + \sum_v c'_v \otimes y_v.$$

Let $f \in I$. Then also $f\delta_{x_u} \in I$ and so

$$0 = \langle f\delta_{x_u}, x\rangle = \sum_{u'} \langle f \otimes \delta_{x_u}, c_{u'} \otimes x_{u'}\rangle = \langle f, c_u\rangle.$$

This holds for all f and so $c_u \in I^\perp$ for all $u \in U$. Similarly $c'_v \in I^\perp$ for all $v \in V$ and so

$$\mu(x) \in I^\perp \otimes C.$$

In the same manner, writing $\mu(x)$ as a sum

$$\mu(x) = \sum_u x_u \otimes d_u + \sum_v y_v \otimes d'_v$$

and multiplying f with δ's on the left, one sees that

$$\mu(x) \in C \otimes I^\perp$$

Now apply lemma 2.6.6 which says that $I^\perp \otimes C \cap C \otimes I^\perp = I^\perp \otimes I^\perp$ and the proof of 4 $\Rightarrow$ is finished.

Proof that none of the implications 4 $\Leftarrow$, 5 $\Leftarrow$, 6 $\Leftarrow$ holds in general. Here is a counter example to all three simultaneously (more or less).

Take again the graded cofree coalgebra $\mathrm{CoF}(k) = kX_0 \oplus kX_1 \oplus kX_2 \oplus \cdots$ and its dual algebra the power series algebra $k[[Y]]$. In $k[[Y]]$ consider the subspace I spanned by polynomials of the form $Y^n - Y^{n^2}$ with $n > 1$. This subspace of polynomials is not multiplicatively closed. Indeed, consider the product of the first two elements and suppose that element of $k[[Y]]$ to be equal to a linear combination of elements of the form $Y^n - Y^{n^2}$,

$$(Y^2 - Y^4)(Y^3 - Y^9) = Y^5 - Y^7 - Y^{11} - Y^{13} = \sum_n a_n(Y^n - Y^{n^2}). \qquad (2.8.46)$$

Then the coefficients of the elements $Y^{2^k} - Y^{2^{k+1}}$ are equal to 1. Indeed this is obvious for $k = 0$. So suppose with induction that it holds for $k \leq r$. Consider the power $Y^{2^{r+1}}$. This one only occurs in the terms

$$a_{2^r}(Y^{2^r} - Y^{2^{r+1}}) \quad \text{and} \quad a_{2^{r+1}}(Y^{2^{r+1}} - Y^{2^{r+2}})$$

from the right hand side of (2.8.46) and so $a_{2^{r+1}} = a_{2^r} = 1$. So for (2.8.46) to hold infinitely many nonzero coefficients are needed proving that the product on the left is not in I and so I is not a subalgebra (not even in the weaker sense where one forgets about existence of a unit element) and it is not a left ideal or right ideal or ideal. The next claim is that

$$I^\perp \quad \text{is equal to } \mathrm{CoF}(k)^{(1)} = kX_0 \oplus kX_1.$$

Indeed, let $x \in \mathrm{CoF}(k)$, $x = \sum\limits_i^{<\infty} a_i X_i$. Then there is a largest n such that $a_n \neq 0$. If $n \leq 1$, there is nothing to prove. If $n > 1$ take the pairing with the element $Y^n - Y^{n^2}$ in I to see that $a_n = 0$, given this take the pairing of x with $Y^{n-1} - Y^{n-1^2}$ to see that also $a_{n-1} = 0$ and so on up to and including taking the pairing of x with $Y^2 - Y^4$ to find that $a_2 = 0$. This proves the claim.

This is therefore a counterexample for both $4 \Leftarrow$ and $6 \Leftarrow$. This is not a counterexample to $5 \Leftarrow$ because $\varepsilon(\mathrm{CoF}(k)^{(1)}) \neq 0$.

To get a counterexample to $5 \Leftarrow$ let $I_1 = I + k$. Then I_1 is not multiplicatively closed and $I_1^{\perp} = kX_1$ as an ideal.

There are obviously many more examples of this kind.

For another counterexample to $5 \Leftarrow$, take for I all power series or polynomials with nonzero constant term and the coefficient of Y equal to zero and add the element 0. This subspace is multiplicatively closed but there is no unit element in it, so it is not a sub algebra. The orthogonal under the pairing is again kX_1, so this is another type of counterexample.

Another, rather more elegant, counterexample to $4 \Leftarrow$ and $6 \Leftarrow$ is the following. Take the same coalgebra and dual algebra as before. Now take for the space I the space of all recursive power series with the first m coefficients equal to zero; that is all power series that can be written in the form

$$\frac{a_{m+1}Y^{m+1} + \cdots + a_{m+r}Y^{m+r}}{1 + b_1 Y + \cdots + b_n Y^n} = c_{m+1}Y^{m+1} + c_{m+2}Y^{m+2} + \cdots \tag{2.8.47}$$

This is multiplicatively closed but not a left ideal, right ideal or ideal. The reason is that recursive power series have at the most exponential growth in the absolute value of their coefficients. Indeed if M is a constant larger than the absolute value of all the a's and b's occurring in (2.8.47), then an easy induction shows that the coefficients of the power series represented by (2.8.47) satisfy

$$|c_r| \leq (nM)^r$$

So for instance the power series

$$1 + 1^1 Y + 2^2 Y^2 + 3^3 Y^3 + \cdots$$

is not recursive and a product of this power series with a recursive power series is also not recursive.

The orthogonal of I is $\mathrm{CoF}(k)^{(m)} = kX_0 \oplus kX_1 \oplus \cdots \oplus kX_m$ which is a sub coalgebra and a left and right ideal of $\mathrm{CoF}(k)$, so this is another counter example.

2.8.48. Algebra-coalgebra duality and subs and quotients. Now consider the situation of an algebra A and its zero dual also called 'finite dual' or 'restricted dual' A^0. As before there are twelve potential implications as follows.

Let $J \subset A$ be a subspace of the algebra A and let $I^{\perp}$ be its orthogonal in A^0. Then there are the following six potential implications.

1. I is a sub algebra of A $\quad \overset{?}{\Longleftrightarrow} \quad$ $I^{\perp}$ is an ideal in A^0
2. I is an ideal in A $\quad \overset{?}{\Longleftrightarrow} \quad$ $I^{\perp}$ is a sub algebra of A^0
3. J is a left (right) ideal in A $\quad \overset{?}{\Longleftrightarrow} \quad$ $I^{\perp}$ is a left (right) ideal in A^0

and sort of dually, let $J \subset A^0$ be a subspace of the algebra A^0 and let $J^{\perp}$ be its orthogonal in A. Then there are the following six more potential implications to consider.

4.	J is a coideal in A^0	$\overset{?}{\Longleftrightarrow}$	$J^\perp$ is a sub algebra of A
5.	J is a sub coalgebra of A^0	$\overset{?}{\Longleftrightarrow}$	$J^\perp$ is a ideal in A
6.	J is a left (right) coideal in A^0	$\overset{?}{\Longleftrightarrow}$	$J^\perp$ is a left (right) ideal in A

Theorem 2.8.49. *The six implications* $1 \Rightarrow$, $2 \Rightarrow$, $3 \Rightarrow$, $4 \Rightarrow$, $5 \Rightarrow$, $6 \Rightarrow$, *all hold.*

Proofs can be found in [**13**]. They are very similar in spirit to the proofs in the case of $C - C^*$ duality.

The situation as regards the reverse potential implications the situation is unclear and largely uninvestigated. The reverse implications $1 \Leftarrow$ and $2 \Leftarrow$ definitely do not hold in full generality. But that is based on such nasty examples like $A = k(t)$, the field of rational functions over k. In this case $A^0 = \{0\}$.

The proper setting to investigate these potential implications might be when A is a proper algebra.

2.8.50. Proper algebras. A subspace $V \subset W^*$ of a dual space is said to be **dense** if

$$\{w \in W : \langle v, w\rangle = 0, \text{ for all } v \in V\} = 0.$$

For instance the algebra of polynomials $k[Y]$ is dense in the algebra of power series $k[[Y]]$ (which is a dual space).

An algebra A is **proper** if $A^0 \subset A^*$ is dense. Algebras that are the dual of a coalgebra are proper. An algebra that is finitely generated (as an algebra over k) is proper.

2.8.51. Remarks on the case where the base ring k is not a field. Suppose that k is not a field and that the algebras and coalgebras we are dealing with are free or duals of free modules. There are some good properties in this case which could help improve the duality situation. Notably the fact that if M is a countable free module over the integers, then the canonical morphism $M \longrightarrow M^{**}$ is an isomorphism (which is a thing that never happens over fields), see 2.11 below.

On the other hand there are nasty examples like the following.

Example 2.8.52. Take $C = \mathrm{CoF}(\mathbf{Z})^{(2)} = \mathbf{Z}X_0 \oplus \mathbf{Z}X_1 \oplus \mathbf{Z}X_2$ and its dual which is $\mathbf{Z}[Y]/(Y^3)$. Take the sub Abelian group J of C generated by $2X_1, X_2$. Then $J^\perp = \mathbf{Z}$ which is a subalgebra. But J is not a coideal.

Here it seems that the notion of a pure submodule could be appropriate.

2.9. Comodules and representations

Algebras have modules, which is the same as representations of that algebra. And one does not really understand an algebra if one does not have a complete description of all its modules (or 'realizations'). Dually coalgebras have comodules or corepresentations and the same remark applies.

Definition 2.9.1. Modules in diagram form. Given an algebra A over k, a **left module** over A is a k-module M together with a morphism of k-modules

$$A \otimes M \xrightarrow{\rho} M$$

such that the following two diagrams commute:

$$\begin{array}{ccc} A \otimes A \otimes M & \xrightarrow{\mathrm{id} \otimes \rho} & A \otimes M \\ {\scriptstyle m_A \otimes \mathrm{id}} \downarrow & & \downarrow {\scriptstyle \rho} \\ A \otimes M & \xrightarrow[\rho]{} & M \end{array} \qquad \begin{array}{ccc} k \otimes M & \xrightarrow{e_A \otimes \mathrm{id}} & A \otimes M \\ {\scriptstyle =} \downarrow & & \downarrow {\scriptstyle \rho} \\ M & = & M \end{array}$$

where the left hand vertical arrow of the second diagram is the canonical identification.

The category of left A modules is denoted ${}_A\mathbf{Mod}$. The category of right A modules is $\mathbf{Mod}_A$.

As expected a comodule over a coalgebra is obtained by reversing all arrows.

Definition 2.9.2. Comodules. Let C be a coalgebra over k. A **left comodule** over C is a k-module N together with a morphism of k-modules

$$N \xrightarrow{\lambda} C \otimes N$$

such that the following two diagrams commute

$$\begin{array}{ccc} N & \xrightarrow{\lambda} & C \otimes N \\ {\scriptstyle \lambda} \downarrow & & \downarrow {\scriptstyle \mu_C \otimes \mathrm{id}} \\ C \otimes N & \xrightarrow{\mathrm{id} \otimes \lambda} & C \otimes C \otimes M \end{array} \qquad \begin{array}{ccc} N & \xrightarrow{\lambda} & C \otimes N \\ {\scriptstyle =} \downarrow & & \downarrow {\scriptstyle \varepsilon_C \otimes \mathrm{id}} \\ N & \xrightarrow{=} & k \otimes N \end{array}$$

The category of left comodules over C is denoted ${}_C\mathbf{Comod}$ and the category of right comodules $\mathbf{Comod}_C$. (It is of course obvious what a morphism of comodules is.)

In the case of algebras a morphism of algebras $A \xrightarrow{\varphi} B$ defines a left and right A module structure on B : $\rho(a \otimes b) = \varphi(a)b$, respectively $\rho'(b \otimes a) = b\varphi(a)$. Similarly a morphism of coalgebras defines comodule structures.

Example 2.9.3. Let $\psi : C \longrightarrow D$ be a morphism of coalgebras. Then C inherits a left and right D comodule structure as follows

$$\lambda(c) = \sum \psi(c_{1,j}) \otimes c_{2,j}, \text{ respectively } \lambda'(c) = \sum c_{1,j} \otimes \psi(c_{2,j}).$$

This includes of course the case that $\psi = \mathrm{id}$ making a coalgebra a left and right comodule over itself (regular corepresentation).

2.9.4. Comodule-module duality. Let C be a coalgebra over k. Given a k-morphism $\rho : N \longrightarrow N \otimes C$ consider the morphism $\alpha_\rho : C^* \otimes N \longrightarrow N$ defined by the composite

$$C^* \otimes N \xrightarrow{\mathrm{id} \otimes \rho} C^* \otimes N \otimes C \xrightarrow{\mathrm{tw} \otimes \mathrm{id}} N \otimes C^* \otimes C \xrightarrow{\mathrm{id} \otimes \langle , \rangle} N \otimes k \cong N. \tag{2.9.5}$$

Proposition 2.9.6. *A morphism $\rho : N \longrightarrow N \otimes C$ defines a right comodule structure on N if and only if $\alpha_\rho : C^* \otimes N \longrightarrow N$ as defined by (2.9.5) is a left C^*-module stucture.*

The proof is straightforward or see [**31**], section 2.1, p.34ff.

Theorem 2.9.7. (Finiteness theorem for comodules.) *Let C be a coalgebra over a field k and N a right (or left) comodule, $n \in N$. Then the submodule of N generated by n is finite dimensional.*

This is of course the comodule analogue of the 'main theorem for coalgebras' of 2.6.6 above. In turns out it can be used, together with duality, to give a completely different proof of the main theorem, see e.g. [**19**].

Proof. The morphism λ takes n into a finite sum

$$\lambda(n) = \sum_{i=1}^{s} n_i \otimes c_i \tag{2.9.8}$$

where the c_i can be taken to be independent. Let M be the subspace generated by the n_i in (2.9.8).

By the co-unit property for a comodule

$$n = (\mathrm{id} \otimes \varepsilon)\lambda(n) = \sum_{i=1}^{s} n_i \varepsilon(c_i) \in M.$$

Let $f_i : C \longrightarrow M$ be such that $f_i(c_j) = \delta_{i,j}$. Then

$$\begin{aligned}\lambda(n_i) &= (\mathrm{id} \otimes \mathrm{id} \otimes f_i)(\lambda(n_i) \otimes c_i) = (\mathrm{id} \otimes \mathrm{id} \otimes f_i)\Big(\sum_{j=1}^{s} \lambda(n_j) \otimes c_j\Big) = \\ &= (\mathrm{id} \otimes \mathrm{id} \otimes f_i)(\lambda \otimes \mathrm{id})\lambda(n) = (\mathrm{id} \otimes \mathrm{id} \otimes f_i)(\mathrm{id} \otimes \mu_C)\lambda(n) = \\ &= \sum_{j=1}^{s} n_j \otimes (\mathrm{id} \otimes f_i)\mu_C(c_j)\end{aligned}$$

so that $\lambda(n_j)$ is in $N \otimes C$, proving that M is a submodule.

2.10. Graded coalgebras

Many coalgebras and bialgebras and ... as they occur in various parts of mathematics are naturally graded. In these chapters only graded objects graded by the nonnegative integers are considered. It is obvious how to generalize things to graded modules etc. graded by any semigroup.

Definition 2.10.1. Graded module. A module M over k is **graded** (by $N \cup \{0\}$) if there is given a direct sum decomposition into submodules

$$M = \bigoplus_{i=0}^{\infty} M_i.$$

A nonzero element of C is homogenous if it is contained in just one summand; 0 is also considered homogeneous. As a rule in these pages only graded modules are considered for which the pieces M_i, are all of finite rank.

Example 2.10.2. Consider again the monoid of words over the positive integers $\mathbf{N}^*$. The weight and length of a word $\alpha = [a_1, \dots, a_m]$ are defined by $\mathrm{wt}(a) = a_1 + a_2 + \cdots + a_m$ and $\lg(\alpha) = m$. The empty word has length and weight zero. Let M be the module $k[\mathbf{N}^*]$ and define M_i, as the submodule spanned by all words of weight i. Then M is a graded module with these pieces and all these pieces are of finite rank.

If length is chosen to define the pieces a new grading is obtained which does not satisfy the "all pieces are of finite rank" condition.

Definition 2.10.3. Graded coalgebra. A coalgebra whose underlying module is graded is a connected **graded coalgebra** if

$$\mu(C_n) \subset \bigoplus_{i+j=n} C_i \otimes C_j \tag{2.10.4}$$

$$\mathrm{Ker}(\varepsilon) = \bigoplus_{i>0} C_i \tag{2.10.5}$$

and so that $C_0 = k$ (the connectedness condition).

Definition 2.10.6. Graded algebra. An algebra whose underlying module is graded is a connected **graded algebra** if

$$m(A_i \otimes A_j) \subset A_{i+j} \tag{2.10.7}$$

$$e(k) = A_0. \tag{2.10.8}$$

One could of course consider graded coalgebras and algebras whose part 0 pieces are larger than k, but such will not occur in these chapters. The property $C_0 = k$ is sometimes referred to as connected because the homology of a connected space has this property.

Definition 2.10.7. Graded dual. The **graded dual** of a graded module $M = \bigoplus_i M_i$, is the module $M^{gr*} = \bigoplus_i M_i^*$.

If C is a graded coalgebra then C^{gr*} inherits a graded algebra structure. And if A is a graded algebra with summands that are free of finite rank, then, obviously, A^{gr*} gets a coalgebra structure. Note that in this case $A^{gr*} = A^0$ (if one forgets about the grading). Indeed, $\bigoplus_{i \geq n} A_i$, is an ideal of finite corank for all n.

Definition 2.10.8. Graded morphism. A morphism of graded modules $M \xrightarrow{\varphi} N$ is homogenous if $\varphi(M_i) \subset N_i$ for all i.

Definition 2.10.9. Cofree graded coalgebra. Let M be a graded k-module with $M_0 = 0$. The **cofree graded coalgebra** over M is a graded coalgebra CM together with a graded morphism of graded modules $\pi : CM \longrightarrow M$ such that for every graded coalgebra C and graded morphism of modules $\mathrm{Ker}(\varepsilon) = \bigoplus_{j \geq 0} C_j \xrightarrow{\psi} M$ there is a unique morphism of graded coalgebras $C \xrightarrow{\widetilde{\psi}} CM$ such that $\pi\widetilde{\psi}$ on $\mathrm{Ker}(\varepsilon)$.

Theorem 2.10.10. (Existence and construction of cofree graded coalgebras). *Give the tensor coalgebra* $\mathrm{CoF}(M)$ *the grading*

$$\mathrm{CoF}(M)_n = \bigoplus_{i_1+\cdots+i_m=n} M_{i_1} \otimes \cdots \otimes M_{i_m}$$

then $\mathrm{CoF}(M)$ *together with the natural projection* $\mathrm{CoF}(M) \longrightarrow M$ *is the cofree graded coalgebra over* M.

Proof (sketch). Let ψ : $\mathrm{Ker}(\varepsilon) \longrightarrow M$ be a morphism of graded modules. Define

$$\nu_1 = (\mathrm{id} - e\varepsilon) : C \longrightarrow C.$$

This is just the projection of C onto its positively graded part; that is, it kills C_0 and is the identity on all other pieces. Define

$$\nu_n = (\nu_1 \otimes \cdots \otimes \nu_1)\mu_n : C \longrightarrow C^{\otimes n}. \tag{2.10.11}$$

For a given $c \in C$, $\nu_n(c)$ gives all decompositions of c into parts of strictly positive degree. It follows from the fact that C is a graded coalgebra that for a given c there is an n (depending on c) such that $\nu_m(c) = 0$ for all $m \geq n$. This is what makes things work.

The required morphism of coalgebras is now defined by

$$\begin{aligned}&\psi : k = C_0 \longrightarrow \mathrm{CoF}(M)_0 = k \text{ is the identity}\\ &\widetilde{\psi} : \mathrm{Ker}(\varepsilon) \longrightarrow \bigoplus_{j>0} M^{\otimes j},\ c \mapsto (c,\ \psi^{\otimes 2}\nu_2(c),\ \psi^{\otimes 3}\nu_3(c), \ldots).\end{aligned}$$

As already remarked $\mathrm{CoF}(M)$ is not the cofree coalgebra over M in the category of all coalgebras.

2.11. Reflexive modules

Let M be a module over k. There is always the natural morphism of modules of M into its algebraic double dual

$$M \longrightarrow M^{**},\ m \mapsto \varphi_m,\ \varphi_m(f) = f(m).$$

The module is called **reflexive** if this morphism is an isomorphism. This happens of course for finite rank free modules, in particular for finite dimensional vector spaces over a field. Note that purely algebraic duals are taken; there are no topological, in particular no continuity, assumptions made.

Reflexivity never happens for infinite dimensional vector spaces over a field. Thus it comes as a surprise to many mathematicians that there are infinite rank cases where reflexivity happens. Here is a short summary.

The matter of when a module over a ring A is reflexive is a delicate one involving higher set theoretic notions. In the case of $A = \mathbf{Z}$ the answer is as follows. Let M be a free Abelian group. Then M is reflexive if and only if the cardinality of (a basis of) M is non-ω-measurable. (A set is ω-measurable if and only if it has a non-principal ultrafilter $\mathcal{D}$, such that for all countable sets of elements D_i, $i \in \mathbf{N}$, $D_i \in \mathcal{D}$, $\bigcap_i D_i \in \mathcal{D}$). It is easy to see that $\mathbf{N}$ is non-ω-measurable.)

The reflexivity of free Abelian groups with countable basis was established by Specker in 1950, [**30**]. For results on higher cardinals see [**3**], and for a general survey of these matters see [**3**], chapter 3.

An A-module M is **slender** iff for every morphism $\prod_{i\in\mathbf{N}} A \xrightarrow{\lambda} M$, $\lambda(e_n^*) = 0$ for all but finitely many n. Here the e_n^* are the dual 'basis' to the standard basis of $\bigoplus_{i\in\mathbf{N}} A$. A ring A is slender if and only if it is slender as an A-module. It turns out that a PID is slender iff A is not a field or a discrete complete valuation ring. So from the point of view of reflexivity properties of modules over a ring, fields and complete discrete valuation rings are exactly the wrong thing to look at.

2.12. Measuring

Construction 2.12.1. Convolution product. Let $(C, \mu, , \varepsilon)$ be a coalgebra over k and (A, m, e) an algebra over k. Then there is a natural product on $\mathbf{Mod}_k(C, A)$ called **convolution**. It is defined as follows. Let $f, g \in \mathbf{Mod}_k(C, A)$ be two k-module morphisms, then

$$f * g = (C \xrightarrow{\mu} C \otimes C \xrightarrow{f \otimes g} A \otimes A \xrightarrow{m} A). \tag{2.12.2}$$

Note that for the case $A = k$ the convolution product is precisely the product of the dual algebra C^* of the coalgebra C.

Example 2.12.3. Convolution product on $\mathrm{Func}(G, k)$. Consider a finite group G and its bialgebra of functions $\mathrm{Func}(G, k)$ seen as the dual $\mathbf{Mod}_k(k[G], k)$. The convolution product of two elements from this module of morphisms is by definition

$$f * g(s) = m(f \otimes g)\mu(s)$$

$$= m(f \otimes g)(\sum_{tt'=s} t \otimes t') = \sum_{tt'=s} f(t)g(t') = \sum_{t \in G} f(t)g(t^{-1}s).$$

So in this case the convolution product $*$ coincides with the usual idea of a convolution product.

There are relations between the idea of representative functions (as discussed in 2.4.13) and the convolution product. Indeed, consider a coalgebra C and an algebra A and the k-module $\mathrm{Mod}_k(C, A)$. Written out explicitly the convolution product formula is

$$(f * g)(c) = \sum_i f(c_i')g(c_i''), \quad \text{where } \mu(c) \text{ is a finite sum } \mu(c) = \sum_i c_i' \otimes c_i''.$$

Now interpret an element $c \in C$ as a linear function

$$\mathrm{Mod}_k(C, A) \xrightarrow{c} A, \quad f \mapsto c(f) = f(c).$$

In these terms the formula above becomes

$$c(f * g) = \sum_i c_i'(f)c_i''(g)$$

where the sum is finite, showing that c is a representative function.

Example 2.12.4. Convolution of functions in the classical analysis sense. Consider the binomial coalgebra $\mathbf{R}[X]$ of example 2.4.22 over the real numbers. The dual $\mathbf{R}[X]^*$ is an algebra under the convolution product. Let f be a function of compact support on $\mathbf{R}$. Define

$$f^*(P) = \int f(x)P(x)dx, \quad P \in \mathbf{R}[X].$$

Then $f \in \mathbf{R}[X]^*$. Note that for $P \in \mathbf{R}[X]$ and writing X for $X \otimes 1$ and Y for $1 \otimes X$

$$\mu(P) = \sum_j P_{1,j} \otimes P_{2,j} = P(X + Y).$$

Now, if g is another function of compact support, the convolution product of f^* and g^* is

$$\begin{aligned} f^* * g^*(P) &= \sum_i \int f(x)P_{1,i}(x)dx \int g(y)P_{2,i}(y)dy \\ &= \int\int \sum_i (f(x)P_{1,i}(x)g(y)P_{2,i}(y))dxdy = \\ &= \int\int f(x)g(y)P(x+y)dxdy = \\ &= \int\int f(x)g(t-x)P(t)dxdt = \\ &= \int \left(\int f(x)g(t-x)dx\right) P(t)dt = \\ &= \int h(t)P(t)dt = \\ &= h^*(P) \end{aligned}$$

where

$$h(t) = \int f(x)g(t-x)dx$$

is the classical convolution of the functions f and g.

Given k-modules A, B, C there is a natural and well known morphism

$$\mathbf{Mod}_k(C \otimes A,\, B) \xrightarrow{\varphi} \mathbf{Mod}_k(A,\, \mathbf{Mod}_k(C,\, B)),$$

$$(\varphi(f)(a))(c) = f(c \otimes a) \tag{2.12.5}$$

which is an isomorphism if all three modules are free of finite rank (and also in somewhat more general circumstances).

If C is a coalgebra, and A and B are algebras, $\mathbf{Mod}_k(C,\, B)$ is an algebra under the convolution product, and it is natural to ask when the image of $\lambda \in \mathbf{Mod}_k(C \otimes A,\, B)$ is a morphism of algebras. When this is the case, by definition, $(C,\, \lambda)$ measures A to B. This works out as

Definition 2.12.6. Measuring. Let C be a coalgebra over k, let A and B be algebras over k and let $\lambda : C \otimes A \longrightarrow B$ be a k-morphism. The pair $(C,\, \lambda)$ **measures** the algebra A to the algebra B iff

$$\begin{aligned} \lambda(c \otimes aa') &= \sum_i \lambda(c_{1,i} \otimes a)\lambda(c_{2,i} \otimes a') \\ \lambda(c \otimes 1_A) &= \varepsilon(c)1_B \end{aligned} \tag{2.12.7}$$

where $c \in C$, $a, a' \in A$ and $\mu(c) = \sum_i c_{i,1} \otimes c_{i,2}$. The first of these conditions says that in a sense 'the action λ of C on A to B is distributive over the multiplication on A' and the second that it is 'unit preserving'. When C is a Hopf algebra these phrases take on a clearer meaning.

Example 2.12.8. Measuring and algebra morphisms. Consider the case that C is a set coalgebra as in example 2.4.24. Suppose that $\lambda : kS \otimes A \longrightarrow B$ measures A to B. Then writing simply $s(a)$ for $\lambda(s \otimes a)$

$$s(aa') = s(a)s(a'),\ s(1) = 1$$

because $\mu(s) = s \otimes s$, $\varepsilon(s) = 1$. Thus in this case a measuring morphism is nothing but a bunch of algebra morphisms.

Example 2.12.9. Measuring and derivations. Now consider a measuring morphism and let $d \in C$ be a primitive element of the coalgebra C with respect to the group like element g. This means that $\mu(d) = g \otimes d + d \otimes g$, $\mu(g) = g \otimes g$. Then

$$d(aa') = d(a)g(a') + g(a)d(a'),\quad d(1_A) = 0$$

so that under a measuring morphism g-primitive elements act as g-derivations. In case C is a bialgebra one can take the group like element 1 and ordinary derivations appear.

Example 2.12.10. Measuring and duality. Let A be an algebra and A^0 its zero dual coalgebra, see 2.8.17 above. Let $\langle , \rangle : A^0 \otimes A \longrightarrow k$ be the duality pairing. Then $(A^0, \langle , \rangle)$ measures A to k.

It is not unlikely that this example is the origin of the terminology 'measuring', the elements of A^0 being functionals on A.

The notion of measuring will play an essential role in two important constructions later on, the smash product and the cross product, see section 3.9 below.

2.13. Addition formulae and duality

This chapter started with addition formulae coming from a group representation and it was shown that there is really a coalgebra behind this, called C_ρ (see section 2.1).

This can be generalized. Instead of a group representation, i.e. a kG-module, one can take any algebra A (instead of kG) and any module M over A (instead of a representation of G).

The construction goes as expected. Let $e_1, \dots, e_t$ be a basis for M (over k). Then

$$ae_j = \sum p_{i,j}(a)e_j$$

giving a number of functions on A, i.e. elements of $A^* = \mathbf{Mod}_k(A, k)$ that span a space C_M (which is not necessarily all of A^*). The duality rule (see also 2.6)

$$\langle \mu(c), a \otimes b \rangle = \langle c, ab \rangle$$

defines a comultiplication on C_M.

This generalization is of relevance because the addition formulae of q-special functions do not come from (representations of) groups but from (representations of) quantum groups, which are not groups but are Hopf algebras. For an initial survey on these matters, see e.g. [**18**].

2.14. Coradical and coradical filtration

Let C be a coalgebra over a field. The coradical $C_0 = \mathrm{CoRad}(C)$ of C is the sum of all simple coalgebras of C. Let $A = C^*$ be the dual algebra of C. This notion is dual to that of the Jacobson radical in the sense that

$$C_0^\perp = \mathrm{JacR}(A).$$

The coradical filtration of C is a series of subcoalgebras

$$C_0 \subseteq C_1 \subset_1 \subset \cdots \subset C_n \subset \cdots$$

obtained as follows

$$C_{n+1} = \mu_C^{-1}(C_n \otimes C + C \otimes C_n).$$

This is an instance of the wedge construction in coalgebras. Quite generally if X and Y are subspaces of a coalgebra C their wedge is

$$X \vee Y = \mu_C^{-1}(X \otimes C + C \otimes Y).$$

See [**31**], chapter IX, for a lot of material on the wedge construction.

Some properties of the coradical filtration are
(i) The C_n are subcoalgebras
(ii) $C_n \subseteq C_{n+1}$
(iii) $\bigcup\limits_{n \geq 0} C_n = C$
(iv) $\mu(C_n) \subseteq \sum\limits_{i=0}^{n} C_i \otimes C_{n-i}$

2.15. Coda to chapter 2

There is quite a bit more to be said on coalgebras and comodules. Quite enough in fact to fill a substantial monograph, and it is a little surprising perhaps that such a book does not yet exist.

Among the topics that have not been discussed here, are

• Rational modules over the dual algebra. They are the modules that come from comodules over C as in 2.9.4 above. They have some nice extra finiteness properties. See e.g. [**13**], section 2.2, p.72ff.

• Quotients of coalgebras and the corresponding 'standard' isomorphism theorems.

• Simple, irreducible, pointed, cosemisimple coalgebras.

• Cotensor product of comodules.

• Umbral calculus. A part of combinatorics largely based on the binomial coalgebra and its dual algebra. See [**17**], [**22**], and also [**25**]. But some aspects of the umbral calculus that are immediately related to Hopf algebras are described in subsection 8.8.7.

• Coalgebras and recursiveness, see [**16**] and the references therein and much more.

Let A be an algebra and consider the category ${}_A\mathbf{Mod}_A$ of (A, A)-bimodules. A coring C is a comonoid object (coalgebra) in this monomial category. That is, there are structure morphisms

$$\mu : C \longrightarrow C \otimes_A C, \quad \varepsilon : C \longrightarrow A$$

that satisfy the natural coassociativity and counit properties.

On the one hand corings are but a mild generalization of coalgebras (in that the algebra A need not be commutative), on the other they are a vast generalization in terms of their relations with and applications to other parts of algebra such as various kinds of sophisticated notions of Hopf algebra modules (entwining structures etc.), matrix problems (BOCSs) in the sense of A. V. Roiter, M. Kleiner and L. A. Nazarova, ring extensions, ...

Unlike in the case of coalgebras, here is a book on corings: [**2**]; see also [**8**].

References

[1] Eichi Abe, *Hopf algebras*, Cambridge Univ. Press, 1977.
[2] Marcelo Aguiar, Nantel Bergeron and Frank Sottile, *Combinatorial Hopf algebras and generalized Dehn-Sommerville relations*, Compos. math. **142**:1 (2006), 1-30; preprint version: `arXiv:math/0310016`.
[3] S. Balcerzyk, *On groups of functions defined on Boolean algebras*, Fund. Math. **50** (1962), 347–367.
[4] L. J. Billera, Ning Jia and Victor Reiner, *A quasisymmetric function for matroids*, 2009; `arXiv:math/0606646 v3`.

[5] Dieter Blessenohl and Manfred Schocker, *Noncommutative character theory of the symmetric group*, Imperial college press, 2005.

[6] Richard E. Block, Pierre Leroux, *Generalized dual coalgebras of algebras with applications to cofree coalgebras*, J. of pure and appl. Algebra **36** (1985), 15–21.

[7] Tomasz Brzezinski, Robert Wisbauer, *Corings and comodules*, Cambridge Univ. Press, 2003.

[8] Tomasz Brzezinski, *Comodules and corings*, in: Michiel Hazewinkel, ed., *Handbook of algebra, Volume 6*, Elsevier, 2009, 237–318.

[9] G. Carnovale, *Le coalgebre sull' anello degli interi*, Master thesis Univ. di Roma I, 1993.

[10] H. Crapo and William R.Schmitt, *A free subalgebra of the algebra of matroids*, 1994; `arXiv:math/0409028 v2`.

[11] H. Crapo and William R.Schmitt, *The free product of matroids*, 1994; `arXiv:math/0409080 v1`.

[12] H. Crapo and William R. Schmitt, *A unique factorization theorem for matroids*, 1994; `arXiv:math/0409099 v1`.

[13] Sorin Dascalescu, Constantin Nastasescu, Serban Raianu, *Hopf algebras. An introduction*, Marcel Dekker, 2001.

[14] P. C. Eklof, A. H. Mekler, *Almost free modules. Set theoretic methods*, North Holland, 1990.

[15] R. Ehrenborg *On posets and Hopf algebras*, Adv. in Math. **119** (1996), 1–25.

[16] Michiel Hazewinkel, *Cofree coalgebras and multivariable recursiveness*, J. of pure and appied Algebra **183** (2003), 61–103.

[17] S. A. Joni, G.- C. Rota, *Coalgebras and bialgebras in combinatorics*, in: Robert Morris, ed., *Umbral calculas and Hopf algebras*, Amer. math. Soc., 1982, 1–47.

[18] Erik Koelink, *Addition formulas for q-special functions*, in: Mourad E. Ismail, David R. Masson, Mizan Rahman eds, *Special functions, q-series and related topic*, Amer. math. Soc., 1997, 109–130.

[19] Dominique Manchon, *Hopf algebras in renormalisation*, In: Michel Hazewinkel, ed., *Handbook of Algebra, Volume 5*, 2008, 365–427.

[20] Walter Michaelis, *Lie coalgebras*, Adv. Math. **38**:1 (1980), 1–54.

[21] W. Miller, *Lie theory and special functions*, Acad. Press, 1968.

[22] Warren Nichols, Moss Sweedler, *Hopf algebras and combinatorics*, in: Robert Morris, ed., *Umbral calculas and Hopf algebars*, Amer. math. Soc., 1982.

[23] J.-Ch. Novelli, Jean-Yves Thibon and N. M. Thiery, *Algèbres de Hopf de graphes*, Comp. rend. math. Acad. Sci. Paris **339** (2004), 607 - 610.

[24] Kathyn L. Nyman, *The peak algebra of the symmetric group*, J. algebraic Combinatorics **17** (2003), 309–322.

[25] S. Roman, *The umbral calculas*, Academic Press, 1984.

[26] S. Shnider, S. Sternberg, *Quantum groups. From coalgebras to Drinfeld algebras*, International Press, 1994.

[27] W. R. Schmitt, *Hopf algebras of combinatorial structures*, Can. J. Math **45** (1993), 412 - 428.

[28] W. R. Schmitt, *Incidence Hopf algebras*, J. pure and applied Algebra **96** (1994), 299 - 330.

[29] W. R. Schmitt, *Hopf algebra methods in graph theory*, J. pure and applied Algebra **101** (1995), 77–91.

[30] E. Specker, *Additive Gruppen von Folgen ganzer Zahlen*, Portugaliae Math. **9** (1950), 131–140.

[31] Moss E. Speedier, *Hopf algebras*, Benjamin, 1969.

[32] J. T. Stembridge, *Enriched P-partitions*, Trans. Amer. math. Soc. **349** (1997), 763-788.

[33] N. Ya. Vilenkin, *Special functions and theory of group representations*, Nauka, 1965. (in Russian); English translation, Amer. Math. Soc., 1968.

[34] N. Ya. Vilenkin, A. U. Klimyk, *Representation of Lie groups and special functions. Vol. 1: Simplest Lie groups, special functions and integral transforms*, KAP, 1991.

[35] N. Ya. Vilenkin, A. U. Klimyk, *Representation of Lie groups and special functions. Vol. 2: Class I representations, special functions and integral transforms*, KAP, 1991.

[36] N. Ya. Vilenkin, A. U. Klimyk, *Representation of Lie groups and special functions. Vol. 3: Classical and quantum groups and special functions*, KAP, 1992.

[37] N. Ya. Vilenkin, A. U. Klimyk, *Representations of Lie groups and special functions. Recent advances*, KAP, 1995.

CHAPTER 3

Bialgebras and Hopf algebras. Motivation, definitions, and examples

In the previous chapter an attempt was made to define and motivate the study of coalgebras. However, the main motivation for the study of coalgebras lies no doubt in the study of and applications of bialgebras and Hopf algebras as well as those special Hopf algebras that go by the name of quantum groups[1].

Bialgebras are k-modules that are both coalgebras and algebras in a compatible way. The precise definition is in section 3.2 below. As in the previous chapter k is a Noetherian commutative and associative integral domain with unit.

3.1. Products and representations

A starting point is the discussion of a most important tool in the representation theory of groups and Lie algebras: the existence of a product of representations.

3.1.1. Construction. Products of representations of groups. Let G be a group. A finite dimensional representation of G over k is of course a group morphism

$$G \xrightarrow{\rho} GL_n(k). \tag{3.1.2}$$

The representation is **faithful** if ρ is injective. In that case ρ exhibits G as a concrete group of invertible $n \times n$ matrices, a realization of G. Whence the importance of the notion of representation. Indeed in the earlier days of group theory finite groups were always thought of in this kind of concrete way, mostly as a group of permutations.

This gives some hint why it might be important to have systematic methods of producing new representations from given (known) ones. Very roughly speaking the more concrete realizations one has of a(n abstract) group the more one knows about it.

Equivalently, a representation ρ is an action of G on a module $V = k^n$, or again, a structure of a kG-module[2] on V:

$$\rho : kG \otimes V \longrightarrow V. \tag{3.1.3}$$

Now let ρ_1, ρ_2 be two representations of a group G. Then with $\otimes$ denoting the Kronecker or tensor product of matrices

$$g \mapsto \rho_1(g) \otimes \rho_2(g) \tag{3.1.4}$$

is another representation of G.

[1]For the time being; there are strong indications that the field of coalgebras and related structures in itself is developing in a substantial and worthwhile specialism, see e.g. [**2**].

[2]Here kG is the group algebra of G over k.

$$\begin{array}{ccc} B\otimes B & \xrightarrow{m} & B \\ \downarrow{\scriptstyle \varepsilon_{B\otimes B}} & & \downarrow{\scriptstyle \varepsilon} \\ k & = & k \end{array} \qquad (3.2.8B)$$

which also means that the counits preserve multiplication. The remaining two diagrams that need to be commutative are

$$\begin{array}{ccc} k & \xrightarrow{e} & B \\ \downarrow{\scriptstyle \cong} & & \downarrow{\scriptstyle \mu} \\ k\otimes k & \xrightarrow{e\otimes e} & B\otimes B \end{array} \quad \text{and} \quad \begin{array}{ccc} k & \xrightarrow{e} & B \\ \downarrow{\scriptstyle =} & & \downarrow{\scriptstyle \varepsilon} \\ k & \xrightarrow{=} & k \end{array} \qquad (3.2.9)$$

The last condition simply says that $\varepsilon e = \mathrm{id}$, so that $B \cong k \oplus \mathrm{Ker}(\varepsilon)$.

The most important compatibility condition, the commutativity of the diagram (3.2.8A) does not say that the comultiplication undoes the multiplication (or vice versa) but rather that they are sort of orthogonal to each other. Think of the comultiplication as giving all ways of composing an object into a left part and a right part and of multiplication as vertical composition. Then in words the condition says that all left parts of the decomposition of a product are all products of left parts of the two factors and the corresponding right parts are the products of the corresponding right parts of the decompositions of the factors involved. In pictures (stacking things vertically is composition; cutting things horizontally into left and right parts is decomposition):

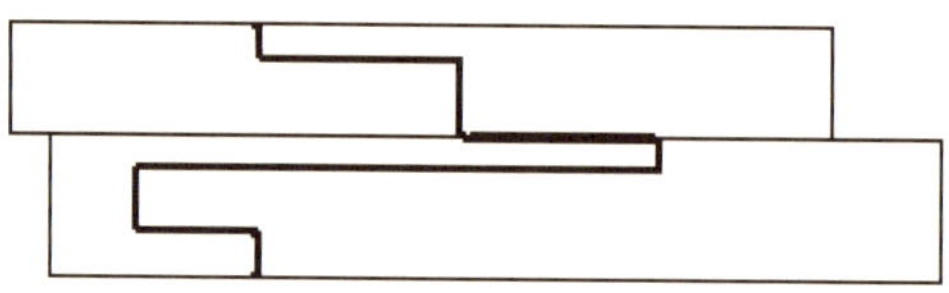

Here the top box is object 1 and the bottom one is object 2. The two boxes together are the composition of object 1 and object 2 (in that order). The first three segments of the jagged heavy line give a decomposition of object 1 into a left part LI and a right part R1 and the last five segments of that heavy jagged line give a decomposition into a left part L2 and a right part R2 of object 2. The complete heavy broken line gives the corresponding decomposition of the product of object 1 and 2 into a left part that is the product of $L1$ and $L2$ and a right part that is the product of $R1$ and $R2$.

In fact there is a most important and beautiful bialgebra (which is in fact a Hopf algebra) that looks almost exactly like this.

Consider stacks of rows of unit boxes.

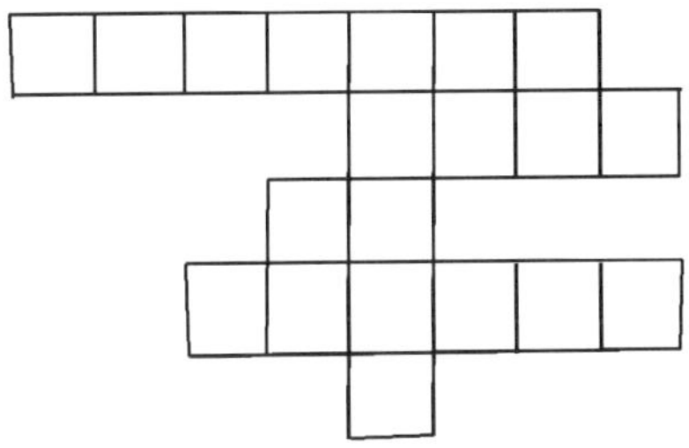

Here two such stacks are considered equivalent if they have the same number of boxes in each layer. Thus it only matters how many boxes there are in each layer. The case depicted is hence given by the word $[7, 4, 2, 6, 1]$ over the positive integers $\mathbf{N}$. The empty word is permitted and corresponds to any stack of empty layers. The possible decompositions of a stack are obtained by cutting each layer into two parts. Two of these for the example at hand are indicated below.

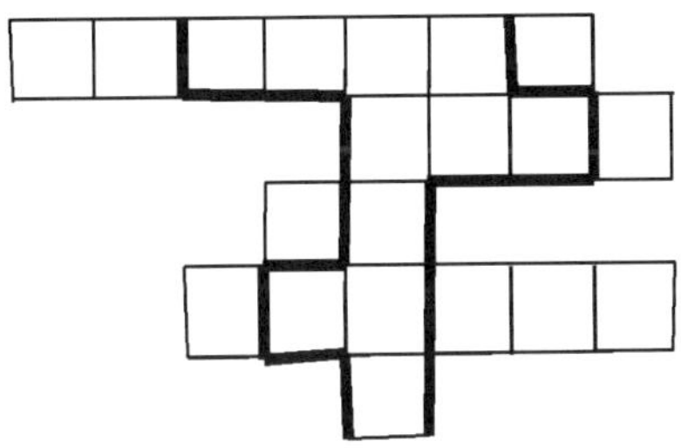

These two correspond to the decompositions $[2, 1, 1] \otimes [5, 4, 1, 5, 1]$ and $[6, 3, 2, 3, 1] \otimes [1, 1, 3]$. A convenient way of encoding this algebraically is to consider the free associative algebra $\mathbf{Z}\langle Z \rangle$ over the integers in the indeterminates $Z_1, Z_2, Z_3, \ldots$. A basis (as a free Abelian group) for this algebra is given by 1 and all (noncommutative) monomials $Z_{i_1} Z_{i_2} \ldots Z_{i_m}$. This monomial encodes the stack with layers of $i_1, i_2, \ldots, i_m$ boxes. Thus the example above corresponds to the monomial $Z_7 Z_4 Z_2 Z_6 Z_1$.

The comultiplication on $\mathbf{Z}\langle Z \rangle$ is given by the algebra homomorphism determined by

$$\mu(Z_n) = \sum_{i+j=n} Z_i \otimes Z_j, \tag{3.2.10}$$

where $Z_0 = 1$, and $i, j \in \mathbf{N} \cup \{0\} = \{0, 1, 2, 3, \ldots\}$
This, together with the counit

$$\varepsilon(Z_n) = 0, \quad n \geq 1 \tag{3.2.11}$$

defines the bialgebra (Hopf algebra) **NSymm** of noncommutative symmetric functions, which will be studied in more detail later.

Example 3.2.12. Monoid bialgebra of a finite monoid. Let G be a finite monoid. Let kG be the monoid algebra. Define

$$\mu : kG \longrightarrow kG \otimes kG \text{ and } \varepsilon : kG \longrightarrow k \text{ by}$$

$$\mu(g) = g \otimes g, \ \varepsilon(g) = 1, \quad \forall g \in G. \tag{3.2.13}$$

It is easy to check that these are algebra morphisms. Indeed it suffices to check this on the basis $\{g : g \in G\}$ and then $\mu(gh) = gh \otimes gh = (g \otimes g)(h \otimes h) = \mu(g)\mu(h)$, $\varepsilon(gh) = 1 = \varepsilon(g)\varepsilon(h)$.

Example 3.2.14. Monoid bialgebra of an arbitrary monoid. The same definitions and proofs work for the infinite rank monoid algebra of an infinite monoid.

Example 3.2.15. Universal enveloping bialgebra. Let $\mathfrak{g}$ be a Lie algebra and $U\mathfrak{g}$ its universal enveloping algebra. Let μ, ε be given by

$$\mu(x) = 1 \otimes x + x \otimes 1, \ \varepsilon(x) = 0, \ \ \forall x \in \mathfrak{g}. \tag{3.2.16}$$

The associative algebra $U\mathfrak{g}$ is generated by the $x \in \mathfrak{g}$ subject to the relations

$$xy - yx = [x, y]$$

where [ ] denotes the bracket of the Lie algebra. Now

$$\begin{aligned}\mu(xy-yx) &= (1\otimes x + x\otimes 1)(1\otimes y + y\otimes 1) - (1\otimes y + y\otimes 1)(1\otimes x + x\otimes 1)\\ &= 1\otimes xy + y\otimes x + x\otimes y + xy\otimes 1 - 1\otimes yx - x\otimes y - y\otimes x - yx\otimes 1\\ &= 1\otimes (xy-yx) + (xy-yx)\otimes 1\\ &= 1\otimes [x, y] + [x, y]\otimes 1\\ &= \mu([x, y])\end{aligned}$$

$$\varepsilon(xy - yx) = 0 = \varepsilon([x, y])$$

so that μ, ε are algebra morphisms.

Example 3.2.17. Bialgebra of representative functions on a monoid. Consider the coalgebra of representative functions on a monoid of example 2.4.13. There is a multiplication on this module, viz pointwise multiplication of functions on G with values in k. Now, if f and g are representative, so that, say,

$$f(xy) = \sum_i^{<\infty} f_{i,1}(x)f_{i,2}(y), \ \ g(xy) = \sum_j^{<\infty} g_{j,1}(x)g_{j,2}(y) \tag{3.2.18}$$

then

$$\begin{gathered}L_x(fg)(y) = (fg)(xy) = f(xy)g(xy) = \\ = \sum_i f_{i,1}(x)f_{i,2}(y)\sum_j g_{i,1}(x)g_{i,2}(y) = \sum_{i,j}(f_{i,1}g_{j,1})(x)(f_{i,2}g_{j,2})(y).\end{gathered} \tag{3.2.19}$$

Thus the pointwise product fg is then also representative. There is also the function that takes the constant value 1 which is manifestly representative. We now claim that the comultiplication and counit defined in example 2.4.13 are algebra morphism for this pointwise multiplication and unit.

Indeed, (3.2.19), disregarding the left most term, says that (see 2.4.17)

$$\mu(fg) = \sum_{i,j} f_{i,1}g_{j,1} \otimes f_{i,2}g_{j,2}$$

which is equal to $\mu(f)\mu(g)$ by (3.2.18) and, again, (2.4.17). And, obviously, if $\mathbf{1}$ denotes the function with constant value 1, $\mu(\mathbf{1}) = \mathbf{1} \otimes \mathbf{1}$.

3.2.20. Nonexample. Words with concatenation and cut. As in section 2.2 consider an alphabet $\mathcal{X}$ and the free k-module $k\langle\mathcal{X}\rangle$ with as basis all words on $\mathcal{X}$, including the empty word. Concatenation of words defines an algebra structure (with the empty word as unit element), and 'cut' defines a coalgebra structure. However, these two do not combine to define a bialgebra structure. Indeed, taking $\mathcal{X} = \mathbf{N}$, consider e.g. the words [2] and [3]. Their concatenation product is [2, 3], and applying cut to that gives

$$[\,] \otimes [2,3] + [2] \otimes [3] + [2,3] \otimes [\,].$$

On the other hand applying cut to [2] and [3] gives, respectively, $[\,] \otimes [2] + [2] \otimes [\,]$ and $[\,] \otimes [3] + [3] \otimes [\,]$ and the product of these two is

$$[\,] \otimes [2,3] + [3] \otimes [2] + [2] \otimes [3] + [2,3] \otimes [\,]$$

which is different.

3.2.21. Nonexample. Consider the coalgebra C_ρ of a representation of a monoid G as discussed in section 2.1. The elements of C_ρ are functions on G and hence can be multiplied pointwise. This, however, usually takes one out of the module C_ρ. For instance the function $\sin(x)\cos(x)$ is not, as a function of x, a linear combination (with constant coefficients) of the functions $\sin(x)$ and $\cos(x)$,

3.2.22. Counterexample and open problem. Consider the k-module $M^{n\times n}(k)$ of $n \times n$ matrices with entries from k. Matrix multiplication turns this into an algebra $M^{n\times n}_{\text{alg}}(k)$. There is also a comultiplication on $M^{n\times n}(k)$, see 2.4.1. However, for $n \geq 2$, the two do not combine to define a bialgebra structure.

In fact there is no comultiplication on $M^{n\times n}_{\text{alg}}(k)$ that turns it into a bialgebra. The reason is trivial. The kernel of the counit morphism of a bialgebra B needs to be a nontrivial ideal of B. But $M^{n\times n}_{\text{alg}}(k)$ is a simple algebra (if k is a field) and hence has no such ideal.

With some effort one can also show that there is no bialgebra structure with underlying algebra $k \oplus M^{2\times 2}_{\text{alg}}(k)$.

This raises the following completely open question. Which algebras can arise as underlying algebras of Hopf algebras? At the moment this seems completely out of reach even for the case of finite dimensional semisimple algebras over the complex numbers, i.e. algebras of the form

$$\mathbf{C} \oplus M^{n_2\times n_2}_{\text{alg}}(\mathbf{C}) \oplus \cdots \oplus M^{n_r\times n_r}_{\text{alg}}\mathbf{C}. \qquad (3.2.22a)$$

The analogous question has been raised for groups (by Bertram Huppert, for instance in a talk at Amsterdam Univ.). That is, what are the conditions on the natural numbers $1, n_1, \ldots n_r$ in order that (3.2.22a) is the group algebra of a finite group. This is also almost completely open.

Definitions 3.2.23. Cocommutative and commutative bialgebras. Let B be a bialgebra and let

$$\text{tw} : B \otimes B \longrightarrow B \otimes B,\ x \otimes y \mapsto y \otimes x \qquad (3.2.24)$$

be the usual twist (switch) morphism of k-modules. The bialgebra B is said to be **commutative** if the left one of the two diagrams below commutes, i.e., if the underlying algebra is commutative; it is said to be **cocommutative** if the right hand one of the diagrams below commutes, i.e., if the underlying coalgebra is cocommutative.

$$\begin{array}{ccc} B\otimes B & \xrightarrow{\text{tw}} & B\otimes B \\ \downarrow{\scriptstyle m} & & \downarrow{\scriptstyle m} \\ B & = & B \end{array} \qquad \text{and} \qquad \begin{array}{ccc} B & = & B \\ \downarrow{\scriptstyle \mu} & & \downarrow{\scriptstyle \mu} \\ B\otimes B & \xrightarrow{\text{tw}} & B\otimes B \end{array} \tag{3.2.25}$$

All examples given so far are either commutative or cocommutative. Examples that are (very) noncommutative and (very) noncocommutative will be described later.

3.3. Hopf algebras

A **Hopf algebra** is a bialgebra H with an extra morphism $\iota : H \longrightarrow H$, called antipode, that plays to some extent the same role as the inverse for groups. [5]

Hopf algebras first turned up in algebraic topology, in the work of Heinz Hopf, as the homology of H-spaces, that is based spaces with a 'multiplication" map that is associative up to homotopy [6] and such that $m(\mathcal{X},*)$ and $m(*,\mathcal{X})$ are homotopic to the identity. Here $*$ is the base point[7].

3.3.1. Recollection. Convolution product. Let C be a coalgebra over k and A an algebra over k. Consider the k-module of morphisms $\mathbf{Mod}_k(C,\, A)$. Recall that there is a natural product structure on $\mathbf{Mod}_k(C,\, A)$ called the **convolution product** defined as follows

$$f*g = (C \xrightarrow{\mu_C} C\otimes C \xrightarrow{f\otimes g} A\otimes A \xrightarrow{m_A} A). \tag{3.3.2}$$

The convolution product is associative and their is a unit element, viz the morphism

$$C \xrightarrow{\varepsilon_C} k \xrightarrow{\varepsilon_A} A. \tag{3.3.3}$$

Definition 3.3.4. Antipode. An **antipode** $\iota : H \longrightarrow H$ for a bialgebra H is a k-morphism that is a two sided inverse to $\mathrm{id} : H \longrightarrow H$ for the convolution product on $\mathrm{End}_k(H)$. A bialgebra complete with antipode is called a Hopf algebra.

The conditions on an antipode in diagram terms say that the two following diagrams are commutative.

$$\begin{array}{ccccc} H & \xrightarrow{\text{tw}} & H & \xrightarrow{\mathrm{id}\otimes\iota} & H \\ \downarrow{\scriptstyle \varepsilon} & & & & \downarrow{\scriptstyle m} \\ H & \xrightarrow{e} & H & = & H \end{array} \tag{3.3.5}$$

$$\begin{array}{ccccc} H & \xrightarrow{\mu} & H & \xrightarrow{\iota\otimes\mathrm{id}} & H \\ \downarrow{\scriptstyle \varepsilon} & & & & \downarrow{\scriptstyle m} \\ H & = & H & \xrightarrow{e} & H \end{array} \tag{3.3.6}$$

---

[5] In the literature the symbol S is often used for the antipode.

[6] The 'H' in H-space also stands for Hopf.

[7] In a substantial part of the algebraic topology of H-spaces "homotopy associativity" is not needed and left out

Example 3.3.7. Antipode for group Hopf algebras. Let G be a group and kG its group algebra. Then

$$\iota : kG \longrightarrow kG : \qquad s \mapsto s^{-1}$$

defines an antipode. Indeed

$$(\mathrm{id} * \iota)(s) = m(\mathrm{id} \otimes \iota)\mu(s) = m(\mathrm{id} \otimes \iota)(s \otimes s) = m(s \otimes s^{-1}) = ss^{-1} = 1 = e\varepsilon(s)$$

and similarly for the other condition.

Proposition 3.3.8. (Antipode on monoid bialgebras). *Let G be a monoid and suppose that the monoid bialgebra has an antipode. Then G is a group.*

Proof. Take an $s \in G$ and let $\iota(s) = r_1 s_1 + \cdots + r_m s_m$, $s_1, \ldots s_m \in G$, $r_1, \ldots, r_m \in k$ with distinct s_i. Then by the definition of antipode $r_1 ss_1 + \cdots + r_m ss_m = 1$ and this can only happen if all but one of the r_i are zero, the remaining r_i is 1 and the corresponding s_i satisfies $ss_i = 1$.

This shows that there is quite a strong relation between antipodes and inverse elements. It also shows that there are many bialgebras that do not admit an antipode.

Theorem 3.3.9. (Properties of the antipode). *Let H be a Hopf algebra, $x, y \in H$, $\mu(x) = \sum x_{i,1} \otimes x_{i,2}$. Then its antipode is unique, and satisfies the following properties*

(1) $\iota(xy) = \iota(y)\iota(x)$
(2) $\iota(1) = 1$
(3) $\mu(\iota(x)) = \sum x_{i,2} \otimes x_{i,1}$
(4) $\varepsilon(\iota(x)) = \varepsilon(x)$.

Proof. Uniqueness follows from the fact that the antipode is a two sided inverse of something (in casu id) in an associative algebra.

Consider the morphisms $F, G, m \in \mathrm{End}_k(H \otimes H,\, H)$ defined by

$$F : x \otimes y \mapsto \iota(y)\iota(x),\ G : x \otimes y \mapsto \iota(xy),\ m : x \otimes y \mapsto xy.$$

It will now be shown that m is a left inverse of F (under convolution on $\mathrm{End}_k(H \otimes H,\, H)$) and that m is a right inverse for G. That then implies that $F = G$ proving (1). Indeed, let $\mu(x) = \sum x_{i,1} \otimes x_{i,2}$, $\mu(y) = \sum y_{j,1} \otimes y_{j,2}$, then

$$\begin{aligned}(m * F)(x \otimes y) &= \sum m_H(m \otimes F)(x_{i,1} \otimes y_{j,1} \otimes x_{i,2} \otimes y_{j,2}) \\ &= \sum x_{i,1} y_{j,1} \iota(y_{j,2})\iota(x_{i,2}) \\ &= e\varepsilon(y) \sum x_{i,1}\iota(x_{i,2}) \\ &= (e\varepsilon(y))(e\varepsilon(x)) = e(\varepsilon(y)\varepsilon(x)) = e(\varepsilon(x)\varepsilon(y)) \\ &= e_H \varepsilon_{H\otimes H}(x \otimes y)\end{aligned}$$

proving the first claim. The other one goes similar. Property (2) is an almost immediate consequence of the defining property of an antipode and so is property (4). Property (3) is proved in a similar way as property (1) by considering the morphisms

$$x \mapsto \mu(\iota(x)),\ x \mapsto \sum \iota(x_{i,2}) \otimes \iota(x_{i,1}),\ x \mapsto \mu(x)$$

from H to $H \otimes H$ and calculating convolutions.

Theorem 3.3.10. (Order of the antipode). *Let H be a Hopf algebra, $x \in H$, $\mu(x) = \sum x_{i,1} \otimes x_{i,2}$. Then the following are equivalent*

(1) $\sum \iota(x_{i,2})x_{i,1} = e\varepsilon(x)$

(2) $\sum x_{i,2}\iota(x_{i,1}) = e\varepsilon(x)$

(3) $\iota^2 = e\varepsilon(x)$

where ι^2 is the composition of ι with itself.

Proof. The implication (1) $\Longrightarrow$ (3) is proved by calculating the convolution $\iota * \iota^2$ to be id using (1) and property (1) of theorem 2.3.9. Given (3), (1) results from applying ι to the defining property of the antipode $\sum x_{i,1}\iota(x_{i,2}) = e\varepsilon(x)$ and using again property 1 of 2.3.9. Using the other defining property one gets in the same way (3) $\Longrightarrow$ (2). Finally (2) $\Longrightarrow$ (3) results from calculating the convolution product $\iota^2 * \iota$.

Corollary 3.3.11. *If the Hopf algebra H is commutative or cocommutative its antipode has order two.*

However, it is not true that the antipode always has order 2. In fact it can have any even order including infinity. See examples 3.4.3-4 below for Hopf algebras with antipodes of any finite even order. The MPR Hopf algebra of chapter 7 below has an antipode of infinite order. The order of the antipode of a finite dimensional Hopf algebra is always finite. This last result is due to David E. Radford, *The order of the antipode of a finite dimensional Hopf algebra is finite*, Amer. J. of Math. **98** (1976), 333 - 335. See also [**12**], page 293.

For an infinite dimensional Hopf algebra the antipode need not even be injective. See Takeuchi, Mitsuhiro, *There exists a Hopf algebra whose antipode is not injective*, Sci. papers, College general Education, University of Tokyo **21** (1971), 127 - 130.

3.4. Some more examples of Hopf algebras

In the beginning, when Hopf algebras were just beginning to be studied, Kaplansky once wrote that they are such rare and beautiful objects, that each new one deserves the most careful study[8]. Nowadays, this is far from being the case: Hopf algebras abound.

In example 2.3.7 we saw already that group algebras are not only bialgebras but even Hopf algebras. Similarly the function algebras of representative functions on a group are Hopf algebras and so are the algebras of (algebraic) functions on an affine algebraic group, see [**1**], chapter 4. Formal groups are another source of Hopf algebras, see [**19**]. Further there is a natural Hopf algebra structure on the universal enveloping algebras of Lie algebras. All these are either commutative or co commutative. Before quantum groups there were very few non commutative and non co commutative Hopf algebras and in this connection it should be remarked that quantum groups (function algebra version) are only mildly non commutative while the universal enveloping algebra version is only mildly non co commutative (as has been remarked several times in the published literature.

Example 3.4.1. Hopf algebra of representative functions on a group. As in 3.2.17 let $\mathrm{Func}(G, k)_{\mathrm{repr}}$ be the bialgebra of representative functions on a group.

[8]See [**24**]

Define

$$\iota(f)(s) = f(s^{-1}),\ f \in \mathrm{Func}(G,\, k)_{\mathrm{repr}},\ s \in G. \tag{3.4.2}$$

Then $\iota(f)$ is again representative (see 2.4.13), and ι is an antipode. To prove the latter statement, by definition, one has to show that $m(\mathrm{id} \otimes \iota)\mu(f) = f(u)$ where u is the unit element of the group G. Let $\mu(f) = \sum g_i(x)h_i(y)$ which, by definition, is equivalent to $f(xy) = \sum g_i(x)h_i(y)$. Thus

$$m(\mathrm{id} \otimes \iota)\mu(f)(s) = m(\mathrm{id} \otimes \iota)\sum g_i \otimes h_i(s) = \sum g_i(s)h_i(s^{-1}) = f(ss^{-1}) = f(u)$$

as required. Similarly $m(\iota \otimes \mathrm{id})\mu(f) = f(u)$.

Example 3.4.3. The Sweedler 4-dimensional Hopf algebra. Let k be a field of characteristic $\neq 2$. As an algebra let H be given by generators c, x and relations $c^2 = 1$, $x^2 = 0$, $xc = -cx$. Then H is a four dimensional vector space with basis $\{1,\, c,\, x,\, cx\}$. Define a comultiplication, counit, and antipode by

$$\mu(c) = c \otimes c,\ \mu(x) = c \otimes x + x \otimes 1,\ \varepsilon(c) = 1,\ \varepsilon(x) = 0,\ \iota(c) = c,\ \iota(x) = -cx.$$

It is simple to check that this is compatible with the relations and that the result is a Hopf algebra. The antipode is of order 4. Indeed, $\iota^2(c) = c$, $\iota^2(x) = -x$.

Example 3.4.4. The Taft Hopf algebras. Let k be a field of characteristic zero with roots of unity; e.g. $k = \boldsymbol{C}$. Let ζ_n be a primitive n th root of unity. Consider the n^2-dimensional algebra H generated by symbols c, x subject to the relations $c^n = 1$, $x^n = 0$, $xc = \zeta_n cx$. Define

$$\mu(c) = c \otimes c,\ \mu(x) = c \otimes x + x \otimes 1,\ \varepsilon(c) = 1,\ \varepsilon(x) = 0,\ \iota(c) = c^{-1},\ \iota(x) = -c^{-1}x.$$

Then this defines a Hopf algebra with an antipode of order $2n$. Indeed $\iota^2(c) = c$, $\iota^2(x) = \zeta_n x$.

Example 3.4.5. The Lie Hopf algebra $LieHopf$. Consider the free associative algebra over the integers, $\mathbf{Z}\langle U\rangle = \mathbf{Z}\langle U_1,\, U_2,\, \ldots\rangle$, in the noncommuting indeterminates U_1, U_2, ... Let μ, ι, and ε be the algebra (and anti-algebra) morphisms determined by

$$\mu(U_n) = 1 \otimes U_n + U_n \otimes 1,$$
$$\iota(U_n) = -U_n,\ \varepsilon(U_n) = 0.$$

This defines a Hopf algebra.

Example 3.4.6. The shuffle Hopf algebra. Consider the monoid $\mathbf{N}^*$ of all words over the alphabet $\boldsymbol{N}$ of positive integers (including the empty word) and let *Shuffle* be the free $\mathbf{Z}$-module (Abelian group) with basis $\mathbf{N}^*$. The **shuffle product** of two words $\alpha = [\alpha_1, \ldots, \alpha_m]$ and $\beta = [\beta_1, \beta_2, \ldots, \beta_n]$ is defined as follows. Take a 'sofar empty' word with $n + m$ slots. Choose m of the available $n + m$ slots and place in it the natural numbers from α in their original order; place the entries from β in their original order in the remaining n slots. The product of the two words α and β is the sum (with multiplicities) of all words that can be so obtained. So, for instance

$$[a,b] \times_{\mathrm{sh}} [c,d] = [a,b,c,d] + [a,c,b,d] + [a,c,d,b] + [c,a,b,d] + [c,a,d,b] + [c,d,a,b]$$

$$[1] \times_{\mathrm{sh}} [1,1,1] = 4[1,1,1,1].$$

This defines a commutative associative multiplication on *Shuffle* for which the empty word is a unit element. Moreover with cut as a comultiplication

$$\mu([\alpha_1,\ldots,\alpha_m]) = \sum_{i=0}^{m}[\alpha_1,\ldots,\alpha_i]\otimes[\alpha_{i+1},\ldots,\alpha_m]$$

counit $\varepsilon([\,]) = 1$, $\varepsilon(\alpha) = 0$ if $\lg(\alpha) \geq 1$, where the length of a word $a = [\alpha_1,\ldots,\alpha_m]$ is $\lg(\alpha) = m$, and antipode

$$\iota([\alpha_1,\ldots,\alpha_m]) = (-1)^m[\alpha_m,\alpha_{m-1},\ldots,\alpha_1]$$

Shuffle becomes a Hopf algebra

It is not difficult to verify this directly. However, as will be discussed later, *Shuffle* is the graded dual of the Lie Hopf algebra and that takes care of things automatically.

Later in chapter 6 it will be shown that *Shuffle* is a free commutative algebra in certain explicitly specified variables. This makes it monomial, something that is certainly not apparent from the defining formula for the multiplication.

Shuffle algebras first came up in the work of Ree, [**36**]. More recently they play a significant role in the world of control theory. This comes about because shuffles play an essential role in the theory of the Chen iterated path integrals, see [**10**], [**11**], and these are just what is needed in studying control systems of the form

$$\dot{x} = u_1V(x) + \cdots + u_mV_m(x),\ x \in M,$$

where M is a smooth manifold, the V_i are vector fields on M, and the u_i, are controls. For a lot of information on how shuffles and related things enter here see [**3**], [**4**], [**5**], [**15**], [**16**], [**26**], [**38**].

Example 3.4.7. The Hopf algebra of noncommutative symmetric functions **NSymm**. This one was already briefly described in 3.2 above, see (3.2.10) and (3.2.11). As an algebra over the integers it is the same as *LieHopf*. That is, it is a free associative algebra in infinitely indeterminates, denoted $Z_1, Z_2, \ldots$ for the occasion: $\mathbf{NSymm} = \mathbf{Z}\langle Z\rangle = \mathbf{Z}\langle Z_1, Z_2, \ldots\rangle$. The coalgebra structure is different:

$$\mu(Z_n) = \sum_{i+j=n} Z_i\otimes Z_j,\ \varepsilon(Z_n) = 0,\ n\geq 1 \tag{3.4.8}$$

where $Z_0 = 1$, and the antipode is given by

$$\iota(Z_n) = \sum_{\substack{i_1+\ldots+i_r=n\\ i_1,\ldots,i_r\in\mathbf{N}}} (-1)^r Z_{i_1}Z_{i_2}\cdots Z_{i_r}. \tag{3.4.9}$$

E.g. $\iota(Z_3) = -Z_3 + Z_1Z_2 + Z_2Z_1 - Z_1^3$. Over the rationals **NSymm** and *LieHopf* are isomorphic. One particularly nice isomorphism is given by

$$1 + Z_1t + Z_2t^2 + Z_3t^3 + \cdots = \exp(U_1t + U_2t^2 + U_3t^3 + \cdots) \tag{3.4.10}$$

where *exp* is defined by its usual power series expression. This works out as

$$\varphi : \mathbf{NSymm_Q} \longrightarrow LieHopf_{\mathbf{Q}} : Z_n \mapsto \sum_{i_1+\cdots+i_r=n}\frac{U_{i_1}U_{i_2}\cdots U_{i_r}}{r!}. \tag{3.4.11}$$

This is an isomorphism of Hopf algebras. See [**20**] for two detailed proofs, and/or chapter 6 below. The inverse isomorphism is

$$U_n \mapsto \sum_{i_1+\cdots+i_r=n,\ i_j\in\mathbf{N}} (-1)^{r+1}\frac{Z_{i_1}Z_{i_2}\cdots Z_{i_r}}{r}. \tag{3.4.12}$$

There are many more isomorphisms (over $\mathbf{Q}$) [9]. Over the integers **NSymm** and *LieHopf* are definitely not isomorphic.

Example 3.4.13. The Hopf algebra of symmetric functions **Symm** [10]. Consider the algebra of polynomials in an infinite number of commuting variables over the integers $\mathbf{Z}[x_1, x_2, \ldots]$ and in it consider the subalgebra of symmetric polynomials in the x_i. The symmetric function theorem says that this is a free polynomial algebra in the elementary symmetric functions e_n or, equivalently, in the complete symmetric functions h_n

$$\mathbf{Symm} = \mathbf{Z}[e_1, e_2, \ldots] = \mathbf{Z}[h_1, h_2, \ldots] \subset \mathbf{Z}[x_1, x_2, \ldots] \tag{3.4.14}$$

$$e_n = \sum_{i_1<i_2<\cdots<i_n} x_{i_1}x_{i_2}\cdots x_{i_n}, \qquad h_n = \sum_{i_1\le i_2\le\cdots\le i_n} x_{i_1}x_{i_2}\cdots x_{i_n}. \tag{3.4.15}$$

There is a well known recursion relation between the e_n and the h_n given by

$$h_n - e_1h_{n-1} + \cdots + (-1)^i e_ih_{n-i} + \cdots + (-1)^{n-1}e_{n-1}h_1 + (-1)^n e_n = 0. \text{ [11]} \tag{3.4.16}$$

There is a Hopf algebra structure on **Symm** given by [12]

$$\mu(h_n) = 1\otimes h_n + h_1\otimes h_{n-1} + \cdots + h_{n-1}\otimes h_1 + h_n\otimes 1 \tag{3.4.17}$$

or, equivalently,

$$\mu(e_n) = 1\otimes e_n + e_1\otimes e_{n-1}, + \cdots + e_{n-1}\otimes e_1 + e_n\otimes 1.$$

Thus

$$\mathbf{NSymm} \longrightarrow \mathbf{Symm}, \quad Z_n \mapsto h_n \tag{3.4.18}$$

is a surjective morphism of Hopf algebras with as kernel the Hopf ideal generated by the commutators in the Z_i.

Symm will be discussed in some depth in chapter 4. But in fact it would take several books to describe all that is known and interesting; meanwhile see [**28**].

Example 3.4.19. The **divided power Hopf algebra**. Consider the divided power coalgebra of example 2.4.8. Recall that as a k-module it has a basis $\{d_0, d_1, d_2, \ldots\}$ and that the comultiplication and counit are given by

$$\mu(d_n) = \sum_{i+j=n} d_i\otimes d_j,\ \varepsilon(d_0) = 1,\ \varepsilon(d_j) = 0 \ \text{ for } j\ge 1.$$

Define a multiplication by

$$m(d_i\otimes d_j) = \binom{i+j}{j} d_{i+j}$$

---

[9] The automorphism group of *LieHopf* is very large, see below.

[10] **Symm** may very well be the most beautiful algebraic object in mathematics.

[11] These are known as the Wronski relations.

[12] There is a good reason to work primarily with the complete symmetric functions rather than the elementary symmetric functions; see chapter 4 below.

with unit d_0. There is also an antipode given by

$$\iota(d_n) = (-1)^n d_n.$$

These extra bits of structure make it a Hopf algebra.

Note that as a coalgebra this is a sub coalgebra of **NSymm**. But it is not a sub algebra, and also not a sub Hopf algebra.

Example 3.4.20. Affine line. Consider the algebra of polynomials in one variable $k[X]$. Define a comultiplication, counit, and antipode on the free algebra $k[X]$ by requiring

$$\mu : X \mapsto 1 \otimes X + X \otimes 1, \quad \varepsilon : X \mapsto 0, \quad \iota : X \mapsto -X \tag{3.4.21}$$

to be k-algebra morphisms. This defines a Hopf algebra, that is called the **affine line** [13].

Example 3.4.22. The Pareigis Hopf algebra. Above there have been two examples of Hopf algebras that are neither commutative nor cocommutative. But they are a bit contrived, constructed for precisely that purpose and to find examples where the antipode is not of order two. However, such Hopf algebras also occur naturally, as answers to questions from somewhere else in mathematics. One very beautiful and important example is the Hopf algebra of permutations (= *MPR* Hopf algebra) that will be discussed in some detail in chapter 7. Meanwhile here is another. It is defined as follows. Let x and y be noncommuting variables.

$$\begin{aligned}
&H = k\langle x, y, y^{-1}\rangle/(xy + yx, x^2) \\
&\mu(x) = x \otimes 1 + y^{-1} \otimes x,\ \mu(y) = y \otimes y \\
&\iota(x) = xy,\ \iota(y) = y^{-1},\ \varepsilon(x) = 0,\ \varepsilon(y) = 1.
\end{aligned}$$

Then, as is easily verified, H is a noncommutative, noncocommutative Hopf algebra with an antipode of order 4.

Of course, to be honest, as an algebra it is only very mildly noncommutative: the x and y anticommute and in some contexts that is just as good as commuting. For an extremely noncommutative and noncocommutative Hopf algebra, see chapter 7.

At first sight this example looks just as 'artificial' as the Sweedler and Taft Hopf algebras. However, it is not. It is the unique Hopf algebra H such that the category of left H-comodules is isomorphic to the category of complexes over k, k - *complex*, as monoidal categories. Moreover the isomorphism is compatible with the forgetful functors ${}_H\mathbf{Comod} \longrightarrow {}_k\mathbf{Mod}$ and k-*complex* $\longrightarrow {}_k\mathbf{Mod}$. To get some impression of how this goes, here is one of the basic constructions. Recall that a complex over k is a sequence of k-modules

$$\mathcal{M} = (\ldots \to M_i \xrightarrow{\partial_i} M_{i-1} \to \ldots) \text{ such that } \partial^2 = 0.$$

Here is the corresponding H-comodule M

$$M = \bigoplus_i M_i,\ \lambda : M \longrightarrow H \otimes M,\ \lambda(m_i) = y^i \otimes m_i + y^{i+1}x \otimes \partial_i(m_i).$$

[13] Taking the spectrum in the sense of algebraic geometry of this Hopf algebra gives one dimensional affine space over k. The comultiplication then gives the usual addition on this line.

The opposite functor is almost as easy to construct. There are a lot of technicalities involved to complete the proof, but this is the basic idea. For details, see [**33**].

Example 3.4.23. The quantum group $\mathbf{SL}_q(2;k)$. The quantum 2×2 matrix function algebra $M_q(2\times 2;k)$ is the algebra generated over k by four indeterminates a,b,c,d subject to the relations

$$\begin{aligned} ab = qba,\quad bd = qdb,\quad ac = qca,\quad cd = qdc\\ bc = cb,\quad ad - da = (q-q^{-1})bc. \end{aligned} \tag{3.4.24}$$

Here q can be seen as either a fifth inteterminate that commutes with the other ones or as a specific element from k such that $q^2\neq -1$, $q\neq 0$.

The comultiplication is the usual matrix one

$$\begin{aligned} \mu(a) = a\otimes a + b\otimes c,\quad \mu(b) = a\otimes b + b\otimes d,\\ \mu(c) = c\otimes a + d\otimes c,\quad \mu(d) = c\otimes b + d\otimes d. \end{aligned} \tag{3.4.25}$$

This Hopf algebra is therefore a deformation of the Hopf algebra that defines the 2×2-matrix valued functor on the category of commutative algebras over k. It is a deformation in that what is defined by (3.4.24) and (3.4.25) can be seen as as a family of Hopf algebras depending on a parameter q that for $q=1$ turns into the classical Hopf algebra (bialgebra) of polynomial functions on matrices.

The quantum determinant is

$$\det_q\begin{pmatrix} a & b\\ c & d\end{pmatrix} = ad - qbc$$

and the quantum group $\mathbf{SL}_q(2;k)$ is now defined by (3.4.24), (3.4.25), and

$$\det_q\begin{pmatrix} a & b\\ c & d\end{pmatrix} = ad - qbc = 1. \tag{3.4.26}$$

The antipode is

$$\iota(a) = d,\quad \iota(b) = -q^{-1}b,\quad \iota(c) = -qc,\quad \iota(d) = a \tag{3.4.26a}$$

This Hopf algebra is therefore to be seen as a deformation of the algebra of functions on the algebraic group $\mathbf{SL}_q(2;k)$ over k.

Example 3.4.27. The quantum universal enveloping algebra $U_q\mathfrak{sl}(2;k)$. This one should be seen as (the universal enveloping algebra of) the 'Lie algebra' of $\mathbf{SL}_q(2;k)$. Fix an invertible element q different from 1 and -1 and such that $(q-q^{-1})^{-1}$ exists in k. Or treat q as an indeterminate commuting with everything in sight.

The algebra $U_q\mathfrak{sl}(2;k)$ is now the algebra generated by four indeterminates K, K^{-1}, E, F subject to the relations

$$\begin{aligned} KK^{-1} = K^{-1}K = 1,\quad KEK^{-1} = q^2E,\\ KFK^{-1} = q^{-2}F,\quad [E,F] = (q-q^{-1})^{-1}(K-K^{-1}). \end{aligned} \tag{3.4.28}$$

The comultiplication and counit on $U_q\mathfrak{sl}(2;k)$ are

$$\mu(E) = 1\otimes E + E\otimes K,\quad \mu(F) = K^{-1}\otimes F + F\otimes 1 \tag{3.4.29}$$

$$\mu(K) = K\otimes K,\quad \mu(K^{-1}) = K^{-1}\otimes K^{-1}$$

$$\varepsilon(E) = \varepsilon(F) = 0,\quad \varepsilon(K) = \varepsilon(K^{-1}) = 1 \tag{3.4.30}$$

and finally the antipode is

$$\iota(E) = -EK^{-1}, \quad \iota(F) = -KF, \quad \iota(K) = K^{-1}, \quad \iota(K^{-1}) = K. \tag{3.4.31}$$

The book [**368**] is almost exclusively devoted to the quantum groups $\mathbf{SL}_q(2;k)$ and $\mathbf{GL}_q(2;k)$ and the quantum universal enveloping algebra $U_q\mathfrak{sl}(2;k)$ and their interelations with and applications to other parts of mathematics and theoretical physics such as knot invariants, monodromy, braid group representations, Yang-Baxter equations, ...

Quite generally there are quantum versions of all semisimple Lie algebras and Lie groups (and many other things such as planes, spheres, homogenous spaces, Minkowski space, ...; see e.g. [**22**] and [**6**].

3.5. Primitive elements

Intuitively, a primitive element should be something like one that admits no non-trivial decompositions. This fits well with the technical definition in the case of bialgebras and Hopf algebras.

Definition 3.5.1. Let B be a bialgebra. An element p in B is called **primitive** if

$$\mu(p) = 1 \otimes p + p \otimes 1.$$

An immediate first property is that

$$p \text{ is primitive} \Longrightarrow \varepsilon(p) = 0.$$

This follows immediately from the counit property. Denote the module of primitive elements of a bialgebra B by $\mathrm{Prim}(B)$.

Proposition 3.5.2. *$\mathrm{Prim}(B)$ is a Lie algebra under the commutator difference product $(x, y) \mapsto [x, y] = xy - yx$.*

Proof. Let $x,\ y \in \mathrm{Prim}(B)$. Then [14]

$$\begin{aligned}\mu(xy - yx) &= (1 \otimes x + x \otimes 1)(1 \otimes y + y \otimes 1) - (1 \otimes x + x \otimes 1)(1 \otimes x + x \otimes 1)\\ &= 1 \otimes xy + y \otimes x + x \otimes y + xy \otimes 1 - 1 \otimes yx - x \otimes y - y \otimes x - yx \otimes 1\\ &= 1 \otimes (xy - yx) + (xy - yx) \otimes 1.\end{aligned}$$

Example 3.5.3. Let G be a monoid, then $\mathrm{Prim}(kG) = 0$.

Example 3.5.4. Let k be a field of characteristic zero. Let $\mathfrak{g}$ be a Lie algebra (over k) and $U\mathfrak{g}$ its universal enveloping algebra. Then $\mathrm{Prim}(U\mathfrak{g}) = \mathfrak{g}$.

This is a consequence of the Poincaré-Birkhoff-Witt theorem which says that if $\{y_i : e \in I\}$ is a totally ordered basis for the k-module $\mathfrak{g}$, then the monomials $y_{i_1}y_{i_2}\cdots y_{i_m}$, with $i_1 \le i_2 \le \cdots \le i_m$ form a basis for $U\mathfrak{g}$.

Example 3.5.5. The Newton primitives of **NSymm**. Define the noncommutative polynomials $P_n(Z)$ by the formula

$$P_n(Z) = \sum_{\substack{i_1+\cdots+i_m=n\\ i_j\in\mathbf{N}}} (-1)^{m+1} i_m Z_{i_1} Z_{i_2} \cdots Z_{i_m}. \tag{3.5.6}$$

---

[14] This is the same calculation as was done in 3.2.15 to show that the universal enveloping algebra is a bialgebra. That is no accident.

Note that if Z_j is given weight j, these are homogeneous of weight n. For instance, writing $P_n = P_n(Z)$

$$\begin{aligned} P_1 &= Z_1,\ P_2 = 2Z_2 - Z_1Z_1,\ P_3 = 3Z_3 - Z_2Z_1 - 2Z_1Z_2 + Z_1Z_1Z_1 \\ P_4 &= 4Z_4 - Z_3Z_1 - 3Z_1Z_3 - 2Z_2Z_2 + Z_2Z_1Z_1 \\ &+Z_1Z_2Z_1 + 2Z_1Z_1Z_2 - Z_1Z_1Z_1Z_1. \end{aligned} \tag{3.5.7}$$

The P_n, satisfy a recursion formula

$$P_n = nZ_n - (Z_{n-1}P_1 + Z_{n-2}P_2 + \cdots + Z_1P_{n-1}). \tag{3.5.8}$$

This a noncommutative version of the Newton formulas [15] expressing the power sum symmetric functions

$$p_n = \sum_i x_i^n \tag{3.5.9}$$

in terms of the complete symmetric functions

$$h_n = \sum_{i_1 \le i_2 \le \cdots \le i_n} x_{i_1}x_{i_2}\cdots x_{i_n}. \tag{3.5.10}$$

For instance

$$p_1 = h_1,\ p_2 = 2h_2 - h_1^3,\ p_3 = 3h_3 - 3h_1h_2 + h_1^3. \tag{3.5.11}$$

Some other primitives of **NSymm** are [16]

$$Z_1Z_2 - Z_2Z_1,\ Z_1Z_3 - Z_3Z_1 - Z_1Z_1Z_2 + Z_1Z_2Z_1. \tag{3.5.12}$$

Here is a proof (with induction) of the fact that the P_n are primitives. Of course $P_1 = Z_1$, is primitive by the definition of the comultiplication. Assume with induction that $P_1, \ldots P_{n-1}$ are all primitive. Then using the recursion formula (3.5.8) one sees that

$$\begin{aligned} \mu(P_n) &= n(1 \otimes Z_n + \cdots + Z_i \otimes Z_{n-i} + \cdots + Z_n \otimes 1) \\ &- (1 \otimes Z_1 + \cdots + Z_{n-1} \otimes 1)(1 \otimes P_1 + P_1 \otimes 1) \\ &\vdots \\ &- (1 \otimes Z_1 + Z_1 \otimes 1)(1 \otimes P_{n-1} + P_{n-1} \otimes 1). \end{aligned} \tag{3.5.13}$$

The terms of bi-weight $(0, n)$ give precisely $1 \otimes P_n$ and those of bi-weight $(n, 0)$ yield $P_n \otimes 1$. Now consider the terms of bi-weight $(i, n-i)$ in (3.5.13) for some

[15] This is an instance where a noncommutative version (of a formula) is more elegant than its commutative counterpart. The formula (3.5.6) fits rather better with its recursion recipe than the commutative version of (3.5.6).

[16] By taking iterated commutators of the P_n one obtains a basis over the rational numbers of the primitives, that is a basis over $\mathbf{Q}$ of Prim($\mathbf{NSymm_Q}$). Over the integers this is far from being the case. Instead one obtains a full subgroup (of quite large index in a suitable sense). This is already apparent from taking the commutator of P_1 and P_2 which yields $2(Z_1Z_2 - Z_2Z_1)$ instead of $Z_1Z_2 - Z_2Z_1$. The matter of writing down an explicit recipe for obtaining a basis of the Abelian group Prim(**NSymm**) was only recently settled, see [**21**].

Prim($\mathbf{NSymm_Q}$) is (isomorphic to) the free Lie algebra in countably many indeterminates over the rationals. Prim(**NSymm**) is definitely not the free Lie algebra in countably many indeterminates over the integers.

$1 \leq i \leq n-1$. These add up to

$$\begin{aligned}
&nZ_i \otimes Z_{n-i} \\
&-(Z_i \otimes Z_{n-i-1}P_1 + Z_{i-1}P_1 \otimes Z_{n-i} \\
&-(Z_i \otimes Z_{n-i-2}P_2 + Z_{i-2}P_2 \otimes Z_{n-i}) \\
&\cdots\cdots\cdots\cdots\cdots\cdots\cdots\cdots\cdots\cdots \\
&-(Z_i \otimes Z_1 P_{n-i-1} + Z_1 P_{i-1} \otimes Z_{n-i}) \\
&-(Z_i \otimes P_{n-i} + P_i \otimes Z_{n-i}) \\
&= Z_i \otimes ((n-i)Z_{n-i} - (Z_{n-i-1}P_1 - Z_{n-i-2}P_2 - \cdots - Z_1 P_{n-i-1} + P_{n-i}) \\
&+(iZ_i - Z_{i-1}P_1 - Z_{i-2}P_2 - \cdots - Z_1 P_{i-1} - P_i)) \otimes Z_{n-i} \\
&= Z_i \otimes 0 + 0 \otimes Z_{n-i} = 0.
\end{aligned}$$

Consider $\mathbf{NSymm_Q} = \mathbf{NSymm} \otimes_{\mathbf{Z}} \mathbf{Q}$. I.e., consider the noncommutative symmetric functions over the rationals. Noting that

$$P_n = nZ_n, \quad \operatorname{mod}(Z_1, Z_2, \ldots, Z_{n-1}) \tag{3.5.14}$$

it follows that

$$\mathbf{NSymm_Q} = \mathbf{Q}\langle P_1, P_2, \ldots \rangle. \tag{3.5.15}$$

Now define

$$\mathbf{NSymm_Q} \longrightarrow LieHopf_{\mathbf{Q}},\ P_n \mapsto U_n. \tag{3.5.16}$$

This is an isomorphism of algebras because $\mathbf{NSymm_Q}$ is the free associative algebra over $\mathbf{Q}$ in the P_n and $LieHopf_{\mathbf{Q}}$ is the free associative algebra over $\mathbf{Q}$ in the U_n. It is also a morphism of Hopf algebras because both the P_n and the U_n are primitive. Hence (3.5.16) is a second isomorphism of Hopf algebras of $\mathbf{NSymm_Q}$ and $LieHopf_{\mathbf{Q}}$.

Example 3.5.17. The power sum primitives of **Symm**. It follows immediately from the fact that $\mathbf{NSymm} \longrightarrow \mathbf{Symm}$, $Z_n \mapsto h_n$ is a morphism of Hopf algebras, that the power sums (3.5.9) are primitives in **Symm**. They are in fact a basis for the Abelian group of primitives $\operatorname{Prim}(\mathbf{Symm_Q})$.

Example 3.5.18. The primitives of *Shuffle*. Prim(*Shuffle*) is the free module spanned by the words of length 1. [17]

Proposition 3.5.19. *Let B be a finite dimensional bialgebra over a field of characteristic zero. Then* $\operatorname{Prim}(B) = 0$.

This is proved by showing with induction that if p is a nonzero primitive then $1, p, p^2, \ldots, p^n$ are linearly independent for all n.

The proposition does not remain true for bialgebras over fields of characteristic > 0. Indeed let k have characteristic $l > 0$ and consider the bialgebra $k[x]/(x^{l^2})$ of polynomials in one variable modulo the ideal (x^{l^2}) with $\mu(x) = 1 \otimes x + x \otimes 1$. Then $\mu(x^{l^2}) = 0$ so that μ is well defined. The elements x and x^l are both nonzero primitives. [18]

Theorem 3.5.20. (The Friedrichs theorem). *Let k be a field of characteristic zero or $k = \mathbf{Z}$. Let $\mathcal{X} = \{X_i : i \in I\}$ be a collection of indeterminates (symbols, letters), and let $Ass_k(\mathcal{X})$ be the free associative algebra over k in the indeterminates*

[17] This fits very well with the idea of primitives as indecomposables.

[18] And, in fact these two elements are a basis for the Lie algebra of primitives in this case.

X_i, $i \in I$. Give $Ass_k(\mathcal{X})$ a Hopf algebra structure by setting $\mu(X_i) = 1 \otimes X_i + X_i \otimes 1$. Then

$$\mathrm{Prim}(Ass_k(\mathcal{X})) = FrLie_k(\mathcal{X}) \tag{3.5.21}$$

where $FrLie_k(\mathcal{X})$ is the free Lie algebra over k in the indeterminates from $\mathcal{X}$.

Proving this is a bit beyond the scope of the present volume. See [**37**]. This makes $LieHopf$ very useful and important in the study of free Lie algebras and related matters. The book loc. cit. is completely devoted to such matters.

Inversely $Ass_k(\mathcal{X})$ is the universal enveloping algebra of $FrLie_k(\mathcal{X})$. [19]

3.6. Group-like elements

As should be abundantly clear by now, Hopf algebras are a far reaching simultaneous generalization of both Lie algebras and groups. On the one hand that suggests that there should be room within the subject for a really good examination of exponentiation: going from Lie algebras to groups. And to some extent that is true; but only up to a point; convergence and completion matters intrude, and much remains to be done.

On the other hand this suggests that there should be traces of both Lie algebras and groups in Hopf algebras. Primitive elements capture the Lie algebra part. Group-like elements, the subject of this section, capture the group part. One could even hope that the two combine to classify. As both group algebras and universal enveloping algebras are cocommutative, one cannot hope for much more than a structure theorem of this kind in the cocommutative case. And in fact there is such a one, in any case when k is a field of characteristic zero. It basically says that a cocommutative Hopf algebra over a field of characteristic zero is a smash product, see 3.9.5 for this notion,

$$H \cong U(\mathrm{Prim}(H^1))\#k\mathrm{Group}(H)$$

where H^1 is a certain sub Hopf algebra of H, the irreducible component of the group like element $1 \in H$. This structure theorem, however, goes beyond the scope of these chapters. See [**39**], section 13.1, see also [**17**]. In case k is a field of positive characteristic things become a good deal more complicated, see [**39**], section 13.2.

Definition 3.6.1. Group-like elements. Let C be a coalgebra. A **group-like element** is an element $g \in C$ different from 0 such that

$$\mu(g) = g \otimes g. \tag{3.6.2}$$

It follows immediately that $\varepsilon(g) = 1$ if g is group-like.

The group like elements of a bialgebra form a monoid. Indeed, $1 \in B$ is always group-like, and if g, $h \in B$ are group-like then so is gh. If C is just a coalgebra, it is possible that there are no group-like elements at all. This happens e.g. for the matrix coalgebra $M^{n\times n}_{\mathrm{coalg}}(k)$ when $n \geq 2$.

If H is a Hopf algebra, so that there is an antipode, group-like elements have inverses. More precisely, by the very definition of an antipode, if g is group like then $\iota(g)$ satisfies

$$\iota(g)g = 1, \ g\iota(g) = 1. \tag{3.6.3}$$

[19] This is an instance of the Milnor-Moore theorem, see below.

Let Group(B) denote the set of group-like elements of a bialgebra B. Then Group(B) is a functorial monoid on the category of bialgebras, and on the category of Hopf algebras it is a functorial group.

Example 3.6.4. Let kG be the group bialgebra of a monoid G. Then

$$\mathrm{Group}(kG) = G. \tag{3.6.5}$$

Example 3.6.6. Consider the shuffle Hopf algebra, **NSymm**, **Symm**, $U\mathfrak{g}$. In all these cases Group$(-) = 0$. In the first three cases this results immediately from degree considerations. In the last case, this is another consequence of the PBW theorem.[20]

The matter is, however, a bit more delicate than the blunt assertion of this example might suggest. Consider the polynomials in one variable over k with the comultiplication $x \mapsto 1 \otimes x + x \otimes 1$, This is an instance of a universal enveloping algebra (for the trivial one dimensional Lie algebra; it is also the algebra of symmetric functions in one variable). Now consider the power series completion $k[[x]]$ and suppose that k is a field of characteristic zero. Then

$$\exp(x) = 1 + \frac{x}{x!} + \frac{x^2}{2!} + \frac{x^3}{3!} + \cdots \tag{3.6.7}$$

is a group like element.

Moreover this trick makes sense for every Hopf algebra over a characteristic zero field for which such exponential series make sense in some kind of completion. Thus in those cases every primitive exponentiates to a group-like.

Definition 3.6.8. Divided power sequences. Let B be a bialgebra. A **divided power sequence** (DPS) in B is a sequence of elements

$$d = (d(0) = 1,\, d(1),\, d(2), \ldots) \tag{3.6.9}$$

such that

$$\mu(d(n)) = \sum_{i+j=n} d(i) \otimes d(j). \tag{3.6.10}$$

Note that $d(1)$ is a primitive.

If we are in a situation where the infinite sum

$$1 + d(1) + d(2) + \cdots \tag{3.6.11}$$

makes sense in some completion (which it does for instance in the graded case with $d(1)$ of positive weight), (3.6.11) is group like.

In the case of a bialgebra over a field of characteristic zero every primitive extends to a DPS: the terms of the exponential series do the job.

In the case of a bialgebra over the integers this need not be the case. For instance in **NSymm** every primitive extends to a DPS, but in *LieHopf* this is definitely not the case. The latter remark is an easy calculation: not even the second term of a DPS extending the primitive U_1 exists. The fact that in **NSymm** every primitive extends to a DPS is a deep theorem and equivalent to the theorem that **NSymm** over the integers is the cofree graded coalgebra over its module of

[20] So this is basically another degree type argument, the difference being that here we are dealing with a filtration rather than a grading.

primitives; see [18] for details. [21] In **Symm** it is also true that every primitive extends to a DPS (over the integers). [22]

Proposition 3.6.12. (Linear independence of group-likes). *Let C be a coalgebra over k. The group-like elements of C are linearly independent over k.*

Proof. Working over the quotient field of k if necessary one may as well assume that k is a field. Let $r_1g_1 + r_2g_2 + \cdots + r_ng_n = 0$ be a minimal length relation between distinct group-like elements. Applying the comultiplication it follows that $r_1g_1 \otimes g_1 + \cdots + r_ng_n \otimes g_n = 0$. On the other hand, $(r_1g_1 + \cdots + r_ng_n) \otimes g_1 = 0$. Subtract these two relations to find $g_2 \otimes r_2(g_2 - g_1) + \cdots + g_n \otimes r_n(g_n - g_1) = 0$. By the minimal length assumption the $g_2, \ldots, g_n$ are linearly independent. So $r_2 = \ldots = r_n = 0$.

Proposition 3.6.13. (Dual of a group-like). *Let C be a finite dimensional coalgebra over a field. Then the group-like elements of C are in bijective correspondence with the algebra morphisms $C^* \longrightarrow k$:*

$$\mathrm{Group}(C) \longrightarrow \mathrm{Alg}_k(C^*, k). \tag{3.6.14}$$

Indeed, practically by definition a group-like is the same as the image of 1 under a coalgebra morphism $k \longrightarrow C$.

There are more group-like aspects of Hopf algebras than just the things involving group-like elements. For instance there is the Nichols-Zoeller theorem.

Theorem 3.6.15. (Freeness over subalgebras; Nichols-Zoeller theorem.) *Let H be a finite dimensional Hopf algebra over a field and let H' be a sub Hopf algebra. Then H is free as a module over H'. In particular* $\dim_k(H')$ *divides* $\dim_k(H)$.

This is a generalization of the Lagrange theorem which says that the order of a subgroup of a finite group divides the order of that group. For a proof of the Nichols-Zoeller theorem see e.g. [**12**], section 7.2 and the original paper Nichols, Warren D. and M.Bettina Zoeller, *Freeness of infinite dimensional Hopf algebras over grouplike subalgebras*, Comm. Algebra **17** (1989), 413 - 422.

Example 3.6.16. Oberst-Schneider example. Theorem 3.6.15 does not necessarily hold in the infinite dimensional case. A simple example (over a field) is due to Oberst and Schneider, [**32**].

---

[21] Indeed, if there were a $d(2) \in \mathit{LieHopf}$ such that $\mu(d(2)) = 1 \otimes d(2) + U_1 \otimes U_1 + d(2) \otimes 1$, then, by degree considerations, $d(2)$ would have to be of the form $d(2) = aU_1^2 + bU_2$ with $a, b \in Z$. A quick calculation than gives that one must have $2a = 1$. On the other hand $(1, Z_1, Z_2, \ldots)$ is a DPS in **NSymm**. This does not yet suffice to prove that **NSymm** and *LieHopf* are not isomorphic over the integers. But it does suffice to show that there is no homogeneous isomorphism.

[22] A basis of the primitives of **Symm** consists of the power sum symmetric functions. These lift to the primitives P_n in **NSymm**. These extend to DPS's and hence the images of these DPS's under the Hopf algebra quotient morphism **NSymm** $\longrightarrow$ **Symm** are DPS's for the power sums. Writing down explicit DPS's extending the power sums in **Symm** is tricky and unsolved except in this roundabout not very explicit way.

3.7. Bialgebra and Hopf algebra duality

For convenience, in this section k is a field. Let B be a bialgebra over k. As in the case of coalgebra-algebra duality in the previous chapter consider the linear dual and the zero dual

$$B^0 \subset B^* = \mathbf{Mod}_k(B,k). \tag{3.7.1}$$

As in the case of the dual of an algebra the best one can hope for is a coalgebra structure on B^0 (not one on B^*) which together with the algebra structure on B^* induces a bialgebra structure on B^0. This does indeed happen.

Theorem 3.7.2. (Dual bialgebra) *Let $B = (B, m, e, \mu, \varepsilon)$ be a bialgebra over a field k and let B^0 be the zero dual of B. Then the coalgebra structure on B^0 together with the restriction to B^0 of the multiplication $\mu^* : B^* \otimes B^* \longrightarrow B^*$ define a bialgebra structure on B^0.*

Proof. Let $f, g \in B^0$. That is the same as saying that the spaces of left translates Lf and Rf are finite dimensional and the same for g, see 2.8.21. Let $x, y \in B$

$$\mu(x) = \sum_i x_{i,1} \otimes x_{i,2}, \quad \mu(y) = \sum_j y_{j,1} \otimes y_{j,2}. \tag{3.7.3}$$

Now consider the left translates of $\mu^*(f \otimes g)$

$$\begin{aligned} L_x(\mu^*(f \otimes g))(y) &= \mu^*(f \otimes g)(xy) \\ &= \sum_{i,j} f(x_{i,1}y_{j,1})g(x_{i,2}y_{j,2}). \end{aligned} \tag{3.7.4}$$

On the other hand

$$\begin{aligned} (L_{x_{i,1}}f)(L_{x_{i,2}}g)(y) &= \sum_{i,j} L_{x_{i,1}}f(y_{j,1})L_{x_{i,2}}g(y_{j,2}) \\ &= \sum_{i,j} f(x_{i,1}y_{j,1})g(x_{i,2}y_{j,2}) \end{aligned}$$

and thus

$$L_x(fg) = L_x(\mu^*(f \otimes g)) = \sum_i (L_{x_{i,1}}f)(L_{x_{i,2}}g)$$

showing that $L(fg)$ is finite dimensional. Note that the fact that μ is an algebra morphism is essentially used in (3.7.4). The verification of all the bialgebra conditions on $(B^0, \mu^0, \varepsilon^0, m^0, e^0)$ is routine.

Corollary 3.7.5. (Dual Hopf algebra). *Let H be a Hopf algebra. Then H^0 is a Hopf algebra with antipode ι^0 (which is the restriction of ι^* to H^0).*

Proof. To see this it must be shown that $\iota^*(H^0) \subset H^0$. Consider left translates

$$\begin{aligned} L_x(\iota^* f)(y) &= f(\iota(xy)) = f(\iota(y)\iota(x)) \\ &= R_{\iota(x)}f(\iota(y)) = (\iota^*(R_{\iota(x)}f))(y). \end{aligned}$$

Now $\iota^*(Rf)$ is finite dimensional because Rf is finite dimensional and thus the space of left translates of $\iota^* f$ is finite dimensional (see 2.8.21), proving the claim. Again, it is routine to verify that ι^0 satisfies the required properties.

Example 3.7.6. Let G be a group and kG its group algebra. Then $(kG)^0 = \mathrm{Func}(G,k)_{\mathrm{repr}}$.

Example 3.7.7. Let k be a field of characteristic zero and $\mathbf{SL}_n$ the special linear group over k. The coordinate ring of $\mathbf{SL}_n$ is the algebra $\mathcal{O}(\mathbf{SL}_n) = k[X_{ij} : i,j = 1,\ldots,n]/(\det(X_{i,j}) - 1)$. This is a Hopf algebra under the comultiplication, counit, and antipode

$$X_{ij} \mapsto \sum_s X_{i,s} \otimes X_{s,j},\ \varepsilon(X_{i,j}) = \delta_{i,j},\ \iota(X_{i,j}) = (-1)^{i+j} M_{j,i}(X)$$

where $M_{j,i}(X)$ is the (j,i)-th minor of the matrix $(X_{i,j})$. (NB This object is not to be confused with the matrix coalgebra of 2.4.1; here the $X_{i,j}$ are functions on $\mathbf{SL}_n$, not elements of k.) The zero dual of the universal enveloping algebra $U(\mathfrak{sl}_n)$ is $\mathcal{O}(\mathbf{SL}_n)$. This comes about through interpreting the elements of the Lie algebra as (left invariant) differential operators on the ring of coordinate functions and then taking values at the identity matrix. See [**1**] for details.

Counterexample 3.7.8. For G a nontrivial group it can happen that $(kG)^0 = k$. This happens when $G = PSL_2(K)$ for K a field of cardinality larger than that of k. See [**7**], section 2.7: see also [**1**], exercise 2.5.

Thus it is not necessarily the case that $H \cong H^{00}$.

Proposition 3.7.9. *Let H be a Hopf algebra or bialgebra over k (not necessarily a field) whose underlying module is free of finite rank over k. Then $H^{00} = H$.*

Proof. In this case $'* = 0'$ and the double dual morphism $H \longrightarrow H^{**}$ is an isomorphism.

3.8. Graded bialgebras and Hopf algebras

Many of the most important and most beautiful Hopf algebras are graded. Though more general gradings also occur naturally, here only gradings by the nonnegative integers shall be considered.

Definition 3.8.1. Graded modules. A **graded module** over k is a module M that comes with a direct sum decomposition into submodules

$$M = \bigoplus_{i=0}^{\infty} M_i. \tag{3.8.2}$$

The elements of the direct summand M_i are called homogeneous of degree i. A morphism φ between graded k-modules M and N is called homogenous if $\varphi(M_i) \subset N_i$ for all i. A graded module M is connected if $M^0 = k$. [23]

The tensor product (over k) of two graded modules M and N is made a graded module by

$$(M \otimes N)_n = \bigoplus_{i+j=n} M_i \otimes N_j. \tag{3.8.3}$$

Definition 3.8.4. Graded algebras and coalgebras. An algebra A whose underlying module is graded is a **graded algebra** if the multiplication morphism $m : A \otimes A \longrightarrow A$ is homogeneous (and $e(k) \subset A_0$). A coalgebra C whose underlying module is graded is a **graded coalgebra** if the comultiplication morphism $\mu : C \longrightarrow C \otimes C$ is homogenous (and $\varepsilon(C_i) = 0$ for all $i > 0$).

[23] This terminology comes from algebraic topology. The cohomology and homology of connected spaces are connected graded modules in this sense.

Definition 3.8.5. Graded Hopf algebras. A Hopf algebra H is a **graded Hopf algebra** if it is a both a graded algebra and a graded coalgebra. [24]

It is possibly a good idea, however, to sort have in the back of the mind when reading this text, that the graded Hopf algebras that will be discussed here as a rule are connected and also have a finite type + freeness property which says that homogeneous component is a free finite rank k-module. This is referred to as a graded Hopf algebra of free finite type.

Examples 3.8.6. Consider the Hopf algebra **Symm** of 3.4.13 above. Give h_n weight n and a monomial $h_{n_1} \cdots h_{n_r}$ weight $n_1 + \cdots + n_r$. Let $\mathbf{Symm}_n$, denote all linear combinations of monomials of weight n. Then with this grading **Symm** becomes a graded Hopf algebra in the sense of definition 3.8.5.

Similarly, consider **NSymm**, see 3.4.7, and give Z_n weight n. Then **NSymm** becomes a graded Hopf algebra in the sense of definition 3.8.5.

Now consider the Lie Hopf algebra of example 3.4.5. Give each U_n weight 1 (as is natural in some contexts). Then this is a graded Hopf algebra in the sense of definition 3.8.5.

The examples **Symm** and **NSymm** are both of free finite type and connected. The example *LieHopf* with $\mathrm{wt}(U_n) = 1$ is not of finite type. The homogeneous components $LieHopf_n$ are not finitely generated if this weight assignment is used. However, if one assigns weight n to U_n then *LieHopf* becomes a connected graded Hopf algebra of free finite type.

In the text below the three Hopf algebras **Symm**, **NSymm**, *LieHopf* will always be given the free finite type graded Hopf algebra structure that results from giving h_n, Z_n, U_n weight n.

Remark 3.8.7. Nonexistence of nontrivial connected graded finite dimensional Hopf algebras over a field of characteristic zero [25]. Let H be a nontrivial connected graded Hopf algebra over a field k of characteristic zero. Nontrivial meaning that $H \neq H_0 = k$. Let $j > 0$ be minimal such that $H_j \neq 0$. Then all nonzero elements of H_j are nonzero primitives. Indeed, by the homogeneity of the comultiplication and connectedness, if $x \in H_j$, $\mu(x)$ must be of the form

$$\mu(x) = 1 \otimes x_1 + x_2 \otimes 1 \text{ for some } x_1, x_2 \in H.$$

The counit property then gives that $x_1 = x = x_2$.

Now let x be a nonzero primitive. If H is finite dimensional then there must be an n such that $x^n = 0$. Let n be the minimal natural number with this property. Note that then $1, x, \ldots, x^{n-1}$ are linearly independent because they live in different H_j. But then

$$0 = \mu(x^n) = (1 \otimes x + x \otimes 1)^n = 1 \otimes x^n + \cdots + \binom{n}{i} x^{n-i} \otimes x^i + \ldots + x^n \otimes 1.$$

The two outer terms of this are zero. But, because k is a field of characteristic zero and the $1, x, \ldots, x^{n-1}$ are linearly independent all the middle terms are nonzero

[24] There is a more general notion of graded Hopf algebra, or, depending on how one looks at things, a different notion of graded Hopf algebra. That one is the original one which came out of algebraic topology. See section 3.13 for some details.

[25] There is an apparent contradiction here with the remark in section 3.3 that Hopf algebras first arose as the cohomology (homology) of H-spaces. For the cohomology of a connected H-space is definitely graded and connected. For an explanation see the historical remarks of section 3.13.

and because they live in different $H_j \otimes H_{j'}$ their sum is also nonzero, giving a contradiction. [26]

Proposition 3.8.8. (Automatic existence of antipodes for connected graded bialgebras). *Let B be a connected graded bialgebra. Then there is an antipode making it a Hopf algebra. Moreover this antipode is homogeneous.*

Proof. On B_0 the antipode is necessarily the identity. Now suppose that the antipode has been constructed on all B_i, $i < n$. By the homogeneity of the comultiplication and the properties of the counit $\mu(x)$, $x \in B_n$ is of the form

$$\mu(x) = 1 \otimes x + \cdots + \sum y_i \otimes y_{n-i} + \cdots + x \otimes 1$$

where the middle terms are sums of elements of the form

$$y_i \otimes y_{n-i},\ y_i \in B_i,\ y_{n-i} \in B_{n-i},\ 0 < i < n.$$

Now define

$$\iota(x) = -x - \sum_{i=1}^{n} y_i \iota(y_{n-i}).$$

Then ι is a right inverse to the identity under convolution. Similarly there is a left inverse. These two are then necessarily equal and are a (unique) antipode. The defining formula shows that the antipode is homogeneous.

Definition 3.8.9. Quasi bialgebra. Let B be a module equipped with a multiplication, comultiplication, unit and counit morphisms. Then it is called a **quasi bialgebra** if the unit and multiplication preserve counit and comultiplication or, equivalently, the counit and comultiplication preserve unit and multiplication.

Thus the difference with a bialgebra is just that it is not assumed that the multiplication and comultiplication are associative and coassociative.

The main reason for introducing this notion is that under suitable conditions quasi bialgebras are automatically Hopf algebras, i.e. the (co)associativity (and (co)commutativity) come automatically as a consequence of other properties., see 3.8.13 below.

Definition 3.8.10. Selfadjoint (quasi) bialgebras. Let B be a connected graded quasi bialgebra and suppose it comes with a nondegenerate bilinear form $<,>$ that reduces to multiplication on $k = B_0$ and that makes B_i and B_j orthogonal to each other if $i \neq j$. The bialgebra B is said to be **selfadjoint** if the multiplication and comultiplication are adjoint to one another, which means that for all x, y, z

$$\langle m(x \otimes y), z \rangle = \langle x \otimes y, \mu(z) \rangle, \qquad (3.8.11)$$

where on the right, the induced tensor product bilinear form is used

$$\langle x \otimes y,\ u \otimes v \rangle = \langle x,\ u \rangle \langle y,\ v \rangle \qquad (3.8.12)$$

and, also counit and unit are adjoint to each other, where, of course, on k the bilinear form is multiplication (as on B_0).

Self adjointness is a very strong property, as recorded in the following theorem.

---

[26]The situation over a field of characteristic $p > 0$ is different. For instance $k[X]/(X^p)$, $\mu(X) = 1 \otimes X + X \otimes 1$, $\deg(X) = 1$ is a graded connected Hopf algebra of finite dimension p.

Theorem 3.8.13. *Let B be a connected selfadjoint graded quasi bialgebra with respect to a positive definite symmetric bilinear form. Then the multiplication on B is commutative and associative and the comultiplication is cocommutative and coassociative, so that B is a commutative and cocommutative Hopf algebra.*

Proof. Let $P(B)$ be the module of primitive elements of B, and let $I = \bigoplus_{i>0} B = \mathrm{Ker}(\varepsilon)$ be the so called augmentation ideal of B. The first step is to show that

$$P(B) \cap I^2 = 0. \tag{3.8.14}$$

This is easy. Indeed let $x = yz$, $x \in P(B)$, $y, z \in I$. Then by selfadjointness

$$\langle x, x\rangle = \langle x, yz\rangle = \langle \mu(x), y \otimes z\rangle = \langle 1 \otimes x + x \otimes 1, y \otimes z\rangle = 0.$$

And as the bilinear form is positive definite this means $x = 0$.

Proof of commutativity. The proof goes by induction. So assume that if $\deg(x) + \deg(y) < n$, then $xy - yx = 0$. The induction starts because $B_0 = k$ is commutative. It also can be assumed that $\deg(x), \deg(y) > 0$. As in the proof of proposition 3.8.8

$$\mu(z) = 1 \otimes z + z \otimes 1 + v(z)$$

with

$$v(z) \in I \otimes I$$

where, as before, I is the augmentation ideal of B. Now calculate

$$\begin{aligned}
&\mu(xy - yx) = (1 \otimes x + v(x) + x \otimes 1)(1 \otimes y + v(y) + y \otimes 1)- \\
&-(1 \otimes y + v(y) + y \otimes 1)(1 \otimes x + v(x) + x \otimes 1) = \\
&= 1 \otimes xy + (1 \otimes x)v(y) + y \otimes x + v(x)(1 \otimes y)+ \\
&+v(x) \otimes v(y) + v(x)(y \otimes 1) + x \otimes y + (x \otimes 1) \otimes v(y)+ \\
&+xy \otimes 1 - 1 \otimes yx - (1 \otimes y)v(x) - x \otimes y - v(y)(1 \otimes x)- \\
&-v(y)v(x) - v(y)(x \otimes 1) - y \otimes x - (y \otimes 1)v(x) - yx \otimes 1 = \\
&= 1 \otimes (xy - yx) + (xy - yx) \otimes 1.
\end{aligned}$$

Thus $xy - yx$ is a primitive. On the other hand if $x, y \in I$, $xy - yx \in I^2$. And thus by (3.8.14) $xy - yx = 0$, proving commutativity.

Similarly for associativity consider the associator $a(x, y, z) = (xy)z - x(yz)$. This is obviously zero if one of the x, y, z has degree zero. With induction on $\deg(x) + \deg(y) + \deg(z)$ one now calculates the associator to be a primitive and so zero.

Given commutativity and associativity the self adjointness and nondegeneracy of the bilinear form now immediately give cocommutativity and coassociativity.

3.8.15. Construction. Graded dual. Let H be a graded Hopf algebra of finite type, meaning that each homogenous component is finitely generated. Define

$$H^{gr*} = \bigoplus_i H_i^*$$

where '$gr*$' should be read as 'graded dual'. The multiplication and comultiplication morphisms of H induce dual morphisms

$$m^* : H_{i+j}^* \longrightarrow H_i^* \otimes H_j^*, \qquad \mu^* : H_i^* \otimes H_j^* \longrightarrow H_{i+j}^*$$

and unit and counit provide morphisms

$$\varepsilon^* : H_0^* \longrightarrow k, \qquad \varepsilon : k \longrightarrow H_0^*$$

which all together make H^{gr*} a graded Hopf algebra (with antipode ι^*), called the graded dual of H.

Please note that as a rule the graded dual of a graded finite type Hopf algebra is different from the zero dual of that Hopf algebra when the grading is disregarded.

Example 3.8.16. Consider the Lie Hopf algebra *LieHopf* with the (for this text) standard grading defined by giving U_n weight n. The shuffle Hopf algebra of 3.4.6 can be given a grading by assigning to each word $a = [a_1, \dots, a_m]$ the weight $\mathrm{wt}(a) = a_1 + \cdots + a_m$. Then *Shuffle* is the graded dual of *LieHopf*. This is a relatively easy and anyway a straightforward exercise.

To conclude this section, here are two famous theorems on connected graded Hopf algebras over fields of characteristic zero.

Let $I(H)$ be the augmentation ideal of a connected graded Hopf algebra, i.e. the kernel of the counit morphism ε. The module $Q(H) = I(H)/I(H)^2$ is called the module of indecomposables.

Theorem 3.8.17. (Leray theorem [27]). *Let H be a commutative graded connected Hopf algebra over a field k of characteristic zero. Then any vector space section $s : Q(H) \longrightarrow I(H)$ induces an isomorphism from the free commutative algebra on $Q(H)$ over k to H.*

Theorem 3.8.18. (Milnor-Moore theorem.) *Let H be a cocommutative connected graded Hopf algebra of finite type over a field of characteristic zero. Let $P(H)$ be the Lie algebra of primitives of H. Then the inclusion $P(H) \subset H$ induces an isomorphism of the universal enveloping algebra of $P(H)$ with H.*

For proofs of both theorems, see [24].

3.8.19. Construction-definition. Primitives and derivations. Let H be a Hopf algebra and H^* its dual in one of the senses discussed above, i.e. zero dual or graded dual where appropriate.

Let

$$\langle\, , \rangle : H \times H^* \longrightarrow k,\ (x, f) \mapsto f(x) \tag{3.8.20}$$

be the duality pairing. The adjoint of multiplication by an element x of H gives an operator on H^* defined by

$$(y,\, x^* f) = \langle xy,\, f\rangle. \tag{3.8.21}$$

There is a formula for how the adjoints of multiplications act on a product:

$$x^*(fg) = \sum_i (x^*_{1,i} f)(x^*_{2,i} g) \quad \text{if} \quad \mu(x) = \sum_j x_{1,j} \otimes x_{2,j}. \tag{3.8.22}$$

Proof. By definition

$$\begin{aligned}\langle y, x^*(fg)\rangle &= \langle xy, fg\rangle = \langle \mu(x)\mu(y),\, f \otimes g\rangle \\ &= \langle \sum_{i,j} x_{1,i}y_{1,i} \otimes x_{2,i}y_{2,i},\, f \otimes g\rangle = \sum_{i,j} \langle y_{1,j},\, x^*_{1,i} f\rangle \langle y_{2,j},\, x^*_{2,i} g\rangle.\end{aligned}$$

On the other hand

$$\langle y, \sum_i (x^*_{1,i} f)(x^*_{2,i} g)\rangle = \sum_i \langle \mu(y),\, x^*_{1,i} f \otimes x^*_{2,i} g\rangle =$$

[27]This theorem has now been vastly generalized to the setting of operands, see [28].

$$= \sum_{i,j} \langle y_{1,j},\, x_{1,i}^* f\rangle \otimes \langle y_{2,j},\, x_{2,i}^* g\rangle$$

proving the two sides of (3.8.22) to be equal.

Corollary 3.8.23. (Primitives and derivations.) *Let $p \in H$ be a primitive, then p^* acts as a derivation on H^*.*

3.8.24. Discussion. 'Duality' of generators and primitives. Finally (for this section) examine the matter of the relation between primitives in H and generators for the dual of H. Or in other words, what does the notion of a set of generators for an algebra dualize to for the dual coalgebra.

For this subsection H is a connected graded Hopf algebra of free finite type and H^{gr*} is its graded dual. Leaving out the direct summands $H_0 = k$ and $H_0^{gr*} = k$, this gives a duality pairing

$$\langle\,,\,\rangle : H_{>0} \otimes J \longrightarrow k \tag{3.8.25}$$

where J is the augmentation ideal $J = (H^{gr*})_{>0}$.

Proposition 3.8.26. *$p \in H$ is primitive if and only if $\langle p,\, fg\rangle = 0$ for all $f, g \in J$.*

Proof. Let p be primitive. Then

$$\langle p,\, fg\rangle = \langle \mu(p),\, f \otimes g\rangle = \langle 1 \otimes p + p \otimes 1,\, f \otimes g\rangle = 0$$

because $f, g \in J$. On the other hand if p is not primitive, then

$$\mu(p) = 1 \otimes p + p \otimes 1 + \sum_{i,j} c_{i,j} x_i \otimes x_j.$$

For some basis $\{x_i\}_{i \in I}$ of $H_{>0}$ with at least one $c_{i,j} \neq 0$, say $c_{i_0,\, j_0}$. Let $\{f_i\}_{i \in I}$ be the dual basis, then

$$\langle p,\, f_{i_0} f_{j_0}\rangle = \langle \mu(p),\, f_{i_0} \otimes f_{j_0}\rangle = c_{i_0, j_0} \neq 0.$$

Thus the duality pairing (3.8.25) induces a duality pairing between the module of primitives $P(H) \subset H$ and the quotient module J/J^2. And of course a set of generators for H^{gr*} is the same as a set of module generators (spanning set) for J/J^2. It is in this sense that primitives and generators are dual.

When H^{gr*} is a free commutative algebra over k there results a duality between sets of free generators (bases of J/J^2) and bases (as a module) of the module of primitives of H. There will be examples of precisely this situation later in chapter 5.

3.8.27. Conilpotency and the Cartier theorem.

The starting point here is the observation that for graded connected coalgebras the cotensor coalgebra $\mathrm{CoF}(M)$ of a module M is cofree in the category of connected graded coalgebras in the following sense.

Consider $\mathrm{CoF}(M)$ together with its canonical projection $\imath$ onto the component M of degree 1. Let C be a graded coalgebra and consider a morphism of modules $\varphi : C \longrightarrow M$ such that $\varphi(C_0) = 0$. Then there is a unique morphism of coalgebras $\widetilde{\varphi} : C \longrightarrow \mathrm{CoF}(M)$ such that $\pi\widetilde{\varphi} = \varphi$. The formula is (2.7.13), which in the present setting contains only finitely many nonzero terms because for every c all summands of $\mu_n(c)$ for n large enough, contain factors from C_0 because C is graded. Thus in this setting the recursive completion is not needed.

This theorem/observation generalizes to the case of conilpotent coalgebras, a notion that will be described now.

Let $(H, m, \mu, \iota, e, \varepsilon)$ be a Hopf algebra over a field k of characteristic zero. Let J be the augmentation ideal and consider the reduced comultiplication

$$v(x) = \mu(x) - x \otimes 1 - 1 \otimes x$$

and its iterates $v_n : J \longrightarrow J^{\otimes n}$:

$$\begin{aligned} v_0 &= 0 \\ v_1 &= \mathrm{id} \\ v_2 &= v \\ &\cdots\cdots\cdots\cdots \\ v_{n+1} &= (v \otimes \mathrm{id} \otimes \cdots \otimes \mathrm{id})v_n \end{aligned} .$$

Let J_n be the kernel of v_{n+1} (so that in particular $J_0 = 0$). Then the filtration

$$J_0 \subset J_1 \subset J_2 \subset \cdots \subset J_n \subset \cdots$$

satisfies

$$J_i J_j \subset J_{i+j}, \quad \mu(J_n) \subset \sum_{i+j=n} J_i \otimes J_j.$$

The comultiplication μ is said to be **conilpotent** if J is the union of the J_n. This notion is the predual of the notion of nilpotency in the theory of algebras.

The **Cartier theorem** (1957) now says:
Let H be a Hopf algebra over a field of characteristic zero with a cocommutative and conilpotent comultiplication. Then $\mathfrak{g} = J_1$ is a Lie algebra and the inclusion of $\mathfrak{g}$ into H extends to an isomorphism of Hopf algebras of $U(\mathfrak{g})$ and H.

There is an obvious family likeness with the Milnor-Moore theorem mentioned above. For more information see [**9**] and [**35**].

3.9. Crossed products

In group theory and Lie algebra theory and ring theory there are a variety of twisted, skew, crossed, ... products that are important for various reasons, for instance as test cases and as sources of counterexamples. This section is devoted to the appropriate generalizations for Hopf algebras.

Definition 3.9.1. Hopf module algebra. Let H be a bialgebra over k and A an algebra over k and let there be given an action

$$\rho : H \otimes A \longrightarrow A \tag{3.9.2}$$

that is, a morphism of k-modules. The morphism (3.9.2) is said to give A the structure of an H **module algebra** if

$$(i) \;\; \rho \text{ makes } A \text{ an } H - \text{module} \tag{3.9.3}$$

$$(ii) \;\; \rho \text{ measures } A \text{ to } A, \text{ see } 2.12.6. \tag{3.9.4}$$

Note that condition (i) only involves the algebra structure of H and that condition (ii) only involves the coalgebra structure.

What is defined above is a left H module algebra. There is of course a corresponding notion of a right H module algebra

3.9.5. Construction. Smash product. Let A be a left H module algebra. Consider the tensor product $A \otimes H$ (as a k-module) and define a product on it by the formula

$$(a\#g)(b\#h) = \sum_i ag_{1,i}(b)\#g_{i,2}h \tag{3.9.6}$$

where $\mu(g) = \sum g_{1,i} \otimes g_{2,i}$ and $a\#g$ is written for $a \otimes g$ to indicate that we are not dealing with the multiplication of the usual tensor product of algebras. This product is called the **smash product** and denoted $A\#H$.

Example 3.9.7. Smash product and skew group rings. Let G be a group and let A be a kG module algebra. The condition that the action $kG \otimes A \longrightarrow A$ be a measuring says that each $g \in G$ is an automorphism of A, see section 2.12. The condition that A in addition be an H module adds to this that the map

$$G \longrightarrow \mathrm{Aut}_{\mathrm{Alg}} M(A)$$

be a morphism of groups. The multiplication on $A\#kG = A\#G$ works out as

$$(ag)(bh) = (ag(b))(gh)$$

and this is known as the **skew group ring** of G over A.

Example 3.9.8. Semidirect product of groups. Let G and H be groups and let there be given an action of G on H, i.e. a morphism of groups $G \longrightarrow \mathrm{Aut}_{\mathrm{Group}}(H)$. As a set the semidirect product of G and H is the Cartesian product $G \times H$ and the multiplication is given by

$$(g, h)(g', h') = (gg', g'(h)h') \tag{3.9.9}$$

see [15], pp 88-90. This semidirect product is denoted $G \times_s H$.

The action $G \longrightarrow \mathrm{Aut}_{\mathrm{Group}}(H)$ defines the structure of an kG module algebra on kH and it is an easy matter to verify that in this case

$$kH\#kG = k(G \times_s H). \tag{3.9.10}$$

Example 3.9.11. Smash product and Lie algebras. Let $\mathfrak{g}$ be a Lie algebra and $U\mathfrak{g}$ its universal enveloping algebra, which is of course a Hopf algebra. Let A be a $U\mathfrak{g}$ module algebra. This implies that the unit element of $U\mathfrak{g}$ acts as the identity. Therefore the elements of $\mathfrak{g} \subset U\mathfrak{g}$ act as derivations, see 2.12.9. Further $\mathfrak{g}$ generates $U\mathfrak{g}$ as an algebra and thus the structure of a $U\mathfrak{g}$ module algebra on an algebra A is exactly the same as a morphism of Lie algebras

$$\mathfrak{g} \longrightarrow \mathrm{Der}(A) \tag{3.9.12}$$

where $\mathrm{Der}(A)$ is the Lie algebra of derivations on A.

The smash product $A\#U\mathfrak{g}$ defined by such a morphism of Lie algebras is called the differential polynomial ring (of $\mathfrak{g}$ over A).

In the particular case that $\mathfrak{g}$ is the one dimensional Lie algebra kx there is just one derivation involved, say δ and one finds the so called Ore extension $A[x, \delta]$ of 'polynomials' over A with the twisted multiplication

$$xa = ax + \delta(a). \tag{3.9.13}$$

In the still more particular case that $A = k[y]$ and the derivation is $\delta = d/dy$ one finds the first Weyl algebra of differential operators in one variable with polynomial coefficients

$$k\langle x,\, \frac{d}{dx}\rangle = k\langle x,\, y\rangle/(xy - yx - 1). \tag{3.9.14}$$

In group theory at least, there are more sophisticated constructions than the semidirect product. Basically, if N is a normal subgroup of a group G with quotient $H = G/N$, then there is an action of H on N and there is the notion of a factor system to describe G in terms of H and N, the action and the factor system, see [**18**], p 218ff. The question arises whether this can be generalized to the context of Hopf algebras. That can be done and the relevant construction is called the cross product.

In the setting of groups the 'action' of $H = G/N$ on N arises by taking a set of representatives of G/N in G and then conjugation by these representatives gives a bunch of automorphisms of N because N is normal. But unless N is Abelian these automorphisms need not combine to define a group morphism $G/N \longrightarrow \mathrm{Aut}_{\mathrm{group}}(N)$. Thus something more general will be needed than a $k(G/N)$ module algebra structure on kN. Note that there is still a measuring present given by the bunch of automorphisms just mentioned.

3.9.15. Construction. Crossed product. The ingredients needed are

(i) A Hopf algebra H;

(ii) An action $\lambda : H \otimes A \longrightarrow A$ on an algebra A that is a measuring;

(iii) A k-bilinear morphism $\sigma : H \otimes H \longrightarrow A$ that is invertible under convolution with convolutional inverse σ^{-1}.

These ingredients are supposed to satisfy the following conditions (in addition to the measuring condition)

• (normalization)

$$\sigma(1,\, h) = \sigma(h,\, 1) = \varepsilon(h)1_A \tag{3.9.16}$$

• (cocycle) Let $\mu(h) = \sum h_{1,i} \otimes h_{2,i}$, $\mu_3(h) = \sum h_{1,i} \otimes h_{2,i} \otimes h_{3,i}$ and similarly for $\mu(l)$, $\mu(n)$. Note that $h_{1,i}$ and $h_{2,i}$ have different meanings depending on whether they come from $\mu_2(h)$ or $\mu_3(h)$. The context makes clear which meaning applies. Then

$$\begin{aligned} &\sum h_{1,j_1}(\sigma(l_{1,j_2},\, n_{1,i_3}))\sigma(h_{2,i_1}, l_{2,i_2}n_{2,i_3}) \\ &= \sum \sigma(h_{1,i_1}, l_{1,i_2})\sigma(h_{2,i_1}l_{2,i_2}, n) \end{aligned} \tag{3.9.17}$$

• (twisted module)[28]

$$h(l(a)) = \sum \sigma(h_{1,i_1}, l_{1,i_2})((h_{2,i_1}, l_{2,i_2})(a))\sigma^{-1}(h_{3,i_1}, l_{3,i_2}) \tag{3.9.18}$$

where, as before $h(a)$ stand for $\lambda(h \otimes a)$.

Given such a measuring with cocycle satisfying these conditions consider the k-vector space $A \otimes H$ and, writing $a\#h$ for $a \otimes h$, define a product by

$$(a\#h)(b\#l) = \sum a(h_{1,i}(b))\sigma(h_{2,i}, l_{l,j})\#h_{3,i}l_{2,j} \tag{3.9.19}$$

where $\mu_3(h) = \sum h_{1,i} \otimes h_{2,i} \otimes h_{3,i}$, $\mu(l) = \mu_2(l) = \sum l_{1,i} \otimes l_{2,j}$.

Theorem 3.9.20 [**7**], [**13**]. (Crossed product existence theorem). *Let H, A, λ, σ be as above. Then the multiplication* (3.9.19) *turns the module $A \otimes H$ into an associative k algebra with unit element $1_A\#1_H$. This algebra is denoted $A\#_\sigma H$. Moreover the morphism $a \mapsto a\#1_H$ is an injective morphism of algebras of A into $A\#_\sigma H$.*

[28]If the cocycle is trivial, that is $\sigma(h, l) = \varepsilon(h)\varepsilon(l)1_A$, the twisted module condition becomes the usual module condition, so that in this case A is an H module algebra.

Further $A\#_\sigma H$ is isomorphic to $A \otimes H$ as a left A module. If σ is invertible (under convolution) then $A\#_\sigma H$ is isomorphic to $H \otimes A$ as a right A module and $A\#_\sigma H$ is free both as a left and as a right A module.

For a a proof see loc. cit. or [**12**], p. 237ff.

The next step is to characterize those (Hopf) algebras that have a crossed product structure.

Here are the formulation and results of [**7**]. A somewhat more general theory is described in [**31**], chapter 7 [29]. Some preparation is needed.

Example 3.9.21. Adjoint action. Let H be a Hopf algebra. Then it is a Hopf module algebra over itself under the so called **adjoint action** defined by

$$(\mathrm{ad}(h))(l) = \sum h_{1,i}l(\iota(h_{2,i})). \tag{3.9.22}$$

In case $H = kG$ this reduces to $\mathrm{ad}(x)(y) = xyx^{-1}$ and in case $H = U\mathfrak{g}$ to $\mathrm{ad}(x)(h) = xh - hx$ for $x \in \mathfrak{g}$, the familiar adjoint actions of group theory and Lie algebra theory.

Let $H \xrightarrow{\pi} \overline{H}$ be a surjective morphism of Hopf algebras. The left Hopf kernel of π is the set

$$\mathrm{LHKer}(\pi) = \{h \in H : (\mathrm{id} \otimes \pi)\mu(h) = h \otimes 1_H\}.$$

This is an $ad(H)$ stable subalgebra of H. The main theorem of [**7**] is now:

Theorem 3.9.23. (BCM theorem on crossed products). *Let k be a field and let $H \xrightarrow{\pi} \overline{H}$ be a surjective morphism of Hopf algebras with left Hopf kernel A. Suppose moreover that there is a section s of π as a morphism of coalgebras, i.e. a morphism of coalgebras $\overline{H} \xrightarrow{s} H$ such that $\pi s = \mathrm{id}_{\overline{H}}$. Then H is a crossed product $A\#_\sigma \overline{H}$ for the following action and cocycle*:

$$\overline{h}(a) = (\mathrm{ad}(s(\overline{h}))(a))$$
$$\sigma(\overline{h}, \overline{l}) = \sum s(\overline{h}_{1,i})s(\overline{l}_{1,j})(\iota(s(\overline{h}_{2,i}\overline{l}_{2,j}))).$$

For a proof see [**7**], p.694ff.

3.10. Integrals for Hopf algebras

Integrals, that is some systematic technique for taking averages, especially (left or right) invariant integrals, play a vitally important role in the investigation of groups and their representations.

Take for instance a finite group. Then there is an invariant integral on G which simply assigns to a function f on G the complex number 'average of f' or 'integral of f'

$$\int_G f = \frac{1}{\#G} \sum_{g \in G} f(g) \tag{3.10.1}$$

---

[29] This bit of crossed product theory is due independently and simultaneously to both [**7**], [**13**].

For the case of Hopf module algebras there is the further construction of the so called double cross product which can be constructed if H and K are both Hopf algebras (or bialgebras) and H is a left K module algebra and K is a right H module algebra. The construction is important in the theory of quantum groups and their applications in physics; see [23], p. 298ff.

where, as usual $\#G$ is the number of elements of G. The idea is used to construct a scalar product on Func(G, $\mathbf{C}$):

$$\langle f_1, f_2 \rangle = \frac{1}{\#G} \sum_{g \in G} f_1(g)\overline{f_2(g)} \tag{3.10.2}$$

and then there follows without much trouble for instance the theorem on the orthogonality of the matrix coefficients of the irreducible representations of G which says that if $r^i_{j,k}(g)$ is the (j, k)-entry of the i-th irreducible representation of G (over the complex numbers) then the $r^i_{j,k}$ are orthogonal to each other and $\langle r^i_{j,k}, r^i_{j,k} \rangle = n_i^{-1}$ where n_i is the dimension of the i-th representation ρ_i.

Moreover, the whole thing goes through, virtually unchanged for compact groups.

Thus, seeing Hopf algebras as a generalization of groups, one wonders whether there exists something like integrals for Hopf algebras.

Warned by the situation for noncompact groups one asks not for invariant integrals but for left (or right) invariant integrals.

Definition 3.10.3. Integrals. Let B be a bialgebra. Then B^* has an algebra structure dual to the coalgebra structure of B. An element $T \in B^*$ is called a **left integral** of the bialgebra B if for all $f \in B^*$

$$fT = f(1)T. \tag{3.10.4}$$

Note that this involves almost nothing more than just the coalgebra structure of B; the qualifier 'almost' refering to the fact that one needs to know $1 \in B$, i.e. 'what is the special element $1 \in B$'.

Let

$$\mu(x) = \sum_i x_{i,1} \otimes x_{i,2}$$

then $T \in B^*$ is a left integral if and only if for all $x \in B$.

$$\sum_i T(x_{i,2})x_{i,1} = T(x)1. \tag{3.10.5}$$

This is immediate from the definition of the multiplication on B^*. In some more detail, (3.10.5) is an equality in B and it holds if and only if it holds when evaluated at every $f \in B^*$. So (3.10.5) holds if and only if for every f

$$\sum_i T(x_{i,2})f(x_{i,1}) = \langle \sum_i T(x_{i,2})x_{i,1}, f) \rangle = \langle T(x)1, f \rangle = T(x)f(1).$$

But indeed by the definition of the multiplication the left most side of this is equal to Tf evaluated at x and thus (3.10.4) is equivalent to (3.10.5).

It is a **right integral** if left-right symmetrically

$$\sum_i T(x_{i,1})x_{i,2} = T(x)1. \tag{3.10.6}$$

These correspond to left integrals on B^{cop} which is the same bialgebra B with the comultiplication reversed (but the multiplication left intact).

3.10.7. Justification and interpretation. Hopf algebra integrals for the Hopf algebras of compact groups. Let I be the (left) invariant Haar integral on the compact group G which assigns to a function on G its integral with respect to Haar measure, which is normalized so that the integral of the function which is constant 1 is 1.

The left invariance means

$$I(xf) = I(f)$$

where xf is the left translate function $L_x f(y) = (xf)(y) = f(xy)$. Now restrict f to the space $\mathrm{Func}(G, \mathbf{C})_{\mathrm{repr}}$ of representative functions on G. By the definition of representative (or the comultiplication)

$$\mu(f) = \sum_i f_{i,1} \otimes f_{i,2} \tag{3.10.8}$$

with the right hand side of (3.10.8) characterized by

$$L_x f(y) = \sum_i f_{i,1}(x) f_{i,2}(y) \tag{3.10.9}$$

as functions of $x, y \in G$. As functions of y (for a temporarily fixed x)

$$L_x f = \sum_i f_{i,1}(x) f_{i,2}.$$

Applying the integral to this it follows that as a function of x

$$I(L_x f) = \sum_i f_{i,1}(x) I(f_{i,2}). \tag{3.10.10}$$

By (3.10.5) a functional I on the coalgebra of representative functions on G is a left integral iff for each $f \in \mathrm{Func}(G, \mathbf{C}_{\mathrm{repr}})$

$$\sum_i f_{i,1} I(f_{i,2}) = I(f)\mathbf{1}$$

where $\mathbf{1}$ is the constant function $x \mapsto 1$ on G. This holds iff it holds for every $x \in G$ and so, using (3.10.10), this holds iff $I(L_x f) = I(f)$. Thus a functional on $\mathrm{Func}(G, \mathbf{C})_{\mathrm{repr}}$ is an integral in the Hopf algebra sense if and only it is a left invariant integral in the usual sense.

A scalar multiple of a bialgebra left integral is again a left integral. (So normalization is not incorporated).

The zero dual, i.e. the bialgebra dual, of $\mathrm{Func}(G, \mathbf{C})_{\mathrm{repr}}$ is the group bialgebra $\mathbf{C}[G]$, but, unless G is finite, there is no element of the group algebra that represents a left invariant Haar integral on $\mathrm{Func}(G, \mathbf{C})_{\mathrm{repr}}$.

Corollary 3.10.11. (The $\mathrm{Func}(G, \mathbf{C})_{\mathrm{repr}}$ example of a left integral in the bialgebra sense.) *A (not necessarily normalized) Haar integral on a compact group gives rise to left integral on the bialgebra of representative functions.*

When the group is finite the scalar multiples of the element

$$\sum_{x \in G} x \in kG = \mathrm{Func}(G, k)^*$$

are left integrals.

Indeed in this case the functions δ_x, $x \in G$ are a basis,

$$\mu(x) = \sum_{yz=x} \delta_y \otimes \delta_z, \quad I(\delta_x) = 1, \quad \mathbf{1} = \sum_{x\in G} \delta_x$$

and so

$$\sum_{yz=x} \delta_y I(\delta_z) = \sum_{yz=x} \delta_y = \sum_{y\in G} \delta_y = \mathbf{1} = I(\delta_x)\mathbf{1}.$$

This fits of course perfectly with the idea that for functions on a finite group, $f \mapsto \sum_{x\in G} f(x)$ is a left invariant integral.

The presence of an inverse is not really needed to make this type of example work.

Example 3.10.12. Consider a monoid G (not necessarily a group) and the monoid bialgebra, kG of 3.2.14. Let

$$T : kG \longrightarrow k, \ \sum_{<\infty} a_x x \mapsto a_e$$

where e is the unit element of the monoid G. Then T is a left and right integral for kG. Indeed, $\mu(x) = x \otimes x$ for any $x \in G$, $\mathbf{1} = e \in G \subset kG$,

$$T(x) = \begin{cases} 0, & \text{if } x \neq e \\ 1, & \text{if } x = e \end{cases}$$

and so for any $x \in G$, $xI(x) = I(x)\mathbf{1}$, which suffices. T is also a right integral because kG is cocommutative.

However, there are Hopf algebras, which do not have nonzero integrals. As it turns out they need to be infinite dimensional for that to happen. This does not come as all that much of a surprise in view of the fact that for noncompact groups integration of functions introduces all kinds of additional questions.

Counterexample 3.10.13. No integrals for the divided power bialgebra. Consider the divided power Hopf algebra of example 3.4.19. Recall that as a free module over k it has a basis $\{d_0, d_1, d_2, \ldots\}$, that the comultiplication is

$$\mu(d_n) = \sum_{i+j=n} d_i \otimes d_j$$

and that the unit element is d_0. Let T be a candidate left integral. Then by (3.10.4) it must be the case that for every n

$$\sum_{i+j=n} d_i T(d_j) = T(d_n) d_0$$

and this implies immediately that $T(d_j) = 0$ for every $j < n$ and hence $T = 0$.

Let H be a finite dimensional Hopf algebra over a field. Then an integral for the dual Hopf algebra H^* is an element of H^{**} with certain properties. But H^{**} can be identified in this case with H and thus an integral for H^* is an element of H with certain properties which surely can be expressed in terms of H itself. Here is the outcome of this exercise.

Definition 3.10.14. Second kind of integrals for Hopf algebras. Let H be a Hopf algebra. Then an element $t \in H$ is a **left integral** in H iff

$$ht = \varepsilon(h)t, \quad \text{for all } h \in H. \tag{3.10.15}$$

Left-right symmetrically, a right integral (second kind) in a Hopf algebra H is an element $t' \in H$ such that

$$t'h = \varepsilon(h)t', \quad \text{for all } h \in H. \tag{3.10.16}$$

Note that in this definition one says left integral **in** a Hopf algebra instead of left integral **for** a Hopf algebra. Still it seems better to always specify exactly which integral one has in mind.

Here is the argument that a left integral (second kind) in a Hopf algebra H that is finite dimensional over a field is the same as a left integral (first kind) for the Hopf algebra H^*.

So let $t \in H$ and interpret it as a functional on H^*, $f \mapsto (t,\, f) = f(t)$. Recall that the unit element in H^* is ε. Let

$$\mu(f) = \sum_i f_{i,1} \otimes f_{i,2}$$

then for f to be a left integral (first kind) for H^* one needs

$$\sum_i f_{i,1} f_{i,2}(t) = f(t)\varepsilon. \tag{3.10.17}$$

Evaluating this at $h \in H$ gives

$$\sum_i f_{i,1}(h) f_{i,2}(t) = f(t)\varepsilon(h).$$

And this is equivalent to (3.10.15) because

$$\langle ht,\, f\rangle = \langle h \otimes t,\, \mu(f)\rangle = \sum_i f_{i,1}(h) f_{i,2}(t),\ \ f(t)\varepsilon(h) = \langle \varepsilon(t)h,\, f\rangle.$$

Theorem 3.10.18. (The Larson-Sweedler theorem on Hopf algebra integrals.) *Let H be a finite dimensional Hopf algebra over a field. Then*

(i) *The space of left (resp. right) Hopf algebra integrals (of the second kind) is one dimensional. In particular there exist nonzero ones.*

(ii) *The antipode is bijective and takes a left Hopf algebra integral into a right Hopf algebra integral and vice versa.*

(iii) *H is a cyclic left and right H^*-module.*

(iv) *H is a Frobenius algebra.*

Here, if H is an algebra, it acts on the dual space H^* on the left by left translation and on the right by right translation: $(x, f) \mapsto L_x f,\ L_x f(y) = f(xy)$ and $(f, x) \mapsto R_x f,\ R_x f(y) = f(yx)$. This explains the meaning of statement (iii), interpreting H as H^{**}.

A Frobenius algebra over k is an algebra A that comes with a nondegenerate bilinear form $\langle\, ,\, \rangle : A \otimes A \longrightarrow k$ that is associative, meaning $\langle xy, z\rangle = \langle x, yz\rangle$.

For a proof see [**27**], [**13**], [**39**], or [**31**]. Applying the theorem to the dual Hopf algebra H^*, whose integrals of the second kind are integrals of the first kind for H, one obtains a similar theorem for integrals of the first kind for a finite dimensional Hopf algebra H.

The theorem makes it clear that admitting a Hopf algebra structure is quite a strong condition on an algebra.

One nice application of the theory of integrals is the Maschke theorem for Hopf algebras. The classical Maschke theorem for finite groups says the following.

Theorem 3.10.19. (Maschke theorem for finite groups.) *Let G be a finite group and k a field. Then the group algebra kG is semisimple if and only if the characteristic of k does not divide the order of G.*

Here is a version for finite dimensional Hopf algebras.

Theorem 3.10.20. (Maschke theorem for finite dimensional Hopf algebras over a field.) *Let H be a finite dimensional Hopf algebra over a field. Then H is semisimple as an algebra if and only if there is a left integral (of the second kind) t such that $\varepsilon(t) \neq 0$.*

Note that this fits with the Maschke theorem for finite groups, taking the group algebra kG for H. Indeed in that case the only integrals (of the second kind) in H are the scalar multiples of the element $\sum\limits_{x \in G} x$ and applying ε to this gives $\#G$.

For a proof of theorem 3.10.20 see for example [**12**], theorem 5.2.10, page 187. There are more general Maschke type theorems, see loc. cit.

3.11. Formal groups

Formal groups were introduced by Bochner, [6], as an intermediate object in that part of Lie theory [30], that deals with the correspondence between Lie algebras and Lie groups. Going from a Lie group to its Lie algebra is a one step process. But constructing a Lie group from a given Lie algebra is nowadays usually thought of as a multistep process

$$\begin{aligned} \text{Lie algebra} &\mapsto \text{infinitesimal Lie group} = \text{formal group} \\ &\mapsto \text{local Lie group} \mapsto \text{Lie group} \end{aligned}$$

The first step is handled by the Campbell-Baker-Hausdorff formula, the second step concerns convergence matters and the third has much to do with universal covering spaces.

Over a field of positive characteristic the correspondence (algebraic) Lie groups - Lie algebras breaks down. And the break down is at the first step; there can be many different formal groups with the same Lie algebra.

Indeed there is only one one-dimensional Lie algebra over any k of characteristic zero. However, there are countably many different one dimensional formal groups over an algebraically closed field of characteristic $p > 0$ and over the integers there are even uncountably many nonisomorphic formal groups of dimension 1, see [**19**].

---

[30] In some texts the phrase 'Lie theory' precisely means the correspondence between Lie algebras and Lie groups.

The formal group of a Lie group is the power series that expresses the coordinates of the product of two elements near the identity in terms of the coordinates of the two elements. The formal definition of a formal group is as follows. [31]

Definition 3.11.1. Formal group. An n-dimensional **formal group** $F(X,Y)$ over k is an n-tuple of formal power series in $2n$ indeterminates

$$F_i(X_1,\dots,X_n;Y_1,\dots,Y_n) \in K[[X_1,\dots,X_n;Y_1,\dots,Y_n]],\quad i=1,\dots n \tag{3.11.2}$$

such that

$$F_i(X,Y) = X_i + Y_i + (\text{terms of higher degree}) \tag{3.11.3}$$

$$F(F(X,Y),Z) = F(X,F(Y,Z)). \tag{3.11.4}$$

It follows that there is also an "inverse", that is an n-tuple of power series $\iota(X)$ such that

$$F(X,\iota(X)) = F(\iota(X),X) = 0. \tag{3.11.5}$$

Writing $Y_i = X_i \otimes 1$, $X_i = 1 \otimes X_i$ the formal power series F_i define a continuous algebra morphism (using hopefully excusable notation)

$$k[[X]] \longrightarrow k[[X]]\widehat{\otimes}k[[X]],\ X_i \mapsto F_i(1\otimes X,\ X\otimes 1) \tag{3.11.6}$$

where $\widehat{\otimes}$ is a completed tensor product. Except when the F_i themselves are polynomials one needs the completion when considering the 'comultiplication' (3.11.6)[32]. The power series ring $k[[X]]$ with the 'comultiplication' (3.11.6) is referred to as the contravariant bialgebra, $R(F)$, of the formal group F. Its continuous dual, the covariant bialgebra $U(F)$ of F is a Hopf algebra in the usual sense.

The topology on $k[[X]] = k[[X_1,\dots,X_n]]$ is the usual power series one defined by the sequence of ideals

$$J_s = \{f \in k[[X]] \ :\ \text{the coefficients of } X^\alpha \text{ in } f \text{ are } 0 \text{ for } \mathrm{wt}(\alpha) \le s\} \tag{3.11.7}$$

Here $\alpha = [r_1, r_2, \dots, r_n]$, $r_i \in \mathbf{N}\cup\{0\}$, is a word of length n over the nonnegative integers of weight $\mathrm{wt}(\alpha) = r_1 + r_2 + \cdots + r_n$ and X^α is short for $X_1^{r_1}X_2^{r_2}\cdots X_n^{r_n}$.

When taking topological duals k is given the discrete topology. Thus for instance for $n=1$

$$k[[X]]^{\mathrm{top}*} = kY_0 \oplus kY_1 \oplus \dots \oplus kY_n \oplus \dots,\qquad \langle X^r, Y_s\rangle = \delta_{r,s} \tag{3.11.8}$$

For the one dimensional additive formal group

$$\hat{\mathbf{G}}_a(X;Y) = X + Y \tag{3.11.9}$$

and the covariant bialgebra[33] of $\hat{\mathbf{G}}_a(X;Y)$ is therefore the binomial Hopf algebra

$$\begin{aligned} U(\hat{\mathbf{G}}_a) &= kY_0 \oplus kY_1 \oplus \cdots \oplus kY_n \oplus \cdots \\ \mu(Y_n) &= \sum_{i+j=n} Y_i \otimes Y_j \\ Y_rY_s &= \binom{r+s}{r} Y_{r+s}. \end{aligned} \tag{3.11.10}$$

---

[31] A formal group is sometimes called a formal group *law* because it is really a recipe for multiplying things, rather than a group multiplication itself.

[32] The reason is that $k_1[[X_1,\dots,X_n]] \otimes k[[Y_1,\dots,Y_n]]$ is only part of $k[[X_1,\dots,X_n;Y_1,\dots,Y_n]]$.

[33] It is in fact a Hopf algebra, as are all the $U(F)$. But it is established convention to call them bialgebras.

3.12. Hopf modules

Just like a Hopf algebra is simultaneously an algebra and a coalgebra in a compatible way, so a Hopf module over a Hopf algebra H is simultaneously a module (over the underlying algebra of H) and a comodule (over the underlying coalgebra of H) in a compatible way. For the notions of module and comodule see 2.9 above.

Definition 3.12.1. Invariants. Let H be a Hopf algebra and M a left H module. Then the **invariants** of M are the elements of the sub k module

$$\mathrm{Inv}(M) = \{m \in M : hm = \varepsilon(h)m \text{ for all } h \in H\}. \tag{3.12.2}$$

Note that in the case of a group Hopf algebra kG or universal enveloping algebra of a Lie algebra, $U\mathfrak{g}$, this fits exactly with the usual idea of invariants.

Definition 3.12.3. Coinvariants. Let H be a Hopf algebra and N an H comodule. Then the **module of coinvariants** is:

$$\mathrm{Coinv}(N) = \{n \in N : \lambda(n) = n \otimes 1\}. \tag{3.12.4}$$

Note that if $\pi : H \longrightarrow \overline{H}$ is a morphism of Hopf algebras and H is regarded as a (right) $\overline{H}$ comodule the left Hopf kernel of π, see 3.9, is precisely the **module of coinvariants**.

Construction 3.12.5. Tensor product of modules. Let H be a Hopf algebra and let V and W be two modules over H. Then the tensor product (over k (sic)) acquires a module structure by

$$h(v \otimes w) = \sum h_{1,i}v \otimes h_{2,i}w. \tag{3.12.6}$$

This is of course the construction alluded to in the beginning of this chapter and which is a main motivation for considering Hopf algebras.

Definition 3.12.7. Hopf module. Let H be a Hopf algebra over k. Then a **right Hopf module** is a k-module M such that

(i) M is a right module over H

(ii) M is a right comodule via a morphism $\lambda : M \longrightarrow M \otimes H$

(iii) the k module morphism λ is a morphism of modules where $M \otimes H$ has the right module structure of 3.12.5).

This last condition can be written out as

$$\lambda(mh) = \sum m_{0,i}h_{1,j} \otimes m_{1,i}h_{2,j}$$

where

$$\lambda(m) = \sum m_{0,i} \otimes m_{1,i}, \quad \mu(h) = \sum h_{1,j} \otimes h_{2,j}.$$

Example 3.12.8. Let W be a right H module and consider $M = W \otimes H$ with the coaction $\lambda = \mathrm{id} \otimes \mu$. Then M is a Hopf module. This construction can be applied in particular for W a trivial H module, meaning $wh = \varepsilon(h)w$. The M's obtained this way are called trivial Hopf modules.

The main theorem on Hopf modules now says that these are, up to isomorphism, the only Hopf modules.

Theorem 3.12.9. (Fundamental theorem for Hopf modules [34].) *Let H be a Hopf algebra over a field k. Let M be a (right) Hopf module. Then*

$$M \simeq \mathrm{Coinv}(M) \otimes H$$

where $\mathrm{Coinv}(M) \otimes H$ *is a trivial Hopf module. The isomorphism is given by* $m \otimes h \mapsto mh$, $m \in \mathrm{Coinv}(M)$, $h \in H$.

For a proof, see [**31**], p. 15ff, or [**12**], p. 171ff. There is an important connection between the "triviality" of Hopf modules in the sense of theorem 3.12.9, more precisely a particular Hopf module denoted $H^{*\ rat}$, and integrals for Hopf algebras; see [**12**], section 5.2.

3.13. Historical remarks

As has been remarked the first Hopf algebras came out of algebraic topology as the homology or cohomology of H-spaces, that is spaces with a multiplication structure

$$X \times X \longrightarrow X \tag{3.13.1}$$

that is associative up to homotopy and for which a base point serves as a unit, again up to homotopy.

It is easiest to explain things in the case of cohomology, simply because most people know that the cohomology of a space comes equipped with a product, the cupproduct, while fewer are aware that the homology comes with a natural coproduct. Both come from the diagonal map

$$\Delta : X \longrightarrow X \times X. \text{ [35]} \tag{3.13.2}$$

So consider the cohomology, $H^*(X)$, of an H-space X with, say, coefficients in a field k. The vector space $H^*(X)$ is graded

$$H^*(X) = \bigoplus_{i=0}^{\dim(X)} H^i(X) \tag{3.13.3}$$

and connected, $H^0(X) = k$, if X is connected.

Further the Künneth theorem says that the cohomology of a product of spaces is the tensor product (with the usual grading, see (3.8.3)), as a graded module.

$$H^*(X \otimes Y) = H^*(X) \otimes H^*(Y). \tag{3.13.4}$$

Thus, if X is an H-space, the map (3.8.1), induces a comultiplication

$$H^*(X) \longrightarrow H^*(X) \otimes X^*(X). \tag{3.13.5}$$

However, the cupproduct on $H^*(X \times Y)$ is not the usual tensor product of the cupproducts on $H^*(X)$ and $H^*(Y)$ but instead the graded tensor product

$$(x \otimes y) \vee (u \otimes v) \mapsto (-1)^{\deg(u)\deg(y)}(x \vee u) \otimes (y \vee v) \tag{3.13.6}$$

which differs from the usual tensor product by the sign factor. Here x, y, u, v are supposed to be homogeneous elements. See [**14**], section 8.16ff, for details.

---

[34] This is another 'niceness theorem'. It is difficult for a module or algebra or ... to have many compatible structures unless the underlying structure is quite nice.

[35] And this is the origin of the frequently used (but unfortunate) notation Δ for the comultiplication in a Hopf algebra.

Thus the Hopf algebras coming from algebraic topology are graded Hopf algebras in a slightly different sense then the graded Hopf algebras of 3.8. The only difference is that in the bialgebra axioms of 3.2.1 in the diagram (3.2.8) the twist morphism tw is replaced by the graded twist

$$b \otimes c \xrightarrow{grtw} (-1)^{\deg(b)\deg(c)} c \otimes b\,. \tag{3.13.7}$$

These Hopf algebras will be called Hopf algebras of algebraic topological type in this text.

The graded twist map (3.13.7) is an example of a braiding (on the category of graded modules). Quite generally, a braiding on a category with tensor products (a monoidal category) is a collection of functorial isomorphisms $\tau_{V,W} : V \otimes W \longrightarrow W \otimes V$ that satisfy some rather natural compatibility conditions, the so called hexagon conditions, see [23], section 9.2.

A braided Hopf algebra or bialgebra is now a Hopf algebra or bialgebra just like in section 3.2 with as only change that the twist morphism in diagram (3.2.8) is replaced by the braiding.

Thus Hopf algebras of topological type are braided Hopf algebras with respect to the graded twist braiding.

The cup product is graded commutative, also called supercommutative, meaning that for homogeneous elements x, y

$$x \cup y = (-1)^{\deg(x)\deg(y)} y \cup x. \tag{3.13.8}$$

Possibly the first structure theorem concerning Hopf algebras is, inevitably, due to Heinz Hopf. It says the following concerning the underlying algebra structure of a finite dimensional graded supercommutative Hopf algebra of algebraic topological type over a characteristic zero field. Such a Hopf algebra is as an algebra the exterior algebra on a finite number of odd degree generators.

This was later extended to a theorem that is variously called the Hopf-Leray theorem or Borel-Hopf theorem, which says the following.

Theorem 3.13.9. *Let H be a connected supercommutative Hopf algebra of algebraic topological type over a field F of characteristic $p > 2$. Then as an algebra H is a tensor product (over F) of algebras with one homogeneous generator*

$$H = \bigotimes_i A_i \tag{3.13.10}$$

with each A_i the (two dimensional) exterior algebra on one generator if that generator is of odd degree and $A_i = F[x]$ or $F[x]/(x^{p^m})$ if x is of even degree, where $p = \operatorname{char}(F)$.

For a thorough discussion of many applications of the theory of Hopf algebras in algebraic topology in the context of H-spaces, see [**23**].

3.14. The Hopf algebra of an algebra

Let A be an algebra. Two important objects attached to A are its automorphism group and its Lie algebra of derivations. For instance, Galois theory links subfields and automorphism groups in the normal separable case of field extensions. In the case of inseparable extensions of height 1 there also is a Galois theory, this time working with derivations rather than automorphisms.

Hopf algebras being a powerful simultaneous generalization of both groups and Lie algebras, one wonders whether there is a canonical Hopf algebra attached to an algebra that encompassed both automorphisms and derivations. There is.

This short section is purely descriptive (no proofs), and solely intended to alert the reader to this business (which appears to deserve more investigation than it has had so far).

We have already seen that the idea of 'measuring' includes both automorphisms and derivations, see section H1.12.

3.14.1. Construction-description. Universal measuring coalgebra. Let A and B be two algebras. Then there exists a coalgebra $M(A,B)$ that measures A to B by a morphism $\vartheta : M(A,B) \otimes A \longrightarrow B$ that is universal in the following sense. For every coalgebra C with measuring morphism $\varphi : C \otimes A \longrightarrow B$ there is a unique coalgebra morphism $\tilde{\varphi} : C \longrightarrow M(A,B)$ such that $\vartheta \circ (\tilde{\varphi} \otimes \mathrm{id}) = \varphi$.

For construction and proof see [**39**], [**40**].

3.14.2. Construction-description. Hopf algebra of an algebra. When $A = B$ the universal measuring coalgebra $M(A,A)$ also carries a unique algebra structure making it a Hopf algebra and making A an $M(A,A)$-module via ϑ.

This is the canonical Hopf algebra attached to an algebra. For some applications of it to Galois type theory and the theory of descent (also essentially a Galois sort of thing), see [**41**].

References

[1] Eiichi Abe, *Hopf algebras*, Cambridge Univ. Press, 1977.

[2] Tomasz Brzezinski, *Comodules and corings*, In: Michiel Hazewinkel, ed., *Handbook of Algebra*, **6**, Elsevier, 2009.

[3] A. A. Agrachev, R. V. Gamkrelidze, *Chronological algebras and nonstationary vectorfields*, Soviet Math **17**:1 (1979), 1650–1675.

[4] A. A. Agrachev, R. V. Gamkrelidze, *Volterra series and permutation groups*, J. of the Mathematical Sciences **71**:3 (1994), 2409–2433. Translated from: Itogi Nauki i Tekhniki, Seriya Sovremennye Problemy Matematiki, Noveishie Dostizheniya **39** (1991), 3–40.

[5] A. A. Agrachev, R. V. Gamkrelidze, A. V. Sarychev, *Local invariants of smooth control systems*, Acta Appl. Math. **14**:3 (1989), 191–237.

[6] Vyjayanthi Chari, Andrew Pressley, *A guide to quantum groups*, Cambridge Univ. Press, 1994.

[7] R. J. Blattner, M. Cohen, S. Montgomery, *Crossed products and inner actions of Hopf algebras.* Trans. Amer. Math. Soc. **298** (1986), 671–711.

[8] S. Bochner, *Formal Lie groups*, Ann. of Math. **47** (1946), 192–201.

[9] P. Cartier, *A primer of Hopf algebras*, IHES, 2006.

[10] Kuo-Tsai Chen, *Algebras of iterated path integrals and fundamental groups*, Trans. Amer. Math. Soc. **156** (1971), 359–379.

[11] Kuo-Tsai Chen, *Iterated path integrals*, Bull. Amer. Math. Soc. **82**:5 (1977), 831–879.

[12] Sorin Dascalescu, Constantin Nastasescu, Serban Raianu, *Hopf algebras. An introduction*, Marcel Dekker, 2001.

[13] Y. Doi, M. Takeuchi, *Cleft comodule algebras for a bialgebra*, Comm. in Algebra **14** (1986), 801–818.

[14] Albrecht Dold, *Lectures on algebraic topology*, Springer, 1972.

[15] Michel Fliess, *Intégrales iterées de K. T. Chen, bruit blanc Gaussien et filtrage nonlinéaire*, C.R. Acad. Sci. Paris **A-284** (1977), 459–462.
[16] Michel Fliess, *Fonctionelles causales non linéaires et indeterminées non commutatives*, Bull. Soc. Math. France **109** (1981), 3–40.
[17] Luzius Grunenfelder, *Über die Struktur von Hopf-algebren*, 1969.
[18] Marshall Hall, Jr., *The theory of groups*, Macmillan, 1959.
[19] Michiel Hazewinkel, *Formal groups and applications*, Acad. Press, 1978.
[20] Michiel Hazewinkel, *The Leibniz-Hopf algebra and Lyndon words*, preprint, CWI, Amsterdam, 1996.
[21] Michiel Hazewinkel, *The primitives of the Hopf algebra of noncommutative symmetric functions over the integers*, São Paolo J. Math. Sci. **1**:2 (2007), 175–203.
[22] Jens Carsten Jantzen, *Lectures on quantum groups*, Amer. math. Soc., Providence, RI, 1996.
[23] R. M. Kane, *The homology of Hopf spaces*, North Holland, 1988.
[24] Irving Kaplansky, *Bialgebras*, Dept. Math., Univ. of Chikago, 1975.
[25] Christian Kassel, *Quantum groups*, Springer, 1995.
[26] M. Kawski, *Chronological algebras: combinatorics and control*, J. mathematical Sciences **103**:6 (2001), 725–744.
[27] Richard Gustavus Larson, Moss Eisenberg Sweedler, *An associative orthogonal bilinear form for Hopf algebras*, Amer. J. Math. **91** (1969), 75–94.
[28] I. G. Macdonald, *Symmetric functions and Hall polynomials*, Clarendon Press, 2-nd edition, 1995.
[29] Shahn Majid, *Foundations of quantum group theory*, Cambridge University Press, 1995.
[30] John W. Milnor, John C. Moore, *On the structure of Hopf algebras*, Ann. Math. **81**:2 (1965), 211–264.
[31] S. Montgomery, *Hopf algebras and their actions on rings*, Amer. math. Soc., 1993.
[32] Ulrich Oberst, H.-J. Schneider, *Untergruppen formeller Gruppen von endlichen index*, J. of Algebra **31** (1974), 10–44.
[33] Bodo Pareigis, *A non-commutative non-cocommutative Hopf algebra in 'nature'*, J. of Algebra **70** (1981), 356–374.
[34] Frédéric Patras, *A Leray theorem for the generalization to operads of Hopf algebras of divided powers*, J. of Algebra **218** (1999), 528–542.
[35] Frédéric Patras, *L'algèbre des descentes d'une bigèbre graduée*, J. of Algebra **170** (1994), 547 – 566.
[36] R. Ree, *Lie elements and an algebra associated with shuffles*, Annals of Math. **68** (1958), 210–220.
[37] Chr. Reutenauer, *Free Lie algebras*, Oxford University Press, 1993.
[38] H. Sussmann, *A product expansion of the Chen series*. In: C. I. Byrnes and A. Lindquist (eds.). Theory and applications of nonlinear control systems, Elsevier, 1986, 323–335.
[39] Moss E. Sweedler, *Hopf algebras*, Benjamin, 1969.
[40] Moss E. Sweedler, *The Hopf algebra of an algebra applied to field theory*, J. of Algebra **8** (1968), 262–276.
[41] Harry Prince Allen and Moss Eisenberg Sweedler, *A theory of linear descent based on Hopf algebraic techniques*, J. of Algebra **12** (1969), 242–294.

CHAPTER 4

The Hopf algebra of symmetric functions

The (algebra of) symmetric functions are (is) one of the most studied objects in mathematics. Also, arguably, possibly the most beautiful one and the most important one. What is known fills several books. And in spite of all that there are still many open questions.

4.1. The algebra of symmetric functions

Probably every one (who is likely to get his fingers on this monograph) knows, or can easily guess at, what is a symmetric polynomial (over the integers, or any other k).

Say, the polynomial is in n (commuting) variables $x_1, x_2, \ldots, x_n$. Then the polynomial $f = f(x_1, x_2, \ldots, x_n)$ is symmetric iff for every permutation σ of $\{1, 2, \ldots, n\}$

$$f(x_{\sigma(1)}, x_{\sigma(2)}, \ldots x_{\sigma(n)}) = f(x_1, x_2, \ldots x_n). \tag{4.1.1}$$

And, in that case, the main theorem of symmetric functions says that: the polynomial f is a polynomial p_f in the elementary symmetric functions $e_1, e_2, \ldots, e_n$ or, equivalently, in the complete symmetric functions $h_1, h_2, \ldots, h_n$, and "the polynomial p_f is independent of the number of variables involved provided there are enough of them" (meaning more than or equal to the degree of f). The last phrase needs explaining. Also this strongly suggests that the way to work with symmetric polynomials is to take an infinity of variables. Both these things need some notation.

An exponent sequence of length m is simply a word $\alpha = [\alpha_1, \alpha_2, \ldots, \alpha_m]$ over the alphabet of the positive integers $\mathbf{N}$. An index sequence γ of length m over $\{1, 2, \ldots, n\}$ is a sequence of indices $i_1 < i_2 < \cdots < i_m$, $i_1 \geq 1$, $i_m \leq n$. Let $IS(n)$ be the set of all index sequences over $\{1, 2, \ldots, n\}$ and IS the set of all index sequences over $\{1, 2, \ldots\}$. A polynomial f in $x_1, x_2, \ldots, x_n$ is a finite sum

$$f = \sum r_{\alpha,\gamma} x_\gamma^\alpha \tag{4.1.2}$$

(where the coefficients $r_{\alpha,\gamma}$ are uniquely determined by f). Here $\gamma \in IS(n)$, $\lg(\gamma) = \lg(\alpha)$ and

$$x_\gamma^\alpha = x_{\gamma_1}^{\alpha_1} x_{\gamma_2}^{\alpha_2} \cdots x_{\gamma_m}^{\alpha_m}. \tag{4.1.3}$$

The polynomial f is quasisymmetric if $r_{\alpha,\gamma} = r_{\alpha,\gamma'}$ for all exponent sequences α and any two index sequences. It is symmetric if moreover

$$r_{[\alpha_1,\ldots,\alpha_m],\gamma} = r_{[\alpha_{\sigma(1)},\ldots\alpha_{\sigma(m)}],\gamma},$$

for all permutations of $\{1, \ldots, m\}$. Thus for instance

$$x_1 x_2^5 + x_1 x_3^5 + x_2 x_3^5 \tag{4.1.4}$$

is a quasisymmetric function in three variables that is not symmetric, and

$$x_1x_2^5 + x_1x_3^5 + x_2x_3^5 + x_1^5x_3 + x_2^5x_3 + x_1^5x_2 \tag{4.1.5}$$

is a symmetric function in three variables.

For quasisymmetric functions the notion of 'the same function in more variables' makes sense. For a given exponent sequence $\alpha = [\alpha_1, \ldots, \alpha_m]$ define the monomial quasisymmetric function in n variables, $n \geq \lg(\alpha) = m$, as the sum

$$M_\alpha = M_\alpha(x_1, \ldots, x_n) = \sum_{1 \leq i_1 < i_2 < \cdots < i_m \leq n} x_{i_1}^{\alpha_1} x_{i_2}^{\alpha_2} \cdots x_{i_m}^{\alpha_m}. \tag{4.1.6}$$

So for instance (4.1.4) is $M_{[1,5]}$ in three variables. The 'same' quasisymmetric function in two variables and four variables is, respectively,

$$x_1x_2^5, \quad x_1x_2^5 + x_1x_3^5 + x_1x_4^5 + x_2x_3^5 + x_2x_4^5 + x_3x_4^5$$

and it also makes perfect sense to write down the monomial quasisymmetric functions in an infinity of variables

$$M_\alpha = \sum_{i_1 < i_2 < \ldots < i_m} x_{i_1}^{\alpha_1} x_{i_2}^{\alpha_2} \cdots x_{i_m}^{\alpha_m} \in \mathbf{Z}[[x_1, x_2, \ldots, x_n, \ldots]].$$

Quite generally, let f be a power series in the $x_1, x_2, \ldots$ over k. This is simply a formal sum with coefficients in k:

$$f = \sum_{\substack{\alpha \in \mathbf{N}^*, \gamma \in IS \\ \lg(\alpha) = \lg(\gamma)}} r_{\alpha,\gamma} x_\gamma^\alpha. \tag{4.1.7}$$

Such a power series is quasisymmetric if $r_{\alpha,\gamma} = r_{\alpha,\gamma'}$ for all α, γ, γ'. It is symmetric if moreover

$$r_{[\alpha_1,\ldots,\alpha_m],\gamma} = r_{[\alpha_{\sigma(1)},\ldots,\alpha_{\sigma(m)}],\gamma} \text{ for all permutations } \sigma \text{ of } \{1, 2, \ldots, m\}. \tag{4.1.8}$$

Quasisymmetric, resp. symmetric, power series can be multiplied and added and multiplied with a scalar to yield again quasisymmetric, resp. symmetric, power series.

This yields the k-algebras

$$\mathbf{QSymm}_k \text{ and } \mathbf{Symm}_k \tag{4.1.9}$$

of quasisymmetric and symmetric polynomials (functions) which are the quasisymmetric and symmetric power series of bounded degree. A power series (4.1.7) is of bounded degree if there is an n such that

$$r_{\alpha,\gamma} = 0 \text{ if } \mathrm{wt}(\alpha) > n.$$

We shall write

$$\mathbf{QSymm} = \mathbf{QSymm}_{\mathbf{Z}}, \quad \mathbf{Symm} = \mathbf{Symm}_{\mathbf{Z}}.$$

The monomial quasisymmetric functions M_α, $\alpha \in \mathbf{N}^*$, form a basis for the free k-module $\mathbf{QSymm}_k$.

Definition 4.1.10. Compositions and partitions. Let $n \in \mathbf{N}$. A **composition** of n is a word over $\mathbf{N}$ of weight n. A **partition** of n is a sequence $\lambda = (\lambda_1, \lambda_2, \ldots, \lambda_t)$ of positive integers such that $\lambda_1 + \lambda_2 + \cdots + \lambda_t = n$ and $\lambda_1 \geq \lambda_2 \geq \cdots \geq \lambda_t$. Let *Part* be the set of all partitions including the empty partition ( ). The length of such a partition is t and its weight is n. The empty partition has weight and length zero.

Given a partition $\lambda = (\lambda_1, \lambda_2, \ldots, \lambda_t)$, the **dual partition** is defined by

$$\lambda^d = (\lambda_1^d, \lambda_2^d, \ldots, \lambda_s^d),\ \lambda_j^d = \#\{i : \lambda_i \geq j\}.$$

In particular $\lambda_1^d = t$, the length of λ, and $s = \lambda_1$. For instance if $\lambda = (5,3,3,2,1,1)$, $\lambda^d = (6,4,3,1,1)$. Note that $(\lambda^d)^d = \lambda$. If, as is often done, $\lambda = (\lambda_1, \lambda_2, \ldots, \lambda_t)$ is depicted as a diagram with λ_j, boxes in the j-th row, numbering from top to bottom, the diagram of the dual partition is obtained by reflection with respect to the diagonal.

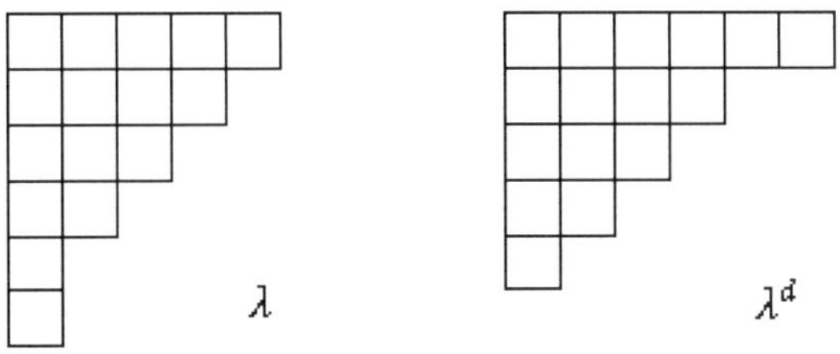

4.1.11. Definition and construction. Monomial symmetric functions. Given a partition $\lambda = (\lambda_l, \lambda_2, \ldots, \lambda_t)$ the associated **monomial symmetric function** is defined by

$$m_\lambda = \sum_{s_i \neq s_j \text{ if } i \neq j} x_{s_1}^{\lambda_1} \cdots x_{s_t}^{\lambda_t} \tag{4.1.12}$$

Note that the monomial symmetric function m_λ is the sum of all quasi monomial functions M_α with distinct composition indices (=exponent sequences) α which are permutations of $(\lambda_1, \ldots, \lambda_t)$. For instance

$$m_{(1,1,3)} = M_{[1,1,3]} + M_{[1,3,1]} + M_{[3,11]}$$

$$m_{(2,4,5)} = M_{[2,4,5]} + M_{[2,5,4]} + M_{[4,2,5]} + M_{[4,5,2]} + M_{[5,2,4]} + M_{[5,4,2]}. \tag{4.1.13}$$

By convention $m_{()} = M_{[]} = 1$.

The monomial symmetric functions form a basis for the free module $\mathbf{Symm}_k$.

4.1.14. Projective limit description. For the finicky (or pertinicky) the following projective limit construction is perhaps more congenial. For each n consider the algebra morphism

$$\pi_{n+1,n} : k[x_1, \ldots, x_n, x_{n+1}] \longrightarrow k[x_1, \ldots, x_n],$$
$$x_i \mapsto x_i \text{ for } i \in \{1, \ldots, n\},\ x_{n+1} \mapsto 0.$$

Let $\mathbf{Symm}_k^{(n)}$, resp. $\mathbf{QSymm}_k^{(n)}$ be the subalgebra of symmetric polynomials, resp. quasisymmetric polynomials, in $k[x_1, \ldots, x_n]$. This gives projective systems of graded algebras

$$\pi_{n+1,n} : \mathbf{Symm}_k^{(n+1)} \longrightarrow \mathbf{Symm}_k^{(n)},\ \pi_{n+1,n} : \mathbf{QSymm}_k^{(n+1)} \longrightarrow \mathbf{QSymm}_k^{(n)}$$

and **Symm** and **QSymm** are the graded projective limits of these systems[1].

4.1.15. Construction and theorem. The elementary symmetric function basis. The elementary symmetric functions are the monomial symmetric functions of the partitions

$$\underbrace{(1,1,\dots,1)}_{n}.$$

That is

$$e_n = \sum_{i_1<\cdots<i_n} x_{i_1}x_{i_2}\cdots x_{i_n}. \tag{4.1.16}$$

The basic theorem on symmetric functions says that each symmetric function is a unique polynomial in the elementary symmetric functions. For each partition $\lambda = (\lambda_1, \lambda_2, \dots \lambda_t)$ define

$$e_\lambda = e_{\lambda_1}e_{\lambda_2}\cdots e_{\lambda_t}. \tag{4.1.17}$$

Then the basic theorem on symmetric functions says that the e_λ, $\lambda \in Part$ (where $e_{(\,)} = 1$) form another basis of **Symm**.

4.1.18. Construction and theorem. Define the complete symmetric functions by

$$h_n = \sum_{i_1\le\cdots\le i_n} x_{i_1}x_{i_2}\cdots x_{i_n}, \quad h_\lambda = h_{\lambda_1}h_{\lambda_2}\cdots h_{\lambda_n}, \quad h_{(\,)} = 1. \tag{4.1.19}$$

The elementary symmetric functions and the complete symmetric functions are related by the Wronski relations

$$h_n - e_1h_{n-1} + \cdots + (-1)^{n-1}e_{n-1}h_1 + (-1)^n e_n = 0. \tag{4.1.20}$$

It follows that the h_λ, $\lambda \in Part$, form a third basis of **Symm**.

Definition 4.1.21. Inner product on **Symm**. A bilinear form is defined on **Symm** by declaring the bases $\{h_\lambda\}$ and $\{m_\lambda\}$ indexed by the partitions to be biorthonormal to each other

$$\langle h_\kappa, m_\lambda\rangle = \delta_{\kappa,\lambda}. \tag{4.1.22}$$

The next theorem implies that this is symmetric, and theorem 4.1.31 or formula 4.1.35 below then show that this is positive definite.

Theorem 4.1.23. *Orthogonality theorem for* **Symm**. *Let* $\{a_\lambda\}$, $\{b_\lambda\}$ *be two bases of the symmetric functions (as a free Abelian group), indexed by all partitions. Then the two bases are biorthogonal with respect to each other, meaning* $(a_\kappa, b_\lambda) = \delta_{\kappa,\lambda}$, *if and only if for each two sets of different independent (commuting) indeterminates* x_i, y_i

$$\sum_\lambda a_\lambda(x)b_\lambda(y) = \prod_{i,j}(1 - x_iy_i)^{-1}. \tag{4.1.24}$$

The first part of this is the next theorem; viz that this holds for the defining pair of biorthogonal bases.

---

[1]This is the projective limit in the category of graded algebras (graded modules). But they are not the projective limits without the word "graded". E.g. $e_1 + e_2^2 + e_3^3 + \cdots$ is in $\underleftarrow{\lim}\,\mathbf{Symm}_k^{(n)}$ (because $e_r = 0$ in $\mathbf{Symm}_k^{(n)}$ for $r > n$) but is not in $\underleftarrow{\text{grlim}}\,\mathbf{Symm}_k^{(n)}$.

Theorem 4.1.25. *Biorthogonality of the monomial symmetric functions and the complete symmetric functions. The bases of monomial functions and complete symmetric functions are biorthogonal to each other (which is by definition) and*

$$\sum_{\lambda} h_\lambda(x)m_\lambda(y) = \prod_{i,j}(1 - x_i y_j)^{-1}. \tag{4.1.26}$$

Proof. Let $H(t)$ be the generating function of the complete symmetric functions

$$H(t) = 1 + h_1(x)t + h_2(x)t^2 + h_3(x)t^3 + \cdots .$$

Then

$$H(t) = \prod_i (1 - x_i t)^{-1}$$

as is easily seen either directly or from the Wronski relations combined with

$$E(-t) = 1 - e_1(x)t + e_2(x)t^2 - e_3(x)t^3 + \ldots = \prod_i (1 - x_i t)$$

and $H(t)E(-t) = 1$.

Hence

$$\begin{aligned}\prod_{i,j}(1 - x_i y_i)^{-1} = \prod_j H(y_j) = \prod_j \left(\sum_{r=0}^{\infty} h_r(x) y_j^r\right) &= \sum_\alpha h_\alpha(x) M_\alpha(y) \\ = \sum_\lambda h_\lambda(x) \left(\sum_{Part(\alpha)=\lambda} M_\alpha(y)\right) &= \sum_\lambda h_\lambda(x) m_\lambda(y)\end{aligned}$$

where α runs over all words over the the natural numbers, $M_\alpha(y)$ is the monomial quasisymmetric function associated to α and $Part(\alpha)$ is the unique partition that as a word is a permutation of α.

Proof of theorem 4.1.23 (continued). The rest of the proof is straightforward and routine. Let $\{a_\lambda\}$, $\{b_\lambda\}$ be two other bases. Write them out in the basis of the complete symmetric functions and the basis of the monomial symmetric functions

$$a_\lambda = \sum_\rho c_{\lambda,\rho} h_\rho, \qquad b_\mu = \sum_\sigma d_{\mu,\sigma} m_\sigma.$$

Then the biorthogonality of $\{a_\lambda\}$, $\{b_\lambda\}$, using the biorthogonality of $\{h_\lambda\}$, $\{m_\lambda\}$, is equivalent to

$$\sum_\rho c_{\lambda,\rho} d_{\mu,\rho} = \delta_{\lambda,\mu} \qquad \text{(Kronecker delta).} \tag{$*$}$$

On the other hand, using (4.1.26), $\sum_\lambda a_\lambda(x)b_\lambda(y) = \prod_{i,j}(1 - x_i y_j)^{-1}$ is equivalent to

$$\sum_\lambda a_\lambda(x)b_\lambda(y) = \sum_\lambda h_\lambda(x)m_\lambda(y). \tag{$**$}$$

Plugging in the expressions for a_λ and b_λ, and using that the $\{h_\lambda(x)\}$ is a basis as functions of the x's, and the $\{m_\lambda(y)\}$ is a basis as functions of the y's, it follows that (**) is equivalent to

$$\sum_\rho c_{\lambda,\rho} d_{\lambda,\sigma} = \delta_{\rho,\sigma}$$

which is the transpose matrix identity of (*) and hence equivalent to it. So biorthogonality is equivalent to (4.1.24) finishing the proof. [2]

Definition 4.1.27. Forgotten symmetric functions. Let $\{f_\lambda : \lambda \in Part\}$ be the dual basis to the elementary function basis of **Symm**. That is the f_λ are defined by $\langle e_\lambda, f_\kappa\rangle = \delta_{\lambda,\kappa}$. These are termed the **forgotten symmetric functions**. There appear to be no 'nice formulas' for the forgotten symmetric functions.

Definition 4.1.28. Schur functions. At first the number of variables $x_1, \dots, x_n$ is supposed to be finite. Let $\alpha = (\alpha_1, \alpha_2, \dots, \alpha_n)$ (same n) be a partition [3] with strictly descending parts, $\alpha_1 > \alpha_2 > \dots > \alpha_n \geq 0$ and consider the following matrix and its determinant.

$$A_\alpha = \begin{pmatrix} x_1^{\alpha_1} & x_1^{\alpha_2} & \dots & x_1^{\alpha_n} \\ x_2^{\alpha_1} & x_2^{\alpha_2} & \dots & x_2^{\alpha_n} \\ \vdots & \vdots & \ddots & \vdots \\ x_n^{\alpha_1} & x_n^{\alpha_2} & \dots & x_n^{\alpha_n} \end{pmatrix}, \qquad a_\alpha = \det(A_\alpha).$$

This determinant is divisible in $\mathbf{Z}[x_1, \dots, x_n]$ by each of the differences $(x_i - x_j)$, $i < j$ and hence by the Vandermonde determinant

$$a_\delta = \prod_{i<j}(x_i - x_j), \qquad \delta = (n-1, n-2, \dots, 1, 0).$$

Now let $\lambda = (\lambda_1, \lambda_2, \dots, \lambda_n)$, $\lambda_1 \geq \lambda_2 \geq \dots \geq \lambda_n \geq 0$ be a partition. The corresponding **Schur function** is defined as

$$s_\lambda(x) = a_{\lambda+\delta}/a_\delta. \tag{4.1.29}$$

Both $a_{\lambda+\delta}$ and a_δ are alternating (= antisymmetric) functions of the x_i, meaning that under a permutation of the variables they change sign according to the sign of that permutation, and so the Schur functions are symmetric and hence can be written as polynomials in the complete symmetric functions. The formula is [4]

$$s_\lambda = \det \begin{pmatrix} h_{\alpha_1} & h_{\alpha_1+1} & \dots & h_{\alpha_1+n-1} \\ h_{\alpha_2-1} & h_{\alpha_2} & \dots & h_{\alpha_2+n-2} \\ \vdots & \vdots & \ddots & \vdots \\ h_{\alpha_n-n+1} & h_{\alpha_n-n+2} & \dots & h_{\alpha_n}. \end{pmatrix}. \tag{4.1.30}$$

This is proved as follows. Let $E^{(k)}(t)$ be obtained from the generating function $E(t) = 1 + e_1 t + \dots + e_n t^n = \prod_i (1 + x_i t)$ of the elementary symmetric functions by setting $x_k = 0$, and, correspondingly, let $e_j^{(k)}$ be thus obtained from e_j. Note

[2]The last part of the proof really takes place in **Symm** $\otimes$ **Symm** of which $\{h_\lambda \otimes m_\mu\}_{\lambda,\mu}$ is a basis.

[3]There is a slightly more general notion of partition being used here: a number of trailing zeros is allowed, and nonincreasing sequences which differ only in the number of trailing zeros are identified. It would be more elegant to define partitions as infinite nonincreasing sequences of nonnegative integers of which only finitely many are nonzero. That is notationally more difficult, however.

[4]These formulas are known as the Jacobi-Trudi formulae. In spite of their name Schur functions were first considered by Jacobi in 1841; Issai Schur lived from 1895-1941.

that $e_n^{(k)} = 0$ (everything is still taking place in the ring of polynomials in the finite number of variables $x_1, \ldots, x_n$). Of course $E^{(k)}(t) = E(t)(1 + x_k t)^{-1}$ and thus, as $H(t)E(-t) = 1$ by the Wronski relations, where $H(t) = 1 + h_1 t + h_2 t^2 + \cdots$ is the generating power series of the complete symmetric functions,

$$H(t)E^{(k)}(-t) = (1 - x_k t)^{-1} = 1 + x_k t + x_k^2 t^2 + \cdots$$

Comparing the coefficients of t^{a_i} on the left and the right (and using that $E^{(k)}(t)$ is a polynomial of degree $n-1$), this gives

$$x_k^{a_i} = \sum_{s=0}^{n-1} h_{a_i - s}(-1)^s e_s^{(k)} = \sum_{j=1}^{n} h_{a_i - n + j}(-1)^{n-j} e_{n-j}^{(k)}$$

which translates to the matrix identity

$$A_\alpha = H_\alpha M \tag{4.1.30A}$$

where A_α is the matrix appearing in the definition of the antisymmetric functions a_α and H_α and M are the $n \times n$ matrices

$$H_\alpha = (h_{a_i - n + j}), \qquad M = ((-1)^{n-i} e_{n-i}^{(k)}).$$

Note that $H_{\lambda+\delta}$ is the matrix appearing on the right of (4.1.30). Note also that H_δ is upper triangular with h_0's on the diagonal. So that $\det(H_\delta) = 1$ and by the matrix identity (4.1.30A), $\det(M) = a_\delta$ and $a_{\lambda+\delta} = \det(H_{\lambda+\delta})a_\delta$, proving formula (4.1.30).

The a_α are obtained by antisymmetrizing (with respect to all of the group of permutations of n letters) the monomials $x_1^{\alpha_1} x_2^{\alpha_2} \cdots x_n^{\alpha_n}$. Antisymmetrizing a monomial with two equal exponents gives zero. It follows immediately that the a_α with $a_1 > a_2 > \cdots > a_n > 0$ form a basis for the Abelian group of antisymmetric polynomials in the $x_1, \ldots, x_n$ and that multiplication by a_δ is an isomorphism of the module of symmetric polynomials to that of the antisymmetric polynomials. Hence the s_λ with λ of length $\leq n$ form a basis for the Abelian group of symmetric polynomials in these n variables.

Theorem 4.1.31. (Orthogonality of Schur functions). [5]

$$\langle s_\lambda, s_\kappa \rangle = \delta_{\lambda,\kappa}, \quad \sum_\lambda s_\lambda(x) s_\lambda(y) = \prod_{i,j} (1 - x_i y_j)^{-1}.$$

Proof. For the moment let the number of x variables and y variables be equal to n. In 4.1.28, in the definition of Schur functions, for an $\alpha = [a_1, \ldots, a_n]$, $a_l >$

[5]Over the real numbers for a positive definite inner product there always is an orthonormal basis: a positive definite matrix can be brought into diagonal form with positive diagonal entries and then a square root can be taken. Things are otherwise over the integers. A free finite rank Abelian group with positive definite inner product has a biorthogonal pair of bases if and only if the matrix defining the inner product (with respect to any chosen basis) has determinant one. A free finite rank Abelian group with positive definite inner product is called irreducible if it cannot be written as a direct sum of of two such animals of lower rank. It turns out that in ranks which are a multiple of 8 there are irreducible positive definite inner products and their number grows very rapidly with rank. See **[11]**, p.28.

Thus the fact that the positive definite inner product defined on *Symm* does have an orthonormal basis goes far beyond standard orthonormalization such as Gramm-Schmidt.

$a_2 > \ldots > a_n > 0$, a_α was defined as

$$a_\alpha = \sum_{w \in S_n} \operatorname{sign}(w) y^{w(a)}$$

where $w(\alpha) = [\alpha_{w(1)}, \ldots, \alpha_{w(n)}]$ for a permutation $w \in S_n$. This results from expanding the determinant in the definition of Schur functions in 4.1.28. This formula still makes sense for any $\alpha = [a_1, \ldots, a_n]$, $a_i \in \{0, 1, 2, \ldots\}$. Note that

$$\begin{array}{ll} a_\alpha = 0 & \text{if two (or more) of the entries of } \alpha \text{ are equal} \\ a_{w(\alpha)} = \operatorname{sign}(w) a_\alpha & \text{for a permutation } w \in S_n \end{array}$$

Also recall that for n variables δ stands for the word $[n-1, n-2, \ldots, 1, 0]$. Further, to end these recollections, the determinantal formula (4.1.30) says that

$$a_\alpha = a_\delta \sum_{w \in S_n} \operatorname{sign}(w) h_{\alpha - w(\delta)}$$

which is equivalent to the formula $a_\alpha = a_\delta \det(H_\alpha)$. Now calculate

$$a_\delta(x) a_\delta(y) \prod_{i,j} (1 - x_i y_j)^{-1} = a_\delta(x) a_\delta(y) \sum_\lambda h_\lambda(x) m_\lambda(y) \quad \text{(by theorem 4.1.23)} =$$

$$= a_\delta(x) \sum_\lambda \left(\sum_{Part(\alpha) = \lambda} h_\alpha(x) y^\alpha a_\delta(y) \right)$$

(because $h_\alpha(x) = h_\lambda(x)$ for any α such that $Part(\alpha) = \lambda$ and because $m_\lambda(y) = \sum_{Part(\alpha)=\lambda} y^\alpha$.) So the expression above in turn is equal to

$$\begin{aligned} &= a_\delta(x) \sum_\alpha h_\alpha(x) y^\alpha a_\delta(y) \\ &= a_\delta(x) \sum_{\alpha, w} h_\alpha(x) y^{\alpha + w(\delta)} \operatorname{sign}(w) \quad \text{(by the formula for } a_\delta) \\ &= a_\delta(x) \sum_{\beta, w} \operatorname{sign}(w) h_{\beta - w(\delta)}(x) y^\beta \quad \text{(change of variable)} \\ &= \sum_\beta a_\beta(x) y^\beta \quad \text{(by the formula just recalled for } a_\alpha). \end{aligned}$$

Now, as also just recalled, $a_\beta = 0$ if two or more of the entries of β are zero, and $a_{w(\beta)} = \operatorname{sign}(w) a_\beta$. For each β there is precisely one partition μ such that $Part(\beta) = \mu$. And so the expression just above is equal to

$$\begin{aligned} &= \sum_{w, \mu} a_\mu(x) \operatorname{sign}(w) y^{w(\mu)} \quad \text{(where } \beta = w(\mu)) \\ &= \sum_\mu a_\mu(x) a_\mu(y) \quad \text{(using again the formula for } a_\lambda). \end{aligned}$$

This proves the theorem for a finite number of variables. The passage to an infinity of variables is routine. Or argue as follows. If the second statement of the the theorem were not true, clearly this would show up for some finite number of variables. That is not the case by what has just been proved and theorem 4.1.23 [6]. So the second part holds and another application of theorem 4.1.23 (in the opposite direction) then gives the orthogonality of Schur functions

Definition 4.1.32. The power sums. The **power sums** are the monomial symmetric functions of the one part partitions (n):

$$p_n = \sum_i x_i^n, \quad p_\lambda = p_{\lambda_1} p_{\lambda_2} \cdots p_{\lambda_t}, \quad p_0 = 1. \tag{4.1.33}$$

The power sums are related to the elementary symmetric functions by the Newton relations

$$p_n - e_1 p_{n-1} + \cdots + (-1)^{n-1} e_{n-1} p_1 + (-1)^n n e_n = 0. \tag{4.1.34}$$

It follows that $\{p_\lambda : \lambda \in Part\}$ is a basis for the $\mathbf{Q}$-vector space $\mathbf{Symm}_{\mathbf{Q}}$ (but definitely not a basis for **Symm**). The values of the inner products of the p_λ are

$$\langle p_\lambda, p_\kappa \rangle = \delta_{\lambda,\kappa} ||\lambda|| \tag{4.1.35}$$

where if λ is of the form

$$\lambda = (\underbrace{1, 1, \ldots, 1}_{n_1}, \underbrace{2, 2, \ldots, 2}_{n_2}, \ldots, \underbrace{m, m, \ldots, m}_{n_m}) \tag{4.1.36}$$

which is often written

$$\lambda = 1^{n_1} 2^{n_2} \cdots m^{n_m}, \quad n_j \in \mathbf{N} \cup \{0\} \tag{4.1.37}$$

$$||\lambda|| = n_1! 1^{n_1} n_2! 2^{n_2} \cdots n_m! m^{n_m}. \tag{4.1.38}$$

Definition 4.1.39. Lexicographic order. Let $\alpha = [a_1, \ldots, a_m]$ and $\beta = [b_1, \ldots, b_n]$ be two words over $\mathbf{N}$. Then α is said to be **lexicographically equal to or larger** than β, denoted $\alpha \geq_{\text{lex}} \beta$, iff there is a $j \leq \min\{m, n\}$ such that $a_1 = b_1, \ldots, a_{j-1} = b_{j-1}$ and $a_j > b_j$, or $m \geq n$ and $a_1 = b_1, \ldots, a_n = b_n$. This is the total order used for dictionaries, whence the name.

In spite of its seemingly artificial nature, not much related to any kind of algebraic thinking, this order plays a most important role in various parts of algebra and algebraic combinatorics.

4.1.40. Construction and proof technique. Generating functions. It is often useful and illustrative (and notationally convenient) to work with generating functions. The **generating function** of the elementary symmetric function is

$$E(t) = \sum_{r \geq 0} e_r t^r = \prod_{i \geq 1} (1 + x_i t) \tag{4.1.41}$$

and that of the complete symmetric functions is

$$H(t) = \sum_{r \geq 0} h_r t^r = \prod_{i \geq 1} (1 - x_i t)^{-1}. \tag{4.1.42}$$

The Wronski relations are then equivalent to the statement that

$$H(t)E(-t) = 1. \tag{4.1.43}$$

[6]What is used here is the finite number of variables version of theorem 3.1.23, which is obtained from the theorem as stated by setting for a given n all x's and y's zero with index larger than n, the formulas look just the same.

The generating function of the power sums is [7]

$$P(t)=\sum_{r\geq 1}p_rt^{r-1}=\sum_{i,r\geq 1}x_i^rt^{r-1}=\sum_{i\geq 1}\frac{x_i}{1-x_it}=\sum_{i\geq 1}\frac{d}{dt}\log(\frac{1}{1-x_it}) \tag{4.1.44}$$

so that

$$P(t)=\frac{d}{dt}\log H(t)=\frac{H'(t)}{H(t)} \tag{4.1.45}$$

which translates into the Newton relations

$$nh_n=p_1h_{n-1}+p_2h_{n-2}+\cdots+p_{n-1}h_1+p_n. \tag{4.1.46}$$

From (4.1.46) one gets

$$\begin{aligned} H(t)&=\exp(\sum_{r\geq 1}\frac{p_rt^r}{r})=\prod_{r\geq 1}\exp(r^{-1}p_rt^r)\\ &=\prod_{r\geq 1}\sum_{m_r\geq 1}p_r^{m_r}t^{rm_r}(r^{m_r}m_r!)^{-1}=\sum_{\lambda\in Part}||\lambda||^{-1}p_\lambda t^{\mathrm{wt}(\lambda)}\end{aligned} \tag{4.1.47}$$

where λ stands for the partition $(1^{m_1},2^{m_2},\ldots,n^{m_n})$ with m_i parts of weight i.

This translates into

$$h_n=\sum_{\mathrm{wt}(\lambda)=n}||\lambda||^{-1}p_\lambda \tag{4.1.48}$$

a formula that is a bit more difficult to establish directly from (4.1.36) thus illustrating the convenience of working with generating functions.

4.2. The Hopf algebra structure

The algebra of symmetric functions carries a Hopf algebra structure:

$$\mu(h_n)=\sum_{i+j=n}h_i\otimes h_j,\quad (h_0=1) \tag{4.2.1}$$

or, equivalently, using the Wronski relations

$$\mu(e_n)=\sum_{i+j=n}e_i\otimes e_j,\quad (e_0=1). \tag{4.2.2}$$

It follows, see 3.5.5 (where a noncommutative version of this is proved), that

$$\mu(p_n)=1\otimes p_n+p_n\otimes 1. \tag{4.2.3}$$

I.e. the power sums are primitives. Formula (4.2.3) is a perfectly good characterization/description of the comultiplication provided one knows in some other way that what is defined is defined over the integers.

This is the natural comultiplication? The question mark is intended to convey the question of whether this is the only 'good' Hopf algebra structure on this algebra. Or, stronger, that there are compelling reasons to define the comultiplication just this way and not, for instance, by $\mu(h_n)=1\otimes h_n+h_n\otimes 1$, which looks simpler (and over the rationals is isomorphic).

[7] Note the small shift in degree. It would perhaps be more elegant to work with the operator $t\frac{d}{dt}\log$ instead of $\frac{d}{dt}\log$.

Theorem 4.2.4. (Auto duality of the Hopf algebra **Symm**). *Let* $\langle , \rangle$ *and* p_λ *be given by* (4.1.22) *and* (4.2.1). *Then for all* $x, y, z \in$ **Symm**

$$\langle xy,\, z\rangle = \langle x \otimes y, \mu(z)\rangle. \tag{4.2.5}$$

Remark. This is much stronger than just saying that **Symm** and $\mathbf{Symm}^{gr*}$ are isomorphic.

Proof. It suffices to check this for elements of the power sum basis of $\mathbf{Symm}_{\mathbf{Q}}$. Each p_λ can be written in the form

$$p_\lambda = p_1^{a_1} \dots p_r^{a_r}, \quad a_i \in \mathbf{N} \cup \{0\}.$$

Thus it suffices to check that always

$$\begin{aligned} &\langle (p_1^{a_1} \cdots p_r^{a_r})(p_1^{b_1} \cdots p_r^{b_r}),\, p_1^{c_1} \cdots p_r^{c_r}\rangle \\ &= \langle p_1^{a_1} \cdots p_r^{a_r} \otimes p_1^{b_1} \cdots p_r^{b_r},\, \mu(p_1^{c_1} \cdots p_r^{c_r})\rangle. \end{aligned} \tag{4.2.6}$$

Now the left hand side is zero unless $a_i + b_i = c_i$ and then it is equal to

$$1^{c_1} c_1! 2^{c_2} c_2! \cdots r^{c_r} c_r! \tag{4.2.7}$$

Now

$$\begin{aligned} \mu(p_1^{c_1} \dots p_r^{c_r}) &= (1 \otimes p_1 + p_1 \otimes 1)^{c_1} \cdots (1 \otimes p_r + p_r \otimes 1)^{c_r} \\ &= \sum_{j_1, \dots, j_r} \left(\tbinom{c_1}{j_1} p_1^{j_1} \otimes p_1^{c_1 - j_1}\right) \cdots \left(\tbinom{c_r}{j_r} p_r^{j_r} \otimes p_r^{c_r - j_r}\right) \end{aligned} \tag{4.2.8}$$

and thus the right hand side is zero unless $a_i + b_i = c_i$ and then the only term that gives a nonzero inner product with $p_1^{a_1} \cdots p_r^{a_r} \otimes p_1^{b_1} \cdots p_r^{b_r}$ has $j_1 = a_1, \dots, j_r = a_r$. Thus in that case the right hand side of (4.2.5) is equal to

$$\binom{c_1}{a_1} 1^{a_1} a_1! 1^{b_1} b_1! \binom{c_2}{a_2} 2^{a_2} a_2! 2^{b_2} b_2! \cdots \binom{c_r}{a_r} r^{a_r} a_r! r^{b_r} b_r!$$

which is equal to (4.2.7).

Theorem 4.2.9. (Positivity of the multiplication and comultiplication on **Symm**). *The product of two Schur functions is a positive linear combination of Schur functions.*

This will be a consequence of later results. It does not seem easy to establish this directly.

4.3. *PSH* algebras

The acronym '*PSH*' stands 'for positive selfadjoint Hopf'. This implies sort of that there is also a positive definite inner product, and that we are working over something like the integers or the reals, where positive makes sense. What is not mentioned is that these Hopf algebras are also supposed to be connected and graded. As will be seen the assumptions 'positive', 'selfadjoint', 'graded' are all three very strong; so strong that these algebras can actually be classified. Indeed they are all products of one example (possibly degree shifted) and that example is the Hopf algebra of the symmetric functions. The notion is due to Zelevinsky, [**15**], and the classification theorem is his.

Definition 4.3.1. *PSH* algebra. A ***PSH* algebra** is a connected graded Hopf algebra over the integers, so that

$$H = \bigoplus_n H_n, \quad H_0 = \mathbf{Z}, \quad \mathrm{rk}(H_n) < \infty \tag{4.3.2}$$

which is free as an Abelian group and which comes with a given, 'preferred' homogeneous basis $\{\omega_i : i \in I\} = \mathcal{B}$. Define an inner product $\langle\, ,\, \rangle$ on H by declaring this basis to be orthonormal. Then the further requirements are:

$$\text{Selfadjointness: } \langle xy,\, z\rangle = \langle x \otimes y,\, z\rangle \tag{4.3.3}$$

$$\text{Positivity: let } \omega_i\omega_j = \sum_r a^r_{i,j}\omega_r, \quad \mu(\omega_r) = \sum_{i,j} b^{i,j}_r \omega_i \otimes \omega_j, \tag{4.3.4}$$

then $a^r_{i,j},\ b^{i,j}_r \geq 0$.

Example 4.3.5. At this stage is is not yet clear that there exists such a thing as a *PSH* algebra. However, as will be shown in the next chapter, it is not all that difficult to see that the Hopf algebra *RS* of representations of the symmetric groups is a *PSH* algebra with precisely one preferred basis element that is primitive. Actually much comes for free, and the remaining bits are taken care of by Frobenius reciprocity (the duality part) and the Mackey double coset theorem (the Hopf algebra property). Thus there is at least one example.

The Hopf algebra of symmetric functions will turn out to be isomorphic to RS, and hence is also *PSH*. At this stage it comes close: there is an inner product with respect to which it is auto adjoint; there is a preferred orthonormal basis, the Schur functions; what is not yet clear at this stage is that the multiplication (and hence the comultiplication) is positive. Nor does it seem that that is easy to establish directly.

Of course the selfadjointness says that $a^r_{i,j} = b^{i,j}_r$ so that it suffices to require positivity for one of m and μ. One of the basis elements must be a basis for $H_0 = \mathbf{Z}$, and the positivity requirement then says that this one must be $1 \in H_0 \subset H$. Denote this one ω_0.

By gradedness, the ω's from the basis $\mathcal{B}$ which have minimal degree > 0 are all primitive. So assuming that H is not trivial, as will de done from now on, there are primitive elements among the basis. The simplest nontrivial case therefore is that there is precisely one primitive among the preferred basis.

Proposition 4.3.6. A unique primitive spreads out to involve everything. *Let H be a (nontrivial) PSH algebra and suppose that only one of the (preferred) basis elements is primitive. Call this one $p = \omega_1$. Then for every $i > 0$ there is an n such that ω_i, occurs in the expression for p^n in terms of the preferred basis with coefficient > 0.*

Proof. If $\omega \in \mathcal{B}$ is primitive it is equal to p by hypothesis. By the positivity condition if ω is not primitive and of degree > 0, there are basis elements ω_i, ω_j $1 < \deg(\omega_i), \deg(\omega_j) < \deg(\omega)$ such that the coefficient of ω in $\omega_i\omega_j$ is positive. Indeed, this holds because

$$\langle \mu(\omega),\, \omega_i \otimes \omega_j\rangle = \langle \omega,\, \omega_i\omega_j\rangle$$

and by positivity and the fact that ω is not primitive, at least one of the coefficients $a_{i,j}$ in

$$\mu(\omega) = 1 \otimes \omega + \sum_{\substack{\deg(\omega_i>0) \\ \deg(\omega_j>0)}} a_{i,j}\omega_i \otimes \omega_j + \omega_i \otimes 1$$

is strictly positive.

By induction we can assume that ω_i, and ω_j occur in p^{n_1} and p^{n_2} with positive coefficients. It then follows, again by positivity, that ω occurs in $p^{n_1+n_2}$ with a positive coefficient.

Corollary 4.3.7. (Adjusting the degrees). *Let H be a nontrivial PSH algebra with one primitive element p in its preferred basis. Then the degree of every preferred basis element is a multiple of the degree of p.*

This follows from the gradedness of H, because by the previous proposition every basis element occurs with positive coefficient in some p^n and hence has degree $n\deg(p)$.

Thus, by dividing all degrees by $\deg(p)$ if necessary one may as well assume that $\deg(p) = 1$ as will be done from now on.

As was done, in a more general setting in chapter 2 above, define for each $x \in H$ a linear operator x^* on H by

$$\langle x^*y,\, z\rangle = \langle y,\, xz\rangle. \tag{4.3.8}$$

In the more general setting alluded to x^* is the operator L_x (left translation) on functions on H. Note that $x^*x = 1$ if x is an element of the preferred basis (because $\deg(xz) > \deg(x)$ if z is homogenous of degree > 0). If x is primitive x^* is a derivation [8]. More generally, as in the case of left translates,

$$\text{if } \mu(x) = \sum_i x_{i,1} \otimes x_{i,2}, \quad x^*(yz) = \sum_i (x^*_{i,1}y)(x^*_{i,2}z). \tag{4.3.9}$$

Proposition 4.3.10. (First step towards the Zelevinsky theorem). *Let H be a PSH algebra with one primitive preferred basis element which is written p and which can be assumed of degree 1. Then p^2 is the sum of exactly two different preferred basis elements*

$$p^2 = x_2 + y_2. \tag{4.3.11}$$

Moreover, writing $p = x_1 = y_1$, $1 = x_0 = y_0$

$$\begin{aligned} x_1^*x_1 = x_0,\ x_1^*x_2 = x_1,\ x_2^*x_2 = x_0,\ x_2^*y_2 = 0 \\ y_1^*y_1 = y_0,\ y_1^*y_2 = y_1,\ y_2^*y_2 = y_0,\ y_2^*x_2 = 0 \end{aligned} \tag{4.3.12}$$

$$\begin{aligned} \mu(x_2) = 1 \otimes x_2 + x_1 \otimes x_1 + x_2 \otimes 1 \\ \mu(y_2) = 1 \otimes y_2 + y_1 \otimes y_1 + y_2 \otimes 1 \end{aligned} \tag{4.3.13}$$

and finally $\mathrm{rk}(H_2) = 2$.

Indeed $\langle p^2,\, p^2\rangle = \langle p,\, p^*(p^2)\rangle = \langle p,\, 2(p^*p)p\rangle = 2$ because p^* is a derivation and p is a preferred basis element so that $1 = \langle p, p\rangle = \langle 1_H, p^*p\rangle$. As p^2 is positive and must involve all degree 2 basis elements and there is only one way to write 2 as

[8] This fits well with the duality of universal enveloping algebras and coordinate algebras, see 3.7.7.

a sum of squares, it must be the case that (4.3.11) holds and that H_2 has rank 2. Equations (4.3.12) follow. For instance $x_1^* x_2 = p^* x_2 = ap$ for some positive a because it must be of degree 1. Then $a = \langle ap, p \rangle = \langle p^* x_2, p \rangle = \langle x_2, p^2 \rangle = 1$.

To get (4.3.13), note that one must have $\mu(x_2) = 1 \otimes x_2 + a x_1 \otimes x_1 + x_2 \otimes 1$ for some (positive) a. And then

$$a = \langle \mu(x_2), p \otimes p \rangle = \langle x_2, p^2 \rangle = 1$$

and similarly for $\mu(y_2)$.

In the following the phrase 'basis element' means an element of the preferred basis $\mathcal{B}$.

Theorem 4.3.14. (First part of the Zelevinsky theorem.) *Let H, $1 = x_0 = y_0$, $p = x_1 = y_1$, x_2 and y_2 be as above.*

(1) *For every n there is a unique basis element x_n of degree n such that*

$$y_2^* x_n = 0. \tag{4.3.15}$$

Moreover these x_n satisfy

$$x_r^* x_n = x_{n-r}, \quad \omega^* x_n = 0 \text{ for all } \omega \in \mathcal{B},\, \omega \neq x_r \tag{4.3.16}$$

(where $x_s = 0$ for $s < 0$), *and*

$$\mu(x_n) = 1 \otimes x_n + x_1 \otimes x_{n-1} + \cdots + x_{n-1} \otimes x_1 + x_n \otimes 1. \tag{4.3.17}$$

(2) *For every n there is a unique basis element y_n of degree n such that*

$$x_2^* y_n = 0. \tag{4.3.18}$$

Moreover these y_n satisfy

$$y_r^* y_n = y_{n-r}, \quad \omega^* y_n = 0 \text{ for all } \omega \in \mathcal{B},\, \omega \neq y_r \tag{4.3.19}$$

(where $y_s = 0$ for $s < 0$), and

$$\mu(y_n) = 1 \otimes y_n + y_1 \otimes y_{n-1} + \cdots + y_{n-1} \otimes y_1 + y_n \otimes 1. \tag{4.3.20}$$

Proof. First note that H is commutative by proposition 3.8.13. Proposition 4.3.10 establishes (1) and (2) for $n < 3$. So proceed by induction.

$$\begin{aligned}\langle p x_{n-1}, p x_{n-1} \rangle &= \langle x_{n-1}, p^*(p x_{n-1}) \rangle = \langle x_{n-1}, (p^* p) x_{n-1} + p(p^* x_{n-1}) \rangle \\ &= \langle x_{n-1}, x_{n-1} \rangle + \langle x_{n-1}, p x_{n-2} \rangle = 1 + \langle p^* x_{n-1}, x_{n-2} \rangle \\ &= 1 + \langle x_{x-2}, x_{n-2} \rangle = 2.\end{aligned}$$

It follows that $p x_{n-1}$ is the sum of precisely two basis elements, say ω and ω'. Further, using (4.3.13) and (4.3.9)

$$y_2^*(p x_{n-1}) = (y_2^* p) x_{n-1} + (p^* p)(p^* x_{n-1}) + p(y_2^* x_{n-1}) = 0 + x_{n-2} + 0 = x_{n-2}.$$

Now both $y_2^* \omega$ and $y_2^* \omega'$ are nonnegative linear combinations of basis elements. It follows that one of them must be zero and the other one equal to x_{n-2}. So define x_n as that one from ω, ω' that is taken to zero by y_2^*. This establishes existence of x_n. Say $x_n = \omega$. For uniqueness observe that, using commutativity

$$y_2^* p^* x_n = p^* y_2^* x_n = 0.$$

But $p^* x_n$ is a nonnegative linear combination of degree $n-1$ basis elements and x_{n-1} is the unique basis element of degree $n-1$ that is taken to zero by y_2^*. Thus $p^* x_n$ is a nonnegative scalar multiple of x_{n-1}. But

$$\langle x_{n-1}, p^* x_n \rangle = \langle p x_{n-l}, x_n \rangle = \langle x_n + \omega', x_n \rangle = 1$$

proving that the scalar is one and taking care of (4.3.16) for basis elements of degree 1. By induction, hence

$$(p^r)^* x_n = (p^*)^r x_n = x_{n-r}.$$

Writing out p^r as a sum of basis elements, because every basis element of degree r occurs with a positive coefficient in p^r, it follows that for all basis elements of degree r, $0 \leq r \leq n$, $\omega^* x_n = 0$, except for precisely one. For $r = 0$ this exceptional basis element is $x_0 = 1$ obviously and for $n = r$ it is x_n because x_n is a basis element, so that $\langle x_n^* x_n, 1 \rangle = \langle x_n, x_n \rangle = 1$. For the remaining r using commutativity

$$(p^{n-r})^* x_r^* x_n = x_r^* (p^{n-r})^* x_n = x_r^* x_r = 1.$$

This proves (4.3.16).

Further

$$\mu(x_n) = 1 \otimes x_n + \sum_{i,j} a_{i,j} \omega_i \otimes \omega_j + x_n \otimes 1$$

where the $a_{i,j}$ are nonnegative and the sum ranges over all pairs of basis elements of strictly positive degree with $\deg(\omega_i) + \deg(\omega_j) = n$. If $\omega_i \neq x_i$

$$a_{i,j} = \langle \mu(x_n), \omega_i \otimes \omega_j \rangle = \langle x_n, \omega_i \omega_j \rangle = \langle \omega_i^* x_n, \omega_j \rangle = \langle 0, \omega_j \rangle = 0$$

and if $\omega_i = x_i$

$$a_{i,j} = \langle \mu(x_n), x_i \otimes \omega_j \rangle = \langle x_n, x_i \omega_j \rangle = \langle x_i^* x_n, \omega_j \rangle = \langle x_{n-i}, \omega_j \rangle$$

which is zero if $\omega_j \neq x_{n-i}$ and 1 otherwise. This proves (4.3.17).

Part (2) of the theorem is proved in exactly the same way, interchanging x's and y's. [9]

Theorem 4.3.21. (The second part of the Zelevinsky theorem). *Let H be a PSH algebra with one primitive preferred basis element p, and let x_n and y_n be as above. Then as a ring $H = \mathbf{Z}[x_1, x_2, \ldots]$, the polynomial ring in the infinitely many indeterminates $\{x_i : i \in \mathbf{N}\}$. It is also the polynomial ring $\mathbf{Z}[y_1, y_2, \ldots]$ and*

$$\sum_{j=0}^{n} (-1)^j x_i y_{n-i} = 0. \tag{4.3.22}$$

The comultiplication is

$$\mu(x_n) = 1 \otimes x_n + x_1 \otimes x_{n-1} + \cdots + x_{n-1} \otimes x_1 + x_n \otimes 1 \tag{4.3.23}$$

or, equivalently,

$$\mu(y_n) = 1 \otimes y_n + y_1 \otimes y_{n-1} + \cdots + y_{n-1} \otimes y_1 + y_n \otimes 1 \tag{4.3.24}$$

[9] Commutativity is only used twice: $py_2 = y_2 p, px_k = x_k p$. It is perhaps amusing to see this directly by calculating, as in the text, all three inner products $\langle py_2, py_2 \rangle$, $\langle py_2, y_2 p \rangle$, $\langle y_2 p, y_2 p \rangle$ to be equal to 2. The Cauchy-Schwarz-Bunyakovsky lemma then says that the two vectors py_2 and $y_2 p$ are equal. Similarly for $px_k = x_k p$. Then, using the theorem, in a similar way, using induction, one shows that the three inner products $\langle x_m x_n, x_m x_n \rangle$, $\langle x_m x_n, x_n x_m \rangle$, $\langle x_n x_m, x_n x_m \rangle$ are equal, which then proves commutativity of the x's. And similarly commutativity of the y's and of x's and y's.

(so that H is uniquely determined as a Hopf algebra). The antipode is given by

$$\iota(x_n) = (-1)^n y_n, \quad \iota(y_n) = (-1)^n x_n. \tag{4.3.25}$$

Proof. Because H is commutative the antipode has order 2 and is a Hopf algebra automorphism, and because H is selfadjoint, ι is also selfadjoint, that is

$$\langle \iota(x),\, y\rangle = \langle x,\, \iota(y)\rangle \tag{4.3.26}$$

and so

$$\langle \iota(x),\, \iota(y)\rangle = \langle x,\, \iota^2(y)\rangle = \langle x,\, y\rangle. \tag{4.3.27}$$

An element ω of degree n is a preferred basis element if and only if $\langle \omega, \omega\rangle = 1$ and $(p^n, \omega) > 0$. Also $\iota(p) = -p$ because p is primitive. Hence if ω is a preferred basis element of degree n then so is $(-1)^n\iota(\omega)$. As $p^2 = x_2 + y_2$ there are precisely two basis elements of degree 2, viz $x_2 \neq y_2$ and we must have $\iota(x_2) = y_2$ or $\iota(x_2) = x_2$. The second possibility would give $2x_2 - p^2 = 0$ which is not the case. Hence

$$\iota(x_2) = y_2, \quad \iota(y_2) = x_2. \tag{4.3.28}$$

Now for any $y \in H$

$$\begin{aligned}\langle x_2^*\iota(x_n), y\rangle &= \langle \iota(x_n), x_2 y\rangle = \langle x_n, \iota(x_2 y)\rangle = \langle x_n, y_2\iota(y)\rangle = \\ &= \langle y_2^* x_n, \iota(y)\rangle = \langle 0, \iota(y)\rangle = 0.\end{aligned}$$

Hence $x_2^*((-1)^n\iota(x_n)) = 0$ and because y_n is the unique preferred basis element of degree n such that $x_2^* y_n = 0$ it follows that

$$\iota(x_n) = (-1)^n y_n. \tag{4.3.29}$$

This takes care of (4.3.25) and hence also (4.3.22), using the definition of antipode and the formulas for the comultiplication (which were established in the previous theorem).

What is left now is to show that the x's generate all of H and that there are no relations between them (except commutativity of course). To this end examine the matrices of inner products

$$\langle x_\kappa, x_\lambda\rangle \text{ and } \langle x_\kappa, y_\lambda\rangle$$

where κ, λ range over all partitions of (weight) n, see (4.1.10), and if $\lambda = (\lambda_1, \dots, \lambda_t)$, $x_\lambda = x_{\lambda_1}x_{\lambda_2}\cdots x_{\lambda_t}$ (and similarly for the y's). The first step is

$$y_i^* x_n = 0 \quad \text{for all } \; i \geq 2, \quad y_1^* x_n = x_{n-1}. \tag{4.3.30}$$

The second part of (4.3.30) was established before. The first part is proved by induction, the case $i = 2$ being part of the definition of the x_n. Now

$$\begin{aligned} y_{i+1}^*(x_1 x_{n-1}) &= \sum_{s+t=i+1} y_s^*(x_1) y_t^*(x_{n-1}) \\ &= x_1 y_{i+1}^*(x_{n-1}) + y_1^*(x_1) y_i^*(x_{n-1}). \end{aligned} \tag{4.3.31}$$

The first term on the right hand side of (4.3.31) is zero because y_{i+1} and x_{n-1} are different elements of the preferred basis and the second term is zero by induction. On the other hand $x_1 x_{n-1}$ is the sum of x_n and one other preferred basis element and y_{i+1}^* is positive. Thus $y_{i+1}^*(x_1 x_{n-1}) = 0$ implies $y_{i+1}^*(x_n) = 0$.

The next thing to establish is triangularity of the matrix of inner products $\langle y_\kappa,\, x_\lambda\rangle$, $\mathrm{wt}(k) = \mathrm{wt}(\lambda) = n$. This takes the form

$$\kappa > \lambda^d \Longrightarrow y_\kappa^* x_\lambda = 0, \quad y_{\lambda^d}^* x_\lambda = 1. \tag{4.3.32}$$

Repeated use of (4.3.9) and the formula for the comultiplication gives

$$y_m^*(x_{j_1}x_{j_2}\cdots x_{j_s}) = \sum_{i_1+\cdots+i_s=m} y_{i_1}^*(x_{j_1})y_{i_2}^*(x_{j_2})\cdots y_{i_s}^*(x_{j_s}). \tag{4.3.33}$$

Now, $\kappa \geq_{lex} \lambda^d$ implies that $\kappa_1 \geq t = \lambda_1^d$ the number of factors of x_λ. So if $\kappa_1 > t$, $y_{\kappa_1}^*(x_\lambda) = 0$ by (4.3.33) and (4.3.32) and if $\kappa_1 = t$, $y_{\kappa_1}^*(x'_\lambda) = x_\lambda$, with $\lambda' = (\lambda_1-1, \lambda_2-1, \ldots, \lambda_t-1)$ (with trailing zeros omitted), and then $(\kappa_2, \ldots, \kappa_s) = (\lambda')^d$, so that induction (with respect to the length of κ) finishes the proof of (4.3.32).

Thus the matrix of inner products $(\langle y_\kappa, x_\lambda\rangle), \mathrm{wt}(\kappa) = \mathrm{wt}(\lambda) = n$ is invertible. From (4.3.22) it follows that the y_κ are integer linear combinations of the x_λ of equal weight and vice versa, so it follows that the matrix of inner products

$$T = (\langle x_\kappa, x_\lambda\rangle),\ \mathrm{wt}(\kappa) = \mathrm{wt}(\lambda) = n \text{ is invertible.} \tag{4.3.34}$$

This suffices of course to show that the $x_\lambda, \lambda \in \mathit{Part}$ are independent. It remains to show that they span everything.

To this end let x be a preferred basis element of degree n. Let V be the $\mathbf{Q}$-vector space spanned by the x_λ, $\mathrm{wt}(\lambda) = n$, inside $H_n \otimes \mathbf{Q}$. Write x as an orthogonal sum $x = y + z$, $y \in V, \langle y, z\rangle = 0$. This can be done over $\mathbf{Q}$ (but, a priori, not necessarily over the integers). Let

$$y = \sum_{\mathrm{wt}(\lambda)=n} b_\lambda x_\lambda.$$

Then, of course

$$(b_\lambda)_\lambda = T^{-1}((\langle x, x_\lambda\rangle)_\lambda)$$

which, as T is invertible over the integers, means that all the b_λ are integral. Thus x is a length one vector which is the sum of two integer vectors y and z. This means that either $y = 0$ or $z = 0$. The first case is not possible because by proposition (4.3.10)

$$\langle x, p^n\rangle = \langle x, x_{\underbrace{[1, \ldots, 1]}_{n}}\rangle = 0$$

for every preferred basis element x. This finishes the proof of theorem 4.3.21.

Theorem 4.3.35. (Part 3 of the Zelevinsky theorem). *Let H be a PSH algebra with one primitive preferred basis element. Let x_n and y_n be as before. For every n there is a unique primitive z_n of degree n such that $\langle x_n, z_n\rangle = 1$. Moreover $\langle z_n, z_n\rangle = n$.*

Proof. Define, inductively, $z_1 = x_1$ and

$$z_n = nx_n - (x_{n-1}z_1 + x_{n-2}z_2 + \cdots + x_1z_{n-1}). \tag{4.3.36}$$

Then, as in the case of the Hopf algebra of symmetric functions, the z_n are primitives. Also, with induction,

$$x_n^*(x_{n-i}z_i) = \sum_{j=0}^{n} x_j^*(x_{n-i})x_{n-j}^*(z_i) = x_{n-i}^*(x_{n-i})x_i^*(z_i) = 1$$

and so, applying this to (4.3.36), we get $\langle x_n, z_n\rangle = x_n^*(z_n) = 1$. Further because $H = \mathrm{Prim}(H) \oplus I^2$, and the x_λ, $\lambda \in \mathit{Part}$ are a basis, an element z of degree n is primitive if and only if it is orthogonal to all x_λ except possibly x_n. That makes the group of primitive elements of degree n of rank 1 and as $\langle z_n, x_n\rangle = 1$ they are all integer multiples of z_n. Finally, for $i \geq 1$, $z_n^*(x_{n-i}z_i) = (z_n^* x_{n-i})z_i + x_{n-i}(z_n^* z_i) = 0$

for degree reasons using that z_n is primitive. So applying z_n^* to (4.3.36) gives $\langle z_n, z_n\rangle = n$.

For a partition λ let $z_\lambda = z_{\lambda_1}\cdots z_{\lambda_t}$. Write $\lambda = (1^{m_1}, 2^{m_2}, \dots, n^{m_n})$ to indicate that there are m_i, entries (parts) of λ equal to i. For example, with natural ink-saving conventions,

$$(6,4,3,3,1,1) = (1^2, 2^0, 3^2, 4^1, 5^0, 6^1) = (1^2, 3^2, 4, 6).$$

Corollary 4.3.37. (Inner product of products of primitives).

$$\langle z_\kappa, z_\lambda\rangle = \begin{cases} 1^{m_1} m_1! 2^{m_2} m_2! \cdots n^{m_n} m_n! & \text{if } \kappa = \lambda \\ 0 & \text{otherwise.} \end{cases} \tag{4.3.38}$$

Proof. As z_j^* is a derivation and $\langle z_i, z_j\rangle = 0$ if $j \neq i$,

$$z_j^* z_\lambda = n_j z_{\lambda'}$$

where $n_j = 0$ if z_j does not occur in z_λ and $n_j = jm_j$ if z_j does occur in z_λ and then $z_{\lambda'}$ is z_λ with one factor z_j removed.

Construction 4.3.39. Bernstein morphism. Let H be any graded commutative and associative Hopf algebra. Let

$$\mu_n(x) = \sum_i x_{i,1} \otimes x_{i,2} \otimes \cdots \otimes x_{i,n}$$

be the n-fold comultiplication written as a sum of tensor products of homogenous components. Now define

$$\beta_n : H \longrightarrow H[\xi_1, \xi_2, \dots] = H \otimes \mathbf{Z}[\xi_1, \xi_2, \dots]$$

by

$$\beta_n(x) = \sum_i x_{i,1} x_{i,2} \cdots x_{i,n} \xi_1^{\deg(x_{i,1})} \xi_2^{\deg(x_{i,2})} \cdots \xi_n^{\deg(x_{i,n})}. \tag{4.3.40}$$

Because H is coassociative and cocommutative this is symmetric in the variables $\xi_1, \dots, \xi_n$ so that there is an induced algebra morphism [10]

$$\beta_n : H \longrightarrow H \otimes \mathbf{Z}[h_1, \dots, h_n]$$

and because H is graded this stabilizes in n giving the **Bernstein morphism** [11]

$$\beta : H \longrightarrow H \otimes \mathbf{Symm}. \tag{4.3.41}$$

Theorem 4.3.42. *Let H be a PSH algebra and let x_n be as before. Then*

$$\beta(x_n) = \sum_{\substack{j_1+\cdots+j_s=n \\ j_1 \geq \cdots \geq j_s}} x_{j_1} x_{j_2} \cdots x_{j_s} m_{(j_1,\dots,j_s)}(\xi) \tag{4.3.43}$$

[10] The fact that this is a morphism of algebras uses commutativity; otherwise multiplication is not an algebra morphism

[11] In the case H is the Hopf algebra of symmetric functions, the Bernstein morphism will turn out to be the second comultiplication on **Symm** which makes **Symm** a coring object in the category of algebras and which defines the multiplication on the big Witt vectors, see below. For a general graded commutative Hopf algebra the Bernstein morphism defines an coaction of **Symm** on H and then by duality also an action of **Symm** on H^{gr*}

where the $m_{(j_1,\dots,j_s)}(\xi)$ *are the monomial symmetric functions in the* ξ.

This is immediate from the definition.

Construction 4.3.44. The multiplicative counit morphism. Let H be a *PSH* algebra and let the x_n be as before. Define a ring morphism

$$\varepsilon_{\Pi} : H \longrightarrow \mathbf{Z}, \quad x_n \mapsto 1 \tag{4.3.45}$$

or, equivalently, by requiring that ε_{Π} be equal to x_n^* on each H_n.

Theorem 4.3.46. (Fourth part of the Zelevinsky theorem.) *Let H be a PSH algebra with one primitive preferred basis element, then the composition of the Bernstein morphism with ε_{Π} (strictly speaking $\varepsilon_{\Pi} \otimes \mathrm{id}$) followed by the canonical identification $\mathbf{Z} \otimes_{\mathbf{Z}} \mathbf{Symm} \cong \mathbf{Symm}$) is an isomorphism of Hopf algebras.*

Proof. It follows immediately from (4.3.44) that the composed morphism takes x_n into h_n and the theorem follows because it is an algebra morphism and because the comultiplication laws on x_n and h_n correspond.

Of course, it was already clear from the structure theory so far developed that a *PSH* algebra with one basis primitive had to be isomorphic to **Symm** as a Hopf algebra. This is a particularly nice way of writing down the isomorphism.

Also this means that if there exists a *PSH* algebra with one basis primitive than **Symm** is a *PSH* algebra. The preferred basis elements will turn out to be the Schur functions. This will be dealt with in section 5.4 below.

In case $H = \mathbf{Symm}$ the Bernstein morphism is the second comultiplication, see 4.5 below, and ε_{Π} is the counit for this (the multiplicative counit). So the composite has to be the identity as is indeed the case.

4.4. Automorphisms of Symm

Consider homogeneous automorphisms of the Hopf algebra **Symm** over the integers. There are very few of them, just four, and they form the Klein four group. This fact was discovered surprisingly late. It is due to A. Liulevicius, [6].

Theorem 4.4.1. The Liulevicius theorem. *The only homogeneous Hopf algebra automorphisms of the graded Hopf algebra* **Symm** *are the identity, the antipode, the automorphism given by $t(h_n) = e_n$ (and, hence $t(e_n) = h_n$ by the Wronski relations) and the composite of t and the antipode. They form the Klein four group.*

Proof. First of all t as defined and the antipode ι are indeed different automorphisms of Hopf algebras and they are different from the identity. Indeed on the power sum primitives they are given by

$$\iota(p_n) = -p_n, \quad t(p_n) = (-1)^{n+1}p_n \tag{4.4.2}$$

and it follows that id, ι, t, $\iota t = t\iota$ form the Klein four group $\mathbf{Z}/(2) \otimes \mathbf{Z}/(2)$. Now let φ be any homogeneous automorphism of **Symm**. It must take primitives into primitives of the same degree and so

$$\varphi(p_n) = a_n p_n \tag{4.4.3}$$

for certain integers a_n. Moreover, because φ is an automorphism (so that an inverse exists) it must be the case that $a_n \in \{1, -1\}$ for all n. Compose φ with ι or t or both, if necessary, to get an automorphism with $a_1 = 1 = a_2$. Now look at the Newton relations

$$p_n = nh_n - (p_1 h_{n-1} + \ldots + p_{n-1} h_1)$$

and suppose with induction it has already been shown that $a_1 = a_2 = \ldots = a_{n-1} = 1$ so that $\varphi(p_i) = p_i$, $\varphi(h_i) = h_i$, $i = 1, \ldots, n-1$. Then $\varphi(p_n) = -p_n$ would give

$$n\varphi(h_n) = -2p_n + nh_n$$

making $2p_n$ (as a polynomial in the h_i) divisible by $n \geq 3$ which is impossible because $p_n \equiv h_1^n \mod (h_2, \ldots, h_n)$.

Note that this is very much a theorem over the integers. Over the rationals there are very many homogeneous automorphisms of the Hopf algebra **Symm**. They are given by (4.4.3) where the only requirement is that the a_n are nonzero.

The automorphism t interchanges the elementary symmetric functions and the complete symmetric functions. In section 4.1 above the complete symmetric functions were used in preference [12]. They simply work out better for the Witt vectors and also in relation with the noncommutative symmetric functions **NSymm** and the quasisymmetric functions **QSymm** which will be studied to some extent in chapter 6.

Applying t to the various formulas in section 4.1 gives various useful additional formulas.

4.5. The functor of the Witt vectors

The functor of the big Witt vectors, $A \mapsto W(A)$, takes commutative rings into commutative rings. And there is very much more extra structure floating around. The functor is represented by **Symm** and so in a way the theory of the big Witt vectors is just a part of the theory of the symmetric functions. Even this part alone of the theory of **Symm** could easily fill a whole monograph. [13]

It is probably impossible to overestimate the importance of the big Witt vector construction. The functor has several different universality properties which makes it important in several different parts of algebra.

The p-adic Witt vectors, W_{p^∞}, are a certain quotient functor, and were studied first. They satisfy $W_{p^\infty}(\mathbf{F}_p) = \mathbf{Z}_p$, where $\mathbf{F}_p$ is the field of p elements and $\mathbf{Z}_p$ is the discrete valuation ring of the p-adic integers. This construction (and generalizations to arbitrary finite fields) is due to Witt, [10], whence the name, and the importance of the Witt vectors in algebraic number theory.

One striking property of the Witt vector functor is the structure on it of a comonad (cotriple) structure $W(-) \longrightarrow W(W(-))$. In the p-adic quotient case this is the so called Artin-Hasse exponential of great importance in algebraic number theory, especially class field theory.

---

[12] Usually the elementary symmetric functions are preferred. But in this setting it is definitely the complete symmetric functions which are better to work with.

[13] And, in fact such a monograph is in the process of being (slowly) written; meanwhile there is the partial survey, [**5**].

Construction 4.5.1. The functorial ring structure on $W(A)$. Define the functor of the big Witt vectors by

$$W(A) = \mathbf{Alg}(\mathbf{Symm}, A) \tag{4.5.2}$$

the set of all algebra morphisms of **Symm** to an algebra A. The comultiplication and counit on the Hopf algebra **Symm**

$$\mu_\Sigma : \mathbf{Symm} \longrightarrow \mathbf{Symm} \otimes \mathbf{Symm}, \quad \varepsilon_\Sigma : \mathbf{Symm} \longrightarrow \mathbf{Z} \tag{4.5.3}$$

then give a functorial Abelian group structure on $W(A)$ which is also easily described directly.

Indeed an algebra morphism $\mathbf{Symm} \longrightarrow A$ is the same thing as specifying an infinite sequence of elements $a_1, a_2, \ldots$ of A, the images under the morphism of the free polynomial generators $h_1, h_2, \ldots$ of **Symm**. And in turn such a sequence can be encoded as a power series with constant term 1:

$$a(t) = 1 + a_1 t + a_2 t^2 + \cdots \in 1 + tA[[t]] \tag{4.5.4}$$

and in this interpretation, obviously, the Abelian group structure is that of multiplication of power series.

Now let $a(t), b(t) \in W(A) = 1 + tA[[t]]$ be two such power series and formally write

$$a(t) = 1 + a_1 t + a_2 t^2 + \cdots = \prod_i (1 - x_i t)^{-1},$$

$$b(t) = 1 + b_1 t + b_2 t^2 + \cdots = \prod_j (1 - y_j t)^{-1}$$

so that the a_i, b_i are seen as the complete symmetric functions in the $x_1, x_2, \cdots$ and $y_1, y_2, \cdots$. Now consider the expression

$$\prod_{i,j} (1 - x_i y_j t)^{-1} = \sum_{\lambda \in Part} s_\lambda(x) s_\lambda(y) t^{\mathrm{wt}(\lambda)}. \tag{4.5.5}$$

This is symmetric in the x's and y's and thus there are polynomials

$$\pi_1(a_1; b_1),\ \pi_2(a_1, a_2;\ b_1, b_2), \ldots \tag{4.5.6}$$

such that

$$\prod_{i,j} (1 - x_i y_j t)^{-1} = 1 + \pi_1(a_1; b_1) t + \pi_2(a_1, a_2;\ b_1, b_2) t^2 + \cdots \tag{4.5.7}$$

and this defines a (functorial) multiplication on $W(A)$ by setting $a(t) * b(t) =$ right hand side of (4.5.7). There is a unit, viz. the power series

$$1 + t + t^2 + \cdots = (1 - t)^{-1}. \tag{4.5.8}$$

The polynomials $\pi_n(a_1, \ldots, a_n; b_1, \ldots, b_n)$ are easily written down in terms of the Schur functions. Indeed, by (4.5.5)

$$\pi_n(a_1, \ldots, a_n; b_1, \ldots, b_n) = \sum_{\mathrm{wt}(\lambda) = n} s_\lambda(a_1, \ldots, a_n) s_\lambda(b_1, \ldots, b_n) \tag{4.5.9}$$

where the s_λ are the Schur functions written as polynomials in the complete symmetric functions. The first two, π_1, π_2, are

$$\pi_1(a_1;\ b_1) = a_1 b_1, \quad \pi_2(a_1, a_2;\ b_1, b_2) = a_1^2 b_1^2 - a_2 b_1^2 - a_1^2 b_2 + 2a_2 b_2. \tag{4.5.10}$$

The multiplication $'*'$ on $W(A)$ is distributive over the 'addition', i.e.

$$a(t) * (b(t)c(t)) = (a(t) * b(t))(a(t) * c(t)).$$

It really suffices (because of functoriality) [14] to prove this for power series of the form $a(t) = (1 - xt)^{-1}$, $b(t) = (1 - yt)^{-1}$, $c(t) = (1 - zt)^{-1}$. And in this case by the definitions

$$\begin{aligned} a(t) * (b(t)c(t)) &= (1 - xt)^{-1} * ((1 - yt)^{-1}(1 - zt)^{-1}) = \\ &= (1 - xyt)^{-1}(1 - xzt)^{-1} = \\ &= ((1 - xt)^{-1} * (1 - yt)^{-1})((1 - xt)^{-1} * (1 - zt)^{-1}) = \\ &= (a(t) * b(t))(a(t) * c(t)). \end{aligned}$$

Not that the general case is much more difficult. Still another proof of distributivity will be given in the next section.

In terms of structure on **Symm** the multiplication and multiplicative unit of the Witt vector functor are given by a second comultiplication and corresponding multiplicative counit, both algebra morphisms,

$$\begin{aligned} &\mu_\Pi : \mathbf{Symm} \longrightarrow \mathbf{Symm} \otimes \mathbf{Symm}, \\ &h_n \mapsto \sum_{wt(\lambda)=n} s_\lambda(1 \otimes x)s_\lambda(x \otimes 1) \\ &\varepsilon_\Pi : \mathbf{Symm} \longrightarrow \mathbf{Z},\ h_1 \mapsto 1. \end{aligned} \tag{4.5.11}$$

The Bernstein morphism [15] of subsection 4.3.40 is precisely this second comultiplication morphism, and the multiplicative counit also occurs there.

The second comultiplication μ_Π is distributive over the first one in the Hopf algebra sense, which means that the following diagram is commutative.

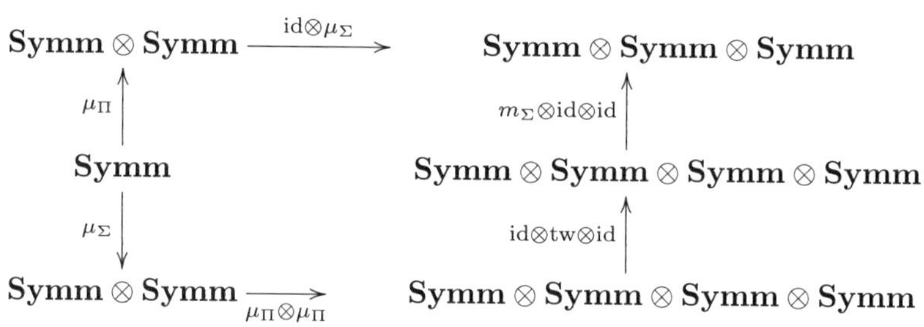

This is the same as saying that the functorial multiplication on $W(-)$ is distributive over the functorial addition (on the right). There is a similar diagram for distributivity on the left. One of the (natural) noncommutative generalizations of the Witt vectors is in fact distributive on only one side and not on the other.

4.5.12. Characterization of the second comultiplication. On the primitives of **Symm**, more generally on the basis of the $\{p_\lambda\}$, $\lambda \in Part$, of $\mathbf{Symm}_\mathbf{Q}$ the second comultiplication is given by

---

[14] This is an instance of what in algebraic topology, especially topological K-theory, is called the splitting principle. For functorial identities it suffices to verify things for line bundles.

[15] In **[15]**, Zelevinsky writes on page 85 that the Bernstein morphism makes sense for any graded commutative Hopf algebra and "Its meaning is not yet well understood." Certainly the fact that for the Hopf algebra **Symm** it is the algebra morphism that defines the Witt vector multiplication goes some way. In general there is an action (or coaction) involved. Moreover, in the noncommutative case something similar happens.

$$\mu_\Pi(p_\lambda) = p_\lambda \otimes p_\lambda. \tag{4.5.13}$$

This is an immediate consequence of the fact that the functorial morphisms $s_n : W(A) \longrightarrow A$, which are given by the p_n, see just below in subsection 4.6.2, turn the Witt vector multiplication into component wise multiplication on $A^{\mathbf{N}}$.

4.6. Ghost components

Symm is a coring object in the category of rings, which is the same as saying that $\mathbf{Alg}(\mathbf{Symm}, -) = W(-)$ is a ring valued functor on the category of rings. There do not seem to be all that many examples and perhaps Kaplansky's remark on Hopf algebras should be modified to say that each of these deserves the most intensive study. There is the functor of Witt vectors and its many quotients (of which the affine line is one), there are the ring valued functors that assign to a ring the ring of $n \times n$ matrices over that ring and there are also quite a few more examples from algebraic topology such as (co)homology of classifying spaces and algebras of operations on cohomology theories [16]. One striking property of all these examples is that their underlying rings are remarkably regular and uncomplicated. This is a so far little understood phenomenon: highly structured objects tend to be remarkably nice. [17]

Construction 4.6.1. The affine line. As a ring, i.e. $\mathbf{Z}$-algebra, this is the ring of polynomials in one variable $\mathbf{A}^1 = \mathbf{Z}[t]$. It has an (additive) comultiplication $\mu_\Sigma(t) = 1 \otimes t + t \otimes 1$ with corresponding (additive) counit $\varepsilon_\Sigma(t) = 0$. There is also a second comultiplication with corresponding (multiplicative) counit, $\mu_\Pi(t) = t \otimes t$, $\varepsilon_\Pi(t) = 1$. The second comultiplication is distributive over the first. All in all this is an unnecessarily complicated way of describing the ring valued functor $A \mapsto \mathbf{A}^1(A) = \mathbf{Alg}(\mathbf{Z}[t], A) = A$.

Construction 4.6.2. Ghost components (version 1). There are a number of important morphisms of ring valued functors

$$\begin{aligned} &s_n : W(-) \longrightarrow \mathbf{A}^1(-), \\ &a(t) = 1 + a_1 t + a_2 t^2 + \cdots \mapsto \text{coefficient of } t^n \text{ in } t\frac{d}{dt}\log(a(t)). \end{aligned}$$

Writing $a(t) = \prod_i (1 - x_i t)^{-1}$, $b(t) = \prod_j (1 - y_j t)^{-1}$ one finds

$$\begin{aligned} t\tfrac{d}{dt}\log(a(t)) &= -t\tfrac{d}{dt}\sum_i \log(1 - x_i t) \\ &= -t\tfrac{d}{dt}\sum_i (-x_i t - \tfrac{x_i t^2}{2} - \tfrac{x_i t^3}{3} - \cdots) \\ &= p_1(x)t + p_2(x)t^2 + p_3(x)t^3 + \cdots \end{aligned} \tag{4.6.3}$$

---

[16] Landweber-Novikov algebras, Steenrod algebras. These ring objects in categories of coalgebras are termed Hopf rings in algebraic topology. See [**13**] and [**12**] for some information on these things. The term 'Hopf ring' as compared to 'Hopf algebra' clashes with the distinction between 'ring' and 'algebra' in ring theory.

[17] And objects which have some universality property tend to have lots of other nice properties as well. See [**6**].

and thus the functorial ring morphisms s_n are described by the expressions for the power sums in terms of the complete symmetric functions. Also

$$s_n(a(t)b(t)) = s_n(a(t)) + s_n(b(t)) \tag{4.6.4}$$

and for the Witt vector product one finds

$$t\frac{d}{dt}\log(a(t) * b(t)) = t\frac{d}{dt}\sum_{i,j}\log(1 - x_i y_j t)^{-1}$$
$$= \sum_{i,j}(x_i y_j t + x_i^2 y_j^2 t^2 + \cdots)$$

of which the coefficient of t^n is $\sum_{i,j} x_i^n y_j^n = p_n(x)p_n(y)$, and thus

$$s_n(a(t) * b(t)) = s_n(a(t))s_n(b(t)) \tag{4.6.5}$$

so that the s_n are indeed functorial ring morphisms. (It is easy to check that things work out fine for the zero and unit Witt vectors.)

At the level of the representing objects of $W(-)$ and $\mathbf{A}^1$, that means **Symm** and $\mathbf{Z}[t]$, the s_n are given by the coring object algebra morphisms

$$t \mapsto p_n$$

(and so it would be natural, but perhaps confusing, to denote these functorial ring morphisms also by p_n).

It also follows immediately from (4.6.5) that the second comultiplication μ_Π on **Symm** is characterized by

$$\mu_\Pi(p_n) = p_n \otimes p_n. \tag{4.6.6}$$

4.6.7. Construction. Ghost components (second version). Most discussion in the published literature on the (big) Witt vectors involve the famous Witt polynomials. These are

$$w_n(u_1, u_2, \ldots, u_n) = \sum_{d|n} d u_d^{n/d}. \tag{4.6.8}$$

The first few Witt polynomials are

$$w_1 = u_1,\ w_2 = u_1^2 + 2u_2,\ w_3 = u_1^3 + 3u_3,\ w_4 = u_1^4 + 2u_2^2 + 4u_4$$
$$w_5 = u_1^5 + 5u_5,\ w_6 = u_1^6 + 2u_2^3 + 3u_3^2 + 6u_6.$$

These enter the picture if a different 'coordinatization' of $W(-)$ is used. As follows. Every element of $W(A) = 1 + tA[[t]]$ can be uniquely written in the form

$$a(t) = 1 + a_1 t + a_2 t^2 + \cdots = (1 - u_1 t)^{-1}(1 - u_2 t^2)^{-1}(1 - u_3 t^3)^{-1} \cdots$$

and vice versa. A simple calculation now shows that in these coordinates

$$s_n(a(t)) = w_n(u_1, \ldots, u_n). \tag{4.6.9}$$

There are often advantages in using these coordinates. Mainly because the formulas for the w_n in terms of the u_n are nice closed formulas and in any case a good deal simpler than the formulas for the p_n in terms of the h_n.

The 'miracle' of the Witt polynomials is now that if polynomials

$$\Sigma_n(u_1, \ldots, u_n; v_1, \ldots, v_n) \text{ and } \Pi_n(u_1, \ldots, u_n; v_1, \ldots, v_n)$$

are defined by requiring that

$$w_n(\Sigma_1,\dots,\Sigma_n)=w_n(u)+w_n(v),\quad w_n(\Pi_1,\dots,\Pi_n)=w_n(u)w_n(v)$$

then these polynomials have integer coefficients. Of course, this 'miracle' is proved by the construction used above. It is also not difficult to prove these integrality statements directly, see [**2**], section 17.1. These polynomials Σ_n, Π_n are then used to define addition and multiplication of Witt vectors. Unit, additive inverse, and zero element can be dealt with in the same way.

4.6.10. Proof technique. The art of using ghost components [18]. Because of functoriality (mainly), when working with the Witt vectors and needing to prove some identities, properties, or characterizations, one often can assume that the ring A is an integral domain or even a field of characteristic zero. And then things often become very easy because in that case

$$s: W(A)\longrightarrow A^{\mathbf{N}},\ a(t)\mapsto (s_1(a(t)),s_2(a(t)),s_3(a(t)),\dots) \tag{4.6.11}$$

is a ring isomorphism (because over the rationals the h_n are polynomials in the p_n).

Here is an example of how this works. Suppose that we are trying to prove the distributivity of the Witt vector multiplication over the Witt vector addition. So takc

$$\begin{aligned} a&=1+a_1t+a_2t^2+\cdots,\\ b&=1+b_1t+b_2t^2+\cdots,\\ c&=1+c_1t+c_2t^2+\cdots \end{aligned}$$

in $W(A)$. Let $\widetilde{A}$ be a characteristic zero integral domain together with a surjective morphism of rings $\pi:\widetilde{A}\longrightarrow A$ and let $\widetilde{a}$, $\widetilde{a}_i$, etc. be lifts of a, a_i, etc. Because of functoriality there is the following commutative diagram of rings, where $Q(\widetilde{A})$ is the quotient field of $\widetilde{A}$

$$\begin{array}{ccccc} W(A) & \xleftarrow{\ \pi\ } & W(\widetilde{A}) & \xrightarrow{\ \subset\ } & W(Q(\widetilde{A}))\\ \downarrow{\scriptstyle s} & & \downarrow{\scriptstyle s} & & \downarrow{\scriptstyle s}\\ A^{\mathbf{N}} & \xleftarrow{\ \pi\ } & \widetilde{A}^{\mathbf{N}} & \xrightarrow{\ \subset\ } & Q(\widetilde{A})^{\mathbf{N}} \end{array}$$

So, $a*(bc)=(a*b)(a*c)$ holds in $W(A)$ if $a*(bc)=(a*b)(a*c)$ holds in $W(\widetilde{A})$ and that is the case if and only if this holds in $W(Q(\widetilde{A}))$ because the right top arrow is an injective ring morphism. Finally the twiddle distributive identity holds in $W(Q(\widetilde{A}))$ because the right most s is an isomorphism of rings and on $Q(\widetilde{A})^{\mathbf{N}}$ addition and multiplication are component-wise so that distributivity holds there.

4.7. Frobenius and Verschiebung endomorphisms

There are two famous families of Hopf algebra endomorphism of **Symm** which hence give rise to two families of group endomorphisms of the functor of the big Witt

[18] This is far from unrelated to the technique of using formal variables as in e.g. the definition of the Witt vector multiplication; nor is this unrelated to the 'splitting principle' of footnote 14 above.

vectors. They are called the Frobenius and Verschiebung endomorphisms. Actually the Frobenius endomorphism are also compatible with the second comultiplication and hence define ring endomorphisms of $W(-)$.

Construction 4.7.1. The Verschiebung endomorphisms. These are defined functorially by

$$\mathbf{v}_m(a(t)) = a(t^m),\ a(t) \in W(A) \tag{4.7.2}$$

On the level of the representing object, **Symm**, this works out as

$$\mathbf{v}_m(h_n) = \begin{cases} 0 & \text{if } m \text{ does not divide } n \\ h_{n/m} & \text{if } m \text{ does divide } n. \end{cases} \tag{4.7.3}$$

Technically, of course, one should not use the same symbol for the functorial morphism on $W(A)$ induced by an endomorphism of the representing object **Symm**. If $\mathbf{v}_m$: **Symm** $\longrightarrow$ **Symm** is defined by (4.7.3) then the functorial endomorphism (4.7.2) should be written

$$W(\mathbf{v}_m) = \mathbf{Alg}(\mathbf{v}_m, A).$$

It is immediate from (4.7.2) that the functorial Verschiebung morphism respects the Witt vector addition which is the same as saying that the morphism defined by (4.7.3) is a Hopf algebra morphism, which is also immediate. On the power sums $\mathbf{v}_m$ works out as

$$\mathbf{v}_m : \mathbf{Symm} \longrightarrow \mathbf{Symm},\ p_n \mapsto \begin{cases} 0 & \text{if } m \text{ does not divide } n \\ mp_{n/m} & \text{if } m \text{ does divide } n. \end{cases} \tag{4.7.4}$$

Construction 4.7.5. The Frobenius endomorphisms. Write

$$a(t) = \prod_i (1 - x_i t)^{-1}. \tag{4.7.6}$$

Then

$$\mathbf{f}_m(a(t)) = \prod_i (1 - x_i^m t)^{-1} \tag{4.7.7}$$

Note that the right hand side of (4.7.7) is symmetric in the x_i. So there are some polynomials $F_{m,n}(h_1, h_2, \dots, h_{nm})$ such that

$$\prod_i (1 - x_i^m t)^{-1} = 1 + F_{m,1}(h_1, \dots, h_{mn})t + F_{m,2}(h_1, \dots, h_{mn})t^2 + \cdots \tag{4.7.8}$$

and then the technically correct definition of the Frobenius endomorphisms is

$$\mathbf{f}_m(1 + a_1 t + a_2 t^2 + \cdots) =$$
$$= 1 + F_{m,1}(a_1, \dots, a_m)t + F_{m,2}(a_1, \dots, a_{2m})t^2 + \cdots \tag{4.7.9}$$

On the level of the representing object, **Symm**, the Frobenius endomorphisms are determined by

$$\mathbf{f}_m : \mathbf{Symm} \longrightarrow \mathbf{Symm},\ h_n \mapsto F_{m,n}(h_1, \dots, h_{mn}). \tag{4.7.10}$$

It is not so easy to write down nice formulas for the $F_{m,n}$ except for $F_{m,1} = p_m$.

Note that applying the ghost component formula to (4.7.7) gives

$$s_n(\mathbf{f}_m(a(t)) = s_{mn}(a(t)) \tag{4.7.11}$$

and because the s_n are given by the power sums this means that on the representing object **Symm** the Frobenius endomorphisms are given by

$$\mathbf{f}_m : \mathbf{Symm} \longrightarrow \mathbf{Symm},\ p_n \mapsto p_{mn}. \tag{4.7.12}$$

Both (4.7.11) and (4.7.12) are perfectly good characterizations of the $\mathbf{f}_m$ provided it is known in some other way that they are well defined over the integers.

It follows immediately from (4.7.12) (or (4.7.11)) (using the ghost component technique) that (4.7.12) defines a Hopf algebra endomorphism that is also second comultiplication preserving (see (4.6.6)) so that it defines functorial ring endomorphisms of the Witt vectors.

If one is in a situation where it makes sense to talk about m-th roots of unity, (4.7.7) can be written

$$\mathbf{f}_m(a(t)) = \prod_{i=1}^{m} a(\zeta_m^i t^{1/m}) \tag{4.7.13}$$

where ζ_m is a primitive m-th root of unity (which sees to it that (4.7.13) is actually a power series in t).

Theorem 4.7.14. (Relations between the Frobenius and Verschiebung morphisms). *The following relations hold between the Frobenius and Verschiebung endomorphisms of the Hopf algebra* **Symm**.

$$\mathbf{f}_m\mathbf{f}_n = \mathbf{f}_{mn} \tag{4.7.15}$$

$$\mathbf{v}_m\mathbf{v}_n = \mathbf{v}_{mn} \tag{4.7.16}$$

$$\text{if } \gcd(m,n) = 1 \quad \mathbf{f}_m\mathbf{v}_n = \mathbf{v}_n\mathbf{f}_m \tag{4.7.17}$$

$$\mathbf{f}_1 = \mathbf{v}_1 = \mathrm{id} \tag{4.7.18}$$

$$\mathbf{v}_m\mathbf{f}_m = d_m \tag{4.7.19}$$

where d_i is the composition of the i-fold comultiplication μ_i : $\mathbf{Symm} \longrightarrow \mathbf{Symm}^{\otimes i}$ *with the i-fold multiplication m_i :* $\mathbf{Symm}^{\otimes i} \longrightarrow \mathbf{Symm}$ *which at the functorial level amounts to adding a Witt vector (under Witt addition, i.e. multiplication of power series) to itself i times.*

Proof. All of this is really immediate from the characterizations of what the Frobenius and Verschiebung morphisms do to the power sums in **Symm**.

Theorem 4.7.20. (Congruence property of the Frobenius morphisms). *In the commutative ring $W(A)$, for each prime number $p, \mathbf{f}_p$ is congruent modulo p to taking $p-th$ powers. That is, using $*$ to denote Witt multiplication, for each $a(t) \in W(A)$*

$$\mathbf{f}_p(a(t)) \equiv (a(t))^{*p} \mod p. \tag{4.7.21}$$

This, of course, takes place in the ring of Witt vectors. So in terms of power series (4.7.21) says that there is a power series $b(t)$ with constant term 1 such that, as power series

$$\mathbf{f}_p(a(t)) = (a(t))^{*p} b(t)^p.$$

Proof. Consider $\mathbf{Symm} \subset \mathbf{Z}[x_1, x_2, \ldots]$ and consider

$$(1 - x_i t)^{-1} \in W(\mathbf{Z}[x_1, x_2, \ldots]).$$

Then

$$\mathbf{f}_p(1-x_it)^{-1}=(1-x_i^pt)^{-1}=((1-x_it)^{-1})^{*p}.$$

In any commutative ring R, $(a_1+a_1+\cdots+a_n)^p\equiv a_1^p+a_2^p+\cdots+a_n^p \mod p$. It follows that for all finite n

$$\mathbf{f}_p(\prod_{i=1}^{n}(1-x_it)^{-1})\equiv(\prod_{i=1}^{n}(1-x_it)^{-1})^{*p} \mod p$$

and a standard passage to an infinite number of variables finishes the proof.

Theorem 4.7.22. (Duality of Frobenius and Verschiebung).

$$\langle \mathbf{f}_n a, b\rangle=\langle a,\mathbf{v}_n b\rangle \quad \text{for all } a,b\in \mathbf{Symm}. \tag{4.7.23}$$

Proof. As both Verschiebung and Frobenius are ring homomorphisms on **Symm**, and as everything is defined over the integers it suffices to verify this over the rationals. That is, it suffices to check this for $a=p_i$, $b=p_j$. In that case

$$\langle \mathbf{f}_n p_i,p_j\rangle=\langle p_{ni},p_j\rangle=\begin{cases}0, & \text{if } j\neq ni,\\ ni, & \text{if } j=ni.\end{cases}$$

On the other hand

$$\langle p_i,\mathbf{v}_n p_j\rangle=\begin{cases}0, & \text{if } n \text{ does not divide } j\\ \langle p_i, np_{j/n}\rangle, & \text{if } n \text{ divides } j \text{ and } i=j/n.\end{cases}$$

And so things fit.

There is something to get used to about this. Verschiebung is very easy to define, while Frobenius takes some thought and technique. Yet they are really the same up to duality.

The Frobenius morphisms are coring object morphisms of the algebra **Symm**; the Verschiebung morphism are ring object morphisms of the coalgebra **Symm**. With the definitions used here the two can be distinguished by saying that the Frobenius endomorphisms of **Symm** are degree increasing and the Verschiebung morphisms on **Symm** are degree decreasing[19].

4.8. The second multiplication of Symm

By duality there is a second multiplication on **Symm**, dual to the second comultiplication which makes **Alg(Symm**, −) a ring valued functor. This second multiplication, m_Π, is of course defined by the formula

$$\langle m_\Pi(a\otimes b,c)\rangle=\langle a\otimes b,\mu_\Pi(c)\rangle. \tag{4.8.1}$$

Theorem 4.8.2. (Description of the second multiplication on **Symm**).

$$m_\Pi(p_\kappa\otimes p_\lambda)=\begin{cases}||\lambda||p_\lambda, & \text{if } \kappa=\lambda\\ 0, & \text{otherwise.}\end{cases} \tag{4.8.3}$$

[19]It can be proved without much difficulty at this stage that the Frobenius morphisms are the only coring object endomorphisms of the coring object **Symm** in the category of algebras.

If a is homogeneous of degree m

$$\begin{aligned} m_\Pi(a \otimes h_n) &= a = m_\Pi(h_n \otimes a) \text{ if } \deg(a) = m = n \\ m_\Pi(a \otimes h_n) &= 0 \text{ if } \deg(a) \neq n. \end{aligned} \tag{4.8.4}$$

Proof. Formula (4.8.3) is an immediate consequence of the description of the second comultiplication in 4.5.12. Formula (4.8.4) then follows because

$$h_n = \sum_{\mathrm{wr}(\lambda)=n} ||\lambda||^{-1} p_\lambda.$$

(see (4.1.48)).

4.8.5. Colloquial description. Let $\mathbf{Symm}^{(n)}$ be the part of **Symm** of elements of degree n. Then each $\mathbf{Symm}^{(n)}$ is a ring under the usual addition and the second multiplication. The element h_n is the unit for this ring structure. Thus, ignoring the first multiplication, from this point of view **Symm** is the direct sum of the countably many rings $\mathbf{Symm}^{(n)}$.

4.8.6. Distributivity. The second multiplication is distributive over the first in the Hopf algebra sense. This simply means that the reversed arrow diagram just above (4.5.12) holds. In explicit formulas this says

$$m_\Pi(a \otimes bc) = \sum_i m_\Pi(a_{1,i} \otimes b) m_\Pi(a_{1,2} \otimes c)$$

where

$$\mu(a) = \sum_i a_{i,1} \otimes a_{i,2}.$$

The distributivity formula implies that that the description of the second multiplication by 4.8.2 is complete. For instance

$$\begin{aligned} &m_\Pi(p_1^2 \otimes p_1^2) = \\ &= m_\Pi(p_1^2 \otimes p_1) m_\Pi(1 \otimes p_1) + 2 m_\Pi(p_1 \otimes p_1) m_\Pi(p_1 \otimes p_1) + m_\Pi(1 \otimes p_1) m_\Pi(p_1^2 \otimes p_1) \\ &= 2p_1^2. \end{aligned}$$

More generally, in degree n all elements are sums of products with the exception of h_n and so the distributivity formula 4.8.6 takes care of everything once it is known that in degree n h_n acts as the unit.

Strictly speaking there is no unit for the second multiplication on all of **Symm**. It does exist as $\sum_{i=0}^{\infty} h_n$ but that is not something that lives in **Symm** itself, only in a certain completion.

4.9. Lambda algebras

A **lambda algebra**, λ-algebra, λ-ring, is a commutative ring A (with 1) with additional operations

$$\lambda^i : A \longrightarrow A \tag{4.9.1}$$

that behave much like exterior powers (of vector spaces, vector bundles, modules, representations, ...). They were first systematically studied by Grothendieck in his investigations concerning the Riemann-Roch theorem, [**1**]. The first requirements on the operations λ^i are

$$\lambda^0(a) = 1,\ \lambda^1 = \mathrm{id},\ \lambda^n(a+b) = \sum_{i+j=n} \lambda^i(a)\lambda^j(b). \tag{4.9.2}$$

The main part of (4.9.2) is conveniently formulated as the requirement that

$$a \mapsto \lambda_t(a) = 1 + at + \lambda^2(a)t^2 + \lambda^3(a)t^3 + \cdots, \quad A \longrightarrow W(A) \tag{4.9.3}$$

is a morphism of Abelian groups.

The other requirements have to do with the behavior of the 'exterior powers' λ^j on products and with how exterior powers compose.

A morphism of λ -rings $\varphi : A \longrightarrow B$ is a morphism of rings that commutes with the exterior product operations: i.e. $\varphi(\lambda^i_A(x)) = \lambda^i_B(\varphi(x))$. With this terminology a λ-ring is basically a ring with exterior product operations such that (4.9.3) is a morphism of λ-rings. This of course requires the definition of a λ-ring structure on $W(A)$. It is also not 100% correct because the way in which the multiplication on the Witt vectors is handled intervenes. [20]

It turns out to be convenient to introduce the corresponding 'symmetric product operations' σ^i

$$\sum_{i=0}^{n} (-1)^i \sigma^i \lambda^{n-i} = 0. \tag{4.9.4}$$ [21]

and

$$\sigma_t(x) = 1 + xt + \sigma^2(x)t^2 + \sigma^3(x)t^3 + \ldots \tag{4.9.5}$$

and the requirement will be that $\sigma_t : A \longrightarrow W(A)$ be a morphism of λ-rings (or, equivalently, σ-rings).

Construction 4.9.6. Exterior and symmetric product operations on the big Witt vectors. Let

$$a(t) = 1 + a_1t + a_2t^2 + \cdots = \prod_i (1 - x_it)^{-1} \in W(A).$$

Define

$$\begin{aligned} \lambda^m a(t) &= \textstyle\prod_{i_1 < i_2 < \cdots < i_m} (1 - x_{i_1}x_{i_2}\cdots x_{i_m}t)^{-1} \\ \sigma^m a(t) &= \textstyle\prod_{i_1 \le i_2 \le \cdots \le i_m} (1 - x_{i_1}x_{i_2}\cdots x_{i_m}t)^{-1}. \end{aligned} \tag{4.9.7}$$

Strictly speaking, of course, it needs to be remarked that the right hand sides of (4.9.7) are symmetric in the x's and hence there are polynomials ... . Also it should be shown that the relations (4.9.4) hold. (Exercise.)[22]

---

[20]There are basically four choices, corresponding to the four automorphisms of the Hopf algebra **Symm**. The four units for the four functorial ring structures are $(1 \pm t)^{\pm 1}$. Three of these occur in the literature, the exception being $(1+t)^{-1}$. In the context of lambda rings mostly the version that corresponds to the unit $1 + t$ (and hence the elementary symmetric functions) is used. And that is natural because in a very real sense the lambda operations correspond to the elementary symmetric functions. We have, however, for good reasons, elected to work with the complete symmetric functions, which basically means working with sigma-rings rather than lambda rings. It is all the same conceptually, but the formulae change a bit.

[21]Thus the exterior product operations relate to the symmetric product operations exactly as the elementary symmetric functions to the complete symmetric functions. This is far from an accident as will be seen later.

[22]This basically amounts to the Wronski relations again.

Definition 4.9.8. Lambda rings, sigma rings. A **lambda ring** [23], equivalently a **sigma ring** [24], is a commutative ring with unit element which comes equipped with exterior product operations $\lambda^i : A \longrightarrow A$, and, equivalently, symmetric product operations $\sigma^i : A \longrightarrow A$, related by (4.9.4) such that $\sigma_t : A \longrightarrow W(A)$ is a morphism of lambda rings (sigma rings). [25]

Example 4.9.9. Simplest example of a λ-ring. The simplest example of a λ-ring is probably the ring of integers with the operations

$$\lambda^n : x \mapsto \binom{x}{n}.$$

Note that in this case $\lambda_t(x) = (1+t)^x$. So λ_t is some kind of exponential and that is a good way of thinking about it.

Theorem 4.9.10. The Artin-Hasse exponential [26]. *The symmetric product operations* σ^i *of* (4.9.7) *define a functorial morphism of* λ*-rings* [27]

$$\begin{aligned} &AH : W(A) \longrightarrow W(W(A)), \\ &a(t) \longrightarrow 1 + (a(t))u + (\sigma^2 a(t))u^2 + (\sigma^3 a(t))u^3 + \cdots \end{aligned} \tag{4.9.11}$$

---

[23] In the older literature one finds the term 'special lambda ring' for what is here called a lambda ring. A ring with operations such that (4.9.2) holds was called a lambda ring. These objects are called pre-lambda rings in the more recent literature. There are many pre-lambda structures that are not lambda but, so far, they seem not to be important in any way.

[24] This notion of a sigma ring has nothing to do with what is called a sigma-algebra in measure theory and set theory.

[25] Because addition, multiplication, and lambda operations on $W(A)$ are given by universal polynomials, meaning independent of A, there are universal polynomials S_n, P_n, Q_{nm} such that for any λ-ring and elements x, y in it $\lambda^n(x+y) = S_n(\lambda^1(x), \lambda^2(x), \ldots; \lambda^1(y), \lambda^2(y), \ldots)$, $\lambda^n(xy) = P_n(\lambda^1(x), \lambda^2(x), \ldots; \lambda^1(y), \lambda^2(y), \ldots)$, $\lambda^m(\lambda^n(x)) = Q_{mn}(\lambda^1(x), \lambda^2(x), \ldots)$. The polynomial S_n is of course $X_n + X_{n-1}Y_1 + \ldots X_{n-2}Y_2 + \ldots + X_1Y_{n-1} + Y_n$

[26] The Artin Hasse exponential first appeared in number theory, more precisely class field theory, as a map $W_{p^*}(\mathbf{F}_p) = \mathbf{Z}_p \longrightarrow 1 + t\mathbf{Z}_p[[t]] = W(\mathbf{Z}_p)$ where $\mathbf{Z}_p$ is the ring of p-adic integers.

[27] Consider a category C and an endofunctor T of C that comes together with functorial morphisms $T(-) \xrightarrow{\mu} T(T-))$ and $T(-) \xrightarrow{\varepsilon} (-)$ such that μ is coassociative and that ε is a counit. These statements mean that the following diagrams are commutative

$$\begin{array}{ccc} T(-) & \xrightarrow{\mu} & T(T(-)) \\ \downarrow \mu & & \downarrow T(\mu) \\ T(T-)) & \xrightarrow{\mu_{T(-)}} & T(T(T(-))) \end{array} , \quad \begin{array}{ccc} T(-) & \xrightarrow{\mu} & T(T(-)) \\ \downarrow = & & \downarrow T(\varepsilon) \\ T(-) & \xrightarrow{=} & T(-) \end{array} , \quad \begin{array}{ccc} T(-) & \xrightarrow{\mu} & T(T(-)) \\ \downarrow = & & \downarrow \varepsilon_{T(-)} \\ T(-) & \xrightarrow{=} & T(-). \end{array}$$

Such a functor is called a cotriple or comonad. It has become increasingly clear that these are very important structures, see [**10**]. A coalgebra for a cotriple is an object C together with a morphism $\sigma : C \longrightarrow T(C)$ such that the following two diagrams are commutative

$$\begin{array}{ccc} C & \xrightarrow{\sigma} & T(C) \\ \downarrow \sigma & & \downarrow \mu_C \\ T(C) & \xrightarrow{T(\sigma)} & T(T(C)) \end{array} , \quad \begin{array}{ccc} C & \xrightarrow{\sigma} & T(C) \\ \downarrow = & & \downarrow \varepsilon_C \\ C & \xrightarrow{=} & C. \end{array}$$

The functor of the big Witt vectors, together with the Artin-Hasse exponential and the morphism $s_1 : W(-) \longrightarrow (-)$ is a very important example of a cotriple on the category of commutative unital rings. The coalgebras for this triple are precisely the λ-rings. The commutativity of diagrams (4.9.12) and (4.9.13), for $n = 1$, give the two counit properties for the cotriple $W(-) \longrightarrow W(W(-))$. For a proof of the coassociativity see [**5**]. The commutativity of diagram (4.9.20) below is the main part of the statement that lambda-rings are the coalgebras for the big Witt vectors cotriple.

that makes the following diagram commutative. Moreover it is uniquely characterized by this property.

$$\begin{array}{ccc} W(A) & \xrightarrow{AH_A} & W(W(A)) \\ \downarrow = & & \downarrow s_{n,W(A)} \\ W(A) & \xrightarrow{\mathbf{f}_{n,A}} & W(A) \end{array} \tag{4.9.12}$$

The functorial commutativity of (4.9.12) *is equivalent to the functorial commutativity of*

$$\begin{array}{ccc} W(A) & \xrightarrow{AH_A} & W(W(A)) \\ \downarrow = & & \downarrow W(s_{n,A}) \\ W(A) & \xrightarrow{\mathbf{f}_{n,A}} & W(A) \end{array} \tag{4.9.13}$$

Proof. The central thing is to establish the commutativity of (4.9.12). This then immediately gives that AH is a ring morphism, because the Frobenius morphisms are ring morphisms and (4.9.12) tells what happens to the ghost components of $AH(a(t))$. So let

$$a(t) = \prod_i (1 - x_i t)^{-1}.$$

Then

$$\mathbf{f}_n a(t) = \prod_i (1 - x_i^n t)^{-1} = 1 + h_1(x_1^n, x_2^n, \ldots)t + h_2(x_1^n, x_2^n, \ldots)t^2 + \cdots \tag{4.9.14}$$

On the other hand

$$AH(a(t)) = 1 + (\prod_i (1 - x_i t)^{-1})u + \cdots + (\prod_{i_1 \leq \cdots \leq i_m} (1 - x_{i_1} \cdots x_{i_m} t)^{-1})u^m + \cdots$$

so what needs to be shown is that

$$s_n(\prod_{i_1 \leq \cdots \leq i_m} (1 - x_{i_1} \cdots x_{i_m} t)^{-1}) = h_m(x_1^n, x_2^n, \ldots).$$

But the s_n are defined by the power sums. Thus

$$s_n(\prod_{i_1 \leq \cdots \leq i_m} (1 - x_{i_1} \cdots x_{i_m} t)^{-1}) = \sum_{i_1 \leq \cdots \leq i_m} (x_{i_1} \cdots x_{i_m})^n \tag{4.9.15}$$

which is manifestly the same as the right hand side of (4.9.14). This proves the commutativity of (4.9.12) and establishes that AH is a morphism of rings. That it is in fact a morphism of λ-rings is left as a notationally somewhat complicated but otherwise straightforward exercise. This will also follow from the fact to be proved later that the Frobenius morphisms commute with the λ and σ operations.

The equivalence of the commutativity of the diagrams (4.9.12) and (4.9.13) comes from the fact that the Frobenius morphisms are characterized by

$$s_n \mathbf{f}_m = s_{mn} \tag{4.9.16}$$

and that the s_n are functorial morphisms $W(-) \longrightarrow (-)$ so that applying W to them gives a commutative diagram [28]

$$\begin{array}{ccc} W(W(A)) & \xrightarrow{s_{m,W(A)}} & W(A) \\ \downarrow{\scriptstyle W(s_{n,A})} & & \downarrow{\scriptstyle s_{n,A}} \\ W(A) & \xrightarrow[s_{m,A}]{} & A \end{array} \tag{4.9.17}$$

Corollary 4.9.18. $W(A)$ is a functorial λ-ring.

Corollary 4.9.19. In diagram form the fact that A is a λ-ring, i.e. that σ_t, is a ring homomorphism that commutes with the symmetric (or exterior) product operations on A on the one side and on $W(A)$ on the other side, says that the following diagram is commutative

$$\begin{array}{ccc} A & \xrightarrow{\sigma_t} & W(A) \\ \downarrow{\scriptstyle \sigma_t} & & \downarrow{\scriptstyle AH_A} \\ W(A) & \xrightarrow{W(\sigma_t)} & W(W(A)) \end{array} \tag{4.9.20}$$

Construction 4.9.21. Adams operations [29]. Given a λ-ring A the corresponding **Adams operations** Ψ^i are defined by

$$\sum_{i \geq 1} \Psi^i(x)t^i = -t\frac{d}{dt}\log(\lambda_{-t}(x))$$

or, equivalently,

$$\sum_{i \geq 1} \Psi^i(x)t^i = t\frac{d}{dt}\log(\sigma_t(x)). \tag{4.9.22}$$

Theorem 4.9.23. *Let A be a ring with operations λ^i (or σ^i). If it is a λ-ring the corresponding Adams operations are ring morphisms. Moreover they satisfy*

$$\Psi^1 = id, \ \Psi^i \circ \Psi^j = \Psi^{ij} \tag{4.9.24}$$

$$\Psi^n \circ \sigma^j = \sigma^j \circ \Psi^n. \tag{4.9.25}$$

Inversely, if A is of characteristic zero and the Ψ^i are ring morphisms such that (4.9.25) *holds then A is a λ-ring.*

Proof. Practically by definition

$$s_n \circ \sigma_t = \Psi^n.$$

Thus the Ψ^n are ring morphisms. Inversely, if A is of characteristic zero, one can injectively pass to the tensor product $A \otimes \mathbf{Q}$ and then the fact that the Ψ^n are ring

[28] The commutativity of this diagram can also be easily seen directly. It amounts to no more than the observation that $p_n(x_1^m, x_2^m, \ldots) = x_1^{mn} + x_2^{mn} + \cdots = p_m(x_1^n, x_2^n, \ldots)$.

[29] These operations are named for J. Frank Adams, who first used them intensively in algebraic topology. They are related to the exterior product operations exactly as the power sums are related to the elementary symmetric functions, and therefore are also frequently called 'power operations'.

morphisms implies that σ_t is a ring morphism which moreover commutes with the symmetric product operations because of (4.9.24).

Finally the identity (4.9.24) follows from the commutativity of (4.9.20).

Theorem 4.9.26. The Clarence Wilkerson theorem. *Let A be a characteristic zero ring with ring morphisms Ψ^n such that* (4.9.24) *holds* [30]. *Suppose moreover that for every prime number p*

$$\Psi^p(x) \equiv x^p \mod p.$$

Then A is a λ-ring whose Adams operations are the given Ψ^n.

Note the difference of this theorem with 4.9.23. Here it is not supposed that there are some (potential) exterior (symmetric) power operations already defined.

The rings $W(A)$ are (functorial) λ-rings. Thus they have associated (functorial) Adams operations, which are ring endomorphisms. Things are as nice as they possibly could be: these are precisely the Frobenius endomorphisms $\mathbf{f}_n$.

Theorem 4.9.27. First Adams=Frobenius theorem. *The Adams operations associated to the functorial λ-ring structure on the $W(A)$ defined by* (4.9.7), *or, equivalently, the Artin-Hasse exponential AH, are the Frobenius endomorphisms.*

This follows immediately by comparing the characterization of the Frobenius endomorphisms of (4.7.10) with the definition of the Adams operations by (4.9.22) (and the definition of the s_n morphisms (see 4.6.2)).

Construction 4.9.28. λ-ring structure on the ring **Symm**. There is a very simple and obvious λ-ring structure on the ring $\mathbf{Z}[x_1, x_2, \ldots]$ of polynomials over the integers, viz the one determined by

$$\lambda^1(x_i) = x_i, \ \lambda^j(x_i) = 0, \ \text{if } j \geq 2. \tag{4.9.29}$$

The associated Adams operators are the ring endomorphisms given by

$$\Psi^n(x_i) = x_i^n. \tag{4.9.30}$$

The relations between the Adams operations and the exterior product operations are the usual ones linking power sums and elementary symmetric functions. i.e.

$$\Psi^n = \det \begin{pmatrix} \lambda^1 & 1 & 0 & \cdots & 0 \\ 2\lambda^2 & \lambda^1 & 1 & \ddots & \vdots \\ 3\lambda^3 & \lambda^2 & \ddots & \ddots & 0 \\ \vdots & \vdots & \ddots & \lambda^1 & 1 \\ n\lambda_n & \lambda^{n-1} & \cdots & \lambda^2 & \lambda^1 \end{pmatrix} \tag{4.9.31}$$

[30] Such a ring is often called a Ψ-ring.

$$n!\lambda^n = \det \begin{pmatrix} \Psi^1 & 1 & 0 & \cdots & 0 \\ \Psi^2 & \Psi^1 & 2 & \ddots & \vdots \\ \Psi^3 & \Psi^2 & \ddots & \ddots & 0 \\ \vdots & \vdots & \ddots & \Psi^1 & n-1 \\ \Psi_n & \Psi^{n-1} & \cdots & \Psi^2 & \Psi^1. \end{pmatrix} {}^{31)} \tag{4.9.32}$$

Formulae (4.9.31) and (4.9.32) together show that $\mathbf{Symm}_{\mathbf{Q}}$ is stable under the exterior product operations specified by (4.9.29). The defining formula (4.9.29) itself shows that the exterior product operations are defined over the integers. Thus **Symm** has a λ-ring structure and the associated Adams operations are the power sum operations (4.9.30).

4.9.33. Discussion. The many different (?) operations on **Symm**. There are by now some five potentially different unary operations on **Symm** and it is perhaps wise to list them

(a) the exterior product operations that define the λ-ring structure on **Symm**. These are not additive of course.

(b) the corresponding Adams operations as described by (4.9.30)

(c) the rings $W(A)$ are functorial λ-rings. The functor $A \mapsto W(A)$ is represented by **Symm** : $W(A) = \mathbf{Alg}(\mathbf{Symm}, A)$. The functorial exterior product operations on $W(A)$ must therefore come from ring endomorphisms of **Symm**. These endomorphism cannot be Hopf algebra endomorphisms (because otherwise the functorial exterior product operations on $W(A)$ would be additive.

(d) the rings $W(A)$ have functorial Adams operations. These also must come from ring endomorphisms of **Symm**. Moreover these must be Hopf algebra endomorphisms and even coring object morphisms in the category of rings.

(e) the Frobenius endomorphisms as defined in section 4.7.

The first Adams=Frobenius theorem says that the morphisms (d) and (e) are the same. By the remarks already made the only one which could conceivable also be the same is (b). And this is indeed the case

Theorem 4.9.34. Second Adams=Frobenius theorem. *The Adams operations Ψ^n coming from the λ-ring structure on* **Symm** *are the same as the endomorphisms $\mathbf{f}_n$ of* **Symm** *that induce the Frobenius endomorphisms of the functor $W(A)$ of the big Witt vectors.*

Proof. This follows immediately from the fact that $\mathbf{f}_n(p_m) = p_{nm}$ compared with (4.9.30).

There is of course still more structure on **Symm**. **Symm** being auto dual there are also all the duals of (a)-(e). It is not evident what all these dual operations are.

---

[31]) These determinantal formulae are completely equivalent to the definition by power series of the Adams operations.

Theorem 4.9.35. Universal λ-ring on one generator. *The ring of symmetric polynomials with the λ-ring structure defined above is the universal λ-ring on one generator.* [32]

This means the following. For each λ-ring A and element $a \in A$ there is a unique morphism of λ-rings $\varphi : \mathbf{Symm} \longrightarrow A$ such that $\varphi(e_1) = a$.

Proof. As is easily verified from e.g. (4.9.32) the λ-ring structure on **Symm** satisfies $\lambda^n(e_1) = e_n$. It follows that the only ring morphism that could possibly work is defined by $\varphi : e_n \mapsto \lambda^n(a)$. That this is actually a morphism of λ-rings requires a bit more work, as follows. Let x be an element of **Symm**, i.e. a polynomial in the elementary symmetric functions

$$x = P_x(e_1, e_2, \ldots) = P_x(\lambda^1(e_1), \lambda^2(e_1), \ldots).$$

Now consider $\lambda^n(x)$. Because composition of lambda operations, a lambda operation applied to a product, and a lambda operation applied to a sum are given by 'universal polynomials', that means the same polynomials for any λ-ring, see 4.9.8, more precisely the corresponding footnote, there is a universal polynomial Q_{n,P_x} (with coefficients in the integers) [33] such that for any λ-ring and any element a in it

$$\lambda^n(P_x(\lambda^1(a), \lambda^2(a), \ldots)) = Q_{n,P_x}(\lambda^1(a), \lambda^2(a), \ldots).$$

Also φ is a ring homomorphism and so commutes with polynomials

$$\varphi(Q(x_1, x_2, \ldots) = Q(\varphi(x_1), \varphi(x_2), \ldots).$$

It follows that φ commutes with the lambda operations, so that it is a morphism of λ-rings.

4.10. *Exp* algebras

The universal lambda ring on one generator is **Symm** is also a Hopf algebra and even both a ring object in the category of coalgebras and a coring object in the category of algebras. Is this an accident or is there something systematic behind it. As it turns out, at least partially, this is not an accident. There is a good construction of rings from Abelian groups that are automatically Hopf algebras and which can be proved to be universal lambda rings if the group is free. Moreover, if the starting object is a ring (instead of just an Abelian group) the construction yields a Hopf algebra with a second comultiplication. All this is in [5], and it seems that much more should be done about this. Here, just a short outline is given.

Definition 4.10.1. The functor 'completed 1-units'. Let R be a graded ring, $R = \bigoplus_i R_i$. Define

$$CU(R) = 1 + \prod_{i \geq 1} R_i \tag{4.10.2}$$

---

[32]There is a far reaching generalization. As will be shown in the next chapter **Symm** is isomorphic to RS the direct sum of the representation rings of the symmetric groups. For a fixed finite group G let $G \times_{\mathrm{wr}} S_n$ be the wreath product of G and S_n. The direct sum of the representation rings $R(G \times_{\mathrm{wr}} S_n)$ is the free lambda ring on the irreducible representations of G as generators, [**9**].

[33]This polynomial is in fact the plethysm $e_n \circ P_x$, see section 4.11 below.

thus the elements of $CU(R)$ are expressions of the form $1 + r_1 + r_2 + \cdots, r_i \in R_i$. And they are indeed units in the ring $R = \prod_{i=0}^{\infty} R_i$, the graded completion of R. They are 1-units in the sense that the degree zero component is 1. [34]

This definition makes sense for any graded ring. However, from now on in this subsection, it is supposed that R is commutative so that the value of CU at R is an Abelian group.

Note that if CU is applied to a graded ring of the form $A[t]$, where A is a commutative ring and the degree of t is 1, then $CU(A[t]) = W(A)$, the group of the big Witt vectors over A.

Construction 4.10.3. The exponentiation functor. The crucial idea is now that the functor

$$CU : \mathbf{GrCRing} \longrightarrow \mathbf{Ab}$$

from commutative graded rings to Abelian groups has an adjoint

$$Exp : \mathbf{Ab} \longrightarrow \mathbf{GrCRing}.$$

Here **Ab** is the category of Abelian groups and **GrCRing** is the category of graded commutative rings. That is, there is a functorial isomorphism

$$\mathbf{Ab}(G, CU(R)) \xrightarrow{\cong} \mathbf{GrCRing}(Exp(G), R).$$

Another way of saying is as follows. For each Abelian group G, there exist a (so-called universal exponential) morphism of Abelian groups $\varepsilon^G : G \longrightarrow CU(Exp(G))$ such that for each graded ring R and morphism of Abelian groups $G \xrightarrow{\varphi} CU(R)$, there is a unique morphism of graded rings $\widetilde{\varphi} : Exp(G) \longrightarrow R$ such that $CU(\widetilde{\varphi}) \circ \varepsilon^G = \varphi$.

The situation is entirely analogous to that of the universal enveloping algebra of a Lie algebra. There we have the functor *Lie*: **AssAlg** $\longrightarrow$ **LieAlg** that associates to an associative algebra A the Lie algebra $Lie(A)$ which has the same underlying module with as bracket product the commutator difference $[x, y] = xy - yx$. The adjoint is the universal enveloping algebra functor U : **LieAlg** $\longrightarrow$ **AssAlg**, $\mathfrak{g} \mapsto U\mathfrak{g}$, which comes with a canonical morphism $e^* : \mathfrak{g} \longrightarrow Lie(U(\mathfrak{g}))$, that satisfies the adjointness equation

$$\mathbf{LieAlg}(\mathfrak{g}, Lie(A)) \xrightarrow{\cong} \mathbf{AssAlg}(U\mathfrak{g}, A)$$

and the universality property that for each morphism of Lie algebras $\mathfrak{g} \xrightarrow{\varphi} Lie(A)$, there is a unique morphism of associative algebras $\widetilde{\varphi} : U\mathfrak{g} \longrightarrow A$ such that $Lie(\widetilde{\varphi}) \circ \varepsilon^* = \varphi$. The only difference is that, Lie being truly a forgetful functor (which CU is not really) things in the universal enveloping algebra case are usually not written down quite so formally.

Proof of the existence and uniqueness of the functor Exp. Uniqueness is the usual yoga of adjoint functors (or a useful exercise). Existence goes as follows. Let G be an Abelian group and present it by means of generators $\{a_i : i \in I\}$ and relations. Now, form the commutative polynomial ring

$$\mathbf{Z}[x_{ij} : i \in I,\ j > 0]$$

[34] This terminology is borrowed from local algebraic number theory.

with one indeterminate of degree $j > 0$ for each j and each generator a_i, $i \in I$. Now let $Exp(G)$ be the quotient graded ring obtained by imposing one homogeneous relation in each positive degree for each relation in the presentation of G obtained by requiring that

$$x_i \mapsto 1 + \sum_{j=1}^{\infty} x_{i,j}$$

be a morphism of Abelian groups. This works. Again the construction is very similar to that in the case of universal enveloping algebras.

For instance in the case that $G = \mathbf{Z}/(2)$

$$Exp(\mathbf{Z}/(2)) = \mathbf{Z}[x_1, x_2, \ldots]/I$$

where I is the ideal generated by $2x_1$, $2x_2 + x_1^2$, $2x_3 + 2x_1x_2$, $2x_4 + 2x_1x_3 + x_2^2$, ..., $2x_n + x_1x_{n-1} + \cdots + x_{n-1}x_1$ if n is odd, and $2x_n + x_1x_{n-1} + \cdots + x_{n-1}x_1 + x_{n/2}^2$ if n is even, the relations resulting from the requirement that

$$(1 + x_1 + x_2 + \cdots)^2 = 1.$$

4.10.4. Construction and theorem. Hopf algebra structure on $Exp(G)$. Again, as in the case of universal enveloping algebras, it is easy to show that

$$Exp(G \oplus G') = Exp(G) \otimes Exp(G')$$

and thus the diagonal morphism of Abelian groups $G \longrightarrow G \oplus G$ induces a comultiplication $Exp(G) \longrightarrow Exp(G) \otimes Exp(G)$. And (straightforward exercise) this turns $Exp(G)$ into a Hopf algebra.

It is not usually presented that way but in fact the Hopf algebra structure on a universal enveloping algebra arises precisely in this manner.

4.10.5. Second multiplication on $Exp(G)$ There is much more. Let $Exp_n(G)$ denote the degree n component of $Exp(G)$. Now suppose that G is a ring. Then the multiplication on G induces in a natural way a second multiplication on $Exp(G)$ in the form of morphisms

$$Exp_n(G) \otimes Exp_n(G) \longrightarrow Exp_n(G)$$

for each n. which is distributive over the first one in the Hopf algebra sense. This makes $Exp(G)$, G a ring, a ring object in the category of (graded) coalgebras. Just like **Symm**.

4.10.6. Lambda ring structure on $Exp(G)$. It also turns out that for any Abelian group $Exp(G)$ has a natural lambda ring structure which makes it the free lambda ring over the Abelian group G.

The construction goes via a notion called ω-ring, which is just a λ-ring is disguise, but with the advantage that the axioms are linear (permitting, perhaps, dualization)

For details of 4.10.5 and 4.10.6, see [**7**]. As far as we know there has been no follow up on Hoffman's 1983 paper. It seems to us that there is here a rich source of interesting things to do. For instance dualizing the notion of a lambda ring. Also the Hopf algebras $Exp(\mathbf{Z}/(n))$ complete with second multiplication deserve further investigation. Finitely generated Abelian groups being as simple as they are, the Exp's of them are just tensor products. Things could become very much more interesting when rings are considered whose underlying Abelian group is not finitely generated.

One also wonders what happens when one attempts to exponentiate noncommutative groups.

Remark 4.10.7. Note that $Exp(\mathbf{Z}) = \mathbf{Symm}$. More generally if $R(G)$ is the representation ring (over the complex numbers) of a finite group G, then

$$Exp(R(G)) = \bigoplus_n R(G \otimes_{\mathrm{wr}} S_n)$$

as Hopf algebras with second multiplication.

4.11. Plethysm

Let A be a λ-ring and a an element of A. Because **Symm** is the universal λ-ring on one generator there is a unique morphism of λ-rings $\varphi : \mathbf{Symm} \longrightarrow A$ such that $\varphi(e_1) = a$. Letting a vary this defines a mapping

$$\mathbf{Symm} \times A \longrightarrow A, \quad (f, a) \mapsto f \circ a \tag{4.11.1}$$

that for each fixed a is a morphism of rings $\mathbf{Symm} \longrightarrow A$.

This can be applied in particularly to the case $A = \mathbf{Symm}$ itself giving a composition law

$$\mathbf{Symm} \times \mathbf{Symm} \longrightarrow \mathbf{Symm}, (f, g) \mapsto f \circ g \tag{4.11.2}$$

that is called **plethysm**.

An explicit description is as follows. Write g as a sum of monomials (with integer coefficients)

$$g = \sum_\alpha c_\alpha x_\alpha \tag{4.11.3}$$

and formally write

$$\prod_i (1 + y_i t) = \prod_\alpha (1 + x_\alpha t)^{c_\alpha}. \tag{4.11.4}$$

Then

$$f \circ g = f(y_1, y_2, \ldots). \tag{4.11.5}$$

In particular if g is a monomial symmetric function $f \circ g$ is obtained from f by substituting the terms of g for the variables of f. In this form plethysm has already occurred above in the discussion of the Artin-Hasse exponential, where the $h_m \circ p_n$ appeared and in fact the characterization of the Artin-Hasse exponential by means of the Frobenius endomorphisms of the big Witt vectors amounts to the fact that $h_m \circ p_n = p_n \circ h_m$, a special case of theorem 4.11.6 below.

Theorem 4.11.6. Properties of plethysm.

(a) *For each* $g \in \mathbf{Symm}$, $f \mapsto f \circ g$ *is a ring endomorphism of* **Symm** (4.11.7)

(b) $f \circ e_1 = e_1 \circ f$ *for all* $f \in \mathbf{Symm}$ (4.11.8)

(c) $p_n \circ g = g \circ p_n = g(x_1^n, x_2^n, \ldots)$ (4.11.9)

(d) *For each* $n \in \mathbf{N}$, $g \mapsto p_n \circ g$ *is a ring endomorphism of* **Symm** (4.11.10)

(e) *For each* $n \in \mathbf{N}$, $e_n \circ f = \lambda^n(f)$ (4.11.11) (*f*) *Associativity:* $(f \circ g) \circ h = f \circ (g \circ h)$ (4.11.12)

Proof. (a) and (b), are immediate from the definition; (c) follows by taking the logarithm of (4.11.4) and then (d) follows. Further (e) follows from (a) and the standard recursion between the p_n and e_n which is the same as that between the Adams operators Ψ of **Symm** and the exterior product operations, plus the observation that the morphisms of (d) are precisely the Adams operations. Finally, it suffices to verify (f) over the rationals and there the power sums form a set of generators so that (a) and (d) entail (f).

4.11.13. Construction and proposition. Composition of natural operations. As already remarked each element $f \in$ **Symm** gives rise to a natural operation on the category of λ-rings:

$$F_A : A \longrightarrow A,\ a \mapsto f \circ a = f(\lambda^1(a), \dots, \lambda^n(a)). \tag{4.11.14}$$

I.e. write f as a polynomial in the elementary symmetric functions and then substitute $\lambda^i(a)$ for e_i. Or, equivalently, $f \circ a$ is the image of f under the unique morphism of λ-rings that takes e_1 to a.

The operation is natural, i.e. functorial, in the sense that if $\varphi : A \longrightarrow B$ is a morphism of λ-rings, then $\varphi(f \circ a) = f \circ \varphi(a)$.

Under this correspondence plethysm fits with composition. That is the natural operation corresponding to $f \circ g$ is the composition FG of F and G [35].

This comes about because of the following commutative naturality diagram

$$\begin{array}{ccccc}
\mathbf{Symm} & \xrightarrow{G} & \mathbf{Symm} & \xrightarrow{F} & \mathbf{Symm} \\
\downarrow & & \downarrow & & \downarrow \\
A & \xrightarrow{G} & A & \xrightarrow{F} & A
\end{array}$$

Here the three vertical arrows all are the unique morphism of λ-rings that takes e_1 into a (for a given element a of A), and F and G are the natural transformations defined by the elements f and g of **Symm**. Then going from the top left hand corner along the top two horizontal arrows and then down gives

$e_1 \mapsto g$ (substituting e_1 in g) $\mapsto f \circ g$ (substituting g in f) $\mapsto (f \circ g) \circ a$ (substituting a in $f \circ g$ (or taking the image of $f \circ g$ under the vertical morphism of λ-rings).

And going down first and then along the bottom row gives

$e_1 \mapsto a$ (taking the image of e_1 under the vertical morphism) $\mapsto g \circ a$ (substituting a in g) $\mapsto f \circ (g \circ a)$ (substituting $g \circ a$ in f). [36]

4.12. The many incarnations of Symm

The symmetric functions, as a ring, as a Hopf algebra, as a λ-ring, as ... turn up in many parts of mathematics; not always with all structure elements, so far

[35] This is like the associativity of part (f) of theorem 4.11.6.

[36] Incidentally this gives another proof of property (f) of theorem 4.11.6.

described, present. Here are some of its manifestations.

Symm $=\mathbf{Z}[h_1, h_2, \ldots]$, the algebra of symmetric functions

$\simeq \bigoplus_n R(S_n)$, the direct sum of the representation rings of the symmetric groups

$\simeq R_{rat}(GL_\infty)$, the ring of rational representations of the infinite linear group

$\simeq H^*(BU)$, the cohomology of the classifying space BU

$\simeq H_*(BU)$, the homology of the classifying space BU

$\simeq R(W)$, the representative ring of the functor of the (big) Witt vectors

$\simeq U(\Lambda)$, the universal λ-ring on one generator

$\simeq HS(\mathbf{Z})$, the Hall-Steinitz algebra of finitely generated Abelian groups, see section 8.7 below

$\simeq \ldots$

$\simeq \ldots$

$\simeq \ldots$

The ellipses here are not just for show. There are actually a good many more incarnations such as $Exp(\mathbf{Z})$, where Exp is exponentiation, (see above in section 4.10), an interpretation in terms of the K-theory of endomorphisms, [**3**], an interpretation in terms of endomorphisms of polynomial functors, [**9**], and finally as the free algebra over the cofree graded coalgebra over one generator and the graded cofree coalgebra [37] over the free algebra on one generator, [**4**].

Most of these will not be discussed further here. The exceptions are

$$RS = \bigoplus_n RS_n$$

the direct sum of the rings of representations of the symmetric groups which is the subject of the next chapter, and the Hall-Steinitz algebra.

References

[1] A. Grothendieck, *La théorie des classes de Chern*, Bull. Soc. Math. France **86** (1958), 137–154.

[2] Michiel Hazewinkel, *Formal groups and applications*, Acad. Press, 1978.

[3] Michiel Hazewinkel, *Operations in the K-theory of endomorphisms*, J. of Algebra **84** (1983), 285–304.

[4] Michiel Hazewinkel, *Cofree coalgebras and multivariable recursiveness*, J. pure and appied Algebra **183** (2003), 61–103. Revised version of a 1999 preprint.

[5] Michiel Hazewinkel, *Witt vectors. Part 1*, `arXiv: 0804.3888`, in: Michiel Hazewinkel (ed.) *Handbook of Algebra*, **6**, 319–472, Elsevier, 2009.

[6] Michiel Hazewinkel, *Niceness theorems*, in: Boris Kruglikov, Valentin Lychagin, Eidar Straume, eds., *Differential equations - geometry, symmetries and integrability*, Springer, 2009, 107–150, `arXiv: 6810.5691`.

[37] But not the nongraded cofree coalgebra over the free algebra on one generator, see loc. cit.

[7] P. Hoffman, *Exponential maps and λ-rings*, J. pure and appl. Algebra **27** (1983), 131–162.

[8] A. Liulevicius, *Representation rings of the symmetric groups – a Hopf algebra approach*, Preprint, Matematisk Institut, Aarhus Universitet, 1976; *Arrows, symmetries and representation rings*, J. pure and applied Algebra **19** (1980), 259–273.[38]

[9] Ian G. Macdonald, *Polynomial functors and wreath products*, J. of pure and applied Algebra **18** (1980), 173–204.

[10] E. Manes, *Monads of sets.* In: Michiel Hazewinkel (ed.), *Handbook of Algebra. Volume 3*, Elsevier, 2003, 67–154.

[11] J. Milnor, D. Husemoller, *Symmetric bilinear forms*, Springer, 1973.

[12] Douglas C. Ravenel, W. Stephen Wilson, *The Hopf ring for complex cobordiam*, J. pure and applied Algebra **9** (1977), 241–280.

[13] W. Stephen Wilson, *Hopf rings in algebraic topology*, Expositiones mathematicae **18** (2000), 369–388.

[14] E. Witt, *Zyklische Körper und Algebren der Charakteristik p vom Grad p^n*, J. reine und angew. Math. **176** (1937), 126–142.

[15] Andrey V. Zelevinsky, *Representations of finite classical groups*, Springer, 1981.

[38]As far as we know the first two papers on the symmetric functions as a Hopf algebra were [8] and Ladnor Geissinger, Hopf algebras of symmetric functions and class functions, In: D. Foata (ed), Combinatoire et représentation du groupe symétrique, Springer, 1977, 168–181.

CHAPTER 5

The representations of the symmetric groups from the Hopf algebra point of view

As already remarked before, the complex representations of the symmetric groups form a natural PSH algebra. Much of this comes for free; the remainder comes from a few basic general theorems from group representation theory. These will be discussed first.

5.1. A little bit of finite group representation theory

Let A be an associative algebra with unit; let M be a simple A-module. Then

Lemma 5.1.1. Schur lemma. *The ring of A-endomorphisms of a simple module M is a division ring.*

For the case of a finite group G and irreducible representations of G over an algebraically closed field k, taking for A the group algebra kG there is the version

Lemma 5.1.2. Schur lemma for irreducible representations. *Let G be a finite group and let V, W be two irreducible representations of G over the field of complex numbers (or any other algebraically closed field). Then*

$$\mathbf{Mod}_G(V,W) = \begin{cases} 0 & \text{if } V \text{ and } W \text{ are not isomorphic} \\ \mathbf{C} & \text{if } V \text{ and } W \text{ are isomorphic.} \end{cases} \tag{5.1.3}$$

Constructions 5.1.4. Restriction and induction. Let G be a finite group and H a subgroup. Let V be a representation of G, i.e. a (left) kG-module. Then restricting the representation $\rho : G \longrightarrow \mathrm{End}_k(V)$ to H, or, equivalently restricting the operations on V to $kH \subset kG$, one obtains a representation of H, called the **restriction** to H and denoted $\mathrm{Res}_H^G(V)$. Other notations in common use are V_H, $V\downarrow_H^G$, $V_{|H}$. This construction is obviously functorial from G-modules to H-modules.

There is an adjoint construction called **induction**. Let W be an H-module, i.e. a kH-module. Via the inclusion $kH \subset kG$ one can see kG as a right kH-module and hence form the tensor product

$$\mathrm{Ind}_H^G(W) \stackrel{\text{def}}{=} kG \otimes_{kH} W. \tag{5.1.5}$$

Induction is both left and right adjoint to restriction:

Theorem 5.1.6. (Frobenius reciprocity, [**3**], p. 232 and p. 243,. [**9**], p. 674.) *Let k be a commutative ring with unit element, and let G be a finite group with* *subgroup H. Then for all kH-modules V and kG-modules W, functorially,*

$$\mathbf{Mod}_{kG}(\mathrm{Ind}_H^G(V), W) \cong \mathbf{Mod}_{kH}(V, \mathrm{Res}_H^G(W)) \tag{5.1.7}$$

$$\mathbf{Mod}_{kG}(W, \mathrm{Ind}_H^G(V)) \cong \mathbf{Mod}_{kH}(\mathrm{Res}_H^G(W), V) \tag{5.1.8}$$

as A-modules [1].

The first of these adjointness theorems is a special case of a much more general adjointness theorem as follows. Let $\varphi : B \longrightarrow A$ be a morphism of associative k-algebras (with unit element). Then every (left) A-module X can be regarded as a (left) B-module via $bx = \varphi(b)x$. This B-module is denoted $\mathrm{Res}_B^A(X)$, and called the restriction of X to B. The terminology 'restriction' of course derives from the case that B is a subalgebra of A.

Let Y be any B-module. The algebra A can be seen as an (A, B)-bimodule (left A-module and right B-module) [2] with the right B-module structure given by

$$ab = a\varphi(b)$$

and then there is the tensor product

$$\mathrm{Ind}_B^A(Y) = A \otimes_B Y$$

with the left A-module structure $a_1(a_2 \otimes y) = a_1 a_2 \otimes y$. With these notations there is the adjointness relation, [**8**], p. 193,

$$\mathbf{Mod}_A(\mathrm{Ind}_B^A(Y), X) \cong \mathbf{Mod}_B(Y, \mathrm{Res}_B^A(X)) \quad \text{as } k\text{-modules.} \tag{5.1.9}$$

And this adjointness relation, in turn, comes from the fundamental adjointness between tensor product and homomorphisms

$$\mathbf{Mod}_A(V \otimes_B W, U) \cong \mathbf{Mod}_B(W, \mathbf{Mod}_A(V, U)) \tag{5.1.10}$$

where V is an (A, B)-bimodule, W is a B-module, U is an A-module, with the B-module structure on $\mathbf{Mod}_A(V, U)$ given by $(bf)(v) = f(vb)$, $f \in \mathbf{Mod}_A(V, U)$. The morphism of (5.1.10) is of course the obvious one: $\rho \mapsto \rho'$, $\rho'(w)(v) = \rho(v \otimes w)$.

To obtain (5.1.9) from (5.1.10), take $V = A$, $U = X$, $W = Y$ and note that $\mathbf{Mod}_A(A, X) = \mathrm{Res}_B^A(X)$ as B-modules via $f \mapsto f(1)$.

The second adjointness relation of theorem 5.1.6 is more group specific. For a proof see e.g. [**8**], p. 387 [3]. For the purposes below either one of the two adjointness results will do.

The next important bit of representation theory to be discussed is the Mackey double coset theorem, also called Mackey decomposition. It describes what happens

---

[1] The 'reciprocity' character of these adjoint functor statements is perhaps not immediately apparent. Here is a consequence, see [**8**], p.691. Let H be a subgroup of a finite group G and let F be a splitting field for G and H whose characteristic does not divide the number of elements of G. Let V be a simple $F[H]$-module and W a simple $F[G]$-module. Then the multiplicity of V as a simple constituent of $\mathrm{Res}_H^G(F)$ is equal to the multiplicity of W as a simple constituent of $\mathrm{Ind}_H^G(V)$.

There is also an inner product version of Frobenius reciprocity for characters and for class functions, see e.g. [**8**], p.760, [**1**], p.271, [**3**], p.233. These are the more classical formulations of Frobenius reciprocity.

[2] An (A, B)-bimodule, where A and B are algebras over k is a k-module M with a left A-module structure and a right B-module structure such that $(am)b = a(mb)$

[3] There is also a right adjoint to restriction in the general case of an arbitrary morphism of algebras. The functor in question is called coinduction; see [**9**], p. 192ff.

to a representation of a subgroup H when it is first induced up to the full group G and than restricted to a (possibly different) subgroup K.

If ρ is a representation of a subgroup H of G and x is an element of G, ρ^x denotes the representation of the subgroup xHx^{-1}, defined by first conjugating with x^{-1} followed by the representation ρ. Thus ρ^x is the representation: $xhx^{-1} \mapsto \rho(h)$.

Theorem 5.1.11. (Mackey double coset theorem, Mackey decomposition). *Let G a a finite group with subgroups H and K. Let $T = \{t_1, \ldots, t_m\}$ be a set of representatives of (K, H) double cosets in G. Then for a representation $\rho : H \longrightarrow GL_n(V)$, i.e. a kH-module V*

$$\mathrm{Res}^G_K(\mathrm{Ind}^G_H(V)) = \bigoplus_{i=1}^{m} \mathrm{Ind}^K_{t_iHt_i^{-1}\cap K}((\mathrm{Res}^H_{H\cap t_i^{-1}Kt_i})^{t_i}) \tag{5.1.12}$$

(as kK-modules).

Before proving this theorem it is useful to know a little more about induction. Let G be a finite group and W a G-module. Suppose that W decomposes as a direct sum

$$W = \bigoplus_{i\in I} W_i \tag{5.1.13}$$

in a way that is compatible with the (left) G-action on W. This means that for each i and g, gW_i is one of the direct summands of (5.1.13), so that, the left G-action induces a permutation of the set $\{W_i : i \in I\}$. Such a decomposition is called a system of imprimitivity, and W is called an imprimitive module.

An induced module from a proper subgroup has this property. Indeed let $\{g_1, \ldots, g_j\}$ be a left transversal for H in G, i.e. $\{g_1, \ldots, g_s\}$ is a full set of representatives for the right cosets of H in G, so that

$$G = \bigcup_{i=1}^{s} g_iH \text{ (disjoint union).} \tag{5.1.14}$$

It follows immediately that

$$kG = \bigoplus_{i=1}^{s} g_ikH \quad \text{(as right } kH\text{-modules) and } \mathrm{Ind}^G_H(V) = \bigoplus_{i=1}^{s} g_i \otimes V \tag{5.1.15}$$

so that indeed $\mathrm{Ind}^H_H(V) = \bigoplus\limits_{i=1}^{s} g_i \otimes V$ is a system of imprimitivity, with moreover the property that the induced permutation of the direct summands is transitive. There is a converse to this.

Proposition 5.1.16. *Imprimitivity systems vs induced modules. Let W be a module with imprimitivity system* (5.1.13) *and such that the induced permutation is transitive. If H is the stabilizer of any $U \in \{W_i : i \in I\}$, then U is a kH-module such that $W = \mathrm{Ind}^G_H(U)$.*

Proof. Obviously U is a kH-module. Let T be a left transversal for H in G. Then $\{W_i : i \in I\} = \{tU : t \in T\}$ and $W = \bigoplus\limits_{t\in T} tU$ and, see (5.1.15), $t \otimes u \mapsto tu$ induces a k-isomorphism $\varphi : \mathrm{Ind}^G_H(U) \longrightarrow W$. Moreover for given $g \in G$, $t \in T$, let

$gt = t'h$ Then

$$\varphi(g(t \otimes u)) = \varphi(t' \otimes hu) = t'hu = gtu = g\varphi(t \otimes u)$$

proving the proposition.

The direct summands occurring in (5.1.15) are not (left) H-modules (except $e \otimes V$ where e is the identity element of the group G) (except when H is a normal subgroup). However, $g \otimes V \simeq gV \subset \mathrm{Ind}_H^G(V)$ is a gHg^{-1}-module with action $(ghg^{-1})(gv) = g(hv)$. Under the isomorphism of k-modules $gV \simeq V$, $gv \mapsto v$ this becomes the representation V^g.

Proof of theorem 5.1.11. Again, let $\{g_1, \ldots, g_s\}$ be a left transversal for H in G, so that (5.1.15) holds. Left multiplication with an element from G permutes the direct summands $g_i \otimes V$ and $g_i \otimes V$ and $g_j \otimes V$ are in the same K-orbit if and only if g_i and g_j are in the same (K, H) double coset. For each $t \in T$ let

$$W_i = \bigoplus_{g_i \in KtH} g_i \otimes V, \quad \text{and } V_t = (\mathrm{Res}^H_{H \cap t^{-1}Kt}(V))^t. \tag{5.1.17}$$

Then it remains to show that

$$\mathrm{Ind}^K_{tHt^{-1} \cap K}(V_t) = W_t. \tag{5.1.18}$$

To this end let $I_t \subset \{1, \ldots, s\}$ be such that

$$KtH = \bigcup_{i \in I_t} g_t H.$$

Now K acts transitively on the set $\{g_i \otimes V : i \in I_t\}$ and the stabilizer of $t \otimes V$ is

$$\{h \in K : ht \otimes V = t \otimes V\} = \{h \in K : t^{-1}ht \in H\} = tHt^{-1} \cap K.$$

It now suffices to apply proposition 5.1.16 and the remark just below it.

The final important bit of representation theory to be discussed in this section is the Mackey tensor product theorem that describes the (inner) tensor product of two induced representations.

Let ρ and σ be two representations of, respectively, two finite groups G and K with modules V and W. Then the tensor product of the two modules affords (carries) a representation $\rho \otimes \sigma$ of the direct product group $G \times K$ given by

$$(\rho \otimes \sigma)(g, h) = \rho(g) \otimes \sigma(h)$$

called the (outer) tensor product of ρ and σ. In case the two groups are equal this can be composed with the diagonal morphism $G \longrightarrow G \times G, g \mapsto (g, g)$ to give a new representation of G called the inner tensor product and denoted in this text by $p \times \sigma$. Both tensor products will play a role in the sequel (whence the difference in notation).

Theorem 5.1.19. (Mackey tensor product theorem.) *Let H and K be subgroups of a finite group G and let V be an H-module and W a K-module. Then*

$$\begin{aligned} &\mathrm{Ind}_H^G(V) \otimes \mathrm{Ind}_K^G(W) \simeq \\ &\simeq \bigoplus_{t \in T} \mathrm{Ind}^G_{tHt^{-1} \cap K}((\mathrm{Res}^G_{H \cap t^{-1}Kt}(V)^t \otimes \mathrm{Res}^G_{tHt^{-1} \cap K}(W)), \end{aligned} \tag{5.1.20}$$

where T is a full set of (K, H) double coset representatives and the G-module structure on the left is that of the inner tensor product of two G-modules.

Actually for the purposes below only a special case of the above is needed (where $K = G$). This takes the form

Theorem 5.1.21. Projection formula, Frobenius axiom [4]. *Let G be a finite group with subgroup H, and let V be an H-module and W a G-module, then*

$$\mathrm{Ind}_H^G(V \otimes \mathrm{Res}_H^G(W)) = \mathrm{Ind}_H^G(V) \otimes W. \tag{5.1.22}$$

Proof. Let T be a left transversal of H in G. Then there is a well-defined k-module morphism

$$\mathrm{Ind}_H^G(V) \otimes W \xrightarrow{\varphi} \mathrm{Ind}_H^G(V \otimes \mathrm{Res}_H^G(W)),\ (t \otimes v) \otimes w \mapsto t \otimes (v \otimes t^{-1}w)$$

and inversely there is a morphism

$$\mathrm{Ind}_H^G(V \otimes \mathrm{Res}_H^G(W)) \xrightarrow{\psi} \mathrm{Ind}_H^G(V) \otimes W,\ t \otimes (v \otimes t^{-1}w) \mapsto (t \otimes v) \otimes w.$$

These two are inverse to each other. To check that they are actually morphism of G-modules is routine.

The Mackey tensor product formula (5.1.20) is now straightforwardly proved by using the Mackey decomposition, the projection formula and transitivity of induction: If $H \subset K \subset G$ and V is an H-module, then

$$\mathrm{Ind}_K^G(\mathrm{Ind}_H^K(V)) = \mathrm{Ind}_H^G(V). \tag{5.1.23}$$

5.2. Double cosets of Young subgroups

Some information will be needed on double cosets of Young subgroups of the symmetric groups S_n of permutations of n letters.

Definition 5.2.1. Young subgroups. Let S_n be the group of permutations on the n symbols $\{1, \ldots, n\}$. A composition of n is a finite sequence $\alpha = [a_1, \ldots, a_m]$, $a_i \in \mathbf{N} = \{1, 2, \ldots\}$ such that $a_1 + \cdots + a_m = n$. For every composition of n the corresponding **Young subgroup** is

$$S_\alpha = S_{\alpha_1} \times S_{\alpha_2} \times \cdots \times S_{\alpha_m} \subset S_n \tag{5.2.2}$$

where S_{α_i} acts on the symbols $\{a_1 + \ldots + a_{i-1} + 1, \ldots, a_1 + \ldots + a_i\}$ and leaves all others untouched. Thus for instance $\sigma \in S_n$ is in $S_i \times S_j \subset S_n$ (where $i + j = n$) if and only if $\sigma(\{1, \ldots, i\}) \subset \{1, \ldots, i\}$ (and hence $\sigma(\{1, \ldots, i\}) = \{1, \ldots, i\}$ because σ is bijective) and (which is a consequence) $\sigma\{i+1, \ldots n\}) \subset \{i+1, \ldots, n\}$.

An important theorem in the representation theory of the symmetric groups and in algebraic combinatorics describes the double cosets of Young subgroups. Let $\alpha = [a_1, \ldots a_m]$ and $\beta = [b_1, \ldots, b_t]$ be two compositions of the same weight n. Let

---

[4] There are many other contexts in which one has two functors like Ind and Res with very like properties. This has led to the development of abstract representation theory and the theories of Mackey functors and Green functors, see [**15**]. In this context property (5.1.12) is called the Mackey decomposition axiom., and (5.1.22) the Frobenius axiom. In the context of algebraic K-theory and direct and inverse images of sheaves (5.1.22) is called the projection formula, see e.g. [**5**], p. 124, exercise 5.1.(d).

S_α, S_β be the corresponding Young subgroups of S_n and let $S_\alpha \pi S_\beta$ be a double coset. Consider the $m \times t$ matrix of nonnegative integers

$$z_{i,j} = \#(\{a_1+\dots+a_{i-1}+1,\dots,a_1+\dots+a_i\}\cap\pi\{b_1+\dots+b_{j-1}+1,\dots,b_1+\dots+b_j\}).$$

Obviously the matrix $(z_{i,j})$ has row sum vector α meaning $\sum_{j=1}^{t} z_{i,j} = a_i$ and column sum vector β, meaning $\sum_{i=1}^{m} z_{i,j} = b_j$.

Theorem 5.2.3. Young subgroup double coset theorem. *The assignment $\pi \mapsto (z_{i,j})$ is a bijection from the set of double cosets $S_\alpha \pi S_\beta$ to all matrices of nonnegative integers with row sum vector α and column sum vector β.*

In the case that $m = t = 2$, the only case that will be needed below, this means that the double cosets of $S_i \times S_{n-i}$ and $S_j \times S_{n-j}$ are labelled by a single nonnegative integer $a = \{1,\dots,i\}\cap\pi\{1,\dots,j\}$, that must be such that $i-a$, $j-a$, $n-i-j+a$, the other three entries of the 2×2 matrix of cardinalities, are all nonnegative. This works out as $\max\{0,\, i+j-n\} \le a \le \min\{i,j\}$.

Proof of the double coset theorem for the case $m = t = 2$. Incidentally, the general proof is just about the same, only notationally fractionally more complicated, see [**7**], pp. 18,19.

First let $\rho = \sigma\pi\tau$, $\sigma \in S_i \otimes S_{n-i}$, $\tau \in S_j \times S_{n-j}$. Then

$$\rho(\{1,\dots,j\}) = \sigma\pi\tau(\{1,\dots,j\}) = \sigma\pi(\{1,\dots,j\})$$

and so

$$\{1,\dots,i\}\cap\rho(\{1,\dots,j\}) = \{1,\dots i\}\cap\sigma\pi(\{1,\dots,j\}) = \sigma(\{1,\dots i\}\cap\pi(\{1,\dots,j\})$$

proving one of the four cardinality statements, because σ is a bijection. The other three follow immediately. For instance

$$\{1,\dots,i\}\cap\rho(\{j+1,\dots,n\})$$

is the complement in $\{1,\dots i\}$ of $\{1,\dots i\}\cap\rho(\{1,\dots,j\})$ and so has cardinality $i-a$ as has $\{1,\dots,i\})\cap\pi(\{j+1,\dots,n\})$ for the same reason.

Now let the four cardinalities be given. Then $\{1,\dots,n\}$ splits up into four subsets of the same sizes in two ways:

$$\begin{gathered}\{1,\dots,i\}\cap\pi\{1,\dots,j\},\ \ \{1,\dots,i\}\cap\pi\{j+1,\dots,n\}\\ \{i+1,\dots,n\}\cap\pi\{1,\dots,j\},\ \ \{i+1,\dots,n\}\cap\pi\{j+1,\dots,n\}\\ \{1,\dots,i\}\cap\rho\{1,\dots,j\},\ \ \{1,\dots,i\}\cap\rho\{j+1,\dots,n\}\\ \{i+1,\dots,n\}\cap\rho\{1,\dots,j\},\ \ \{i+1,\dots,n\}\cap\rho\{j+1,\dots,n\}.\end{gathered}$$

It follows that there is a permutation σ_1 of $\{1,\dots i\}$ such that

$$\sigma_1(\{1,\dots,i\}\cap\pi\{1,\dots,j\}) = \{1,\dots,i\}\cap\rho\{1,\dots,j\}$$

$$\sigma_1(\{1,\dots,i\}\cap\pi\{j+1,\dots,n\}) = \{1,\dots,i\}\cap\rho\{j+1,\dots,n\}$$

and also there is a permutation σ_2 of $\{i+1,\dots,n\}$ such that

$$\sigma_2(\{i+1,\dots,n\}\cap\pi\{1,\dots,j\}) = \{i+1,\dots,n\}\cap\rho\{1,\dots,j\}$$

$$\sigma_2(\{i+1,\dots,n\}\cap\pi\{j+1,\dots,n\}) = \{i+1,\dots,n\}\cap\rho\{j+1,\dots,n\}.$$

And thus $\sigma = \sigma_1\sigma_2 \in S_i \times S_{n-i}$ takes the four 'π subsets' into which $\{1, \ldots, n\}$ splits into the corresponding 'ρ subsets'. It follows that there is a $\tau \in S_j \times S_{n-j}$ such that $\rho = \sigma\pi\tau$.

It follows that to write down a set of permutations that represent the double cosets $(S_j \times s_{n-j})\pi(S_i \times S_{n-i})$ one just needs some permutations π_α such that

$$\#(\{1, \ldots, i\} \cap \pi_\alpha\{1, \ldots, j\}) = a.$$

The following set of involutions, for the case $i \leq j$, does the job.

$$\pi_a = \begin{pmatrix} 1,\ldots,a, & \overbrace{a+1,\ldots,i,}^{i-a \text{ elements}} & \overbrace{\overbrace{i+1,\ldots,j}^{j-i \text{ elements}}, \overbrace{j+1,\ldots,i+j-a,}^{i-a \text{ elements}}}^{j-a \text{ elements}} & i+j-a+1,\ldots,n \\ 1,\ldots,a, & \underbrace{j+1,\ldots,i+j-a,}_{i-a \text{ elements}} & i+1,\ldots,j,\underbrace{a+1,\ldots,i,}_{i-a \text{ elements}} & i+j-a+1,\ldots,n \end{pmatrix}$$

Using these explicit π_α it is a straightforward and easy exercise to check that for $H = S_i \times S_{n-i}$ and $K = S_j \times S_{n-j}$

$$H \cap \pi_a^{-1} K \pi_a = S_a \times S_{i-a} \times S_{j-a} \times S_{n-i-j+a} \tag{5.2.4}$$

$$\pi_a H \pi_a^{-1} \cap K = S_a \times S_{j-a} \times S_{i-a} \times S_{n-i-j+a} \tag{5.2.5}$$

so that conjugation by π_a switches the two middle factors.

5.3. The Hopf algebra

Construction 5.3.1. The Grothendieck group of representations of a finite group. Let G be a finite group. The finite dimensional group algebra, $\mathbf{C}G$ of G over the complex numbers is semisimple by the Maschke theorem, see 3.10.19 and 3.10.20. Thus, by the Wedderburn structure theorem, the group algebra over the complex numbers splits as a direct product (sum) of matrix algebras

$$\mathbf{C}G = M^{i_1 \times i_1}(\mathbf{C}) \times \cdots \times M^{i_r \times i_r}(\mathbf{C}) \tag{5.3.2}$$

of simple matrix algebras, where $M^{i \times i}(\mathbf{C})$ is the simple matrix algebra of all $i \times i$ matrices with complex entries. Each $M^{i_j \times i_j}(\mathbf{C})$ acts of course on $\mathbf{C}^{i_j}$ and $\mathbf{C}^{i_j}$ is a simple $M^{i_j \times i_j}(\mathbf{C})$ module and hence a simple $\mathbf{C}G$ module. These are all the simple modules over $\mathbf{C}[G]$, that is the irreducible representations over the complex numbers of the finite group G. For more details see chapter 1 in [**314**], or [**4**], § 16.4.

The Grothendieck group [5], $R(G)$, of virtual complex representations of G is the free Abelian group with as basis the irreducible representations of G. It is of course of finite rank by what has been said above, or by the simple, and easy to prove,

[5]There is a general construction that assigns an Abelian group $A(S)$ to any Abelian semigroup S, much like the way the integers are obtained from the natural numbers. This object satisfies and is characterized by the following universal property. The Grothendieck group of an Abelian semigroup S, is an Abelian group $A(S)$, together with a morphism of semigroups $\varphi : S \longrightarrow A(S)$, such that for any Abelian group B and any morphism of semigroups $\psi : S \longrightarrow B$, there is a unique morphism of Abelian groups $\widetilde{\psi} : A(S) \longrightarrow B$ such that $\widetilde{\psi}\varphi = \psi$.

More generally there is a Grothendieck group associated to any category with a suitable notion of short exact sequence, see [**3**], §16B, [**2**], §38A. The construction was first systematically used in (algebraic) K-theory.

In the present case take the semigroup of all real (as opposed to virtual) representation for S or the category of finite dimensional $\mathbf{C}[G]$-modules to obtain $R(G)$.

observation that the regular representation $\mathbf{C}G$ of G contains every irreducible representation.

5.3.3. Recollection. Let G and H be two finite groups. Then the irreducible representations of the Cartesian product $G \times H$ are precisely the outer tensor products

$$(\rho \otimes \sigma)(g, h) = \rho(g) \otimes \sigma(h)$$

where ρ and σ are respectively an irreducible representation of G and H. Thus

$$R(G \times H) = R(G) \otimes R(H)$$

Construction 5.3.4. The quasi bialgebra structure on RS. As a free Abelian group, i.e. $\mathbf{Z}$-module

$$RS = \bigoplus_{n \geq 0} R(S_n) \tag{5.3.5}$$

the direct sum of the representation rings of the symmetric groups. Here, by decree, $R(S_0) = \mathbf{Z}$.

Define a product structure on RS by

$$m(\rho, \sigma) = \mathrm{Ind}_{S_i \times S_j}^{S_{i+j}} (\rho \otimes \sigma) \tag{5.3.6}$$

and by having $1 \in \mathbf{Z} = R(S_0)$ act as a unit element. Define a coproduct structure by

$$\mu(\rho) = (1 \otimes \rho + \sum_{i=1}^{n-1} \mathrm{Res}_{S_i \times S_{n-i}}^{S_n} + \rho \otimes 1) \in \bigoplus_{i=0}^{n} R(S_i) \otimes R(S_{n-i}) \tag{5.3.7}$$

and a counit (augmentation)

$$\varepsilon : RS \longrightarrow \mathbf{Z}, \quad \varepsilon = \begin{cases} \mathrm{id} & \text{on } R(S_0) = \mathbf{Z} \\ 0 & \text{on } R(S_n), \quad n \geq 1. \end{cases} \tag{5.3.8}$$

Finally, define an inner product by

$$\begin{aligned} \langle \rho, \sigma \rangle &= \dim(\mathbf{Mod}_{S_n}(\rho, \sigma)) \text{ for } \rho, \sigma \text{ representations of } S_n \\ &= 0 \text{ when } \rho, \sigma \text{ are representations} \\ &\qquad \text{of different symmetric groups} \end{aligned} \tag{5.3.8a}$$

i.e. by declaring $1 \in R(S_0) = \mathbf{Z}$ and the irreducible representations to be an orthonormal basis. In all the above ρ, σ are real (as opposed to virtual) [6] representations and the formulas are to be extended by (bi)linearity to all of RS.

The situation is now as follows

(i) RS is a connected graded $\mathbf{Z}$ module.

(ii) There is a preferred basis consisting of 1 and the irreducible representations with corresponding inner product for which this basis is orthonormal.

(iii) Multiplication is positive (because taking the outer tensor product of true (real) representations and than inducing up gives a true representation).

(iv) Comultiplication is positive (because restricting a true representation yields true representations.

---

[6] I.e. 'real representation' does not mean a representation over the real numbers; it means a positive linear combination with integer coefficients of irreducible (complex) representations.

(v) Comultiplication is multiplication preserving (or, equivalently, the multiplication is comultiplication preserving). This is an immediate consequence of the Mackey double coset theorem combined with the description of double cosets of Young subgroups in 5.2 above.

(vi) Comultiplication and multiplication are dual to each other with respect to the inner product. This follows from Frobenius reciprocity.

(vii) The counit morphism is a morphism of algebras. Trivial.

It now follows from the results of 3.8 that RS is a selfadjoint commutative, cocommutative, associative, coassociative graded, connected Hopf algebra over the integers and that it is a PSH algebra as defined and discussed in 4.3.

Moreover there is just one preferred basis element that is primitive, viz the identity representation of S_1. This is evident from (5.3.7), the definition of the comultiplication, because every true representation of an S_n, $n \geq 2$, restricts to some true representation of $S_1 \times S_{n-1}$. Thus RS is graded isomorphic to **Symm**.

As RS is a PSH algebra with precisely one primitive preferred basis element, there are elements x_n, y_n such that the various parts of the Zelevinsky theorem hold. The next thing to figure out is what these x_n and y_n correspond to in terms of representations.

As before denote the unique preferred primitive basis element by p and set $x_1 = y_1 = p$. This is the trivial representation of S_1, often denoted by [1]. The next step is to look at p^2. This is the regular representation of S_2 which decomposes as the direct sum of the trivial representation (identity representation), often denoted [2], and the sign representation, often denoted [1,1]. Take x_2 to be the trivial representation of S_2 and y_2 to be the sign representation of S_2. The inductive construction of the x_n in section 3.3 now proceeds as follows:

Given x_{n-1} look at px_{n-1}. This representation must necessarily be the sum of two different irreducible representations (= preferred basis elements). Precisely one of these is mapped by y_2^* to zero. That one is chosen to be x_n. Also it is shown in 4.3 that $p^* x_n = x_{n-1}$.

Theorem 5.3.9. (Free polynomial generators of RS).
(i) $x_n = [n]$, *the trivial representation of* S_n
(ii) $y_n = [\underbrace{1, \dots, 1}_{n}]$, *the alternating (= sign) representation of* S_n.

Proof. With induction we can assume that $x_{n-1} = [n-1]$, the trivial representation of S_{n-1}. Now look at

$$px_{n-1} = \mathrm{Ind}_{S_1 \times S_{n-1}}^{S_n}([1] \otimes [n-1]).$$

It is easy to see that this representation is in fact what is called the natural representation of S_n, the one that assigns to a permutation π the corresponding permutation matrix that takes the i-th basis vector to the $\pi(i)$-th basis vector. To see this write S_n as the disjoint union of cosets of $S_1 \times S_{n-1} = S_{n-1}$ (where S_{n-1} acts on $\{2, \dots, n\}$:

$$S_n = S_{n-1} \cup (1,2)S_{n-1} \cup \dots \cup (1, n-1)S_{n-1}$$

so that

$$\mathbf{C}S_n \otimes_{\mathbf{C}S_{n-1}} \mathbf{C} =$$
$$= \mathbf{C}S_{n-1} \otimes_{\mathbf{C}S_{n-1}} \mathbf{C} \oplus (1,2)\mathbf{C}S_{n-1} \otimes_{\mathbf{C}S_{n-1}} \mathbf{C} \oplus \ldots \oplus (1,n)\mathbf{C}S_{n-1} \oplus_{\mathbf{C}S_{n-1}} \mathbf{C} =$$
$$= \mathbf{C} \oplus (1,2)\mathbf{C} \oplus \ldots \oplus (1,n)\mathbf{C}.$$

It is easy to see that left multiplication with π permutes these cosets exactly according to $(1,i)S_{n-1} \mapsto (1,\pi(i))S_{n-1}$ proving the statement. This natural representation visibly contains a copy of the trivial representation because the vector

$$\underbrace{(1,1,\ldots,1)}_{n}$$

is invariant [7]. The claim is that this is x_n. To see this it must be shown that $y_2^*[n] = 0$. To this end first note that

$$\langle p^*[n], [n-1]\rangle = \langle [n], p[n-1]\rangle = 1$$

so that

$$p^*[n] \text{ contains exactly one copy of } [n-1]. \tag{5.3.10}$$

Further,

$$\langle p^*[n], p^*[n]\rangle = \langle [n], p(p^*[n])\rangle = \langle \mu([n]), p \otimes p^*[n]\rangle = \langle p \otimes [n-1], p \otimes p^*[n]\rangle = 1$$

where (5.3.10) is used for the last equality. As p^* is a positive operator it follows that

$$p^*[n] = [n-1].$$

Now

$$x_2^* + y_2^* = (p^2)^* = (p^*)^2$$

and so

$$x_2^*[n] + y_2^*[n] = (p^*)^2[n] = [n-2].$$

Both x_2^*, y_2^* are positive operators and it follows that either $x_2^*[n] = 0$ or $y_2^*[n] = 0$. But

$$\langle x_2^*[n], [n-2]\rangle = \langle [n], x_2[n-2]\rangle = \langle [n], [2][n-2]\rangle > 0$$

because the product of two trivial representations always contains a copy of the trivial representation.

This proves that $y_2[n] = 0$ and hence $[n] = x_n$, and, with induction, the first part of the theorem. The second part is proved similarly.

5.4. Symm as a PSH algebra

It follows by the Zelevinsky theorem that RS is isomorphic as a Hopf algebra with **Symm** and hence that **Symm** inherits the structure of a PSH algebra. It remains to figure out just what the preferred basis elements of **Symm** are and what the induced inner product looks like.

Theorem 5.4.1. **Symm** as a PSH algebra. *The Hopf algebra of symmetric functions is a PSH algebra with as inner product, given by definition* 4.1.21, *and with the Schur functions as preferred basis elements.*

---

[7] The other representation is the one defined by the partition $[n-1,1]$.

Proof. Let $\varphi : RS \longrightarrow \mathbf{Symm}$ be the Bernstein-Zelevinsky isomorphism of section 4.3. This isomorphism takes the inductively constructed elements x_n in the PSH algebra RS to the complete symmetric functions h_n [8].

Now the formula that defines the primitives z_n of section 4.3 in terms of the x_n, see (4.3.36), is exactly the same as the formula that gives the power sum primitives of **Symm** in terms of the complete symmetric functions, see (4.1.46). Thus under the isomorphism φ the z_n go to the p_n. It now follows from the formulas for the inner product of monomials in the z_n, see 4.3.37, compared with the formula for the original inner product of monomials in the power sum primitives in **Symm**, see (4.1.35). that the Bernstein-Zelevinsky isomorphism takes the inner product of the PSH algebra RS to that of **Symm** as defined in 4.1.21.

The Schur functions are an orthonormal basis for **Symm**. So the preferred basis of **Symm** inherited through the isomorphism φ must be a signed permutation of the Schur functions. Further $p = h_1$ spreads out all over. So it remains to check that p^n is a positive sum of Schur functions. With induction this reduces to proving that ps_λ is positive for all partitions λ, or using selfadjointness, that $p^* s_\lambda$ is always positive (or zero). Now

$$p^* h_n = h_{n-1} \tag{5.4.2}$$

and because p is a primitive p^* acts as a derivation. Let $\lambda = (a_1, a_2, \ldots, a_m)$. Then

$$s_\lambda = \det \begin{pmatrix} h_{a_1} & h_{a_1+1} & \cdots & h_{a_1+m-1} \\ h_{a_2-1} & h_{a_2} & \cdots & h_{a_2+m-2} \\ \vdots & \vdots & \ddots & \vdots \\ h_{a_m-m+1} & h_{a_m-m+2} & \cdots & h_{a_m} \end{pmatrix}$$

and because p^* is a derivation it follows that $p^* s_\lambda$ is the sum of the m determinants

$$\det \begin{pmatrix} h_{a_1} & h_{a_1+1} & \cdots & h_{a_1+m-1} \\ \vdots & \vdots & \ddots & \vdots \\ p^* h_{a_i-i+1} & p^* h_{a_i-i+2} & \cdots & p^* h_{a_i-i+m} \\ \vdots & \vdots & \ddots & \vdots \\ h_{a_m-m+1} & h_{a_m-m+2} & \cdots & h_{a_m} \end{pmatrix}$$

and because of (5.4.2) these determinants are either Schur functions themselves or zero (because $a_1 \geq a_2 \geq \cdots \geq a_m$).

5.5. The second multiplication on RS

There is a natural second binary operation on RS. Indeed each $R(S_n)$ is a ring under what is called the inner (tensor) product, which assigns to two representations ρ, σ of a group G the representation

$$(\rho \times \sigma)(g) = \rho(g) \otimes \sigma(g) \tag{5.5.1}$$

[8] At this stage it is not important to what irreducible elements the x_n correspond. As a matter of fact they are the trivial representations of the S_n, and the corresponding y_n of section 4.3 are the sign representations (for $n \geq 2$)

where on the right hand side the tensor product is the Kronecker product of matrices:

$$A \otimes B = \begin{pmatrix} a_{1,1}B & \cdots & a_{1,m}B \\ \vdots & \ddots & \vdots \\ a_{m,1}B & \cdots & a_{m,m}B. \end{pmatrix} \tag{5.5.2}$$

This can be used to define a second multiplication on RS by decreeing for representations ρ of S_m and σ of S_n that

$$\rho \times \sigma = \begin{cases} 0 & \text{if } m \neq n \\ \rho \otimes \sigma & \text{if } m = n. \end{cases} \tag{5.5.3}$$

There is also the natural Zelevinsky-Bernstein isomorphism $RS \longrightarrow$ **Symm** and on **Symm** there is also a second multiplication as discussed in section 4.8. So it is natural to wonder whether the Zelevinsky-Bernstein isomorphism takes the one second multiplication into the other. And in fact it does.

Theorem 5.5.4. *The inner tensor product x as defined by* (5.5.3) *is distributive over the product on RS in the Hopf algebra sense, meaning*

$$x \times (yz) = \sum (x_{1,i}, y) \times (x_{2,i}, y), \text{ if } \mu(x) = \sum x_{1,i} \otimes x_{i,2}. \tag{5.5.5}$$

Moreover, the Zelevinsky-Bernstein isomorphism takes the inner tensor product on RS into the second multiplication on **Symm**.

Proof. The first statement of the theorem is an immediate consequence of the projection formula (5.1.21). Next, the second multiplication on **Symm** satisfies $h_n \times h_n = h_n$, see theorem 4.8.21. Moreover this suffices to characterize the second multiplication given its distributivity over the first one (in the Hopf algebra sense).

Indeed, with induction on the length of monomials in the h's, given distributivity and commutativity, the formula $h_n \times h_n = h_n$ determines everything. To start with

$$h_n \times (h_i h_{n-i}) = \sum_j (h_j \times h_i)(h_{n-j} \times h_{n-i}) = h_i h_{n-i}$$

and so, with induction

$$h_n \times a = \begin{cases} a & \text{if } \deg(a) = n \\ 0 & \text{if } \deg(a) \neq n \end{cases}$$

for all monomials a. Next

$$(h_r h_s) \times a = \sum (h_r \times a_{1,i})(h_s \times a_{2,i})$$

and with induction this determines the product of any monomial with another.

Under the Zelevinsky-Bernstein isomorphism $x_n = [n]$, the trivial representation of S_n, goes to h_n. Obviously the inner tensor product of two trivial representations of S_n is again the trivial representation of S_n, and this proves the second part of the theorem.

5.6. Remarks and acknowledgements

The theory of PSH algebras can be pursued (abstractly) a great deal further. Thus for a PSH algebra with one primitive preferred basis element, the preferred basis elements can be shown to be labelled by partitions. Under the Zelevinsky-Bernstein isomorphism these correspond bijectively to the irreducible representations of the S_n labelled by those partitions (or Young diagrams).

There is a second approach to the representation theory of the symmetric groups developed independently and simultaneously by A. Liulevicius, [**10**], [**11**]. It is perhaps worth a few lines to quote from the introduction to the second of these two papers.

"The aim of this paper is to present a ridiculously simple proof of a theorem on the representation rings of the symmetric groups which concisely presents theorems of Frobenius, Atiyah, and Knutson."

This is exactly true. The power of a Hopf algebra structure is enormous.

Very much is known about the isomorphic objects **Symm** and RS. Enough to take up some 5000 pages of densely written text. But still there are open questions. Here are some.

- What is the representation theoretic interpretation of the second comultiplication on **Symm**, the one that gives rise to the multiplication on the big Witt vectors.

- How are the Frobenius and Verschiebung morphism to be interpreted in representation theoretic terms. Also the λ-operations. There have been investigations in these matters, e.g. [**12**], [**13**], [**14**], but it seems that more needs to be done.

The matter of the dual of the λ-operations as well as the higher functorial λ-operations seems not to have been investigated at all in representation theoretic terms.

References

[1] Charles W. Curtis, Irving Reiner, *Representation theory of finite groups and associative algebras*, Interscience, 1962.

[2] Charles W. Curtis, Irving Reiner, *Methods of representation theory with applications to finite groups and orders. Volume II*, Wiley, 1987.

[3] Charles W. Curtis, Irving Reiner, *Methods of representation theory with applications to finite groups and orders. Volume I*, Wiley, 1990.

[4] Marshall Hall, Jr., *The theory of groups*, Macmillan, 1959.

[5] Robin Hartshorne, *Algebraic geometry*, Springer, 1977.

[6] Michiel Hazewinkel, Nadiya Gubareni, V. V. Kirichenko, *Algebras, rings and modules, Volume 2*, Springer, 2007.

[7] Gordon James, Adelbert Kerber, *The representation theory of the symmetric group*, Addison-Wesley, 1981.

[8] G. Karpilovsky, *Group representations. Volume 1. part A: Background material*, North Holland, 1992.

[9] G. Karpilovsky, *Group representations. Volume 1, part B: Introduction to group representations and characters*, North Holland, 1992.

[10] A. Liulevicius, *Representation rings of the symmetric groups – a Hopf algebra approach*, Preprint, Matematisk Institut, Aarhus Universitet, 1976.

[11] A. Liulevicius, *Arrows, symmetries and representation rings*, J. of pure and appl. Algebra **19** (1980), 259–273.

[12] Chr. Reutenauer, *On symmetric functions related to Witt vectors and the free Lie algebra*, Adv. Math. **110** (1995), 234-246.

[13] Thomas Scharf, Jean-Yves Thibon, *A Hopf algebra approach to inner plethysm*, Adv. Math. **104** (1994), 30–58.

[14] Thomas Scharf, Jean-Yves Thibon, *On Witt vectors and symmetric functions*, Algebra Colloq. **3**:3 (1996), 231–238.

[15] P. Webb, *A guide to Mackey functors.* In: M Hazewinkel, ed., *Handbook of algebra. Volume 2*, Elsevier, 2000, 805–836.

CHAPTER 6

The Hopf algebra of noncommutative symmetric functions and the Hopf algebra of quasisymmetric functions

More or less recently, which in mathematics means something like thirty years ago, two generalizations appeared of the Hopf algebra of symmetric functions. They are the Hopf algebra of noncommutative symmetric functions, **NSymm**, and the Hopf algebra of quasisymmetric functions, **QSymm**. They are dual to each other. One, **NSymm**, is maximally noncommutative; the other is maximally noncocommutative.

These are far from generalizations for generalization's sake: they are most important objects in their own right [1]. Besides their importance in themselves, they are quite important in understanding **Symm**. **Symm** fairly shrieks with universality [2]: It is the universal lambda-ring on one generator; it is the cohomology and homology of the classifying space **BU**, it embodies the essentials of the representation theory of the symmetric groups, see chapter 5, and of the general linear group; it represents the functor of the big Witt vectors which has in turn at least five universality properties, see [**22**].

As such things go (universal objects) **Symm** is rather small and that may be a main reason for all these coincidences, for this squeezed togetherness (of which the two Adams=Frobenius theorems are another example). Thus it is important to examine generalizations where, perhaps, things separate out, to increase understanding. [3]

Here only something like an outline of the theory of these two Hopf algebra will be presented. For much more see the two survey papers [18, 19] and the many references quoted there.

It should be stressed that the game should be played characteristic free, i.e. over the integers. Just as it is important to understand the representation theory of the symmetric groups over the integers and to understand **Symm** over the integers. For instance Liulevicius' beautiful theorem on the automorphisms of **Symm** over integers is not true over the rationals.

So far, most investigations into **NSymm** and **QSymm** have been carried out over a field of characteristic zero. But over rationals **NSymm** is isomorphic (as a

[1] But it must be admitted that the name 'quasisymmetric functions' is suspicious. Practically all things 'quasi' have the taint of some largely useless generalization about them.

[2] "Other sins only speak but murder shrieks out", Dorothy Osborne.

[3] This is, of course, also a main reason for the part of mathematics called deformation theory, see (the introduction to) [**21**].

Hopf algebra) to

$$LieHopf = \mathbf{Z}\langle U_1, U_2, \ldots\rangle, \quad U_n \mapsto 1 \otimes U_n + U_n \otimes 1$$

the free associative algebra in counably many variables (with it simplest coproduct), i.e. $\mathbf{NSymm}_{\mathbf{Q}} \cong LieHopf_{\mathbf{Q}}$. And the latter is a very well studied object, see e.g. [**29**] and also section 6.2 below. So, up to isomorphism, there maybe is not all that much new to say. However, the burden of the studies of **NSymm** alluded to is not to do things 'up to isomorphism' but to find, construct, and study noncommutative analogues of all kinds of things in the symmetric functions, i.e. formulae etc. in the Z_i that reduce to familiar formulae etc. in **Symm** when the Z_i are made to commute. Technically this means, formulae, constructions etc. that are compatible with the canonical projection $Z_i \mapsto h_i$ of **NSymm** onto **Symm**. See section 6.4 below.

In addition, trying to do things over the integers brings in extra interesting combinatorial problems, well worth understanding. Not everything exists over the integers (nor should this be the case). For instance there would appear to be no good family of Frobenious endomorphisms of **NSymm** over the integers while over the rationals this is no problem at all[4].

Quasisymmetric functions arose first in combinatorics, more precisely the theory of P-partitions, which are a simultaneous generalization of both compositions (totally ordered partitions) and partitions (which are totally unordered). According to Richard P. Stanley, *Enumerative combinatorics*, volume 2, Cambridge Univ. Press, 2001, page 401, quasisymmetric functions were first defined by Ira M. Gessel [**11**], though they appeared implicitly in earlier work (including Stanley's own thesis, *Ordered structures and partitions*, Memoirs of the Amer. math. Soc. **119** (1972)).

Below the Hopf algebra of noncommutative symmetric functions is introduced abstractly (as is done in [**14**], the seminal paper on the topic). It has also a concrete realization as the direct sum of the Solomon descent algebras, see Louis Solomon (Tits, Jacques), A Mackey formula in the group of a Coxeter group. With an appendix by J. Tits, *Two properties of Coxeter complexes*, J. of Algebra **41** (1976), 255-268. It is as such that **NSymm** is seen as a sub Hopf algebra of the Hopf algebra of permutations, see chapter 7 below. See also reference 10 of chapter 7 where the direct sum of the Solomon descent algebras is exhibited as the graded dual of the Hopf algebra of quasisymmetric functions.

6.1. The Hopf algebra NSymm

As an algebra the Hopf algebra of noncommutative symmetric functions over the integers, **NSymm**, is simply the free algebra in countably many indeterminates over **Z**:

$$\mathbf{NSymm} = \mathbf{Z}\langle Z_1, Z_2, \ldots\rangle. \tag{6.1.1}$$

It is made into a Hopf algebra by the comultiplication

$$\mu(Z_n) = \sum_{i+j=n} Z_i \otimes Z_j, \text{ where } Z_0 = 1 \tag{6.1.2}$$

[4]This probably indicates that some rethinking is needed in the noncommutative case concerning the notions of Frobenius operators and lambda rings.

and the counit

$$\varepsilon(Z_n) = 0, \quad n \geq 1; \quad \varepsilon(Z_0) = 1. \tag{6.1.3}$$

There is an antipode determined by the requirement that it be an anti-endomorphism of rings of **NSymm** and

$$\iota(Z_n) = \sum_{\mathrm{wt}(\alpha)=n} (-1)^{\mathrm{length}(\alpha)} Z_\alpha. \tag{6.1.4}$$

Here, as before, α is a word over the alphabet $\{1, 2, \ldots\}$, $\alpha = [a_l, a_2, \ldots, a_m]$, $\mathrm{length}(\alpha) = m$, $wt(\alpha) = a_1 + \cdots + a_m$, and, for later use, $\alpha^t = [a_m, \ldots, a_2, a_1]$. Further $Z_\alpha = Z_{a_1} Z_{a_2} \cdots Z_{a_m}$.

A composition $\beta = [b_1, \ldots, b_n]$ is a refinement of the composition

$$\alpha = [a_1, a_2, \ldots, a_m]$$

iff there are integers $l \leq j_1 < j_2 < \cdots < j_m = n$ such that $a_i = b_{j_{i-1}+1} + \cdots + b_{ji}$, where $j_0 = 0$. For instance the refinements of $[3, 1]$ are $[3, 1]$, $[2, 1, 1]$, $[1, 2, 1]$, and $[1, 1, 1, 1]$. An explicit formula for the antipode is then

$$\iota(Z_\alpha) = \sum_{\beta \text{ refines } \alpha^t} (-1)^{\mathrm{length}(\beta)} Z_\beta. \tag{6.1.5}$$

The variable Z_n is given weight n which defines a grading on the Hopf algebra **NSymm** for which $\mathrm{wt}(Z_\alpha) = \mathrm{wt}(\alpha)$.

This Hopf algebra of noncommutative symmetric functions was introduced in the seminal paper [**14**] and extensively studied there and in a slew of subsequent papers such as [**10**], [**12**], [**22**], [**24**], [**25**], [**26**], [**31**]. (But all over a field of characteristic zero).

It is amazing how much of the theory of symmetric functions has natural analogues for the noncommutative symmetric functions. This includes ribbon Schur functions, Newton primitives (two kinds of analogues of the power sums p_n), Frobenius reciprocity, representation theoretic interpretations, determinantal formulae, ... Some of these and others will turn up below. Sometimes these noncommutative analogues are more beautiful and better understandable then in the commutative case (as happens frequently); for instance the recursion formula of the Newton primitives and the properties of ribbon Schur functions.

Construction 6.1.6. The Newton primitives of **NSymm**. The (noncommutative) **Newton primitives** in **NSymm** are defined by

$$P_n(Z) = \sum_{r_1+\ldots+r_k=n} (-1)^{k+1} r_k Z_{r_1} Z_{r_2} \cdots Z_{r_k}, \quad r_i \in \mathbf{N} = \{1, 2, \ldots\} \tag{6.1.7}$$

or, equivalently, by the recursion relation

$$nZ_n = P_n(Z) + Z_1 P_{n-1}(Z) + Z_2 P_{n-2} + \cdots + Z_{n-1} P_1(Z). \tag{6.1.8}$$

Note that under the projection $Z_n \mapsto h_n$ by (6.1.8) and (4.1.33) $P_n(Z)$ goes to p_n. It is easily proved by induction, using (6.1.8), or directly from (6.1.7)), that the $P_n(Z)$ are primitives of **NSymm**, and it is also easy to see from (6.1.8) that over the rationals **NSymm** is the free associative algebra generated by the $P_n(Z)$. Thus, see 3.5.20, over the rationals the Lie algebra of primitives of **NSymm** is simply the free Lie algebra generated by the $P_n(Z)$, giving a description of $\mathrm{Prim}(\mathbf{NSymm}_{\mathbf{Q}})$, but not a description of Prim(**NSymm**) which is a far more complicated object.

There is a second family of Newton primitives, given by

$$Q_n(Z) = \sum_{r_1+\ldots+r_k=n} (-1)^{k+1} r_1 Z_{r_1} Z_{r_2} \cdots Z_{r_k}, \quad r_i \in \mathbf{N} = \{1,2,\ldots\} \tag{6.1.9}$$

and these satisfy the recursion relation

$$nZ_n = Q_n(Z) + Q_{n-1}(Z)Z_1 + Q_{n-2}Z_2 + \cdots + Q_1(Z)Z_{n-1}. \tag{6.1.10}$$

6.2. NSymm over the rationals

As in section 3.4.5 consider the Lie Hopf algebra

$$LieHopf = \mathbf{Z}\langle U_1, U_2, \ldots \rangle, \quad \mu(U_n) = 1 \otimes U_n + U_n \otimes 1 \tag{6.2.1}$$

The claim is that over the rationals this Lie Hopf algebra becomes isomorphic (as Hopf algebras) to **NSymm**. There are in fact very many isomorphisms, even many homogeneous ones. One is $P_n(Z) \mapsto U_n$. Indeed, $\mathbf{NSymm}_{\mathbf{Q}}$ is the free associative algebra over the rationals in the countable many independent variables $P_n(Z)$.

A particularly beautiful isomorphism [5] is given, by the formula

$$1 + Z_1 t + Z_2 t^2 + \cdots = \exp(U_1 t + U_2 t^2 + \cdots) \tag{6.2.2}$$

where t is a counting variable commuting with everything in sight and 'exp' is the usual exponential formula. The explicit formula for Z_n in terms of the U's is

$$Z_n = \sum_{\substack{i_1+\cdots+i_m=n \\ i_j \in \mathbf{N}}} \frac{U_{i_1} U_{i_2} \cdots U_{i_m}}{m!}. \tag{6.2.3}$$

Theorem 6.2.4. *Define* $\varphi : \mathbf{NSymm}_{\mathbf{Q}} \longrightarrow LieHopf_{\mathbf{Q}}$ *by* (6.2.2), *i.e.* $\varphi(Z_n)$ *is equal to the right hand side of* (6.2.3). *Then* φ *is an isomorphism of Hopf algebras.*

Proof. Because **NSymm** is free, φ certainly defines a unique morphism of algebras. Moreover, φ is degree preserving and $\varphi(Z_n) = U_n \mod (U_1, \ldots, U_{n-1})$ so that φ is an isomorphism of algebras. Further φ respects the augmentations. It remains to show that φ respects the comultiplication; it is then automatic that it respects the antipodes.

Thus it remains to show that if $\mu(U_i) = 1 \otimes U_i + U_i \otimes 1$ is applied to the right hand side of (6.2.3), then the result is

$$\sum_{i+j=n} \varphi(Z_i) \otimes \varphi(Z_j).$$

---

[5] It is not clear at this time why this is a much more beautiful isomorphism compared with all the other ones. It just feels that way. However, it fits well with some of the considerations of chapter 4. Indeed, consider for the moment the commutative analogue of (6.2.2), that is

$$u_1 t + u_2 t^2 + \cdots = \log(1 + h_1 t + h_2 t^2 + \cdots) = \log(\prod_i (1 - x_i t)^{-1}) = p_1 t + \frac{p_2 t^2}{2} + \frac{p_3 t^3}{3} + \cdots$$

where the p_i are the power sums. So the isomorphism (6.2.2) is basically the noncommutative analogue of the operator of ghost components $t\frac{d}{dt}\log$ that plays such an important role in the theory of the big Witt vectors. The "beautiful isomorphism" (6.2.2) also appears in A. Garsia, C. Reutenauer, *A decomposition of Solomon's descent algebra*, Advances in Mathematics **77** (1989), 189-262. This may entail a second reason for singling out this particular isomorphism over the rationals over all others.

Now

$$\mu(U_{i_1}\cdots U_{i_m}) = \sum U_{a_1}\cdots U_{a_r}\otimes U_{b_1}\cdots U_{b_s}$$

where $r+s=m$ and the $a_1,\dots,a_r$; $b_1,\dots,b_s$ are all pairs of complementary subsequences of $i_1,\dots,i_m$. In other words $i_1,\dots,i_m$ is one of the shuffles of $a_1,\dots a_r$ and $b_1,\dots,b_s$, see 3.4.6. It remains to figure out how many $i_1,\dots,i_m$ there are that yield the term $U_{a_1}\cdots U_{a_r}\otimes U_{b_1}\cdots U_{b_s}$. This amounts to choosing precisely r places from m places (where the $a_1,\dots,a_r$ are inserted in that order and the $b_1,\dots b_s$ are inserted in the remaining $m-r=s$ in their original order). This number is $\frac{m!}{r!s!}$. Hence

$$\begin{aligned}\mu(\varphi(Z_n)) &= \sum \tfrac{1}{m!}U_{a_1}\cdots U_{a_r}\otimes U_{b_1}\cdots U_{b_s}(\tfrac{m!}{r!s!})\\ &= \sum \tfrac{U_{a_1}\cdots U_{a_r}}{r!}\otimes \tfrac{U_{b_1}\cdots U_{b_s}}{s!} = (\varphi\otimes\varphi)(\textstyle\sum_{r+s=m} Z_r\otimes Z_r)\end{aligned}$$

which proves what is desired.

Another proof, more conceptual but to some perhaps less immediately convincing, is as follows. Note that $U_i\otimes 1$ and $1\otimes U_j$ are commuting variables in $\mathbf{Q}\langle U\rangle\otimes\mathbf{Q}\langle U\rangle$. Hence $A=(1\otimes U_1)t+(1\otimes U_2)t^2+\cdots$ and $B=(U_1\otimes 1)t+(U_2\otimes 1)t^2+\cdots$ are commuting elements in $\mathbf{Q}\langle U\rangle\otimes\mathbf{Q}\langle U\rangle[t]$. Hence $\exp(A+B)=\exp(A)\exp(B)$ and the desired result follows by applying μ to (6.2.2)

The inverse formula to (6.2.3) is

$$U_n = \sum_{i_1+\cdots+i_k=n} (-1)^{k+1}\frac{Z_{i_1}Z_{i_2}\cdots Z_{i_k}}{k} \tag{6.2.5}$$

and so the $n!U_n$ give rise to a family of primitives in **NSymm** which are symmetric in the Z's. One for each degree. They are most probably worth some further study and may well turn out to better analogues of the power sums then the Newton primitives P_n and Q_n of section 6.1 above.

6.3. The Hopf algebra QSymm

The graded dual of the connected graded Hopf algebra **NSymm** of free finite type over the integers is the Hopf algebra of quasisymmetric functions. This demands some explanations. Quasisymmetric functions were defined in (4.1.7).

6.3.1. Recollection. Monomial quasisymmetric functions. Given an exponent sequence (composition) $\alpha=[a_1,\dots,a_m]$, the corresponding **monomial quasisymmetric function** in a number (possibly infinite) of variables is

$$M_\alpha = M_\alpha(x_1,\dots,x_n) = \sum_{i_1<i_2<\dots<i_m\le n} x_{i_1}^{\alpha_1}x_{i_2}^{\alpha_2}\cdots x_{i_m}^{\alpha_m}.$$

So for instance, in three variables

$$M_{[1,5]} = x_1x_2^5 + x_1x_3^5 + x_2x_3^5$$

and this is also an example of a quasisymmetric function that is not symmetric. The monomial quasisymmetric functions are a free basis for the Abelian group of quasisymmetric functions and the product of two monomial quasisymmetric functions is a finite sum of monomial quasisymmetric functions. This product can be described directly in combinatorial terms as follows.

6.3.2. Definition-construction. Overlapping shuffle product. Let $\alpha = [a_1, a_2, \ldots, a_m]$ and $\beta = [b_1, b_2, \ldots, b_n]$ be two compositions or words over the integers $\mathbf{N} = \{1, 2, \ldots\}$. Take a 'sofar empty' word with $n + m - r$ slots where r is an integer between 0 and $\min\{n, m\}$, $0 \le r \le \min\{m, n\}$. Choose m of the available $n + m - r$ slots and place in it the natural numbers from α in their original order; choose r of the now filled places; together with the remaining $n+m-r-m = n-r$ places these form n slots; in these place the entries from β in their original order; finally, for those slots which have two entries, add them. The product of the two words α and β is the sum (with multiplicities) of all words that can be so obtained. So, for instance,

$$\begin{aligned}[a, b] \times_{osh} [c, d] = {} & [a, b, c, d] + [a, c, b, d] + [a, c, d, b] + [c, a, b, d] + [c, a, d, b] + \\ & + [c, d, a, b] + [a + c, b, d] + [a + c, d, b] + [c, a + d, b] + \\ & + [a, b + c, d] + [a, c, b + d] + [c, a, b + d] + [a + c, b + d]\end{aligned}$$

and

$$[1] \times_{osh} [1] \times_{osh} [1] = 6[1, 1, 1] + 3[1, 2] + 3[2, 1] + [3].$$

The empty word (composition), [ ], corresponding to the constant quasisymmetric function 1, serves as the unit element.

Note that the shuffle product of α and β, discussed in section 3.4.6 just consists of the terms without overlap of the overlapping shuffle product.

Proposition 6.3.3. Multiplication of monomial quasisymmetric functions.

$$M_\alpha M_\beta = M_{\alpha \times_{osh} \beta} \tag{6.3.4}$$

where on the left hand side the product is taken as a product of power series and on the right hand side is meant the sum of all the monomial quasisymmetric functions M_γ (with multiplicities) for the compositions that occur in $\alpha \times_{osh} \beta$.

Proof. This is entirely straightforward. The products of monomials

$$x_{i_1}^{a_1} x_{i_2}^{a_2} \cdots x_{i_m}^{a_m} \quad \text{and} \quad x_{j_1}^{b_1} x_{j_2}^{b_2} \cdots x_{j_n}^{b_n}$$

gives a summand of a nonoverlapping term of $\alpha \times_{osh} \beta$ if and only if $\{i_1, \ldots, i_m\} \cap \{j_1, \ldots j_n\} = \varnothing$. And if this intersection is nonempty with cardinality r then it is a summand of a term of $\alpha \times_{osh} \beta$ with precisely r overlaps.

In the following no real distinction will be made between a composition (word over the integers) and the quasisymmetric function it defines.

Definition 6.3.5. Duality pairing between **NSymm** and **QSymm**. For each composition $\alpha = [a_1, \ldots, a_m]$ let $Z_a = Z_{a_1} \cdots Z_{a_m}$, which is, of course, a monomial of weight $\mathrm{wt}(\alpha)$, and set $Z_{[\,]} = Z_0 = 1$. A duality pairing between **NSymm** and **QSymm** is now given by declaring the basis $\{Z_\alpha\}_{\alpha \in \mathbf{N}^*}$ of **NSymm** and the basis $\{\alpha\}_{\alpha \in \mathbf{N}^*}$ of **QSymm** to be orthonormal dual bases:

$$\langle Z_\alpha, \beta \rangle = \delta_\alpha^\beta \quad \text{(Kronecker delta)}. \tag{6.3.6}$$

Now consider the Abelian group, **QSymm** of the quasisymmetric functions, or, equivalently the graded free Abelian group with as basis all words over the natural numbers including the empty word. Equip it with a multiplication by the overlapping shuffle product with the empty word as unit, and give it as comultiplication the operation 'cut'

$$\mu[a_1, \dots, a_m] = \sum_{i=0}^{m} [a_1, \dots, a_i] \otimes [a_{i+1}, \dots, a_m]$$

where of course $[a_1, \dots, a_0] = [\,] = 1$ and also $[a_{m+1}, \dots, a_m] = [\,] = 1$, and a counit

$$\varepsilon(\alpha) = \begin{cases} 1 & \text{if } \alpha = [\,] \\ 0 & \text{if } \mathrm{wt}(\alpha) > 0. \end{cases}$$

Theorem 6.3.7. Duality of **QSymm** and **NSymm**. **QSymm** *with the structures just above, i.e. overlapping shuffle product as multiplication and cut as comultiplication, is the graded dual of the Hopf algebra* **NSymm** (*and hence a Hopf algebra, complete with antipode*).

Proof. There are a number of formulas to be proved. The most difficult one is the duality of comultiplication on **NSymm** and multiplication on **QSymm**, i.e. the formula

$$\langle \mu(Z_\alpha), \beta \otimes \gamma \rangle = \langle Z_\alpha, \ \beta \times_{osh} \gamma \rangle. \tag{6.3.8}$$

This is still entirely straightforward all the same. The overlapping shuffle product $\beta \times_{osh} \gamma$ can also be described as follows. Consider $2 \times s$ matrices M without zero columns such that the first row consists of the entries of β (in their original order) interspersed with zeros and the second row consists of the entries of γ, also in their original order interspersed with zeros. Let $\mathcal{M}_{\beta,\gamma}$ be the set of all such matrices. Then

$$\beta \times_{osh} \gamma \sum_{M \in \mathcal{M}_{\beta,\gamma}} \mathrm{csum}(M) \tag{6.3.9}$$

where 'csum' stands for column sum [6]. For instance if

$$M = \begin{pmatrix} 0 & b_1 & b_2 & b_3 & 0 & b_4 \\ c_1 & 0 & 0 & c_2 & c_3 & 0 \end{pmatrix}$$

then

$$\mathrm{csum}(M) = [c_1, b_1, b_2, b_3 + c_2, c_3, b_4]$$

which is one of the terms of $[b_1, \dots, b_4] \times_{osh} [c_1, c_2, c_3]$.

From this description it is clear that the RHS of (6.3.8) has a contribution 1 for each pair $(\alpha, \beta \otimes \gamma)$ such that α is a column sum of a matrix in $\mathcal{M}_{\beta,\gamma}$.

On the other hand

$$\mu(Z_\alpha) = \prod_{i=1}^{m} \sum_{\substack{r_i + s_i = a_i \\ r_i, s_i \geq 0}} Z_{r_i} \otimes Z_{s_i}.$$

This expression defines an associated matrix

$$M = \begin{pmatrix} r_1 & r_2 & \dots & r_m \\ s_1 & s_2 & \dots & s_m \end{pmatrix}$$

[6] This description also works for iterated overlapping shuffle products. When taking the overlapping shuffle products of r compositions look at all $r \times s$ matrices without zero columns, whose i-th row consists of the entries of the i-th factor (in their original order) interspersed with zeros and take column sums.

with columns sum α which is in $\mathcal{M}_{\beta,\gamma}$, where β is $[r_1, \dots r_m]$ with the zeros removed and, likewise, γ is $[s_1, \dots s_m]$ with the zeros removed. Thus the LHS of (6.3.8) also has a contribution 1 for every matrix $M \in \mathcal{M}_{\beta,\gamma}$ with column sum α. This proves (6.3.8). The other formulas that need to be checked are still easier and are left as an exercise.

6.4. Symm as a quotient of NSymm

There is a natural morphism of Hopf algebras

$$\pi : \mathbf{NSymm} \longrightarrow \mathbf{Symm}, \quad Z_n \mapsto h_n. \text{ [7]} \tag{6.4.1}$$

Now **Symm** is an extremely richly structured object. The question arises which of the many structures have analogues in **NSymm**, which here we shall take to mean that they have similar properties and descend to **Symm** by means of the projection π.

For instance **NSymm** has two families of Newton primitives given by the formulas

$$P_n(Z) = \sum_{r_1+\cdots+r_k=n} (-1)^{k+1} r_k Z_{r_1} Z_{r_2} \cdots Z_{r_k}, \; r_i \in \mathbf{N} = \{1, 2, \dots\} \tag{6.4.2}$$

$$Q_n(Z) = \sum_{r_1+\cdots+r_k=n} (-1)^{k+1} r_1 Z_{r_1} Z_{r_2} \cdots Z_{r_k}, \; r_i \in \mathbf{N} = \{1, 2, \dots\} \tag{6.4.3}$$

or, equivalently, by the recursion relations

$$nZ_n = P_n(Z) + Z_1 P_{n-1}(Z) + Z_2 P_{n-2}(Z) + \cdots + Z_{n-1} P_1(Z) \tag{6.4.4}$$

$$nZ_n = Q_n(Z) + Q_{n-1}(Z) Z_1 + Q_{n-2}(Z) Z_2 + \cdots + Q_1(Z) Z_{n-1} \tag{6.4.5}$$

and these recursion relations descend to the (same) recursion relation for the power sum primitives of **Symm**. [8]

Related to the matter of analogues on **NSymm** of structures on **Symm** is the matter whether there exist section morphisms of one kind or another of the projection π. That is morphisms

$$s : \mathbf{Symm} \longrightarrow \mathbf{NSymm}, \quad \text{such that } \pi s = \mathrm{id}_{\mathbf{Symm}}. \tag{6.4.6}$$

Obviously there are Abelian group morphism sections. Many.

Proposition 6.4.7. Nonexistence of algebra sections of **NSymm** $\longrightarrow$ **Symm**. *There is no morphism of algebras* $s : \mathbf{Symm} \longrightarrow \mathbf{NSymm}$ *such that* $\pi s = \mathrm{id}_{\mathbf{Symm}}$.

Proof. This is quite easy. If there were such an s then necessarily

$$s(h_1) = Z_1 + R_1 \quad \text{and } s(h_2) = Z_2 + R_2 \quad \text{with } R_1, R_2 \in I$$

---

[7] There is a good reason to take the complete symmetric functions rather than the elementary symmetric functions as will become clear in section 6.6. It is for the same reason that Witt vectors in chapter 4 were treated in terms of the complete symmetric functions. There are in fact several more good reasons to prefer the complete symmetric functions over the elementary symmetric functions; for instance in the theory of Witt vectors, see **[22]**.

[8] The explicit formulas (6.4.2), (6.4.3) are a nice example of how things can become more elegant in the noncommutative case compared to the commutative case.

the commutator ideal of **NSymm**, which means that R_1, R_2 are sums of terms of weight 3 and higher. But, if s is to be multiplication preserving one needs

$$s(h_1)s(h_2) = s(h_1h_2) = s(h_2h_1) = s(h_2)s(h_1)$$

and this then gives

$$Z_1Z_2 - Z_2Z_1 = Z_2R_1 - R_1Z_2 + R_2Z_1 - Z_1R_2 + R_2R_1 - R_1R_2$$

which is not possible because the left hand side is nonzero of weight 3 and the right hand side is a sum of elements of weight 4 or more.

More generally there is no algebra injection of **Symm** into **NSymm**. Were there such a one there would be many commuting elements in the free algebra **NSymm**. A little experimentation rapidly gives one the feeling that it is very difficult to find two commuting noncommutative polynomials (unless they are obtainable from single variable polynomials with one and the same noncommutative polynomial substituted. Proving this rigorously is quite another matter. But indeed there is:

Theorem 6.4.8. Bergman centralizer theorem. *Let k be a field and $A = k\langle X\rangle$ a free algebra over k. Then the centralizer of any nonconstant element of A is isomorphic (as an algebra) to $k[t]$, the algebra of polynomials in one variable over k.*

For a proof see [**2**], [**3**], or [**9**], section 6.8; see also M. Lothaire, *Algebraic combinatorics on words*, Cambridge University Press, 2002, Chapter 9 on centralizers of noncommutative series and polynomials.

On the other hand, as will be shown later, there do exist coalgebra sections s : **Symm** $\longrightarrow$ **NSymm**. This brings the BCM theorem, 3.3.29, into play and exhibits, somewhat surprisingly, **NSymm** as a cross product of **Symm** with a certain subalgebra A of **NSymm** [9].

Definition 6.4.9. Divided power sequence. Let H be a Hopf algebra. A **divided power sequence** (DPS) in H is a sequence of elements $d = (d_0 = 1, d_1, d_2, \ldots)$ such that

$$\mu(d_n) = \sum_{i=0}^{n} d_i \otimes d_{n-i} \quad \text{for all } n. \tag{6.4.10}$$ [10]

Note that d_1 is a primitive. It is often convenient to write a divided power sequence d as a power series

$$d(t) = 1 + d_1t + d_2t^2 + \cdots \tag{6.4.11}$$

where t is a counting variable commuting with everything.

Example 6.4.12. Examples of divided power sequences.

$h(t) = 1 + h_1t + h_2t^2 + \cdots$ is a divided power sequence in **Symm**.

$Z(t) = 1 + Z_1t + Z_2t^2 + \cdots$ is a divided power sequence in **NSymm**.

[9] The theory described in [**18**] should give a good description of the algebra A. The matter of the validity of the BCM theorem over a general ring k also needs attention.

[10] The name has much to do with the divided power coalgebra of example 2.4.8. A divided power sequence in a Hopf algebra H is the same as a morphism of coalgebras from the divided power coalgebra into H that takes Z_0 into 1.

C) $\lg(\gamma) < lg(\beta)$, so that $\beta = \beta' * \gamma$ and

$$\alpha * \beta <_{\text{lex}} \beta <_{\text{lex}} \gamma$$

by (i) and the fact that β is Lyndon.

Theorem 6.5.5. Chen-Fox-Lyndon factorization. *For each word α there is a unique concatenation factorization into nonincreasing Lyndon words*

$$\alpha = \lambda_1 * \lambda_2 * \cdots * \lambda_s, \quad \lambda_1 \geq_{\text{lex}} \lambda_2 \geq_{\text{lex}} \cdots \geq_{\text{lex}} \lambda_s. \tag{6.5.6}$$

Proof. Any word can be written in at least one way as a concatenation product of Lyndon words; simply because words of length 1 are Lyndon. So choose a concatenation factorization

$$\alpha = \lambda_1 * \lambda_2 * \cdots * \lambda_s$$

with s minimal. If there were an i with $\lambda_i <_{\text{lex}} \lambda_{i+1}$, then, by lemma 6.5.4, $\lambda_i * \lambda_{i+1}$ is Lyndon, contradicting that s is minimal. This proves the existence of at least one factorization like (6.5.6).

To show uniqueness, suppose that

$$\alpha = \lambda_1 * \lambda_2 * \cdots * \lambda_s = \lambda'_1 * \lambda'_2 * \cdots * \lambda'_{s'}$$

are two factorizations like (6.5.6). Now, if $\lg(\lambda_1) = \lg(\lambda'_1)$ induction (on the length of α) finishes the job. If $\lg(\lambda_1) > \lg(\lambda'_1)$

$$\lambda_1 = \lambda'_1 * \lambda'_2 * \cdots * \lambda'_i * \beta$$

for some prefix β of λ'_{i+1}. And then

$$\lambda_1 <_{\text{lex}} \beta <_{\text{lex}} \lambda'_{i+1} \leq_{\text{lex}} \lambda'_1 <_{\text{lex}} \lambda_1$$

which is a contradiction. Similarly $\lg(\lambda_1) < \lg(\lambda'_1)$ gives a contradiction. This proves the theorem.

Algorithm 6.5.7. *Let α be a nonempty word. Let λ be the smallest nonempty tail of α, and write $\alpha = \alpha' * \lambda$. Continue with α'. This gives the* CFL *factorization of α.*

Proof. To see this it must be shown that if $\alpha = \lambda_1 * \cdots * \lambda_n$ is the CFL concatenation factorization of α, then λ_n is the smallest nonempty tail of α. First of all $\lambda \in LYN$ because it is a smallest nonempty tail. If $\alpha' = [\,]$ things are done. If not let λ' be the smallest nonempty tail of α'. If $\lambda' < \lambda$, $\lambda' * \lambda \in LYN$ and hence $\lambda' * \lambda < \lambda$ contradicting that λ is smallest. So the algorithm gives a factorization in Lyndon words with the required monotonicity property, which hence must be the CFL factorization (because of the uniqueness part of theorem 6.5.5).

This algorithm is conceptually easy but it is not very fast (and it works from right to left which is also a disadvantage). A much faster algorithm from right to left is the block decomposition algorithm of [**29**], see [**16**] for a proof that the block decomposition algorithm gives the CFL factorization. The block decomposition algorithm also gives the recursive structure of Lyndon words. A linear time algorithm that works from left to right is in [**13**].

For later use, in the proof that *Shuffle* is free over the rationals, the following theorem will be used.

Theorem 6.5.8. (Shuffle product and CFL factorization). *Let* $\alpha = \lambda_1 * \cdots * \lambda_m$ *be the* CFL *factorization of a word* α. *Then the lexicographically largest word that occurs in the shuffle product*

$$\lambda_1 \times_{sh} \lambda_2 \times_{sh} \cdots \times_{sh} \lambda_m$$

is α [13].

Proof. The shuffle product of two words is positive in the sense that all coefficients of the words occurring in it are positive. Thus is suffices to show that α occurs and that it is larger or equal than any other (iterated) shuffle. The first is trivial of course. For the second the following lemma is useful.

Lemma 6.5.9. (Compatibility of shuffle product and lexicographic order). *Let* α *and* β *be positive linear combinations of words. Let* $\alpha_{\max}$ (*resp.* $\beta_{\max}$) *be the lexicographically largest word occurring in* α (*resp.* β). *Then the lexicographically largest word occurring in a* $\alpha \times_{sh} \beta$ *comes from* $\alpha_{\max} \times_{sh} \beta_{\max}$.

Proof. Let α' and β' be any two words from α and β and let γ' be a shuffle of α' and β'. Now in γ' replace the i-th letter from α' with the i-th letter from $\alpha_{\max}$ if $\lg(\alpha_{\max}) \leq i$ and otherwise leave it out; similarly replace the j-th letter from β' in γ' with the j-th letter from $\beta_{\max}$ if $\lg(\beta_{\max}) \leq j$ and otherwise remove it. If after these two procedures there are still letters left from $\alpha_{\max}$ and/or $\beta_{\max}$ put them at the end in any order. The result is a shuffle of $\alpha_{\max}$ and $\beta_{\max}$ that is larger or equal than γ'.

Proof of theorem 6.5.8 (*continued*). Let

$$\alpha_1 \geq_{\text{lex}} \cdots \geq_{\text{lex}} \alpha_m, \ \alpha_i = [a_{i,1}, \ldots, a_{i,n_i}]$$

be the Chen-Fox-Lyndon factors of $\alpha = \alpha_1 * \ldots * \alpha_m$. Using the lemma above and induction on the length of α it follows that the lexicographically maximal word in

$$\alpha_1 \times_{sh} \cdots \times_{sh} \alpha_m$$

occurs in

$$\alpha_1 \times_{sh} (\alpha_2 * \cdots * \alpha_m)$$

So let β be any word occurring in this shuffle product. There are a number of different cases to consider.

Case A. The element $a_{i,1}$ from α_i in β occurs after the letter a_{2,n_2}. So we need to compare

$$\alpha = [a_{1,1}, \ldots, a_{1,n_1}, a_{2,1}, \ldots, a_{2,n_2}, a_{3,1}, \ldots, a_{3,n_3}, \ldots] \tag{6.5.10}$$

and

$$\beta = [a_{2,1}, \ldots, a_{2,n_2}, \ldots, a_{i,1}, \ldots, \underset{\uparrow}{a_{1,1}}, \ldots, a_{i,n_i}, \ldots] \tag{6.5.11}$$

There are a number of subcases.

Case Al. $[a_{1,1}, \ldots, a_{1,n_1}] = [a_{2,1}, \ldots, a_{2,n_2}]$. In this case remove these two prefixes from α and β. Let α'' and β'' be the remaining tails. In β'' replace the letters $a_{1,i}$ by the equal letters $a_{2,i}$ to see that β'' is a shuffle of $\alpha_2, \ldots, \alpha_m$. And so, with induction $\alpha'' \geq_{\text{lex}} \beta''$, so that also $\alpha \geq_{\text{lex}} \beta$ in this case.

[13] The coefficient is also known. It is $r_1! \cdots r_k!$ if the r_i are the multiplicities of the Lyndon factors; i.e. $\alpha = \beta_1^{*r_1} * \cdots * \beta_k^{*r_k}$, $\beta_1, \ldots, \beta_k \in LYN$, $\beta_1 >_{\text{lex}} \cdots >_{\text{lex}} \beta_k$.

Case A2. $[a_{1,1},\ldots,a_{1,n_1}] >_{\text{lex}} [a_{2,1},\ldots,a_{2,n_2}]$ and $n_1 \leq n_2$. In this case there is an i such that $a_{1,i} > a_{2,i}$ and $a_{1,j} = a_{2,j}$ for $j = 1,\ldots,i-1$, and thus $\alpha >_{\text{lex}} \beta$.

Case A3. $[a_{1,1},\ldots,a_{1,n_1}] = [a_{2,1},\ldots,a_{2,n_2},b_1,\ldots,b_s]$, $s \geq 1$. As in case A1, remove the equal prefixes $[a_{1,1},\ldots,,a_{1,n}] = [a_{2,1},\ldots,,a_{2,n_2}]$ from α and β and let α'' and β'' be the corresponding tails. Because α_1 is Lyndon $[b_1,\ldots,b_s] >_{\text{lex}} [a_{1,1},\ldots,a_{1,n}] >_{\text{lex}} [a_{2,1},\ldots,a_{2,n_2}]$ so that $\beta_1 * \ldots * \beta_t * \alpha_2 * \ldots * \alpha_m$ is the Chen-Fox-Lyndon factorization of α'' if $\beta_1 * \ldots * \beta_t$ is the CFL factorization of $[\beta_1,\ldots,\beta_s]$ (because as a tail of $\alpha_1, \beta_t >_{\text{lex}} \alpha_1 \geq \alpha_2$). Now in β'' replace the $a_{1,i}$ by the equal letters $a_{2,i}$, $i = 1,\ldots,n_2$ to see that β'' is a shuffle of $\beta_1,\ldots,\beta_t,\alpha_2,\ldots,\alpha_m$. So with induction $\alpha'' \geq_{\text{lex}} \beta''$ and hence $\alpha >_{\text{lex}} \beta$ also in this case.

Case B. The element $a_{1,1}$ from α_1, in β occurs before the letter a_{2,n_2}. So now

$$\alpha = [a_{1,1},\ldots,a_{1,n_1},a_{2,1},\ldots,a_{2,n_2},a_{3,1},\ldots,a_{3,n_3},\ldots]$$

is to be compared with something of the form

$$\beta = [a_{2,1},\ldots,a_{2,i_2},a_{1,1},\ldots,a_{2,n_2},\ldots],\ i_1 \geq 0$$

and, because $\alpha_1 \geq_{\text{lex}} \alpha_2$

$$[a_{1,1},\ldots,a_{1,i_1}] \geq_{\text{lex}} [a_{2,1},\ldots,a_{2,i_1}]. \tag{6.5.12}$$

Again there are subcases.

Case B1. $a_{1,1}$ is the first letter of β. In this case remove the first letter from both α and β and, arguing as in case A3, it follows by induction that $\alpha >_{\text{lex}} \beta$ (except that no relabelling is needed).

Case B2. Strict inequality holds in (6.5.12). Then it follows immediately that $\alpha >_{\text{lex}} \beta$.

Case B3. Equality holds in (6.5.12). Then remove the prefixes $[a_{1,1},\ldots,a_{1,i_1}] = [a_{2,1},\ldots,a_{2,i_1}]$ from α and β and relabel the letters $a_{1,j}$ with the equal letters $a_{2,j}$, $j = 1,\ldots,i_1$, to see with induction that $\alpha >_{\text{lex}} \beta$, again as in case A3.

This finishes the proof.

Theorem 6.5.13. Freeness of *Shuffle*$_{\mathbf{Q}}$. *As an algebra over the rationals the shuffle algebra is commutative free with as many generators of weight n as there are Lyndon words of weight n.*

Proof. For each $\lambda \in LYN$ let x_λ be an indeterminate [14]. Let $\mathbf{Q}[x_\lambda : \lambda \in LYN]$ be the free commutative algebra of polynomials in these indeterminates over the rational numbers. Give the indeterminate x_λ weight equal to $\text{wt}(\lambda)$. This makes $\mathbf{Q}[x_\lambda : \lambda \in LYN]$ a connected graded algebra (of free finite type). The first thing is to figure out the dimension of the homogeneous summands of weight n of $\mathbf{Q}[x_\lambda : \lambda \in LYN]$. This is not difficult. Order the indeterminates by the lexicographic order of their indices. The dimension of $\mathbf{Q}[x_\lambda : \lambda \in LYN]_n$ over $\mathbf{Q}$ is equal to the number of different monomials in the x_λ of total weight n. Because of commutativity each monomial M can be written uniquely as a product

$$M = x_{\lambda_1}x_{\lambda_2}\cdots x_{\lambda_m},\ \lambda_1 \geq_{\text{lex}} \cdots \geq_{\text{lex}} \lambda_m. \tag{6.5.14}$$

This sets up a bijection between monomials of weight n and words of weight n. Indeed to each monomial (6.5.14) associate the word $\lambda_1 * \cdots * \lambda_m$. Inversely, given

[14] This notation is not to be confused with a notation like $Z_\beta = Z_{b_1}Z_{b_2}\ldots Z_{b_m}$ as has been used before. Here the x_λ are separate distinct indeterminates, one for each Lyndon word.

a word α over the natural numbers, take its CFL factorization

$$\alpha = \lambda_1 * \cdots * \lambda_m$$

and assign to it the monomial $x_{\lambda_1} x_{\lambda_2} \cdots x_{\lambda_m}$. These maps are inverse to one another (obviously). And thus

$$\dim_{\mathbf{Q}}(\mathbf{Q}[x_\lambda : \lambda \in LYN]_n) = \dim_{\mathbf{Q}}((\mathit{Shuffle}_{\mathbf{Q}})_n). \tag{6.5.15}$$

Now define a homogeneous morphism of algebras over the rationals as follows

$$\varphi : \mathbf{Q}[x_\lambda : \lambda \in LYN] \longrightarrow \mathit{Shuffle}_{\mathbf{Q}},\ x_\lambda \mapsto \lambda. \tag{6.5.16}$$

This morphism of algebras is surjective. This is proved by induction on weight, being obvious for weight 0 and weight 1. So assume with induction that every word of weight $< n$ is in the image of φ. Let α be of weight n and let $\alpha = \lambda_1 * \cdots * \lambda_m$ be the CFL factorization of α. Then, by the definition of φ and theorem 6.5.8.

$$\begin{aligned}\varphi(x_{\lambda_1} x_{\lambda_2} \cdots x_{\lambda_m}) &= \lambda_1 \times_{sh} \cdots \times_{sh} \lambda_m \\ &= c\alpha + \text{ (terms that are lexicographically smaller than } \alpha).\end{aligned}$$

where c is some positive integer. By induction all terms that are lexicographically smaller than α are in the image of φ and so also $\alpha \in \operatorname{Im}(\varphi)$.

Thus φ is surjective, and being homogeneous, and because of (6.5.15) it is then an isomorphism finishing the proof.

6.5.17. Remark and counterexample. For the proof of theorem 6.5.8, given above, it is absolutely essential that *Shuffle* is considered over the rationals. In fact *Shuffle* is definitely not free over the integers. Indeed $[1] \times_{sh} [1] = 2[1,1]$ and so $\mathit{Shuffle} \otimes_Z \mathbf{Z}/(2)$ has nilpotents which could not happen if *Shuffle* were free over the integers.

Nevertheless *Shuffle* is a free object in another sense; it is free (over LYN) as an algebra with divided powers. Basically an algebra with divided powers is an algebra that comes with operations $\gamma_n : A \longrightarrow A$ that behave like $x \mapsto x^n/n!$. But note, that it is perfectly possible to have an algebra with divided powers over a field of characteristic > 0 where division by a factorial makes no sense. The precise definition follows. [15]

Definition 6.5.18. [16]. Algebras with divided powers. Let A be a commutative connected graded algebra over k. Then it is said to be equipped with divided powers if there are maps

$$\gamma_n : A_r \longrightarrow A_{nr},\ n = 0, 1, 2, \ldots \tag{6.5.19}$$

that satisfy the following five axioms for all $x, y \in A$

$$\gamma_0(x) = 1,\ \gamma_1(x) = x \tag{6.5.20}$$

$$\gamma_n(x)\gamma_m(x) = \binom{n+m}{n}\gamma_{n+m}(x) \tag{6.5.21}$$

$$\gamma_n(x+y) = \sum_{i+j=n} \gamma_i(x)\gamma_j(y) \tag{6.5.22}$$

[15] This notion of divided powers is not to be confused with the notion of a divided power sequence in a Hopf algebra. Though, certainly, the notions have a common origin.

[16] In the world of Hopf algebras of algebraic topological type, which are graded commutative, the gamma operations are only required to exist for even degree elements and one adds the requirement that they are zero on products of two odd degree elements.

$$\gamma_n(xy) = x^n\gamma_n(y) \tag{6.5.23}$$

$$(\gamma_n \circ \gamma_m)(x) = \frac{(nm)!}{n!(m!)^n}\gamma_{nm}(x). \tag{6.5.24}$$

An algebra with divided powers is also called a Γ-algebra. They play, for instance, an important role in cohomological algebraic geometry, [**4**], [**5**].

6.5.25. Construction and example and theorem. The divided power algebra structure on *Shuffle*. The claim is that for each element of *Shuffle* the n-th shuffle power, $x^{\times_{sh} n}$, is divisible by $n!$. This is seen as follows. First let x be a word α. The n-th shuffle power of α can be described as follows. Let the length of α be m. Consider all $n \times mn$ matrices M of the following form:

(i) each row of M consists of zeros and the entries of α (in their original order)
(ii) each column has precisely one nonzero entry

Then the rows of these matrices are all different (because of condition (ii)) and there are $\binom{mn}{n}$ such matrices (which number is divisible by $n!$ (as had better be the case)).

Then the n-th shuffle power of α consists of the sum of all column sums of such matrices. Now each row permutation of such a matrix gives a different such matrix with the same column sum. Thus

$$\underbrace{\alpha \times_{sh} \cdots \times_{sh} \alpha}_{n}$$

is divisible by $n!$. Set $\gamma_n(a\alpha) = a^n(n!)^{-1}\alpha^{\times_{sh} n}$ for α a word and $a \in \mathbf{Z}$. Now let $x = a_1\alpha_1 + \ldots + a_r\alpha_r$ be a linear combination of words. Then

$$\begin{aligned} x^{\times_{sh} n} &= \sum_{i_1+\cdots+i_r=n} \frac{n!}{i_1!\ldots i_r!}(a_1\alpha_1)^{\times_{sh} i_1}\cdots(a_r\alpha_r)^{\times_{sh} i_r} = \\ &= \sum_{i_1+\cdots+i_r=n} n!\gamma_{i_1}(a_1\alpha_1)\cdots\gamma_{i_r}(a_r\alpha_r) \end{aligned}$$

and one can define $\gamma_n(x) = (n!)^{-1}x^{\times_{sh} n}$.

As it turns out, *Shuffle* with this divided power structure is universal for divided power algebras in the following sense. For every divided power algebra A and set of elements $a_\lambda \in A$, $\lambda \in LYN$, there is a unique morphism of divided power algebras *Shuffle* $\longrightarrow A$ taking λ to a_λ.

6.5.26. Some explicit examples of shuffle powers. Here are two explicit examples of shuffle powers

$$[1,2] \times_{sh} [1,2] = 4[1,1,2,2] + 2[1,2,1,2]$$

$$\begin{aligned} [1,2] \times_{sh} [1,2] \times_{sh} [1,2] = {} & 36[1,1,1,2,2,2] + 24[1,1,2,1,2,2] + 12[1,1,2,2,1,2] \\ & + 12[1,2,1,1,2,2] + 6[1,2,1,2,1,2]. \end{aligned}$$

Remark 6.5.27. Divided power algebras and Hopf algebras. A Hopf algebra with divided powers is a Hopf algebra such that the underlying algebra has divided powers and such that the comultiplication and counit are divided power preserving.

The interest is them is, among other things, that the Leray theorem and the Milnor-Moore theorem, which in their original version fail over fields of characteristic $p > 0$ are saved if divided powers are present (on the original algebra or its dual), see [**1**], [**6**], [**7**], [**30**].

6.5.28. Counting Lyndon words. Let $A = k[x_i,\ i \in I]$ be a free commutative algebra over k in the indeterminates x_i, $i \in I$. Let each indeterminate be given a positive integer weight. That makes $k[x_i : i \in I]$ a connected graded algebra. The number of different commutative monomials then has the generating series

$$\prod_{i\in I}(1 - t^{\mathrm{wt}(x_i)})^{-1} \tag{6.5.29}$$

i.e. $\dim(A_n)$ = coefficient of t^n in (6.5.29); more precisely this coefficient is $\mathrm{rank}(A_n)$.

Now consider the case of the free commutative algebra *Shuffle*. Let b_n be the number of Lyndon words of weight n. Because the number of words of weight n is 2^{n-1} (see the lemma below), one sees [17]

$$\prod_{n}(1 - t^n)^{b_n} = 1 + t + 2t^2 + 4t^3 + 8t^4 + \cdots = \frac{1-t}{1-2t}. \tag{6.5.30}$$

Taking logarithms, a simple calculation gives the recursion formula

$$\sum_{d|n} db_d - 2^n - 1. \tag{6.5.31}$$

The first few values of b_n are

$$b_1 = 1,\ b_2 = 1,\ b_3 = 2,\ b_4 = 3,\ b_5 = 6,\ b_6 = 9,\ b_7 = 18,\ b_8 = 30,\ b_9 = 56,\ b_{10} = 99.$$

Lemma 6.5.32. (Counting words of weight n). *The number of words of weight n is 2^{n-1}.*

Proof. Denote the number of words over the the natural numbers of weight n by a_n. Then, looking at the first entry of a word, one sees that the a_n satisfy the recursion formula and initial condition

$$a_1 = 1,\ a_n = 1 + a_1 + a_2 + \cdots + a_{n-1}$$

the terms in the sum corresponding respectively to the cases that the first letter is $n, n-1, \ldots, 2, 1$. The expression $a_n = 2^{n-1}$ satisfies these conditions.

The next thing on the agenda is a description of a basis (over the integers) of the free Lie algebra on countably many generators over **Z**. This also requires a little preparation.

6.5.33. Construction and lemma. *For each Lyndon word α of length > 1 consider the lexicographically smallest proper tail α'' of α. let α' be the corresponding prefix to α''. Then α' and α'' are both Lyndon and $\alpha = \alpha' * \alpha''$ is called the canonical factorization of α.*

Proof. First of all α'' is certainly Lyndon, because it is the smallest proper tail, so each tail of it is larger. Next let β be a proper tail of α'. Then $\beta * \alpha''$ is a proper

[17] See also [**32**].

tail of α and hence (lexicographically) larger than $\alpha = \alpha' * \alpha''$. As it is also shorter it follows that $\beta >_{\text{lex}} \alpha'$.

Now consider the Lie Hopf algebra *LieHopf* of subsection 3.4.5. For every Lyndon word α inductively construct a primitive, $\psi(\alpha)$, of *LieHopf* as follows.

Construction 6.5.34. Primitives of *LieHopf*.
(i) If $\lg(\alpha) = 1$, i.e. $\alpha = [i]$ for some $i \in \mathbf{N}$, $\psi([i]) = U_i$
(ii) If $\lg(\alpha) > 1$, let $\alpha = \alpha' * \alpha''$ be the canonical factorization of α and set $\psi(\alpha) = [\psi(\alpha'), \psi(\alpha'')] = \psi(\alpha')\psi(\alpha'') - \psi(\alpha'')\psi(\alpha')$.

Theorem 6.5.35. (Free Lie algebra on countably many generators).
(i) *$\psi(LYN)$ is a basis for* Prim(*LieHopf*)
(ii) Prim(*LieHopf*) *is the free Lie algebra in the countably many generators* $\{U_i : i = 1, 2, \ldots\}$.

Given the construction this is not overly difficult. A matter of some two pages, see [**27**], pp. 77-79. [18]

6.6. The autoduality of Symm

In the incarnation *RS* of **Symm** the inner product and the autoduality of **Symm** are entirely natural (Frobenius reciprocity). Here it will be shown that the inner product and autoduality are equally natural from the point of view of **Symm** as a quotient of **NSymm** on the one hand and as a sub Hopf algebra of **QSymm** on the other.

Consider the dual Hopf algebras **QSymm** and **NSymm** as discussed in section 6.3 above. Recall that the duality pairing is

$$\langle Z_{a_1} Z_{a_2} \cdots Z_{a_m}, [b_1, b_2, \ldots, b_n] \rangle = \delta_{m,n} \delta_{a_1,b_1} \cdots \delta_{a_m,b_m}.$$

Now consider **Symm** as a quotient of **NSymm** (by dividing out the commutator ideal) and as a sub of **QSymm** (by regarding symmetric functions as special quasisymmetric functions).

Strictly speaking, in the present context it is not yet clear that these two Hopf algebras are isomorphic, so that should really be reflected in the notation. The main theorem of symmetric functions of course says that in any case as algebras they are definitely isomorphic.
Diagrammatically the situation is as follows

$$\begin{array}{ccc} & & \mathbf{Symm} \\ & & \cap \\ \mathbf{NSymm} & \leftrightarrow & \mathbf{QSymm} \\ \downarrow & & \\ \mathbf{Symm} & & \end{array} \tag{6.6.1}$$

[18] Actually what is proved in loc. cit. is that the primitives thus constructed form a basis of the sub Lie algebra generated by the U_i, (as a Lie algebra) and that this Lie algebra is free. It then requires a bit further work to see that this Lie algebra is all of Prim(*LieHopf*). The construction also works for the free Lie algebras over other alphabets, for instance finite ones. See also [**28**] for a great deal more information.

where "$\leftrightarrow$" stands for "there is a duality pairing", and the left hand arrow is the quotient mapping $Z_n \mapsto h_n$.

The first thing to remark is that the duality pairing in the middle induces a duality pairing between the quotient on the left and the sub on the right.

To see this, let I be the commutator ideal in **NSymm** (which is a Hopf ideal), so that on the left **Symm** = **NSymm**$/I$. Then it needs to be shown that (on the right)

$$\mathbf{Symm} = \{x \in \mathbf{QSymm} : \langle y, x\rangle = 0 \text{ for all } y \in I\}. \tag{6.6.2}$$

This is straightforward. Indeed if m_λ is a monomial symmetric function, see 4.1.11, then obviously its pairing with anything of the form

$$Z_\alpha(Z_nZ_m - Z_mZ_n)Z_\beta \tag{6.6.3}$$

is zero. This proves the inclusion $\subset$ of (6.6.2). Inversely let $x \in$ **QSymm** and write it as a sum of monomial quasisymmetric functions

$$x = \sum_\alpha c_\alpha M_\alpha$$

and suppose that $\langle I, x\rangle = 0$. Then, using that every permutation is a product of transpositions, by testing x against suitable elements of the form (6.6.3), one sees that $c_{[a_1,\dots,a_m]} = c_{[a_{\sigma(1)},\dots,a_{\sigma(m)}]}$ for every permutation σ so that x is a sum of monomial symmetric functions and hence in **Symm**.

So there is an induced duality pairing **Symm** $\leftrightarrow$ **Symm** making the quotient Hopf algebra on the left of (6.6.1) dual to the sub Hopf algebra on the right of (6.6.1). Note that the induced pairing is precisely the one which makes the basis of complete symmetric function monomials $\{h_\lambda\}$ and the monomial symmetric functions basis $\{m_\lambda\}$ biorthonormal, as in chapter 4. [19]

Theoretically the two induced Hopf algebra structures could be different. However as algebras they are certainly isomorphic, both being the algebra of polynomials in an infinity of variables. Pick the algebra isomorphism (from the left **Symm** of (6.6.1) to the right **Symm** of (6.6.1) given by

$$h_n \mapsto e_n.$$

The coalgebra structure on the left is

$$h_n \mapsto \sum_{i+j=n} h_i \otimes h_j.$$

The coalgebra structure on the right is induced by cut on **QSymm**. Now, as a monomial quasisymmetric function

$$e_n = \underbrace{[1, 1, \dots, 1]}_{n}$$

so that cut induces the comultiplication

$$e_n \mapsto \sum_{i+j=n} e_i \otimes e_j$$

which fits.

[19]This is one main reason why the commutative quotient of **NSymm** is seen as $\mathbf{Z}[h_l, h_2, \dots]$ rather than $\mathbf{Z}[e_1, e_2, \dots]$.

6.7. Polynomial freeness of QSymm over the integers

Over the rationals **NSymm** is isomorphic to *LieHopf*. Thus their duals, **QSymm** and *Shuffle*, are also isomorphic over the rationals. This means that over the rationals **QSymm** is a free polynomial algebra. The various isomorphisms available do not give a very pleasing set of generators. Fortunately a modification of the proof that *Shuffle* is free, using a different ordering of words gives a proof that **QSymm** is free (still over the rationals) with as generators the Lyndon words. The ordering to be used is the wll-ordering which stands for "weight first, then length, then lexicographic". See [**17**] for details.

But, as it turns out, things are much more beautifully arranged: **QSymm** is free as an algebra over the integers. However, it is not true that the Lyndon words are generators; nor some more or less obvious alternatives.

Construction 6.7.1. Generators of **QSymm**. To start, consider the following example

$$[1,2]\times_{osh}[1,2]=$$

$$=4[1,1,2,2]+2[1,2,1,2]+2[1,3,2]+2[1,1,4]+2[2,2,2]+[2,4]$$

and observe that the expression

$$e_2([1,2])=2^{-1}([1,2]\times_{osh}[1,2]-[2,4]) \tag{6.7.2}$$

has integral coefficients. Also observe that, as the notation suggests, the formula on the right is the expression for the second elementary symmetric function in terms of the power sums; that is $e_2=2^{-1}(p_1^2-p_2)$, where p_n applied to a word is to be interpreted as

$$p_n([a_1,\dots,a_m])=[na_1,\dots,na_m]. \tag{6.7.3}$$

This is impossibly slender evidence of course. But as it turns out it works. Let $e_n(p_1,\dots,p_n)$ be the formula for the elementary symmetric functions in terms of the power sums, interpret the power sums as applied to a word as in (6.7.2) and write for any word $\alpha=[a_1,\dots,a_m]$

$$e_n(\alpha)=e_n(p_1(\alpha),\dots,p_n(\alpha)). \tag{6.7.4}$$

Then the $e_n(\alpha)$ have integral coefficients. Moreover there is a natural free basis for **QSymm** in these terms as follows. A word $\alpha=[a_1,\dots,a_m]$ is called elementary if the greatest common divisor of its cntries is 1, $gcd\{a_1,\dots,a_m\}=1$.

Theorem 6.7.5. *The $e_n(\alpha)$, α an elementary Lyndon word, $n\in\boldsymbol{N}$, form a free commutative polynomial basis for* **QSymm** *over the integers.*

This can be proved rather directly. But the real mechanism behind it is the formalism and theory of lambda rings. These structures were discussed at some length in section 4.9 and in fact most what will be needed has already been described and proved.

6.7.6. Recollections. (From section 4.9). A lambda ring (λ-ring) is a ring equipped with extra operations

$$\lambda^i:R\longrightarrow R$$

that behave just like exterior powers (of vector spaces or representations or modules). There are associated ring endomorphisms called Adams operations (in algebraic topology) or power operations. Given a ring R with operations $\lambda^i : R \longrightarrow R$ define operations $\Psi^i : R \longrightarrow R$ by the formula

$$\frac{d}{dt}\log \lambda_t(a) = \sum_{n=0}^{\infty}(-1)^n \Psi^{n+1}(a)t^n, \tag{6.7.7}$$

where $\lambda_t(a) = 1 + \lambda^1(a)t + \lambda^2(a)t^2 + \cdots$.

Then a set of operations $\lambda^i : R \longrightarrow R$ for a torsion free ring R turns it into a λ-ring if and only if the Adams operations $\Psi' : R \longrightarrow R$ are all ring endomorphisms and in addition satisfy

$$\Psi^1 = \mathrm{id}, \ \Psi^n\Psi^m = \Psi^{mn}, \quad \text{for all } n, m \in \mathbf{N} = \{1, 2, \ldots\} \tag{6.7.8}$$

([**23**], p 49ff), and section 4.9 above.

Note (or recall) also that the relation (6.7.7) between the lambda operations and the Adams operations is precisely the same as that between the elementary symmetric functions and the power sum symmetric functions. Thus there are the following useful determinantal formulae.

$$n!\lambda^n(a) = \det\begin{pmatrix} \Psi^1(a) & 1 & 0 & \ldots & 0 \\ \Psi^2(a) & \Psi^1(a) & 2 & \ddots & \vdots \\ \vdots & \vdots & \ddots & \ddots & 0 \\ \Psi^{n-1}(a) & \Psi^{n-2}(a) & \ldots & \Psi^1(a) & n-1 \\ \Psi^n(a) & \Psi^{n-1}(a) & \ldots & \Psi^2(a) & \Psi^1(a) \end{pmatrix} \tag{6.7.9}$$

$$\Psi^n(a) = \det\begin{pmatrix} \lambda^1(a) & 1 & 0 & \ldots & 0 \\ 2\lambda^2(a) & \lambda^1(a) & 1 & \ddots & \vdots \\ \vdots & \vdots & \ddots & \ddots & 0 \\ (n-1)\lambda^{n-1}(a) & \lambda^{n-2}(a) & \ldots & \lambda^1(a) & 1 \\ n\lambda^n(a) & \lambda^{n-1}(a) & \ldots & \lambda^2(a) & \lambda^1(a). \end{pmatrix} \tag{6.7.10}$$

Finally recall that there is a simple lambda ring structure on $\mathbf{Z}[x_1, x_2, \ldots]$ given by

$$\lambda^i(x_j) = \begin{cases} x_j & \text{if } i = 1 \\ 0 & \text{if } i \geq 2 \end{cases}, \quad j = 1, 2, \ldots \tag{6.7.11}$$

of which the associated Adams endomorphisms are, obviously, the power operations

$$\Psi^n : x_j \mapsto x_j^n. \tag{6.7.12}$$

Theorem 6.7.13. **QSymm** as a lambda ring. **QSymm** *as a subring of* $\mathbf{Z}[x_1, x_2, \ldots]$ *is a lambda ring and the corresponding Adams operations are* $\Psi^n : \alpha = [a_1, \ldots, a_m] \mapsto [na_1, \ldots, na_m]$.

Proof. Indeed, $\lambda^n(\mathbf{QSymm} \otimes_{\mathbf{Z}} \mathbf{Q}) \subset \mathbf{QSymm} \otimes_{\mathbf{Z}} \mathbf{Q}$ by (6.7.9) and also $(\mathbf{QSymm} \otimes_{\mathbf{Z}} \mathbf{Q}) \cap \mathbf{Z}[x_1, x_2, \ldots] = \mathbf{QSymm}$. The formula for the Adams operators is immediate from the definition of quasisymmetric functions and (6.7.12).

Thus for every composition $\alpha = [a_1, \ldots, a_m]$, $\lambda^n(\alpha)$ is well defined (by formula (6.7.9) and in **QSymm**. By definition $e_n(\alpha) = \lambda^n(\alpha)$. Most things are now in place to prove theorem 6.7.5.

Definition 6.7.14. The **wll-ordering**. The acronym 'wll' stands for 'weight first, than length, then lexicographic'. Thus for example

$$[5] >_{wll} [1,1,2] >_{wll} [2,2] >_{wll} [1,3].$$

Lemma 6.7.15. *If α is a Lyndon word (= composition)*

$$\lambda^n(\alpha) = \alpha^{*n} + \text{ (wll-smaller than } \alpha^{*n}) \tag{6.7.16}$$

where $*$ denotes concatenation (of compositions) and (wll-smaller than α^{*n})stands for a $\mathbf{Z}$-linear combination of monomial quasisymmetric functions that are wll-smaller than α^{*n}.

Proof. This follows rather directly from the determinant expression (6.7.9). Indeed, expanding the determinant, one sees that

$$n!\lambda_n(\alpha) = \alpha^n + \text{ (monomials of length } \leq (n-1) \text{ in the } \Psi^i(\alpha)).$$

Further, all terms are of equal weight, and because $\alpha = [a_1, \ldots, a_m]$ is Lyndon the 'length first-lexicographic thereafter' largest term in its n-th (overlapping shuffle) power is the concatenation power α^{*n} and the coefficient is $n!$ (see theorem 6.5.8).

For any λ-ring R there is an associated mapping

$$\mathbf{Symm} \times R \longrightarrow R, (\varphi, a) \mapsto \varphi(\lambda_1(a), \lambda_2(a), \ldots, \lambda_n(a)). \tag{6.7.17}$$

To define it write $\varphi \in \mathbf{Symm}$ as a polynomial in the elementary symmetric functions $e_1, e_2, \ldots$ and then substitute $\lambda_i(a)$ for $e_i, i = 1, 2, \ldots$. For fixed $a \in R$ this is obviously a homomorphism of rings $\mathbf{Symm} \longrightarrow R$. Often one simply writes $\varphi(a)$ for $\varphi(\lambda_1(a), \lambda_2(a), \ldots, \lambda_n(a))$. Another way to see (6.7.17) is to observe that for fixed $a \in R$ $(\varphi, a) \mapsto \varphi(\lambda_1(a), \lambda_2(a), \ldots) = \varphi(a)$ is the unique homomorphism of λ-rings that takes e_1 into a. (**Symm** is the free λ-ring on one generator, see theorem 4.9.35, see also [**23**]). Note that

$$e_n(\alpha) = \lambda_n(\alpha),\ p_n(\alpha) = \Psi^n(\alpha) = [na_1, na_2, \ldots, na_m]. \tag{6.7.18}$$

The first formula of (6.7.18) is by definition and the second follows from (6.7.10) because the relations between the e_n and p_n are precisely the same as between the $\lambda_n(a)$ and the $\Psi^n(a)$.

Recollection 6.7.19. For any $\varphi, \psi \in \mathbf{Symm}$ and $a \in R$, where R is a λ-ring

$$\varphi(\psi(a)) = (\varphi \circ \psi)(a) \tag{6.7.20}$$

where $\varphi \circ \psi$ is the (outer) plethysm of $\varphi, \psi \in \mathbf{Symm}$.

Proof. See section 4.11.

For the purposes here it does not matter just how plethysm is defined. The only thing needed is that there is some element $\varphi \circ \psi \in \mathbf{Symm}$ such that (6.7.20) holds.

Proof of theorem 6.7.5. Let LYN denote the set of Lyndon words and let $eLYN$ be the set of elementary (or reduced) Lyndon words, i.e. the set of those Lyndon words $\alpha = [a_1, a_2, \ldots, a_m]$ for which $g(\alpha) = gcd\{a_1, a_2, \ldots, a_m\} = 1$.

The first and more difficult part is to prove generation, i.e. that every basis element β (in the Abelian group sense) of **QSymm** can be written as a polynomial in the $e_n(\alpha)$, $\alpha \in eLYN$, $n \in \mathbf{N}$. The rest follows by the same counting argument that was used above in section 6.5. to prove polynomial freeness of *Shuffle* over the rationals.

Let R be the ring generated (over the integers) by all the $e_n(\alpha)$, $\alpha \in eLYN$, $n \in \mathbf{N}$. To start with, let $\beta = [b_1, b_2, \ldots, b_n]$ be a Lyndon composition. Then taking $\alpha = \beta_{red} = [g(\beta)^{-1}b_1, g(\beta)^{-1}b_2, \ldots, g(\beta)^{-1}b_n]$ and $n = g(\beta)$ we have, using (6.7.18), $\beta = p_n(\alpha)$ which is an integral polynomial in the $e_n(\alpha)$, and thus $\beta \in R$.

We now proceed with induction for the wll-ordering. The case of weight 1 is trivial. For each separate weight the induction starts because of what has just been said because compositions of length 1 are Lyndon.

So let β be a composition of weight ≥ 2 and length ≥ 2. By the Chen-Fox-Lyndon concatenation factorization theorem, [**8**], and theorem 6.5.5

$$\beta = \beta_1^{*r_1} * \beta_2^{*r_2} * \cdots * \beta_k^{*r_k}, \ \beta_i \in LYN$$

$$\beta_1 >_{\text{lex}} \beta_2 >_{\text{lex}} \cdots >_{\text{lex}} \beta_k \tag{6.7.21}$$

where, as before, the $*$ denotes concatenation and $\beta >_{lex} \beta'$ means that β is lexicographically strictly larger than β'.

If $k \geq 2$, take $\beta' = \beta_1^{*r_1}$ and for β'' the corresponding tail of β so that $\beta = \beta' * \beta''$. Then

$$\beta'\beta'' = \beta' * \beta'' + \text{ (wll-smaller than } \beta) \ = \beta + \text{ (wll-smaller than } \beta)$$

and with induction it follows that $\beta \in R$.
There remains the case that $k = 1$ in the CFL-factorization (2.2). In this case take $\alpha = (\beta_1)_{\text{red}}$ and observe that by lemma 6.7.15 and formula (6.7.18)

$$\beta = e_n(p_{g(\beta_1)}(\alpha)) + \text{ (wll-smaller than } \beta). \tag{6.7.22}$$

On the other hand, by 6.7.19

$$e_n(p_{g(\beta_1)}(\alpha)) = (e_n \circ p_{g(\beta_1)})(\alpha)$$

where $e_n \circ p_{g(\beta_1)}$ is some polynomial with integer coefficients in the e_j, and hence $(e_n \circ p_{g(\beta_1)})(\alpha)$ is a polynomial with integer coefficients in the $e_j(\alpha)$. With induction this finishes the proof of generation.

Now consider the free commutative ring over the integers in the indeterminates $Y_n(\alpha)$, $\alpha \in eLYN$, $n \in \mathbf{N}$ and consider the ring homomorphism

$$\mathbf{Z}[Y_n(\alpha) : \alpha \in eLYN, n \in \mathbf{N}] \longrightarrow \mathbf{QSymm}, \qquad Y_n(\alpha) \mapsto e_n(\alpha). \tag{6.7.23}$$

By what has just been proved this morphism is surjective. Also, giving $Y_n(\alpha)$ weight $n\text{wt}(\alpha)$ it is homogeneous. There are exactly

$$\beta_n = \#\{ \text{ Lyndon words of weight } n\}$$

$Y_i(\alpha)$ of weight n. And so the rank of the weight n component of $\mathbf{Z}[Y_n(\alpha) : \alpha \in eLYN, n \in \mathbf{N}]$ is equal to the rank of the weight n component of **QSymm**. A surjective morphism between equal finite rank free Abelian groups being an isomorphism it follows that (6.7.23) is an isomorphism, proving the theorem.

There are all kinds of consequences. First, note that $e_n([1]) = \underbrace{[1, 1, \ldots, 1]}_{n}$, the n-the elementary symmetric function. Thus there is a morphism of algebras **QSymm** $\longrightarrow$ **Symm** that is a retraction of the inclusion **Symm** $\subset$ **QSymm**. In fact there are very many. Dually this means that there is a coalgebra section of the projection **NSymm** $\longrightarrow$ **Symm** which, as mentioned before, brings the BCM theorem into play. Writing down an explicit coalgebra section of **NSymm** $\longrightarrow$ **Symm** is a very far from trivial task.

Corollary 6.7.24. *The generators $e_n(\alpha)$, $\alpha \in eLYN$, exhibit* **QSymm**, *as a lambda ring, as a tensor product of infinitely many copies of the λ-ring* **Symm**, *one for each $\alpha \in eLYN$.*

Note that the coalgebra structure does not respect this tensor product decomposition.

6.8. Hopf endomorphisms of QSymm and NSymm

Both **QSymm** and **NSymm** have very many Hopf algebra endomorphisms.

Construction 6.8.1. The universal curve (DPS). Consider **NSymm**. The next bit is only the beginning of something that needs to be explored in great detail and much deeper. Take an additional set of commuting indeterminates $x_1, x_2, \ldots$ (which also commute with the Z_j) and consider the ordered product

$$\begin{aligned}&(1 + x_1 Z_1 t + x_1^2 Z_2 t^2 + \cdots)(1 + x_2 Z_1 t + x_2^2 Z_2 t^2 + \cdots)\\ &\cdots (1 + x_n Z_1 t + x_n^2 Z_2 t^2 + \cdots) \cdots\end{aligned} \tag{6.8.2}$$

Now $Z(t) = 1 + Z_1 t + Z_2 t^2 + Z_3 t^3 + \cdots$ is a DPS, also called a curve, in **NSymm**. It follows that the homothetically transformed power series $Z(x_i t) = 1 + x_i Z_1 t + x_i^2 Z_2 t^2 + \ldots$ is also a DPS (over $\mathbf{Z}[x_1, x_2, \ldots]$), and, hence that the ordered product (6.8.2) is also a DPS (curve).

Proposition 6.8.3. *The expression* (6.8.2) *is equal to*

$$1 + \sum_{\alpha} M_\alpha(x_1, x_2, \ldots) Z_\alpha t^{\mathrm{wt}(\alpha)} \tag{6.8.4}$$

where the sum is over all nonempty words $\alpha = [a_1, a_2, \ldots, a_m]$, $a_i \in \{1, 2, \ldots\}$ over the natural numbers, $\mathrm{wt}(\alpha) = a_1 + \cdots + a_m$ is the weight of α and $Z_\alpha = Z_{\alpha_1} Z_{\alpha_2} \cdots Z_{\alpha_m}$, and $M_\alpha(x_1, x_2, \ldots)$ is the quasi-symmetric function defined by the word α.

The proof is straightforward.

Thus (6.8.2), or (6.8.4), can be seen as a curve in **NSymm** with coefficients in **QSymm**.

Now take any algebra homomorphism φ : **QSymm** $\longrightarrow$ **Z**. This gives a new divided power series in **NSymm**, viz:

$$d_0 = 1, d_1, d_2, \dots, d_j = \sum_{\mathrm{wt}(\alpha)} \varphi(M_\alpha) Z_\alpha, \dots$$

and hence a Hopf algebra endomorphism of **NSymm** given by $Z_j \mapsto d_j$. Because **QSymm** is free polynomial there are very many such endomorphisms.

In addition there are all kind of 'Frobenius like' endomorphisms (and of course also the Verschiebung and homothety operations). These 'Frobenius like' endomorphisms are defined as follows. In (6.8.2) replace t by $t^{1/m}$. Instead of (6.8.4) one then finds the same expression but with $t^{\mathrm{wt}(\alpha)/m}$ instead of $t^{\mathrm{wt}(\alpha)}$. Now take a homomorphism φ that is zero on all the free polynomial generators of weight not a multiple of m. Then the resulting series will be in t instead of just $t^{1/m}$ (easy to see) and we obtain a new divided power series and new endomorphisms. Again, because **NSymm** is free graded, there are very many of these.

It remains to see what one can do with all these endomorphisms.

6.9. Verschiebung and Frobenius on NSymm and QSymm

As described in section 4.7 above the Hopf algebra **Symm** comes with two beautiful families of endomorphisms called Frobenius and Verschiebung. Let's start with a description of one reason why these are important.

Definition 6.9.1. Canonical curves (= DPS's) in **Symm** and **NSymm**. Consider the power series

$$h(t) = 1 + h_1 t + h_2 t^2 + \cdots \tag{6.9.2}$$

over **Symm**, and

$$Z(t) = 1 + Z_1 t + Z_2 t^2 + \cdots \tag{6.9.3}$$

over **NSymm**. These are both divided power sequences (=DPS's = curves [20]). They are called the **canonical curves** over **Symm** respectively **NSymm**.

Theorem 6.9.4. Universal properties of **Symm** and **NSymm** with respect to curves.

(i) *For every commutative Hopf algebra H and every curve $d(t) \in H[[t]]$ there is a unique morphism of Hopf algebras* **Symm** $\xrightarrow{\varphi}$ *H such that $\varphi_*(h(t)) = d(t)$.* [21]

(ii) *For every Hopf algebra H and every curve $d(t) \in H[[t]]$ there is a unique morphism of Hopf algebras* **NSymm** $\xrightarrow{\varphi}$ *H such that $\varphi_*(h(t)) = d(t)$.*

Here φ, applied to a power series in t means 'apply φ to the coefficients'.

---

[20]The terminology 'curve' comes from the following. A DPS in a Hopf algebra H is the same as a coalgebra morphism $\mathrm{CoF}(k) \longrightarrow H$, which dually is the same as a morphism of algebras from a dual (such as H^0 or H^{gr*}) to the affine line $k[t]$, which in the setting of (formal) algebraic geometry is a curve, i.e. a morphism of schemes $A^1 \longrightarrow Spec(dual)$.

[21]Reformulated in the context of formal groups this says that the formal group of the big Witt vectors together with a certain canonical curve in it is universal. In that setting this statement is known as Cartier's first theorem. See [**15**], §27.

Proof. Let $\mathrm{CoF}(k)$ be the divided power coalgebra of example 2.4.8[22]. Obviously, by the definition of DPS (= curve) and $\mathrm{CoF}(k)$, a DPS in a Hopf algebra H over k is the same as a morphism of coalgebras

$$\mathrm{CoF}(k) \longrightarrow H$$

given by $Z_i \mapsto d_i$. The theorem is now immediate because **Symm** is the free commutative associative algebra over $\mathrm{CoF}(k)$ (identifying h_i and X_i) and **NSymm**. is the free associative algebra over $\mathrm{CoF}(k)$.

6.9.5. Definition and construction. The functor *Curve*. For each Hopf algebra H over k define

$$\begin{aligned} Curve(H) &= \{DPS' \text{ s in } H\} - \mathbf{Hopf}_k(\mathbf{NSymm}, H) - \\ &= \mathbf{Coalg}_k(\mathrm{CoF}(k), H). \end{aligned} \tag{6.9.6}$$

In case H is commutative every algebra morphism $\mathbf{NSymm} \longrightarrow H$ factors uniquely through the commutative quotient **Symm**, setting up a bijection. So in that case

$$Curve(H) = \mathbf{Hopf}_k(\mathbf{Symm}, H). \tag{6.9.7}$$

The set $Curve(H) = \mathbf{Hopf}_k(\mathbf{NSymm}, H)$ comes with a great deal of structure. First of all it is a group, commutative if H is commutative, otherwise noncommutative in general.

In addition there are the homothety operators $d(t) \mapsto d(at), \forall a \in k$. Further, every Hopf algebra endomorphism of **NSymm** (respectively **Symm** in the commutative case) induces a group operation on the group

$$Curve(H) = \mathbf{Hopf}_k(\mathbf{NSymm}, H).$$

6.9.8. Description. Classification of commutative formal groups. Now concentrate on the commutative case. So let H be a commutative Hopf algebra. In that case the commutative group $Curve(H) = \mathbf{Hopf}_k(\mathbf{Symm}, H)$ comes with three families of operators:

- the homothety operators $d(t) \mapsto d(at)$, $\forall a \in k$
- the Verschiebung operators $\mathbf{v}_n : d(t) \mapsto d(t^n)$
- the Frobenius operators induced by the Frobenius endomorphisms of **Symm**, which can be described by $\mathbf{f}_n : d(t) \mapsto \prod_{i=1}^{n} d(\zeta_n^i t^{1/n})$, where ζ_n is a primitive n-th root of unity, whenever the idea of primitive n-th roots of unity makes sense. [23]

These operators combine to define a ring $Cart(k)$ which contains the ring of big Witt vectors, $W(k)$, as a subring and in suitable sense is a polynomial ring with relations over $W(k)$ [24]. Thus $Curve(H)$, H commutative, becomes a module over $Cart(k)$, and as it turns out the modules that arise from commutative formal groups (more precisely the covariant Hopf algebras of commutative formal groups, see 3.11) can be characterized and are classifying, see [**15**], § 27.

It is unknown to what extent the functor *Curve* is classifying for possibly noncommutative Hopf algebras.

[22]The theorem is true over any k and the proof given works over any k. Thus a better notation would be $\mathbf{Symm}_k$ and $\mathbf{QSymm}_k$

[23]Using functoriality this characterizes the Frobenius operators completely.

[24]When k is a perfect field of positive characteristic a suitable quotient of *Cart(k)* is the famous Dieudonné ring $W_{p^\infty}[\mathbf{f}, \mathbf{v}]/(\mathbf{fv} = \mathbf{vf} = p)$. Modules over this ring classify commutative formal groups over k and commutative algebraic groups over k.

All this immediately raises the question whether the group valued functor *Curve* for noncommutative Hopf algebras, represented by **NSymm** and equipped with homothety operators and (a selection of) the operators coming from the Hopf algebra endomorphisms of **NSymm** [25], can be classifying (for some large collection of Hopf algebras). Whence, taking inspiration from the commutative case, one powerful reason, to inquire about families of Hopf algebra endomorphisms of **NSymm** that behave more or less like Frobenius and Verschiebung on **Symm**.

6.9.9. Recollections. Frobenius and Verschiebung on **Symm**. To fix notations let

$$\mathbf{f}_n^{\mathbf{Symm}} \text{ and } \mathbf{v}_n^{\mathbf{Symm}}$$

be the classical Frobenius and Verschiebung Hopf algebra endomorphisms over the integers of **Symm**, characterized by

$$\mathbf{f}_n^{\mathbf{Symm}}(p_k) = p_{nk},\ \mathbf{v}_n^{\mathbf{Symm}}(p_k) = \begin{cases} np_{k/n} & \text{if } n \text{ divides } k \\ 0 & \text{otherwise.} \end{cases} \tag{6.9.10}$$

These satisfy a number of properties recalled in (6.9.11) below.

6.9.11. Frobenius on QSymm and Verschiebung on NSymm. Now **NSymm** comes with a canonical projection $\mathbf{NSymm} \longrightarrow \mathbf{Symm}$, $Z_n \mapsto h_n$ and there is the inclusion

$$\mathbf{Symm} \subset \mathbf{QSymm}$$

(see section 2). The question arises whether there are lifts $\mathbf{f}_n^{\mathbf{NSymm}}$, $\mathbf{v}_n^{\mathbf{NSymm}}$ on **NSymm** and extensions $\mathbf{f}_n^{\mathbf{QSymm}}$, $\mathbf{v}_n^{\mathbf{QSymm}}$ on **QSymm** that satisfy respectively

(i) $\mathbf{f}_n^{?\mathbf{Symm}}\mathbf{f}_m^{?\mathbf{Symm}} = \mathbf{f}_{mn}^{?\mathbf{Symm}}$

(ii) $\mathbf{f}_n^{?\mathbf{Symm}}$ is homogeneous of degree n, i.e. $\mathbf{f}_n^{?\mathbf{Symm}}(?\mathbf{Symm}_k) \subset ?\mathbf{Symm}_{nk}$

(iii) $\mathbf{f}_1^{?\mathbf{Symm}} = \mathbf{v}_1^{?\mathbf{Symm}} = \mathrm{id}$

(iv) $\mathbf{f}_n^{?\mathbf{Symm}}\mathbf{v}_m^{?\mathbf{Symm}} = \mathbf{v}_m^{?\mathbf{Symm}}\mathbf{f}_n^{?\mathbf{Symm}}$ if $(n, m) = 1$

(v) $\mathbf{v}_n^{?\mathbf{Symm}}\mathbf{v}_m^{?\mathbf{Symm}} = \mathbf{v}_{nm}^{?\mathbf{Symm}}$

(vi) $\mathbf{v}_n^{?\mathbf{Symm}}$ is homogeneous of degree n^{-1}, i.e.

$$\mathbf{v}_n(?\mathbf{Symm}) \subset \begin{cases} ?\mathbf{Symm}_{n^{-1}k} & \text{if } n \text{ divides } k \\ 0 & \text{otherwise} \end{cases} \tag{6.9.12}$$

(vii) $\mathbf{v}_n^{?\mathbf{Symm}}\mathbf{f}_n^{?\mathbf{Symm}} = [n]_{?\mathbf{Symm}}$

Here '?' can be '**N**', or '**Q**'.

---

[25] Actually *Curve(H)* has even more structure. In addition to the noncommutative group operation it has a multiplication that is left (but not right) distributive over the group multiplication. This is also of course the case for $Curve(H)$ for H commutative. But this ring structure, so far, seems to play no role in the classification and has been largely uninvestigated.

Now there exist a natural lifts of the $\mathbf{v}_n^{\mathbf{Symm}}$ to **NSymm** given by the Hopf algebra endomorphisms

$$\mathbf{v}_n^{\mathbf{NSymm}}(Z_k) = \begin{cases} Z_{k/n} & \text{if } n \text{ divides } k \\ 0 & \text{otherwise} \end{cases} \tag{6.9.13}$$

and there exist natural extensions of the Frobenius morphisms on **Symm** to

$$\mathbf{QSymm} \supset \mathbf{Symm}$$

given by the Hopf algebra endomorphisms

$$\mathbf{f}_n^{\mathbf{QSymm}}([a_1, \dots, a_m]) = [na_1, \dots, na_m] \tag{6.9.14}$$

which, moreover, have the Frobenius p-th power property

$$\mathbf{f}_n^{\mathbf{QSymm}}(\alpha) = \alpha^p \mod p \tag{6.9.15}$$

for each prime number p.

These two families are so natural and beautiful that nothing better can be expected and in the following these are fixed as the Verschiebung morphisms on **NSymm** and Frobenius morphisms on **QSymm**. They are also dual to each other.

The question to be examined now is whether there are supplementary families of morphisms $\mathbf{f}_n$ on **NSymm**, respectively $\mathbf{v}_n$ on **QSymm**, such that (6.9.12) holds. The first result is negative

Theorem 6.9.16. *There are no (Verschiebung-like) coalgebra endomorphisms* $\mathbf{v}_n$ *of* **QSymm** *that extend the* $\mathbf{v}_n$ *on* **Symm**, *such that parts* (iii)-(vi) *of* (6.9.12) *hold. Dually there are no (Frobenius-like) algebra homomorphisms of* **NSymm** *that lift the* $\mathbf{f}_n$ *on* **Symm** *such that parts* (i)-(iv) *of* (6.9.12) *hold.*

The proof is technical and not very illuminating. Basically it calculates the consequences of the assumptions in case at least three different prime numbers are involved and finds a contradiction. In case there are only two primes involved, i.e. over a localization $\mathbf{Z}_{p_1,p_2}$, things can be arranged. These calculations heavily rely on the descriptions of Hopf endomorphisms of **NSymm** via the universal curve of section 6.8 and algebra morphisms $\mathbf{QSymm} \longrightarrow \mathbf{Z}$.

In the case of **QSymm** it is the coalgebra morphism property which makes it difficult for (6.9.12) (iii)-(vi) to hold. It is not particularly difficult to find algebra endomorphisms of **QSymm** that do the job. For instance define the $\mathbf{v}_n$ on **QSymm** as the algebra endomorphisms given on the generators (6.7.4) by

$$\mathbf{v}_n(e_k(\alpha)) = \begin{cases} e_{k/n}(\alpha) & \text{if } n \text{ divides } k \\ 0 & \text{otherwise.} \end{cases} \tag{6.9.17}$$

It then follows from (6.7.18) that

$$\mathbf{v}_n(p_k(\alpha)) = \begin{cases} np_{k/n}(\alpha) & \text{if } n \text{ divides } k \\ 0 & \text{otherwise} \end{cases} \tag{6.9.18}$$

and all the properties (6.9.12) follow. But these $\mathbf{v}_n$ are not Hopf algebra endomorphisms.

The last topic to be discussed in this section is whether there are Hopf algebra endomorphisms of **QSymm** (and dually, **NSymm**) such that some weaker versions of (6.9.12) hold.

To this end first consider a filtration by Hopf subalgebras of **QSymm**. Define

$$G_i(\mathbf{QSymm}) = \sum_{\alpha,\, \mathrm{lg}(\alpha)\leq i} \mathbf{Z}\alpha \subset \mathbf{QSymm} \tag{6.9.19}$$

the free subgroup spanned by all α of length $\leq i$, and let

$$F_i(\mathbf{QSymm}) = \mathbf{Z}[e_n(\alpha) : \mathrm{lg}(\alpha) \leq i] \tag{6.9.20}$$

be the subalgebra spanned by those generators $e_n(\alpha)$, $\alpha \in eLYN$, $\mathrm{lg}(\alpha) \leq i$ of length less or equal to i. Note that this does not mean that the elements of $F_i(\mathbf{QSymm})$ are bounded in length. For instance $F_1(\mathbf{QSymm}) = \mathbf{Symm} \subset \mathbf{QSymm}$ contains the elements

$$\underbrace{[1, 1, \ldots, 1]}_{n} = e_n = e_n([1]) \tag{6.9.21}$$

for any n.

Theorem 6.9.22. $G_i(\mathbf{QSymm}) \subset F_i(\mathbf{QSymm})$.

This is a consequence of the proof of the free generation theorem 6.7.5. Moreover,

$$\begin{aligned} F_i(\mathbf{QSymm}) \otimes_{\mathbf{Z}} \mathbf{Q} &= \mathbf{Q}[p_n(\alpha) : \mathrm{lg}(\alpha) \leq i] \\ F_i(\mathbf{QSymm}) &= \mathbf{Z}[p_n(\alpha) : \mathrm{lg}(\alpha) \leq i] \cap \mathbf{QSymm}. \end{aligned} \tag{6.9.23}$$

It follows from (6.9.22) and (6.9.23) that the $F_i(\mathbf{QSymm})$ are not only subalgebras but sub Hopf algebras (because $\mu(p_n(\alpha)) = \sum_{\alpha' * \alpha'' = \alpha} p_n(\alpha') \otimes p_n(\alpha'')$. [26]

Now consider a coalgebra endomorphism of **QSymm**. Because of the graded commutative cofreeness of **QSymm** as a coalgebra over the module $t\mathbf{Z}[t]$ for the projection

$$\mathbf{QSymm} \longrightarrow t\mathbf{Z}[t], [\,] \mapsto 0, [n] \mapsto t^n, \alpha \mapsto 0 \text{ if } \mathrm{lg}(\alpha) \geq 2$$

or, equivalently, because of the freeness of **NSymm** over its submodule $\sum_{i=1}^{\infty} \mathbf{Z}Z_i$, a homogeneous coalgebra morphism of degree n^{-1} of **QSymm** is necessarily given by an expression of the form

$$\mathbf{v}_\varphi(\alpha) = \sum_{\alpha_1 * \cdots * \alpha_r = \alpha} \varphi(\alpha_1) \cdots \varphi(\alpha_r)[n^{-1}\mathrm{wt}(\alpha_1), \ldots, n^{-1}\mathrm{wt}(\alpha_r)] \tag{6.9.24}$$

for some morphism of Abelian groups $\varphi : \mathbf{QSymm} \longrightarrow \mathbf{Z}$. The endomorphism $\mathbf{v}_\varphi$ is a Hopf algebra endomorphism iff φ is a morphism of algebras.

Proposition 6.9.25. $\mathbf{v}_\varphi(F_i(\mathbf{QSymm})) \subset F_i(\mathbf{QSymm})$

One particularly interesting family of φ's is the family of ring morphisms given by

$$\begin{aligned} \tau_n(e_n[1]) &= \tau_n(e_n) = (-1)^{n-1} \\ \tau_n(e_k(\alpha)) &= 0 \text{ for } k \neq n \text{ or } \mathrm{lg}(\alpha) \geq 2\, (\alpha \in eLYN). \end{aligned} \tag{6.9.26}$$

[26] The corresponding Hopf ideals in **NSymm** are most likely the iterated commutator ideals.

Let $\mathbf{v}_n$ be the Verschiebung type Hopf algebra endomorphism defined by τ_n according to formula (6.9.24). Then

Theorem 6.9.27.

(i) $$\mathbf{v}([a_1, \dots a_m]) \equiv \begin{cases} n^m[n^{-1}a_1, \dots n^{-1}a_m] \ \text{mod (length}\, m-1), & \text{if } n|a_i \ \forall i \\ 0 \ \text{mod (length}\, m-1), & \text{otherwise} \end{cases}$$

(ii) $\mathbf{v}_n$ extends $\mathbf{v}_n^{\mathbf{Symm}}$ on $\mathbf{Symm} = F_1(\mathbf{QSymm}) \subset \mathbf{QSymm}$

(iii) $\mathbf{v}_p\mathbf{v}_q = \mathbf{v}_q\mathbf{v}_p$, on $F_2(\mathbf{QSymm})$

(iv) $\mathbf{v}_n\mathbf{f}_n(\alpha) = n^{\lg(\alpha)}\alpha \mod (F_{\lg(\alpha)-1}(\mathbf{QSymm}))$.

And of course there is a corresponding dual theorem concerning Frobenius type endomorphisms of **NSymm**.

This seems about the best one can do. One unsatisfactory aspect of theorem 6.9.27 is that there are also other families $\mathbf{v}_n$ that work.

To conclude we would like to conjecture a stronger version of theorem 6.9.16, viz that there is no family $\mathbf{f}_n$ of algebra endomorphisms of **NSymm** over the integers that satisfies (6.9.12) (i)-(iii) and such that these $\mathbf{f}_n$ descend to the $\mathbf{f}_n^{\mathbf{Symm}}$ on **Symm**.

References

[1] M. André. *Hopf algebras with divided powers*, J. of Algebra **18** (1971), 19–50.

[2] George M. Bergman, *Commuting elements in free algebras and related topics in ring theory*, Thesis, 1967.

[3] George M. Bergman, *Centralizers in free associative algebras*, Trans. Amer. Math. Soc. **137** (1969), 327–344.

[4] Pierre Berthelot, *Cohomoloqie cristalline des schémas de caractéristique* $p > 0$, Springer, 1974.

[5] Pierre Berthelot, Arthur Ogus, *Notes on crystalline cohomology*, Princeton University Press, 1978.

[6] Richard E. Block, *Commutative Hopf algebras, Lie coalgebras, and divided powers*, J. of Algebra **96** (1985), 275–306.

[7] Richard E. Block, *Determination of the irreducible divided power Hopf algebras*, J. of Algebra **96** (1985), 307–317.

[8] K. T. Chen, R. H. Fox, R. C. Lyndon, *Free differential calculus IV*, Ann. Math. **68** (1958), 81 95.

[9] P. M. Cohn, *Free rings and their relations*, Acad. Press, 1971.

[10] Gerard Duchamp, Alexander Klyachko, Daniel Krob, Jean-Yves Thibon, *Noncommutative symmetric functions III: deformations of Cauchy and convolution algebras*. Preprint, Université Marne-la-Vallée, 1996.

[11] Ira M. Gessel, *Multipartite P-partitions and inner products of skew Schur functions*, Contemp. Math. **34** (1984), 289–301.

[12] G. Duchamp, D. Krob, E. A. Vassilieva, *Zassenhaus Lie idempotents, q-bracketing, and a new exponential/logarithmic correspondence*, J algabraic combinatorics **12**:3 (2000), 251–278.

[13] J. P. Duval, *Factorizing words over an ordered alphabet*, J. of Algorithms **4** (1983), 363–381.

[14] Israel M. Gelfand, Daniel Krob, Alain Lascoux, Bernard Leclerc, Vladimir S. Retakh, Jean-Yves-Thibon, *Noncommutative symmetric functions*, Adv. Math. **112** (1995), 218–348.

[15] Michiel Hazewinkel, *Formal groups and applications*, Acad. Press, 1978.

[16] Michiel Hazewinkel, *The Leibniz-Hopf algebra and Lyndon words*, preprint, CWI, Amsterdam, 1996.

[17] Michiel Hazewinkel, *Quasisymmetric functions*, In: D. Krob, A. A. Mikhalev and A. V. Mikhalev, eds., *Formal series and algebraic combinatorics. Proc. of the 12-th international conference*, Moscow, June 2000, Springer, 2000, 30–44.

[18] Michiel Hazewinkel, *The primitives of the Hopf algebra of noncommutative symmetric functions over the integers*, São Paulo J. math. Sci. **1**:2 (2007), 175–203; `arXiv: 0410365`.

[19] Michiel Hazewinkel, *Symmetric functions, noncommutative symmetric functions, and quasisymmetric functions*, Acta Appl. Math. **75** (2003), 55–83; `arXiv:0410468`.

[20] Michiel Hazewinkel, *Symmetric functions, noncommutative symmetric functions, and quasisymmetric functions II*, Acta Appl. Math., **85** (2005), 319–340. Also to appear in: F. van Oystayen, ed., *Proceedings of the Borewicz memorial conference*, Skt Petersburg 2002, Marcel Dekker; `arXiv: 0410470`.

[21] Michiel Hazewinkel, Murray Gerstenhaber, eds., *Deformation theory of algebras and structures and applications*, Reidel, 1988.

[22] Michiel Hazewinkel, *Witt vectors. Part 1*, in: Michiel Hazewinkel, ed., *Handbook of Algebra, Volume 6*, Elsevier, 2009, 319–472; `arXiv: 0904.3888`.

[23] Florent Hivert, *Hecke algebras, difference operators and quasi symmetic functions,* Adv. Math. **155** (2000), 181–238.

[24] D. Knutson, *λ-Rings and the representation theory of the symmetric group*, Springer, 1973.

[25] D. Krob, B. Leclerc, J.- Y. Thibon, *Noncommutative symmetric functions II: transformations of alphabeths*, Int. J. Algebra and Computation **7**:2 (1997), 181–264.

[26] Daniel Krob, Jean-Yves Thibon, *Noncommutative symmetric functions IV: quantum linear groups and Hecke algebras at $q = 0$*, J. of Algebraic Combinatorics **6** (1997), 339–376.

[27] D. Krob, J.- Y. Thibon, *Noncommutative symmetric functions V: a degenerate version of $U_q(\mathfrak{gl}_N)$*, Int. J. Algebra and Computation **9**:3&4 (1997), 405–430.

[28] M. Lothaire (ed.), *Combinatorics on words*, Addison-Wesley, 1983.

[29] Chr. Reutenauer, *Free Lie algebras*, Oxford University Press, 1993.

[30] A. C. J. Scholtens, *S-typical curves in non-commutative Hopf algebras*, Thesis, Free University of Amsterdam, 1996.

[31] G. Sjödin, *Hopf algebras and derivations*, J. of Algebra **64** (1980), 218–229.

[32] J.-Y. Thibon, B-C-V Ung, *Quantum quasisymmetric functions and Hecke algebras*, J. Phys. A: Math. Gen. **29** (1996), 7337–7348.

[33] E. Witt, *Die Unterringe der freien Lieschen Ringe*, Math. Zeitschrift **64** (1956), 195–216.

CHAPTER 7

The Hopf algebra of permutations

7.1. The Hopf algebra of permutations of Malvenuto, Poirier and Reutenauer

There is a beautiful, highly noncommutative and highly noncocommutative Hopf algebra, that generalizes both **NSymm** and **QSymm**. It is highly noncommutative and noncocommutative in that it is both free and cofree. It was invented and studied by Reutenauer, Malvenuto and Poirier, [6,7,8,10,11]. It will usually be referred to it as the ***MPR* Hopf algebra**. It is connected graded and it is self dual up to an isomorphism [1].

Construction 7.1.1. The underlying graded group of the *MPR* Hopf algebra over the integers. The underlying group of the *MPR* Hopf algebra is the free Abelian group with as basis all permutations including the empty one

$$MPR = \mathbf{Z}S = \mathbf{Z}[\,] \oplus \bigoplus_{i=1}^{\infty} \mathbf{Z}S_n \tag{7.1.2}$$

where $\mathbf{Z}S_n$ is (the underlying Abelian group of) the group ring of the group of permutations on n letters, S_n, over the integers. The grading is given by $MPR_n = \mathbf{Z}S_n$.

Often below a permutation of $\{1, 2, \cdots, n\}$

$$\begin{pmatrix} 1 & 2 & \cdots & n \\ a_1 & a_2 & \cdots & a_n \end{pmatrix}$$

will be written as a word $[a_1, a_2, \ldots, a_n]$. Thus permutations are simply special kinds of words over the integers.

Definition 7.1.3. Multiplication on *MPR*. Define a multiplication on these basis elements as follows. Let $\alpha = [a_1, \ldots, a_m] \in S_m$, $\beta = [b_1, \ldots, b_n] \in S_n$. Then

$$m(\alpha \otimes \beta) = [a_1, \ldots, a_m] \times_{sh} [m + b_1, \ldots, m + b_n] \tag{7.1.4}$$

where, of course, $\times_{sh}$ is the shuffle product. Note that all terms on the RHS of (7.1.3) are indeed permutation words. The empty word (permutation) serves as the unit element. Note also that this multiplication respects the grading

The definition of the coproduct needs the notion of the standardization of a word.

[1]So theorem 3.8.13 does not apply! As has better be the case for the *MPR* Hopf algebra is very far from being commutative.

Definition 7.1.5. Standardization of words [2]. Let $\mathcal{A}$ be any totally ordered alphabet (usually a subset of $\mathbf{N}$ with the induced order) and let $\alpha = [a_1, \dots, a_m]$ be a word over $\mathcal{A}$ with all entries different. Then

$$\mathrm{st}(\alpha) = [\theta(a_1), \dots, \theta(a_m)]$$

where θ is the unique order preserving bijection $\{a_1, \dots, a_m\} \longrightarrow \{1, 2, \dots, m\}$. For instance

$$\mathrm{st}([5, 6, 2, 9]) = [2, 3, 1, 4].$$

And for the empty word $\mathrm{st}([\,]) = [\,]$.

Definition 7.1.6. Coproduct for the *MPR* Hopf algebra. The coproduct μ is now defined by

$$\mu(\alpha) = \sum_{\beta * \gamma = \alpha} \mathrm{st}(\beta) \otimes \mathrm{st}(\gamma) \qquad (7.1.7)$$

where the sum is over all concatenation factorizations of α. For instance

$$\begin{aligned} \mu([2,3,5,4,1]) &= \mathrm{st}([\,]) \otimes \mathrm{st}([2,3,5,4,1]) + \mathrm{st}([2]) \otimes \mathrm{st}([3,5,4,1]) + \\ &+ \mathrm{st}([2,3]) \otimes \mathrm{st}([5,4,1]) + \mathrm{st}([2,3,5,]) \otimes \mathrm{st}([4,1]) + \\ &+ \mathrm{st}([2,3,5,4]) \otimes \mathrm{st}([1]) + \mathrm{st}([2,3,5,4,1]) \otimes \mathrm{st}([\,]) = \\ &= [\,] \otimes [2,3,5,4,1] + [1] \otimes [2,4,3,1] + [1,2] \otimes [3,2,1] \\ &+ [1,2,3] \otimes [2,1] + [1,2,4,3] \otimes [1] + [2,3,5,4,1] \otimes [\,]. \end{aligned}$$

The counit is defined by $\varepsilon([\,]) = 1$ and $\varepsilon(\alpha) = 0$ if α has length ≥ 1.

Theorem 7.1.8. *MPR* is a Hopf algebra. *The graded Abelian group with grading, multiplication, unit, comultiplication and counit as defined just above is a connected graded Hopf algebra. It is noncommutative and noncocommutative and its antipode has order infinity.*

Proof. Associativity and coassociativity are evident, and so are the unit and counit properties. The only thing that requires some thought is the 'Hopf algebra axiom', the one that says that the multiplication and comultiplication respect each other (diagram (3.2.8)). So let $\alpha = [a_1, \dots, a_r]$ and $\beta = [b_1, \dots, b_s]$ be two permutations. The following diagram then must be shown to be commutative.

$$\begin{array}{ccc}
\alpha \otimes \beta & \mapsto & \sum\limits_{\alpha' * \alpha'' = \alpha} \mathrm{st}(\alpha') \otimes \mathrm{st}(\beta') \otimes \sum\limits_{\beta' * \beta'' = \beta} \mathrm{st}(\alpha'') \otimes \mathrm{st}(\beta'') \\
\downarrow & & \downarrow \\
\alpha \times_{sk} [r + b_1, \dots, r + b_s] & & \sum\limits_{\substack{\alpha' * \alpha'' = \alpha \\ \beta' * \beta'' = \beta}} \mathrm{st}(\alpha') \otimes \mathrm{st}(\beta') \otimes \mathrm{st}(\alpha'') \otimes \mathrm{st}(\beta'') \\
\downarrow & & \downarrow \\
\sum\limits_{\substack{\text{shuffles} \\ w}} \sum\limits_{u * v = w} \mathrm{st}(u) \otimes \mathrm{st}(v) & = & \sum\limits_{\substack{\alpha' * \alpha'' = \alpha \\ \beta' * \beta'' = \beta}} m(\mathrm{st}(\alpha') \otimes \mathrm{st}(\beta')) \otimes m(\mathrm{st}(\alpha'') \otimes \mathrm{st}(\beta''))
\end{array}$$

Now going down on the left we have to look at all concatenation factorizations of a shuffle $u * v = w$ of $[a_1, \dots, a_r]$ and $[r + b_1, \dots, r + b_s]$. For each factorization there are a unique i and j, $0 \leq i \leq r$ and $0 \leq j \leq s$ such that u is a shuffle of $\alpha' = [a_1, \dots a_i]$ and $[r + b_1, \dots r + b_j]$ and v is a shuffle of $\alpha'' = [a_{i+1}, \dots a_r]$ and $[r + b_{j+1}, \dots r + b_s]$ and vice versa. This is just (a special case of) the Hopf

[2] There is a more general notion of standardization of which this is a special case. That one also deals with words where not all letters are different. See [**12**], reference 12 of chapter 5 (Chr.Reutenauer), and M.Lothaire, *Combinatorics on words*, Cambridge University Press, 2002.

property of the *Shuffle* bialgebra (with has comultiplication cut). Now if u is such a shuffle then (as $r \geq i$) $\mathrm{st}(u)$ is a shuffle of $\mathrm{st}(\alpha')$ and $[i+b'_1, \ldots, i+b'_j]$ where $\beta' = [b_1, \ldots, b_j]$ and $\mathrm{st}(\beta') = ([b'_1, \ldots b'_j])$. Similarly for v. This proves that the lower left and lower right hand corners of the diagram above are equal and hence proves the Hopf property [3].

The comultiplication etc. obviously respect the grading. Then, see proposition 3.8.8, the existence of the antipode is automatic. Calculating it explicitly is another matter. But see [**1**], section 5. The example formula given there also shows that the antipode has infinite order.

Definition 7.1.9. Inner product on *MPR*. Define a positive definite inner product on *MPR* by decreeing the given basis consisting of all permutations to be orthonormal.

Definitions 7.1.10. Another multiplication and another comultiplication on the Abelian group *MPR*. These are given by

$$m'(\alpha \otimes \beta) = \sum_{\substack{alph(u*v)=\{1,\ldots,r+s\} \\ \mathrm{st}(u)=\alpha,\, \mathrm{st}(v)=\beta}} u * v \tag{7.1.11}$$

where $\alpha = [a_1, \ldots, a_r], \beta = [b_1, \ldots, b_s]$ are permutations of $\{1, \ldots, r\}$ and $\{1, \ldots, s\}$ respectively and $alph(\gamma)$ of a word $\gamma = [c_1, \ldots, c_t]$ is the collection $\{c_1, ..., c_t\}$ of letters occuring in γ. For example

$$\begin{aligned} m'([2,1] \otimes [3,1,2]) = {} & [2,1,5,3,4] + [3,1,5,2,4] + [3,2,5,1,4] \\ & + [4,1,5,2,3] + [4,2,5,1,3] + [4,3,5,1,2] \\ & + [5,1,4,2,3] + [5,2,4,1,3] + [5,3,4,1,2] + [5,4,3,1,2]. \end{aligned}$$

For a permutation $\alpha = [a_1, \ldots, a_r]$ and a subset $B \subset alph(\alpha)$ let α_B be the permutation of B that is obtained from α by removing all letters that are not in B. For instance for $\alpha = [6,7,2,1,3,5,4]$ and $B = \{1,2,4,7\}$, $\alpha_B = [7,2,1,4]$.

The comultiplication μ' is now given by

$$\mu'(\alpha) = \sum_{i=0}^{r} \alpha_{\{1,\ldots,i\}} \otimes \mathrm{st}(\alpha_{\{i+1,\ldots,r\}}). \tag{7.1.12}$$

For example

$$\begin{aligned} \mu'([5,3,4,1,2]) = {} & [\,] \otimes [5,3,4,1,2] + [1] \otimes [4,2,3,1] + [1,2] \otimes [3,1,2] \\ & + [3,1,2] \otimes [2,1] + [3,4,1,2] \otimes [1] + [5,3,4,1,2] \otimes [\,]. \end{aligned}$$

Theorem 7.1.13. Duality on *MPR*. *The multiplication m and the comultiplication μ' are dual to each other and so are the comultiplication μ and the multiplication m'.*

This means that the following formulae hold with respect to the inner product of 7.1.9 for all permutations $\alpha = [a_1, \ldots, a_r]$, $\beta = [b_1, \ldots, b_s]$, $\gamma = [c_1, \ldots, c_t]$.

---

[3] Thus the essential reason that the Hopf property holds for *MPR* (apart from simple properties of standardization) is that *Shuffle* is a bialgebra. There is a good reason for this. It comes about because in a very real sense *MPR* can be seen as an endomorphism bialgebra of *Shuffle*, see [**5**].

$$\langle m(\alpha \otimes \beta), \gamma \rangle = \langle \alpha \otimes \beta, \mu'(\gamma) \rangle \tag{7.1.14}$$

$$\langle m'(\alpha \otimes \beta), \gamma \rangle = \langle \alpha \otimes \beta, \mu(\gamma) \rangle . \tag{7.1.15}$$

Proof. There is a contribution 1 to $\langle m(\alpha \otimes \beta), \gamma \rangle$ if and only if γ is a shuffle of $\alpha = [a_1, \dots, a_r]$ and $\beta = [r + b_1, \dots, r + b_s]$. But then $\gamma_{\{1,\dots,r\}} = \alpha$ and $\gamma_{\{r+1,\dots,r+s\}} = [r + b_1, \dots, r + b_s]$, so that then also there is precisely a contribution 1 to the RHS of (7.1.14). This proves (7.1.14).

Further, there is a contribution 1 to the LHS of (7.1.15) if and only if γ is a permutation of $\{1, \dots r + s\}$ with a concatenation factorization $\gamma = u * v$ with $\mathrm{st}(u) = \alpha$, $\mathrm{st}(v) = \beta$. And as by definition $\mu(\gamma) = \sum_{u*v\gamma} \mathrm{st}(u) \otimes \mathrm{st}(v)$, this is also precisely when the right hand side of (7.1.15) picks up a contribution 1.

Corollary 7.1.16. Second Hopf algebra structure on *MPR*. $(MPR, m', \mu', e, \varepsilon)$ *is a second Hopf algebra structure on MPR.*

Theorem 7.1.17. *Let* $\vartheta : MPR \longrightarrow MPR$ *be the isomorphism of graded Abelian groups* $\alpha \mapsto \alpha^{-1}$ *that takes a permutation to its inverse. Then* ϑ *induces an isomorphism of Hopf algebras*

$$\vartheta : (MPR, m, \mu) \longrightarrow (MPR, m', \mu'), \alpha \mapsto \alpha^{-1}. \tag{7.1.18}$$

To prove this use the following observation

Lemma 7.1.19. Inverses, restriction and standardization. Let $\alpha = [a_1, \dots, a_r]$ be a permuation and $I = \{i_1 < i_2 < \dots < i_s\}$ a subset of $\{1, \dots, r\}$. Then

$$\vartheta(\mathrm{st}(\alpha_I)) = \mathrm{st}([\alpha^{-1}(i_1), \dots, \alpha^{-1}(i_s)]). \tag{7.1.20}$$

Proof. To see this it is useful to view permutations slightly more generally as bijections of finite sets where it does not matter in which order the pairs of corresponding elements are written down. Thus

$$\alpha = \begin{pmatrix} 1 & 2 & \cdots & r \\ a_1 & a_2 & \cdots & a_r \end{pmatrix} = \begin{pmatrix} \sigma(1) & \sigma(2) & \cdots & \sigma(r) \\ a_{\sigma(1)} & a_{\sigma(2)} & \cdots & a_{\sigma(r)} \end{pmatrix}$$

for any permutation of $\{1, \dots, r\}$, where the rule is that α takes the top element of a column into the bottom element. With this notation

$$\alpha_I = \begin{pmatrix} \alpha^{-1}(i_1) & \cdots & \alpha^{-1}(i_s) \\ i_1 & \cdots & i_s \end{pmatrix}, \quad \mathrm{st}(\alpha_I) = \begin{pmatrix} \mathrm{st}([\alpha^{-1}(i_1), & \dots, & \alpha^{-1}(i_s)]) \\ 1 & \cdots & s \end{pmatrix}$$

and (7.1.19) follows because taking inverses means going from the bottom to the top instead.

Proof of theorem 7.1.17. Using lemma 7.1.19, calculate

$$(\vartheta \otimes \vartheta)\mu'(\alpha) = (\vartheta \otimes \vartheta)(\sum_{i=0}^{r} \alpha_{\{1,\dots,i\}} \otimes \mathrm{st}(\alpha_{\{i+1,\dots,r\}}))$$

$$= \sum_{i=0}^{r} \mathrm{st}([\alpha^{-1}(1),\dots,\alpha^{-1}(i)]) \otimes \mathrm{st}([\alpha^{-1}(i+1),\dots,\alpha^{-1}(r)])$$

$$= (\mathrm{st} \otimes \mathrm{st})(\sum_{u*v=\alpha^{-1}} u \otimes v)$$

$$= \sum_{u*v=\alpha^{-1}} \mathrm{st}(u) \otimes \mathrm{st}(v)$$

$$= \mu(\alpha^{-1}) = \mu(\vartheta(\alpha)).$$

This proves that ϑ takes μ' into μ. The compatibility with the multiplication follows by duality (theorem 7.1.13). Indeed, using that obviously $\langle \vartheta(\alpha), \beta\rangle = \langle \alpha, \vartheta(\beta)\rangle$, one has

$$\begin{aligned}\langle \vartheta(m(\alpha \otimes \beta)), \gamma\rangle &= \langle m(\alpha \otimes \beta), \vartheta(\gamma)\rangle = \langle \alpha \otimes \beta, \mu'(\vartheta(\gamma))\rangle \\ &= \langle \alpha \otimes \beta, (\vartheta \otimes \vartheta)\mu(\gamma)\rangle = \langle (\vartheta \otimes \vartheta)(\alpha \otimes \beta), \mu(\gamma)\rangle \\ &= \langle \vartheta(\alpha) \otimes \vartheta(\beta), \mu(\gamma)\rangle = \langle m'(\vartheta(\alpha) \otimes \vartheta(\beta)), \gamma\rangle\end{aligned}$$

proving that $\vartheta(m(\alpha \otimes \beta)) = m'(\vartheta(\alpha) \otimes \vartheta(\beta))$ and proving the theorem, the compatibilities of ϑ with unit and counit being trivial.

Summing things up there are two isomorphic Hopf algebra structures on *MPR* which are graded dual to each other with respect to the positive definite inner product 7.1.13.

Another way of saying this is to introduce a new inner product

$$\langle \alpha, \beta\rangle' = \langle \alpha, \vartheta(\beta)\rangle = \langle \alpha, \beta^{-1}\rangle.$$

Then $(MPR, m, \mu, e, \varepsilon)$ is self adjoint with respect to the inner product $\langle\ ,\ \rangle'$. But, fortunately, the Zelevinsky theorem, see section 4.3, does not apply because $\langle\ ,\ \rangle'$ is not positive definite.

7.2. The imbedding of NSymm into *MPR*

The Hopf algebra *MPR* generalizes both **NSymm** and **QSymm**, in the sense that there is an embedding of Hopf algebras $\mathbf{NSymm} \longrightarrow MPR(m', \mu')$ and dually a surjection of Hopf algebras $(MPR, m, \mu) \longrightarrow \mathbf{QSymm}$.

7.2.1. Definitions and constructions. Compositions, descent sets and permutations. Given a composition $\alpha = [a_1, \dots, a_m]$ of weight n define a subset $Desc(\alpha) \subset \{1, 2, \dots, n-1\}$ by

$$Desc(\alpha) = [a_1, a_1 + a_2, \dots, a_1 + \dots + a_{m-1}] \subset \{1, 2, \dots n-1\}. \tag{7.2.2}$$

Inversely, given a subset $D = \{d_1 < d_2 < \dots < d_r\} \subset \{1, 2, \dots, n-1\}$ define a composition of n by

$$comp(D) = [d_1, d_2 - d_1, d_3 - d_2, \dots, d_r - d_{r-1}, n - d_r]. \tag{7.2.3}$$

These two constructions are inverse to one another. Note that, for this to work, it is very essential to specify of which set $\{1, 2, \dots, n-1\}$ $D = \{d_1 < d_2 < \dots < d_r\}$

is a subset of. For instance $D = \{2\} \subset \{1,2,3\}$ corresponds to the composition $[2,2]$, while $D = \{2\} \subset \{1,2\}$ corresponds to the composition $[2,1]$, and $D = \varnothing \subset [1,\dots,n-1]$ corresponds to the composition $[n]$ which depends on n.

For a permutation $\sigma = [s_1,\dots,s_n]$ its **descent set** is defined by

$$desc(\sigma) = \{i \in \{1,2,\dots,n-1\} : \sigma(i) > \sigma(i+1)\} \subset \{1,2,\dots,n-1\}. \quad (7.2.4)$$

Note the difference in notation, $Desc(\alpha)$ and $desc(\sigma)$, depending on whether a word w with distinct entries which are all in $\{1,2,\dots,\lg(w)\}$ is regarded as a composition or as a permutation.

For every subset $D = \{d_1 < d_2 < \dots < d_r\} \subset \{1,2,\dots,n-1\}$ there is a permutation $\sigma \in S_n$ such that $desc(\sigma) = D$, see section 7.3. below.

Construction 7.2.5. Mapping **NSymm** into MPR. Let α be a composition of weight n. Define

$$\varphi : \mathbf{NSymm} \longrightarrow MPR,\ \varphi(Z_\alpha) = D_{\leq\alpha} = \sum_{desc(\sigma)\subset Desc(\alpha)} \sigma. \quad (7.2.6)$$

This, obviously, is an injective morphism of Abelian groups. (Because S_n splits into disjoint descent classes, that is the classes of permutations with the same descent set, and because of the bijective correspondence between descent sets and compositions.)

Lemma 7.2.7. Garcia-Remmel description of the inverses of descent classes. *Let $\alpha = [a_1,\dots,a_r]$ be a composition of length r and weight n. Then*

$$\vartheta(D_{\leq\alpha}) = u_1 \times_{sh} \dots \times_{sh} u_r \quad (7.2.8)$$

where $u_1 * u_2 * \dots * u_r$ is the unique r factor concatenation factorization of $[1,2,\dots,n]$ *such that* $\lg(u_i) = a_i$.

Proof. This is really quite simple [4]. First observe that if the alphabets of $u_1,\dots,u_r$ are disjoint then all shuffles of them are different (no multiplicities). This comes about because, in this case, given a shuffle, for every entry in it one can tell which word it came from.

Let

$$\begin{aligned} D = Desc(\alpha) &= \{a_1, a_1+a_2,\dots,a_1+\dots+a_{r_1}\} \\ &= \{d_1 < d_2 < \dots < d_{r-1}\} \subset \{1,2,\dots,n\}. \end{aligned}$$

Then

$$u_1 = [1,\dots,d_1],\ u_2 = [d_1+1,\dots,d_2],\dots,u_r = [d_{r-1}+1,\dots,n].$$

Now let σ be a shuffle of $u_1,\dots,u_r$ seen as a permutation

$$\sigma = \begin{pmatrix} 1 & 2 & \cdots & n \\ \multicolumn{4}{c}{\text{shuffle of } u_1,\dots,u_r} \end{pmatrix}. \quad (7.2.9)$$

The inverse of σ is of course given by reading (7.2.9) from bottom to top. Now let i be an element of the j-th interval $d_{j-1}+1,\dots,d_j$, i.e. an entry of u_j, but unequal to the endpoint d_j. Then because we are dealing with a shuffle product the element

[4] But in the complicated paper **[2]** the result is buried deep as a very special case of much more difficult and sophisticated results.

$i+1$ occurs to the right of i in the bottom row of σ above and so $i \notin desc(\sigma^{-1})$. This shows $desc(\sigma^{-1}) \subset D$.

Inversely, let σ be a permutation with the descent set of σ^{-1} contained in D. Then none of the $d_{j-1}+1, \ldots, d_j - 1$ is in the descent set of σ^{-1} and so the elements $d_{j-1}+1, \ldots, d_j$ must occur in their original order in the bottom row of (7.2.9). This shows that these permutations are shuffles of $u_1, \ldots, u_r$. As there are no multiplicities either on the right or an the left of (7.2.8) this proves the lemma.

Theorem 7.2.10. Imbedding of **NSymm** into MPR. *The morphism* φ *of* (7.2.6) *is an imbedding of Hopf algebras* **NSymm** *into* (MPR, m', μ').

By way of preparation for the proof of this consider 'compositions with zeros', which are simply words over $\mathbf{N} \cup \{0\}$. Let $\beta = [b_1, \ldots, b_s]$ be such a word. Then, by definition

$$Z_\beta = Z_{b_1} Z_{b_2} \cdots Z_{b_k}, \quad \text{where } Z_0 = 1.$$

Thus $Z_\beta = Z_{\beta_{cleaned}}$ where $\beta_{cleaned}$ is the compostion obtained by removing all zeros from β. With this notation the comultiplication on **NSymm** has the following simple formulaic expression on each basis element

$$\mu_{\mathbf{NSymm}}(Z_\alpha) = \sum_{\substack{\beta,\gamma \in (\mathbf{N} \cup \{0\})^* \\ \beta+\gamma=\alpha}} Z_\beta \otimes Z_\gamma. \tag{7.2.11}$$

Proof of theorem 7.2.10. Write $\otimes_{ssh}$ for the product m on MPR as defined by (7.1.4). The acronym 'ssh' stands for 'shifted shuffle'. With this notation the Garsia-Remmel formula becomes: if $\alpha = [a_1, \ldots, a_r]$ is a composition, then

$$\vartheta(D_{\leq\alpha}) = [1, \ldots, a_1] \times_{ssh} [1, \ldots, a_2] \times_{ssh} \cdots \times_{ssh} [1, \ldots, a_r]. \tag{7.2.12}$$

Note also that this still works for compositions with zeros. Indeed interspersing some zeros in a composition α does not alter the set $Desc(\alpha)$, and a zero a_i in the right hand side of (7.2.12) just gives a factor $[\ ] = 1$.

Thus

$$\begin{aligned}
\vartheta(D_{\leq\alpha}) \times_{ssh} \vartheta(D_{\leq\beta}) &= ([1, \ldots, a_1] \times_{ssh} [1, \ldots, a_2] \times_{ssh} \cdots \times_{ssh} [1, \ldots, a_r]) \\
&\quad \times_{ssh} ([1, \ldots, b_1] \times_{ssh} [1, \ldots, b_2] \times_{ssh} \cdots \times_{ssh} [1, \ldots, b_s]) \\
&= \vartheta(D_{\leq\alpha*\beta})
\end{aligned}$$

by the associativity of $\times_{ssh}$. This proves that the composite $\vartheta \circ \varphi :$ **NSymm** $\longrightarrow (MPR, m, \mu)$ respects the multiplications.

As to the comultiplications:

$$
\begin{aligned}
\mu(\vartheta(D_{\leq\alpha})) &= \mu([1,\ldots,a_1] \times_{ssh} [1,\ldots,a_2] \times_{ssh} \cdots \times_{ssh} [1,\ldots,a_r]) \\
&= \mu([1,\ldots,a_1]) \times_{ssh} \mu([1,\ldots,a_2]) \times_{ssh} \cdots \times_{ssh} \mu([1,\ldots,a_r]) \\
&= (\sum_{i_1=0}^{a_1} [1,\ldots,i_1] \otimes [1,\ldots,a_1 - i_1]) \times_{ssh} \cdots \\
&\times_{ssh} (\sum_{i_r=0}^{a_r} [1,\ldots,i_r] \otimes [1,\ldots,a_r - i_r]) \\
&= ((\sum_{i_1=0}^{a_1} [1,\ldots,i_1]) \times_{ssh} \cdots \times_{ssh} (\sum_{i_r=1}^{a_r} [1,\ldots,i_r])) \\
&\otimes ((\sum_{i_1=0}^{a_1} [1,\ldots,a_1 - i_1]) \times_{ssh} \cdots \times_{ssh} (\sum_{i_r=1}^{a_r} [1,\ldots,a_r - i_r])) \\
&= \sum_{\beta+\gamma=\alpha} \vartheta(D_{\leq\beta}) \otimes \vartheta(D_{\leq\gamma})
\end{aligned}
$$

where β is the composition with zeros $\beta = [i_1,\ldots,i_r]$ and γ is the composition with zeros $\gamma = [a_1 - i_1,\ldots,a_r - i_r]$. Zeros must occur because i_j takes the values 0 and a_j. In view of (7.2.9) this proves that $\vartheta \circ \varphi$: **NSymm** $\longrightarrow (MPR, m, \mu)$ also respects the comultiplications so that $\vartheta \circ \varphi$: **NSymm** $\longrightarrow (MPR, m, \mu)$ is an imbedding of Hopf algebras. Here it is the Hopf property of (MPR, m, μ) that provides the essential step. As ϑ is an isomorphism of Hopf algebras, φ : **NSymm** $\longrightarrow (MPR, m', \mu')$ is an imbedding of Hopf algebras.

Definition 7.2.13. Refinements. Let $\alpha = [a_1,\ldots,a_r]$ be a composition. A **refinement** of α is a composition $\beta = [b_1,\ldots,b_s]$ of the same weight such that there are integers $1 \leq i_1 < i_2 < \ldots < i_r = s$ such that

$$a_1 = b_1 + \cdots + b_{i_1},\ a_2 = b_{i_1+1} + \cdots + b_{i_2}, \ldots a_r = b_{i_{r-1}+1} + \cdots + b_{i_r}$$ [5]

For example the refinements of [1,3] are [1,3], [1,2,1], [1,1,2], and [1,1,1,1].

Lemma 7.2.14. Refinements and set inclusion. *Let a composition β be a refinement of a composition α, then Desc(β) $\supset$ Desc(α). Inversely if $D \supset D'$ are two subsets of $\{1, 2, \ldots, n-1\}$, then comp(D) is a refinement of comp(D').*

This is immediate from the definitions.

Theorem 7.2.15. **QSymm** as a quotient of *MPR*. *For each composition $\alpha = [a_1,\ldots,a_r]$ defines $F_\alpha \in$* **QSymm** *by*

$$F_\alpha = \sum_{\beta \text{ refines } \alpha} \beta \quad \in \textbf{QSymm}. \tag{7.2.16}$$

Then the morphism

$$\pi : (MPR, m, \mu) \longrightarrow \textbf{QSymm}, \quad \sigma \mapsto F_{comp(desc(\sigma))} \tag{7.2.17}$$

exhibits **QSymm** *as a quotient Hopf algebra of (MPR, m, μ).*

---

[5]Putting $i_0 = 0$ makes this formula completely regular.

This comes about because, using lemma 7.2.14, the morphism π is the graded dual of the imbedding of Hopf algebras $\varphi :$ **NSymm** $\longrightarrow (MPR, m', \mu')$.

7.3. *LSD* permutations

The acronym '*LSD*' stands for 'Lexicographically Smallest in its Descent set. As will be shown these permutations have useful and interesting properties as elements of *MPR*.

Construction 7.3.1. Permutations with prescribed descent sets. Let D be a subset of $\{1, 2, \ldots, n-1\}$. Divide D up into 'runs'

$$D = \{d_1, d_1{+}1, \ldots, d_1{+}j_1, d_2', d_2'{+}1, \ldots, d_2'{+}j_2, \ldots, d_r', d_r'{+}1, \ldots, d_r'{+}j_r\}. \quad (7.3.2)$$

For instance if $D = \{1, 3, 4, 5, 7, 8, 11, 12, 13, 14, 15\}$ the division into runs is $D = \{1; 3, 4, 5; 7, 8; 11, 12, 13, 14, 15\}$. Let $\mathcal{A} = \{a_1 < \cdots < a_n\}$ be a totally ordered alphabet of size n. Now define

$$\begin{aligned}\sigma_D = [&a_1, \ldots, a_{d_1-1}; a_{d_1+j_1}, a_{d_1+j_1-1}, \ldots, a_{d_1}; a_{d_1+j_1+1}, a_{d_1+j_1+2}, \ldots, a_{d_2'-1};\\ &a_{d_2'+j_2}, a_{d_2'+j_2-1}, \ldots, a_{d_2'}; a_{d_2'+j_2+1}, a_{d_2'+j_2+2}, \ldots, a_{d_3-1}; a_{d_3+j_3}\\ &\ldots, a_{d_r'+j_r}, a_{d_r'+j_r-1}, \ldots, a_{d_r'}; a_{d_r'+j_r+1}, a_{d_r'+j_r+2}, \ldots, a_n]. \end{aligned} \quad (7.3.3)$$

where, for convenience of reading, the 'up runs' (omitting starting and endpoints except, possibly, at the beginning and the end) and 'down runs' (including starting and end points) have been separated by semicolons. The starting and ending 'up runs' may be empty. Here is an example for the alphabet $\{1, 2, \ldots, 17\}$. Let $D = \{2, 4, 5, 6, 8, 12, 14, 15, 16\}$, then σ_D is the following permutation

$$\begin{pmatrix} 1 & 2 & 3 & 4 & 5 & 6 & 7 & 8 & 9 & 10 & 11 & 12 & 13 & 14 & 15 & 16 & 17 & 18 \\ 1 & 3 & 2 & 7 & 6 & 5 & 4 & 9 & 8 & 10 & 11 & 13 & 12 & 17 & 16 & 15 & 14 & 18 \\ & \uparrow & & \uparrow & \uparrow & \uparrow & & \uparrow & & & & \uparrow & & \uparrow & \uparrow & \uparrow & & \end{pmatrix} \quad (7.3.4)$$

In this case $d_1 = d_1' = 2$, $j_1 = 0$, $d_2' = 4$, $j_2 = 2$, $d_3' = 8$, $j_3 = 0$, $d_4' = 12$, $j_4 = 0$, $d_5' = 14$, $j_5 = 2$, The descent set is indicated by the up arrows in the third row. So the up and down runs are (1) (up); (3,2) (down); (up); (7,6,5,4) (down); (up); (9,8) (down); (10,11) (up); (13,12) (down); (up); (17,16, 15, 14) (down); (18) (up).

The up down pattern of this permutation can be depicted as follows

$$\begin{array}{cccccccccccccccccc} & 3 & & 7 & & & & & & & & 13 & & 17 & & & & \\ 1 & & 2 & & 6 & & & & & & 11 & & 12 & & 16 & & & \\ & & & & & 5 & & 9 & & 10 & & & & & & 15 & & 18 \\ & & & & & & 4 & & 8 & & & & & & & & 14 & \end{array} \quad (7.3.5)$$

and written with semicolons separating the up and down runs it is the word

$$[1; 3, 2; 7, 6, 5, 4; 9, 8; 10, 11; 13, 12; 17, 16, 15, 14; 18].$$

Theorem 7.3.6. Properties of *LSD* permutations. *Let D be a subset of $\{1, 2, \ldots, n-1\}$ and let σ_D be the corresponding permutation on the alphabet $\{1, 2, \ldots, n\}$ as defined by* (7.3.3). *Then*

(i) $desc(\sigma_D) = D$

(ii) *σ_D is the lexicographically smallest word (permutation) of all permutations with descent set D. I.e. it is an LSD permutation.*

(iii) *If* $u * v = \sigma_D$ *is a concatenation factorization then both* u *and* v *are LSD permutations (on their respective alphabets).*

(iv) *For all* i, $(\sigma_D)_{\{1,\dots,i\}}$ *and* $(\sigma_D)_{\{i+1,\dots,n\}}$, *are LSD permutations.*

(v) *All LSD permutations are of the form* σ_D *for some* D.

(vi) *An LSD permutation is an involution.*

(vii) *The property LSD is invariant under standardization, meaning that a permutation a is LSD if and only if* $\mathrm{st}(\sigma)$ *is LSD.*

Proof. (i) is immediate from the explicit formula (7.3.3) for σ_D. As to (ii) if $1 \notin D$, σ_D starts with 1, which is the smallest possible start and induction takes care of things. If $1 \in D$, let i be the smallest integer such that $\{1, \dots, i\} \subset D$ and $i + 1 \notin D$, Then any permutation with descent set D must start with an integer that is at least i. Removing this one induction does the job. (v) follows because for every D there is such a σ_D and because the lexicographic smallest element from a set of permutations is unique, (vi) is immediate from the defining formula, (vii) is immediate from the definitions, (iii) is seen by direct inspection from the defining formula. And so is (iv). Alternatively (iii) says that the comultiplication μ' takes an LSD permutation into a sum of tensor products of LSD permutations. The isomorphism ϑ takes (MPR, m', μ') into (MPR, m, μ) and so (iv) follows from (iii) because ϑ is the identity on involutions and because of the defining formulae of μ and μ'.

Corollary 7.3.7. Subcoalgebras of *MPR*. *The LSD permutations span isomorphic sub coalgebras of* (MPR, μ) *and* (MPR, μ').

7.3.8. Definition and theorem. The ideal of *nonLSD* permutations. *Let* I_{nonLSD} *be the subgroup of MPR spanned by all permutations that are not LSD. Then* I_{nonLSD} *is an ideal in the algebra* (MPR, m').

Proof. Let α and β be two permutations and let at least one of them be *nonLSD*. The product $m'(\alpha \otimes \beta)$ is a sum of concatenations $u * v$ such that $\mathrm{st}(u) = \alpha$, $\mathrm{st}(v) = \beta$. So none of these summands can be *LSD* because of theorem 7.3.6, part (iii).

Construction 7.3.9. Ribbon Schur functions [6]. For each composition α define the ribbon Schur function

$$R_\alpha = \sum_{\alpha \text{ refines } \beta} (-1)^{\lg(\alpha) - \lg(\beta)} Z_\beta. \tag{7.3.10}$$

Ribbon Schur functions multiply in **NSymm** almost as nicely as the monomials Z_α (which multiply by concatenation of their indices).

Theorem 7.3.11. Multiplication of ribbon Schur functions. *For a composition* $\alpha = [a_1, \dots, a_m]$ *and a composition* $\beta = [b_1, \dots, b_n]$ *let*

$$\alpha \bullet \beta = [a_1, \dots, a_{m-1}, a_m + b_1, b_2, \dots, b_n].$$

Then

$$R_\alpha R_\beta = R_{\alpha * \beta} + R_{\alpha \bullet \beta}.$$

[6]There is a skew Schur function in *NSymm* attached to any skew Young diagram. Ribbons are skew Young diagrams which do not contain a 2×2 block. The ribbon Schur functions are the skew Schur functions that are attached to ribbons. See [**3**].

Proof. Consider $R_{\alpha*\beta}$. Among the coarsenings of $\alpha*\beta$ there are those that add a_m and b_1 and those that do not. All of the first type occur in $R_{\alpha\bullet\beta}$ but with opposite sign because the length of $R_{\alpha\bullet\beta}$ differs by one from the length of $R_{\alpha*\beta}$. Thus the only coarsenings that remain are those that are a concatenation of a coarsening of α and a coarsening of β, proving the theorem. [7]

Description 7.3.12. Let for each composition α

$$D_{=\alpha} = \sum_{comp(desc(\sigma))=\alpha} \sigma.$$

Then the imbedding $\varphi : \mathbf{NSymm} \longrightarrow (MPR, m', \mu')$ of theorem 7.2.10 is (also) given by

$$R_\alpha \mapsto D_{=\alpha} \tag{7.3.13}$$

This is immediate from the original definition combined with lemma 7.2.14 relating refinements of compositions and inclusions of descent sets.

Construction 7.3.14. Define a morphism of Abelian groups

$$\psi : MPR \longrightarrow \mathbf{NSymm},\ \sigma \mapsto \begin{cases} R_{comp(desc(\sigma))} & \text{if } \sigma \text{ is} LSD \\ 0 & \text{otherwise.} \end{cases} \tag{7.3.15}$$

Theorem 7.3.16. Retraction algebra morphism $(MPR, m') \longrightarrow \mathbf{NSymm}$. *The morphism ψ of* (7.3.15) *is an algebra morphism that is a retraction of the imbedding φ (meaning $\psi \circ \varphi = \mathrm{id}$). This exhibits the algebra* $\mathbf{NSymm}$ *as the quotient of* (MPR, m') : $\mathbf{NSymm} = MPR/I_{nonLSD}$.

Proof. The fact that $\phi \circ \varphi = \mathrm{id}$ is immediate from the definition (7.3.15) and the description (7.3.13) of the imbedding. It is also obvious that the kernel of ϕ is I_{nonLSD}. So it just remains to be proved that ψ respects the multiplications.

So let $\alpha = [a_1, \dots, a_m]$ and $\beta = [b_1, \dots, b_n]$ be two permutations. By definition the product $m'(\alpha \otimes \beta)$ is the sum of all concatenations $u * v$ which are permutations and satisfy $\mathrm{st}(u) = \alpha$ and $\mathrm{st}(v) = \beta$. Let $desc(\alpha) = \{d_1, \dots, d_r\}$, $desc(\beta) = \{f_1, \dots, f_s\}$. Then there are only two possibilities for $desc(u * v)$, viz.

$$desc(u * v) = \{d_1, \dots, d_r\} \cup \{m + f_1, \dots, m + f_s\} \tag{7.3.17}$$

or

$$desc(u * v) = \{d_1, \dots, d_r\} \cup \{m\} \cup \{m + f_1, \dots, m + f_s\} \tag{7.3.18}$$

depending on whether the last letter of u is larger then the first letter of v or not. If either α or β is $nonLSD$ the product is a sum of $nonLSD$ permutations and so goes to zero under ψ as should be the case. It remains to examine the case that α and β are both LSD permutations. And what needs to be shown is that both descent set possibilities actually occur. (7.3.16) is easy; that descent set occurs as the descent set of $u * v = [a_1, \dots, a_m, m + b_1, \dots, m + b_n]$. For the second observe that α is of

[7] In the seminal reference [**3**] the ribbon Schur functions are defined differently (but the same up to an obvious isomorphism). There the theorem is proved by means of quasi-determinants and determinental expressions for the ribbon Schur functions.

the form $[a_1, \dots, a_{r-1} < a_r = m, m-1, \dots, r]$ and $\beta = [s, s-1, \dots, 1, b_{s+1}, \dots, b_n]$. Now consider

$$(\underbrace{a_1, \dots, a_{r-1}}_{\substack{\text{alphabeth}\\ \{1,2,\dots,r-1\}}}, \underbrace{m+s, m+s-1, \dots, r+s, r+s-1, \dots, r}_{\substack{\text{alphabeth}\\ \{r,\dots,m+s\}}}, \underbrace{m+b_{s+1}, \dots, m+b_n}_{\substack{\text{alphabeth}\\ \{m+s+1,\dots,m+n\}}})$$

which is a permutation of the form $u * v$ with $\mathrm{st}(u) = \alpha$ and $\mathrm{st}(v) = \beta$ and which has descent set (7.3.18). This concludes the proof.

Taking graded duals, the previous theorem says that there is a coalgebra section of the projection $\pi : (MPR, m, \mu) \longrightarrow$ **QSymm** which again brings the BCM theorem into play saying that MPR is some crossed product of **QSymm** with a subalgebra of MPR. [8]

Remark 7.3.19. For more on LSD permutations see the very recently published paper [**5**].

7.4. Second multiplication and second comultiplication

In this section we report on where the second multiplication on **NSymm** (and of course the dual second comultiplication on **QSymm**) come from. There are no proofs; for those see the references quoted (and even those are incomplete in this matter).

Description 7.4.1. Solomon descent algebras of the symmetric groups. For each natural number n consider the group algebra $\mathbf{Z}S_n$ of the symmetric group on n letters. In it consider the elements $D_{=\alpha} = \sum\limits_{comp(desc(\sigma))=\alpha} \sigma$. It is a discovery of Louis Solomon that these span a subalgebra of $\mathbf{Z}S_n$. That means that for all compositions α, β of weight n there are (nonnegative) integers $c^{\gamma}_{\alpha,\beta}$ such that $D_{=\alpha}D_{=\beta} = \sum c^{\gamma}_{\alpha,\beta} D_{=\gamma}$. Denote this algebra by $D(S_n)$. It is called the **Solomon descent algebra**. There is such a thing for each Coxeter group, [**13**].

Description 7.4.2. Second multiplication on **NSymm**. On (MPR, m', μ') consider a second multiplication defined by

$$\sigma \times \tau = \begin{cases} \text{usual composition of permutations} & \text{if } \sigma, \tau \text{ are in the same } S_n \\ 0 & \text{otherwise.} \end{cases} \tag{7.4.3}$$

[8] This of course asks for a version of the BCM theorem over the integers rather than over a field. The whole situation begs for a dual version of the BCM theorem dealing with the structure of a Hopf algebra H that contains a Hopf algebra K such that there is an algebra retraction $H \longrightarrow K$.

As far as we know the first coalgebra section of $(MPR, \mu) \longrightarrow$ **QSymm** was defined by Aguiar and Sottile, [**1**]. That one is different from the present one.

At this stage it is also not difficult to prove that the projection $(MPR, \mu) \longrightarrow$ **QSymm** induces an isomorphism of coalgebras from the subcoalgebra of MPR spanned by the LSD permutations to the coalgebra **QSymm**. This gives yet another coalgebra section of the projection different from both the previous ones, [M.Hazewinkel, unpublished].

On *MPR* this is just a second multiplication without extra properties (as far as we know) [9]. For instance it is not distributive over the first multiplication in the Hopf algebra sense.

By what has just been said in 7.4.1 this multiplication preserves descent set sums, i.e. the image of **NSymm** in (MPR, m') is closed under this second multiplication. This induces a second multiplication on **NSymm** and this one is distributive over the first one in the Hopf algebra sense on the left (but not on the right), for some more detail (not much), see [**3**].

Description 7.4.4. Second comultiplication on **QSymm**. Dually, of course, there is a second comultiplication on **QSymm** that is also distributive over the first one (in the Hopf algebra sense) on the left but not on the right. An explicit formulaic description exists, see [**3**], [**6**], [**7**].

This turns the functor $H \mapsto \mathbf{Hopf}(\mathbf{QSymm}, H)$ into something that is noncommutative group valued, with a second binary operation that is left distributive over the group operation and which has an enormous number of operations coming from the many Hopf endomorphism of **QSymm**. One wonders whether this functor might be classifying for suitable classes of Hopf algebras.

The second comultiplication on **QSymm** restricts to a second comultiplication on $\mathbf{Symm} \subset \mathbf{QSymm}$ and this is the familiar second comultiplication that accounts for the multiplication operation on the functor of the big Witt vectors. A byproduct of this analysis is a good formula for the multiplication of the big Witt vectors, something which has always been a bit elusive. See loc.cit.

7.5. Rigidity and uniqueness of MPR

The Hopf algebra **Symm** of symmetric functions has the uniqueness property embodied in the Zelevinsky theorem (see section 4.3). It is also almost rigid according to the Liulevicius theorem 4.4.1 in that there are only four homogenous automorphisms.

One wonders whether there are similar theorems for *MPR*. It turns out that *MPR* is completely rigid: there is no other automorphism than the identity. See [**9**].

It seems likely that in a suitable sense it is also unique. This is proved up to and including level 4, i.e. for the part

$$\mathbf{Z} \oplus \bigoplus_{n=1}^{4} \mathbf{Z}S_n \subset MPR$$

of rank 34. See [**8**].

The Hopf algebra *MPR* can be seen as arising from the consideration of endomorphisms of another Hopf algebra, viz *LieHopf* or its dual *Shuffle*, see [**5**], and,

[9] This does not necessarily mean that this second multiplication is not a valuable thing to have. There may have been, in algebra, too much stress on multiplications and the like that are distributive and such. The mere fact that there is a systematic way of obtaining new objects of a certain kind from two (or more) old ones can be immensally valuable. See for example [**4**], where two methods of obtaining a DPS from other DPS's are of crucial importance, even though these two methods have no standard algebraic niceness properties.

as far as the multiplication is concerned, also [**10**]. It is almost certainly worth the trouble to investigate this type of construction for endomorphisms of other Hopf algebras. A start in this direction is [**5**].

References

[1] Marcelo Aguiar, Frank Sottile, *Structure of the Malvenuto-Reutenauer Hopf algebra of permutations*, Adv. in Math, **191** (2002), 225–275.

[2] A. M. Garsia, J. Remmel, *Shuffles of permutations and the Kronecker product, Graphs and Combinatorics* **1** (1985), 217–263.

[3] Israel M. Gelfand, Daniel Krob, Alain Lascoux, Bernard Leclerc, Vladimir S. Retakh, Jean-Yves Thibon, *Noncommutative symmetric functions*, Adv. Math. **112** (1995), 218–348.

[4] Michiel Hazewinkel, *The primitives of the Hopf algebra of noncommutative symmetric functions over the integers*, São Paulo J. of Math., **1** (2007), 175–203, arXiv: 0410365.

[5] Michiel Hazewinkel, *Hopf algebras of endomorphisms of Hopf algebras*, Arab. J. math. Sci. **33**:(2C) (2008), 239–272, arXiv:0410364.

[6] Michiel Hazewinkel, *Symmetric functions, noncommutative symmetric functions, and quasisymmetric functions*, Acta Appl. Math. **75** (2003), 55–83, arXiv:0410468.

[7] Michiel Hazewinkel, *Symmetric functions, noncommutative symmetric functions, and quasisymmetric functions II*, Acta Appl. Math. **85** (2005), 319–340. Also to appear in: F. van Oystayen, ed., Proceedings of the Borewicz memorial conference, Skt Petersburg 2002, arXiv:0410470.

[8] Michiel Hazewinkel, *Towards uniqueness of the Malvenuto-Poirier-Reutenauer Hopf algebra of permutations*, Honam math. J. **29** (2007), 109–192.

[9] Michiel Hazewinkel, *Rigidity for MPR, the Malvenuto-Poirier-Reutenauer Hopf algebra of permutations*, Honam math. J. **29** (2007), 495–510; Addenda et corrigenda, ibib. **30** (2008), 205.

[10] C. Malvenuto, Chr. Reutenauer, *Duality between quasi-symmetric functions and the Solomon descent algebra*, J. of Algebra **177** (1994), 967–982.

[11] Stephane Poirier, Christophe Reutenauer, *Algèbres de Hopf de tableaux*, Ann. Sci. Math. Quebec **19**:1 (1995), 79–90.

[12] C. Schensted, *Longest increasing and decreasing subsequences*, Canadian J. Math. **13** (1961), 179–191.

[13] L. Solomon, *A Mackey formula in the group ring of a Coxeter group*, J. of Algebra **41** (1976), 255–268.

CHAPTER 8

Hopf algebras: Applications in and interrelations with other parts of mathematics and physics

The number of published papers on Hopf algebras and their applications in and interrelations with other parts of mathematics and (theoretical) physics is rather large. In October 2004 a biographic search was done and resulted in 10589 hits. It also appeared that Hopf algebras and quantum groups appear in virtually all parts of mathematics. See [**316**]. Since then at least 1348 more papers have appeared (2005-2008).

These numbers underestimate the real situation. Nowadays it happens frequently that papers appear on Hopf algebras and/or quantum groups that do not mention these phrases in their abstract/review. For example papers on Yangians, Hall algebras, path and quiver algebras, quantum calculi, renormalization, noncommutative geometry, Galois theory, Bethe Ansatz and quantum integrable systems, knot invariants and invariants of three manifolds, ...

Thus only a very selective set of applications and interrelations can be presented here; severely limited by the taste and knowledge of the authors.

8.1. Actions and coactions of bialgebras and Hopf algebras

To start with here are various notations of Hopf algebra and bialgebra modules and comodules. The notion of an action (representation) of an algebra, bialgebra or Hopf algebra has already been formulated and so has the notion of a coaction (corepresentation). If the action is on a coalgebra or algebra it is natural to consider actions that are compatible with these extra structures in various ways. This leads to a rather large collection of notion of "modules" for Hopf algebras and coalgebras. Below they are listed systematically.[1]

8.1.1. Sweedler sigma notation. Consider a coalgebra C with comultiplication given by

$$\mu_C(c) = \sum_i c_{1,i} \otimes c_{2,i}. \tag{8.1.2}$$

This is a careful but cumbersome notation and the number of indices and such can get rapidly out of hand. For instance applying $\mathrm{id} \otimes \mu_C$ to express something like coassociativity gives

$$(\mathrm{id} \otimes \mu_C)\mu_C(c) = \sum_{i,j} c_{1,i} \otimes c_{2,i,1,j} \otimes c_{2,i,2,j}. \tag{8.1.3}$$

[1]But there are still more.

To get around these masses of indices Moss Eisenberg Sweedler invented a special notation. It is elegant and useful but carries certain dangers (which is why it has been avoided so far). In this notation (8.1.2) is written

$$\mu_C(c) = \sum_{(c)} c_{(1)} \otimes c_{(2)}$$

so that e.g. the coassociativity condition can be written as

$$\sum_{(c),(c_{(1)})} c_{(1)(1)} \otimes c_{(1)(2)} \otimes c_{(2)} = \sum_{(c),(c_{(2)})} c_{(1)} \otimes c_{(2)(1)} \otimes c_{(2)(2)}. \tag{8.1.4}$$

It is also useful to use a shorthand for iterated coproducts and to write the two sides of (8.1.4) as

$$\mu_3(c) = \sum_{(c)} c_{(1)} \otimes c_{(2)} \otimes c_{(3)}.$$

For a right coaction $\gamma_V : V \to V \otimes C$ of a coalgebra, C, on a k-module V, there is the notation

$$\gamma_V(\nu) = \sum_{(\nu)} \nu_{(0)} \otimes \nu_{(1)}.$$

Here the zero index is there to remind the reader that a coaction is involved and which elements live in the comodule. Thus the main right comodule property can be written

$$\sum_{(\nu),(\nu_{(0)})} \nu_{(0)(0)} \otimes \nu_{(0)(1)} \otimes \nu_{(1)} = \sum_{(\nu),(\nu_{(1)})} \nu_{(0)} \otimes \nu_{(1)(1)} \otimes \nu_{(1)(2)} = \sum_{(\nu)} \nu_{(0)} \otimes \nu_{(1)} \otimes \nu_{(2)}.$$

For a left comodule $\gamma_V : V \to H \otimes V$ one writes

$$\gamma_V(\nu) = \sum_{(\nu)} \nu_{(-1)} \otimes \nu_{(0)}$$

and

$$(\mu_C \otimes \mathrm{id})\gamma_V(\nu) = (\mathrm{id} \otimes \gamma_V)\gamma_V(\nu) = \sum_{(\nu)} \nu_{(-2)} \otimes \nu_{(-1)} \otimes \nu_{(0)}.$$

There are several variants of this notation used in the literature. For instance upper indices instead of lower for coactions, possibly decorated in some way; leaving out the subscripts under the summation sign; leaving out the various brackets. So for instance coassociativity can be and has been written

$$\sum c_{1,1} \otimes c_{1,2} \otimes c_2 = \sum c_1 \otimes c_{2,1} \otimes c_{2,2}$$

(which perhaps carries notational simplification a bit to far).

8.1.5. Modules and comodules, representations and corepresentations. These things have been defined and are repeated here for convenience and systematics. Let H be a bialgebra over a ground ring k (as usually commutative with unit element). A left **H-module** is a k-module V together with a morphism of k-modules

$$\alpha_V : H \otimes V \to V, \quad \alpha_V(h \otimes \nu) = h \cdot \nu \tag{8.1.6}$$

such that

$$h \cdot (g \cdot \nu) = (hg) \cdot \nu \quad \text{and } 1_H \cdot \nu = \nu. \tag{8.1.7}$$

One also says that this is a representation of H in V or an action[2] of H on V.

This is purely an algebra notion. The role of the comultiplication lies in the fact that it can be used to multiply representations as follows

$$H \otimes V \otimes W \overset{\mu_H \otimes \mathrm{id} \otimes \mathrm{id}}{\longrightarrow} H \otimes H \otimes V \otimes W \cong H \otimes V \otimes H \otimes W \overset{\alpha_V \otimes \alpha_W}{\longrightarrow} V \otimes W$$

thus defining a ring structure on the (Grothendieck) group of representations. (The addition comes of course from the direct sum of representations.)

In diagram form the conditions (8.1.7) say that the following diagrams must commute.

$$\begin{array}{ccc} H \otimes H \otimes V & \xrightarrow{m_H \otimes \mathrm{id}} & H \otimes V \\ {\scriptstyle \mathrm{id} \otimes \alpha_V} \downarrow & & \downarrow {\scriptstyle \alpha_V} \\ H \otimes V & \xrightarrow{\alpha_V} & V \end{array} \qquad \begin{array}{ccc} k \otimes V & \xrightarrow{\cong} & V \\ {\scriptstyle e_H \otimes \mathrm{id}} \downarrow & & \downarrow {\scriptstyle =} \\ H \otimes V & \xrightarrow{\alpha_V} & V \end{array} \tag{8.1.8}$$

Sort of dually a **right comodule** for H (or a corepresentation, or a coaction) is a k-module V together with a homomorphism of k-modules

$$\gamma_V : V \longrightarrow V \otimes H \tag{8.1.9}$$

such that

$$\sum v_{(0)(0)} \otimes v_{(0)(1)} \otimes v_{(1)} = \sum v_{(0)} \otimes v_{(1)(1)} \otimes v_{(1)(2)}$$

$$\sum v_{(0)} \varepsilon_H(v_{(1)}) = v \tag{8.1.10}$$

which is the formula version of the commutativity of the following diagrams

$$\begin{array}{ccc} V \otimes H \otimes H & \xleftarrow{\mathrm{id} \otimes \mu_H} & V \otimes H \\ {\scriptstyle \gamma_V \otimes \mathrm{id}} \uparrow & & \uparrow {\scriptstyle \gamma_V} \\ V \otimes H & \xleftarrow{\gamma_V} & V \end{array} \qquad \begin{array}{ccc} V \otimes k & \xleftarrow{\cong} & V \\ {\scriptstyle \mathrm{id} \otimes \varepsilon_H} \uparrow & & \uparrow {\scriptstyle =} \\ V \otimes H & \xleftarrow{\gamma_V} & V \end{array} \tag{8.1.11}$$

Note that apart from the left right switch the diagrams for comodules are obtained from the ones for modules by reversing all arrows.

The notion of a comodule over a bialgebra involves only the coalgebra part of the structure. The multiplication enters only in that it can be used to define a multiplication of comodules via the composite

$$V \otimes W \overset{\gamma_V \otimes \gamma_W}{\longrightarrow} V \otimes H \otimes W \otimes H \cong V \otimes W \otimes H \otimes H \overset{\mathrm{id} \otimes \mathrm{id} \otimes m_H}{\longrightarrow} V \otimes W \otimes H.$$

Now let A be a k-algebra (with unit, not necessarily commutative). Then a left H-module algebra structure on A is an H-module structure $\alpha_A : H \otimes A \to A$ that in addition satisfies the following properties embodying the required extra compatibilities.

$$h \cdot (ab) = \sum (h_{(1)} \cdot a)(h_{(2)} \cdot b) \quad \text{if } \mu_H(h) = \sum h_{(1)} \otimes h_{(2)}$$

$$\text{and } h \cdot 1_A = \varepsilon_H(h) 1_A. \tag{8.1.12}$$

[2]Actions on a module V are denoted α_V where the alpha is a mnemonic for "action"; similarly coactions are denoted by γ_V as a mnemonic for "coaction" in that γ is the third letter of the Greek alphabet.

This can also be stated as the commutativity of the following diagrams

$$\begin{array}{ccccc}
H\otimes A\otimes A & \xrightarrow{\mathrm{id}\otimes m_A} & H\otimes A & \xrightarrow{\alpha_A} & A \\
\Big\downarrow{\scriptstyle \mu_H\otimes\mathrm{id}\otimes\mathrm{id}} & & & & \Big\uparrow{\scriptstyle m_A} \\
H\otimes H\otimes A\otimes A & \xrightarrow[\cong]{} & H\otimes A\otimes H\otimes A & \xrightarrow[\alpha_A\otimes\alpha_A]{} & A\otimes A
\end{array}$$

$$\begin{array}{ccc}
H\otimes A & \xrightarrow{\alpha_A} & A \\
{\scriptstyle e_H\otimes e_A}\Big\uparrow & & \Big\uparrow{\scriptstyle e_A} \\
k\otimes k & \xrightarrow{\cong} & k
\end{array} \tag{8.1.13}$$

This notion has also already occurred before in this text. See section 3.9. Given an H-module algebra there is the very useful construction of the smash product, see loc. sit., and for some more stuff on the smash product, see below.

Further, let C be a k-coalgebra. Then a left H-module coalgebra structure on C is a module structure $\alpha_C : H\otimes C \to C$ that in addition satisfies the compatibility properties

$$\mu_C(h\cdot c) = \sum (h_{(1)}\cdot c_{(1)})\otimes(h_{(2)}\cdot c_{(2)}) \text{ and } \varepsilon_C(h\cdot c) = \varepsilon_H(h)\varepsilon_C(c). \tag{8.1.14}$$

In diagram for (8.1.14) says that the following diagrams are commutative

$$\begin{array}{ccccc}
H\otimes C & \xrightarrow{\mu_H\otimes\mu_C} & H\otimes H\otimes C\otimes C & \xrightarrow{\cong} & H\otimes C\otimes H\otimes C \\
\Big\downarrow{\scriptstyle \alpha_C} & & & & \Big\downarrow{\scriptstyle \alpha_C\otimes\alpha_C} \\
C & & \xrightarrow{\mu_C} & & C\otimes C
\end{array}$$

$$\begin{array}{ccc}
H\otimes C & \xrightarrow{\varepsilon_H\otimes\varepsilon_C} & k\otimes k \\
{\scriptstyle C}\Big\downarrow & & \Big\downarrow{\scriptstyle \cong} \\
C & \xrightarrow{\varepsilon_C} & k
\end{array} \tag{8.1.15}$$

Some examples of module algebras and module coalgebras will be given below.

The next step in order to complete the picture is to consider comodule structures on algebras and coalgebras.

Consider an algebra A over k. A right H-comodule algebra structure on A is a comodule structure $\gamma_A : A \longrightarrow A\otimes H$ such that in addition

$$\gamma_A(ab) = \sum (a_{(1)}b_{(1)})\otimes(a_{(2)}b_{(2)}) \text{ and } \gamma_A(1_A) = 1_A\otimes 1_H \tag{8.1.16}$$

i.e. such that γ_A is an algebra morphism. In diagram terms this amounts to the commutativity of

$$\begin{array}{ccccc}
A\otimes H & \xleftarrow[m_A\otimes m_H]{} & A\otimes A\otimes H\otimes H & \xleftarrow[\cong]{} & A\otimes H\otimes A\otimes H \\
{\scriptstyle \gamma_A}\Big\uparrow & & & & \Big\uparrow{\scriptstyle \gamma_A\otimes\gamma_A} \\
A & & \xleftarrow[m_A]{} & & A\otimes A
\end{array}$$

$$\begin{array}{ccc} A \otimes H & \xleftarrow{\;e_A \otimes e_H\;} & k \otimes k \\ {\scriptstyle \gamma_A}\uparrow & & \uparrow{\scriptstyle \cong} \\ A & \xleftarrow{\;e_A\;} & k \end{array} \tag{8.1.17}$$

Note that apart from left-right switching at the nodes this diagram is exactly diagram (8.1.5) with all arrows reversed.

Next let C be a coalgebra over k. Then a right H-comodule coalgebra structure on C is a comodule structure such that in addition

$$\sum (c_{(0)(1)} \otimes c_{(0)(2)} \otimes c_{(1)}) = \sum (c_{(1)(0)} \otimes c_{(2)(0)} \otimes c_{(1)(1)} c_{(2)(1)}$$
$$\text{and} \sum \varepsilon_C(c_{(0)}) \varepsilon_H(c_{(1)}) = \varepsilon_C(c). \tag{8.1.18}$$

In diagram form this is the commutativity of the diagrams

$$\begin{array}{ccccc} C \otimes C \otimes H & \xleftarrow{\;\mu_C \otimes \mathrm{id}\;} & C \otimes H & \xleftarrow{\;\gamma_C\;} & C \\ {\scriptstyle \mathrm{id} \otimes \mathrm{id} \otimes m_H}\uparrow & & & & \downarrow{\scriptstyle \mu_C} \\ C \otimes C \otimes H \otimes H & \xleftarrow{\;\cong\;} & C \otimes H \otimes C \otimes H & \xleftarrow{\;\gamma_C \otimes \gamma_C\;} & C \otimes C \end{array}$$

$$\begin{array}{ccc} C \otimes H & \xleftarrow{\;\gamma_C\;} & C \\ {\scriptstyle \varepsilon_C \otimes \varepsilon_H}\downarrow & & \downarrow{\scriptstyle \varepsilon_C} \\ k \otimes k & \xleftarrow{\;\cong\;} & k \end{array} \tag{8.1.19}$$

Apart from left-right switching at the nodes these diagrams are exactly the ones obtained from (8.1.13) by reversing all arrows.

These four kinds of modules over a Hopf algebra by no means exhaust the interesting and important module like structures that occur in the world of Hopf algebras, quantum groups and their applications.

In addition there are for instance Hopf modules as already discussed in section 3.12, relative Hopf modules, entwined modules, Doi-Koppinen modules, Yetter-Drinfel'd modules. See e.g. [**111**], chapter 2, for a discussion of these.

8.1.20. Action-coaction duality. From now on, in this subsection 8.1, k will be a field and H will a Hopf algebra (or bialgebra if the antipode is not needed) that is finite dimensional over k.

Quite generally a left action of H on a module V is the same as a right coaction of the dual H^* on the same module V (and vice versa).

As is often the case it is easier to go from comodules to modules than the other way. So let

$$\gamma_V : V \to V \otimes H^*, \qquad \gamma_V(v) = \sum v_{(0)} \otimes \varphi_{(1)} \tag{8.1.21}$$

be a comodule structure on V over H^*. Now take

$$h \cdot v = \sum v_{(0)} \langle \varphi_{(1)}, h \rangle \tag{8.1.22}$$

as the definition of an action of H on V.

To go the other way take biorthogonal bases $h_1, \dots, h_n$ of H and $\varphi_1, \dots, \varphi_n$ of H^* so that $\langle \varphi_i, h_j \rangle = \delta_{i,j}$. Then given a left action of H on the k-module V define

$$V \longrightarrow V \otimes H^*, \qquad v \mapsto \sum_i (h_i \cdot v) \otimes \varphi_i. \tag{8.1.23}$$

These constructions are inverse to each other. Indeed, going from formula (8.1.23) to the definition of a coaction according to a formula (8.1.22) gives

$$\sum_i (h_i \cdot v)\langle \varphi_i, h\rangle = \sum_i (\langle \varphi_i, h\rangle h_i \cdot v) = h \cdot v.$$

Theorem 8.1.24. *The construction above takes left modules over H into right comodules over H^* (and vice versa) (and, switching H^* and H, modules over H^* into comodules over H). Under this correspondence module algebras go to comodule algebras and module coalgebras go to comodule coalgebras (and vice versa).*

The proof is routine.

8.1.25. Examples of (co)algebra (co)modules. Just as in the case of, say, groups and Lie algebras the regular and adjoint representations are nice important examples, so there are in the present case (co)regular and (co)adjoint (co)modules of a Hopf algebra acting on its dual or on itself, providing eight examples, two each of the four (co)module (co)algebra structures. Here they are in detail.

Example A1. The left regular action of H on itself is given by

$$h \cdot a = ha. \tag{8.1.26}$$

This turns H into a coalgebra module over itself. The commutativity of the diagrams (8.1.15) are the same as occur in the definition of the structure of a bialgebra.

Example A2. The left coregular action of H on H^* is given by

$$(h \cdot \varphi)(a) = \langle \varphi, ah \rangle. \tag{8.1.27}$$

(As $h \cdot \varphi$ is to be an element oh H^* it can be specified by giving its values on the elements of H.) This turns H^* into a left module algebra over H.

For this particular action the notation $h \rightharpoonup \varphi$ is frequently used in the literature. There are three more 'half arrow' notations that are employed. They are $\langle \psi, h \leftharpoonup \varphi \rangle = \langle \varphi\psi, h \rangle$ (right action of H^* on H); $\varphi \leftharpoondown h = \iota(h) \rightharpoonup \varphi$ (right action of H on H^*) and $\varphi \rightharpoondown h = h \leftharpoonup \iota(\varphi)$ (left action of H^* on H).

Example A3. The left adjoint action of H on itself is given by

$$h \cdot a = \sum h_{(1)} a \iota(h_{(2)}). \tag{8.1.28}$$

This action makes H into a module algebra over itself.

Example A4. The left coadjoint action of H on H^* is given by

$$h \cdot \varphi = \sum \varphi_{(2)} \langle \iota(\varphi_{(1)})\varphi_{(3)}), h \rangle. \tag{8.1.29}$$

This makes H^* into a module coalgebra over H.

Example C1. The right coregular coaction of H^* on H is given by

$$\gamma_H(h) = \sum_i h_i h \otimes \varphi_i \tag{8.1.30}$$

where as in 8.1.20 the $\{h_i\}$ and $\{\varphi_i\}$ are biorthogonal bases of H and H^* respectively. This makes H a comodule coalgebra over H^*.

As is regularly the case this coaction is not easy to write down without using bases. The element $\gamma_H(h)$ is in $H \otimes H^*$ and can thus be specified be giving its evaluations on elements $a \in H$. This gives

$$\gamma_H(h)(a) = \sum_i h_i h \otimes \langle \varphi_i, a \rangle = ah$$

which gets rid of the explicit use of bases and also makes clear why this is called a coregular coaction.

Example C2. The right regular coaction of H^* on itself is

$$\gamma_H = \mu_H. \tag{8.1.31}$$

This makes H^* into a comodule algebra over itself. As is to be expected the commutativity of the relevant diagrams is part of the definition of the structure of a bialgebra.

Example C3. The right coadjoint coaction of H^* on H is given by

$$\gamma_H(h)(\varphi) = \sum \langle \varphi_{(2)}, h \rangle \varphi_{(1)} \iota(\varphi_{(3)}).$$

This one makes H into a comodule algebra over H^*.

Example C4. The right adjoint coaction of H^* on itself is

$$\gamma_{H^*}(h)(\varphi) = \sum \varphi_{(2)} \otimes \iota(\varphi_{(1)}) \varphi_{(3)}.$$

This turns H^* into a comodule coalgebra over itself.

The examples have been so numerated that the examples Ci correspond to the examples Ai for $i = 1, 2, 3, 4$ under theorem 8.1.24.

The proofs are fairly routine. One of the trickier ones is the proof of the algebra property in the case of example A3. To see this consider

$$\mu_3(h) = \sum h_{(1)} \otimes h_{(2)} \otimes h_{(3)}$$

as obtained by applying $(\mu_H \otimes \mathrm{id})\mu_H$ to h and see

$$\mu_4(h) = \sum h_{(1)} \otimes h_{(2)} \otimes h_{(3)} \otimes h_{(4)}$$

both as obtained from applying $\mu_H \otimes \mu_H$ to $\mu_H(h)$ and as result of applying $\mathrm{id} \otimes \mu_H \otimes \mathrm{id}$ to $\mu_3(h)$. Then

$$\begin{aligned}(h_{(1)} \cdot a)(h_{(2)} \cdot b) &= \sum h_{(1)} a \iota(h_{(2)}) h_{(3)} b \iota(h_{(4)}) = \sum h_{(1)} a (e\varepsilon(h_{(2)}) b \iota(h_{(3)}) \\ &= \sum h_{(1)} (e\varepsilon(h_{(2)}) ab \iota(h_{(3)}) = \sum h_{(1)} ab \iota(h_{(2)}).\end{aligned}$$

Here the first equality is the definition of the action (twice), the second equality is part of the axioms defining an antipode, the third equality holds because $e\varepsilon(h_{(2)})$

is a scalar, the fourth equality comes from the properties of unit and counit and the fifth equality is again the definition of the action.

All these examples, complete with proofs, can be found in [**481**], section 1.6, p.15-27. (Not all the more standard monographs on Hopf algebras and quantum groups describe all 8 examples.)

Of course switching right to left and interchanging H^* and H there are 24 more examples. These do not bring anything really new and it is not worth the trouble to list them.

8.1.32. Group graded algebras. Let G be a finite group of n elements. A G-group graded algebra is an algebra over k that decomposes as a direct sum

$$A = \bigoplus_g A_g \quad \text{such that } A_g A_h \subset A_{gh} \quad \text{and } 1_A \subset A_e$$

where e is the identity element of G. Let $\mathrm{Func}(G)$ be the algebra of k-valued functions on G. This is a Hopf algebra, the dual of the group Hopf algebra kG. A basis of $\mathrm{Func}(G)$ is given by the orthogonal idempotents p_g, $g \in G$ defined by $p_g(h) = \delta_{g,h}$. On this basis the comultiplication on $\mathrm{Func}(G)$ is

$$\mu(p_g) = \sum_{h \in G} p_{gh^{-1}} \otimes p_h. \tag{8.1.33}$$

There is now the following not unimportant remark. A group grading on an algebra A is the same thing as a module algebra structure on A over the Hopf algebra $\mathrm{Func}(G)$. This gives another example of module algebras.

Indeed, suppose there is a G-grading. Then every element of A can be uniquely written as a sum

$$a = \sum_g a_g.$$

Now define the action of $\mathrm{Func}(G)$ on A by $p_g \cdot a = a_g$. This is indeed an action (trivial) and it satisfies the extra module algebra conditions. Indeed the unit element of $\mathrm{Func}(G)$ is the sum of the p_g so that $1_{\mathrm{Func}(G)} \cdot a = a$ and as $ab = \sum_{g,h} a_g b_h$ it follows that $(ab)_g = \sum_h a_{gh^{-1}} b_h$ so that

$$p_g \cdot ab = (ab)_g = \sum_h a_{gh^{-1}} b_h = \sum_h (p_{gh^{-1}} \cdot a)(p_h \cdot b)$$

as it should be in view of (8.1.33).

Inversely, if A is a $\mathrm{Func}(G)$-module algebra, then $A_g = p_g \cdot A$ defines a group grading on A.

8.1.34. Smash products and group algebras and their duals. Now let A be a module algebra over kG. Then one can form the smash product $A\#kG = A * G$ which is the skew group ring of G over A with its multiplication given by $(a\#g)(b\#h) = ag(b)\#gh$, see example 3.9.7. This algebra is obviously G-graded and so is a $\mathrm{Func}(G)$-module algebra and one can form the further smash product $(A\#kG)\#\mathrm{Func}(G)$. It is now a theorem that

$$(A\#kG)\#\mathrm{Func}(G) \cong A^{n \times n} \tag{8.1.35}$$

the algebra of $n \times n$ matrices with entries in A.

Second let A be a Func(G)-module algebra. Then one can form the smash product $A\#\mathrm{Func}(G)$ whose elements are sums of elements of the form $a\#p_g$. There is a natural (left) group action by algebra automorphism of G on this smash product defined by

$$(a\#p_g)^h = a\#p_{gh^{-1}}$$

turning $A\#\mathrm{Func}(G)$ into a kG-module algebra so that the further smash product with kG can be taken. A dual theorem now says that also

$$(A\#\mathrm{Func}(G))\#kG \cong A^{n\times n}. \tag{8.1.36}$$

These theorems are from [**151**], and they are (elementary) Hopf algebra versions of results for von Neumann algebras as in [**426**]; [**532**]. They have a common generalization to the effect that under suitable hypotheses

$$(A\#H)\#H^0 \cong A \otimes (H\#H^0).$$

See [**71**] for this. For a great deal more information on smash products and their uses see [**152**]; [**153**]; [**522**]; [**520**].

8.2. The quantum groups $GL_q(n, \mathbf{C})$ and multiparameter generalizations

The phrase 'quantum groups' was coined by Vladimir G. Drinfel'd in the middle 1980's. And it stuck as an attractive and evocative term even if the objects in question are not necessarily quantum and they are most definitely not groups. Since then 'quantum' versions of most everything in mathematics have made their appearance.

Quantum groups are a special kind of Hopf algebra. One important bit of motivation came from the work of Ludwig D. Faddeev and the Leningrad school (Leon A. Takhtajan, Nicolai Yu. Reshetikhin, Evgeny K. Sklyanin, Michael Semenov Tian-Shansky, Petr P. Kulish, Vladimir G. Turaev, ...) on the quantum inverse scattering method (Bethe Ansatz, quantum R-matrices, Quantum Yang-Baxter equation (QYBE), see also section 8.4 below).

Both the "deformations of algebras of functions on a group approach", Vladimir G. Drinfel'd, [**212**], [**211**]; also Yuri I. Manin, [**496**], and "the deformations of universal enveloping algebras approach", Nicolai Yu. Reshetikhin - Petr P. Kulish (the algebara $U_q(\mathfrak{sl}_2)$, [**420**], Evgeny K. Sklyanin (the Hopf algebra $U_q(\mathfrak{sl}_2)$, [**645**]), Michio Jimbo and Vladimir G. Drinfel'd, (deformations of universal enveloping algebras of semisimple Lie algebras and more generally symmetrizable Kac-Moody algebras, [**350**], [**211**], [**212**], [**213**], started from certain solutions of the QYBE. These objects are dual each other just as in the non-quantum, i.e. classical case. Note, however that is argued by Vaughan F.R. Jones, [**354**], that there are solutions of the QYBE that do not fit into the quantum group picture.

A totally different bit of motivation came from Tannaka-Krein reconstruction theory (Tannaka-Krein duality): a compact group can be recovered from the category of its representations. The question S. Lev Woronowich asked himself was: what happens if the symmetry condition on the categories involved is relaxed. This led to what are called compact matrix pseudogroups, [**688**], and they are, remarkably, the same as quantum groups, showing that quantum groups were something whose time had come.

See [**358**] for an introduction to both classical Tannaka-Krein reconstruction and an introduction to quantum groups from this point of view; see also [**181**] for classical Tannaka-Krein theory.

The papers quoted by the Leningrad school and Drinfel'd, Jimbo, Woronowicz are generally considered as the papers that started the business of quantum groups (independently of each other). There is now a vast literature on quantum groups including more that a score of monographs.

In this chapter the "deformation of functions on a Lie group" point of view will have preference. (But see the next section for a brief account of the deformed universal enveloping algebras).

8.2.1. The idea of symmetry. The idea of symmetry is usually understood as referring to a situation where one has a group, G, and a space, X, on which it acts. In such a situation, among other things one is interested in how things decompose into indecomposables. That is the orbits or subspaces that are transitive under the given action. However, for e.g. a group acting on itself by, say, left, translation there is just one orbit and little can be done. Things change drastically if instead of the space X one considers the functions on it. The group acts on the algebra of functions on X by

$$g(f)(x) = f(g^{-1}x) \tag{8.2.2}$$

and as a rule $\mathrm{Func}(X)$ is not indecomposable and decomposing it can be a very useful tool, witness e.g. Fourier analysis: the case of the circle group acting on itself (decomposing periodic functions). Instead of an action $G \times X \longrightarrow X$ there is now an action $G \times \mathrm{Func}(G) \longrightarrow \mathrm{Func}(G)$ or, more algebraically, an action

$$\mathbf{R}G \otimes \mathrm{Func}(X) \longrightarrow \mathrm{Func}(X).$$

This opens up, for instance, the possibility of considering more general symmetries by replacing the Hopf algebra $\mathbf{R}G$ by a more general bialgebra or Hopf algebra.

One can also, better, take functions on both the group G and the space X. This then gives a coaction

$$\mathrm{Func}(X) \longrightarrow \mathrm{Func}(X) \otimes \mathrm{Func}(G). \tag{8.2.3}$$

There are more reasons to consider such an algebraic picture. Here is one. In classical (Liouville) completely integrable system theory (soliton equations) and also in many other situations of classical physics there are large groups of (hidden) symmetries.

The question is (still largely open) what happens to these symmetries under the dressing transformations which make soliton theory work or during quantization. Mostly the classical symmetry gets lost[3]; one needs to deform (quantize) the symmetry group along with the space to retain symmetry[4]. This is what leads to quantum groups. It is also what makes them important in noncommutative geometry.

8.2.4. Functions on an algebraic group or Lie group. Consider the group of invertible $n \times n$ matrices over k and the affine space on which it acts (on

---

[3]This seems to have been remarked first by Michael Semenov Tian-Shansky, [**635**].

[4]The point was overlooked for quite some time. Possibly because the representation theory of classical quantum groups is very like the representation theory of the corresponding classical groups themselves, so that the classical groups still worked.

the left).

$$\begin{pmatrix} a_1^1 & \cdots & a_n^1 \\ \vdots & \ddots & \vdots \\ a_1^n & \cdots & a_n^n \end{pmatrix} \begin{pmatrix} x^1 \\ \vdots \\ x^n \end{pmatrix} = \begin{pmatrix} \sum_i a_i^1 x^i \\ \vdots \\ \sum_i a_i^n x^i \end{pmatrix}. \tag{8.2.5}$$

Consider the coordinate functions t_j^i on the (algebraic) group $\mathbf{GL}(n;k)$ over a field k. These form an algebra (under pointwise multiplication) and the comultiplication that induces matrix multiplication (of function values) is

$$\mu(t_j^i) = \sum_{r=1}^{n} t_r^i \otimes t_j^r, \qquad i, j = 1, 2, \dots, r.$$

This can be written succintly

$$\mu \begin{pmatrix} t_1^1 & \cdots & t_n^1 \\ \vdots & \ddots & \vdots \\ t_1^n & \cdots & t_n^n \end{pmatrix} = \begin{pmatrix} t_1^1 & \cdots & t_n^1 \\ \vdots & \ddots & \vdots \\ t_1^n & \cdots & t_n^n \end{pmatrix} \tilde{\otimes} \begin{pmatrix} t_1^1 & \cdots & t_n^1 \\ \vdots & \ddots & \vdots \\ t_1^n & \cdots & t_n^n \end{pmatrix}$$

where the tilde over the tensor product sign is there to distinguish this expression from the Kronecker product of matrices. The (i,j)-th element of the expression on the right is by definition $\sum\limits_r t_r^i \otimes t_j^r$.

Add a counit $\varepsilon(t_j^i) = \delta_{i,j}$, an extra variable z, and the relation $z \det(t_j^i) = 1$ to capture the idea that the determinant is nonzero and one has the Hopf algebra that defines the algebraic group (functor) of the general linear group

$$H_{\mathbf{GL}(n;k)} = k[t_j^i, z]/(z \det(t_j^i) = 1), \quad \mu(t_j^i) = \sum_r t_r^i \otimes t_j^r, \quad \mu(z) = z \otimes z. \tag{8.2.6}$$

There is an antipode given by the familiar minor formula $\iota(t_j^i) = (-1)^{i+j} z M_i^j$.

A quantum group is now obtained by deforming the commutative algebra $k[t_j^i, z]/(z \det(t_j^i) = 1)$ to something (mildly) noncommutative while retaining the comultiplication as is.

One way to do this is as follows. First consider the bialgebra of the $n \times n$ matrix valued functor with the usual matrix multiplication. This is

$$H_{\mathbf{M}_{n\times n}(k)} = k[t_j^i], \quad \mu(t_j^i) = \sum_r t_r^i \otimes t_j^r, \quad \varepsilon(t_j^i) = \delta_{i,j}. \tag{8.2.7}$$

The Hopf algebra coacts on the left on the underlying algebra $k[X^1, X^2, \dots, X^n]$ of n-dimensional affine space by the formula

$$X^i \mapsto \sum_r t_r^i \otimes X^r \tag{8.2.8}$$

making $k[X^1, X^2, \dots, X^n]$ a left comodule algebra over $H_{\mathbf{M}_{n\times n}(k)} = k[t_j^i]$. This formula of course comes from (8.2.5). It also coacts on the right on the (same) underlying algebra $k[Y_1, Y_2, \dots, Y_n]$ by the formula

$$Y_j \mapsto \sum_s Y_s \otimes t_j^s \tag{8.2.9}$$

making $k[Y_1, Y_2, \dots, Y_n]$ a right comodule algebra over $H_{\mathbf{M}_{n\times n}(k)}$. The idea is now to deform the affine space algebras $k[X^1, X^2, \dots, X^n]$ and $k[Y_1, Y_2, \dots, Y_n]$ to quantum affine space algebras and to see what relations are needed on the free associative algebra $k\langle t^i_j;\, i,j \in \{1,2,\ldots,n\}\rangle$ in order that these two coactions remain well defined.

Probably the most natural quantum deformation of $k[X^1, X^2, \ldots, X^n]$ is to choose parameters (or indeterminates that commute with all the X's) $q^{a,b}$ such that $q^{a,b}q^{b,a} = 1$, $q^{a,a} = 1$ and to consider the algebra

$$k\langle X^1, X^2, \ldots, X^n\rangle/(X^iX^j - q^{i,j}X^jX^i) \tag{8.2.10}$$

Now require that the left action of a suitable quotient of $k\langle t^i_j;\, i,j \in \{1,2,\ldots,n\}\rangle$ coacts on the left on (8.2.10) according to formula (8.2.8).

8.2.11. The case $n = 2$. Here is how this works out in the case $n = 2$. Write q^L for $q^{1,2}$ so that

$$X^1X^2 = q^LX^2X^1 \tag{8.2.12}$$

$$X^1 \mapsto t^1_1 \otimes X^1 + t^1_2 \otimes X^2 \quad \text{and } X^2 \mapsto t^2_1 \otimes X^1 + t^2_2 \otimes X^2.$$

Thus

$$X^1X^2 \mapsto t^1_1t^2_1 \otimes X^1X^1 + t^1_1t^2_2 \otimes X^1X^2 + t^1_2t^2_1 \otimes X^2X^1 + t^1_2t^2_2 \otimes X^2X^2$$
$$X^2X^1 \mapsto t^2_1t^1_1 \otimes X^1X^1 + t^2_1t^1_2 \otimes X^1X^2 + t^2_2t^1_1 \otimes X^2X^1 + t^2_2t^1_2 \otimes X^2X^2$$

and plugging in $X^1X^2 = q^LX^2X^1$ there result, as necessary and sufficient condition for (8.2.8) to be well defined, the three relations

$$t^1_1t^2_1 = q^Lt^2_1t^1_1, \quad t^1_2t^2_2 = q^Lt^2_2t^1_2, \quad [t^1_1, t^2_2] = q^Lt^2_1t^1_2 - (q^L)^{-1}t^1_2t^2_1. \tag{8.2.13}$$

Compared with the quantum group $\mathbf{SL}_q(2;k)$, apart from the determinant condition, this is only half of the relations needed; see 3.4.23 above. The other three result from letting a suitable quotient of $k\langle t^1_1, t^1_2, t^2_1, t^2_2\rangle$ coact on the right on $k\langle Y_1, Y_2\rangle/(Y_2Y_1 - q_RY_1Y_2)$ which results in the additional three relations

$$\begin{aligned} t^1_1t^1_2 &= q_Rt^1_1t^1_2,\\ t^2_1t^2_2 &= q_Rt^2_2t^2_1,\\ [t^1_1, t^2_2] &= q_Rt^1_2t^2_1 - (q_R)^{-1}t^2_1t^1_2. \end{aligned} \tag{8.2.14}$$

If q_R is taken to be equal to $q^L = q$ (and $q \neq -q^{-1}$) it follows from (8.2.12) and (8.2.13) together that

$$t^2_1t^1_2 = t^1_2t^2_1 \quad \text{and } [t^1_1, t^2_2] = (q - q^{-1})t^2_1t^1_2 \tag{8.2.15}$$

so that apart from the determinant condition the relations of 3.4.24 reappear. The six relations

$$\begin{gathered} t^1_1t^1_2 = qt^1_2t^1_1, \quad t^2_1t^2_2 = qt^2_2t^2_1\\ t^1_1t^1_2 = qt^2_1t^2_1, \quad t^1_1t^1_2 = qt^1_2t^1_1\\ t^2_1t^1_2 = t^1_2t^2_1, \quad [t^1_1, t^2_2] = (q - q^{-1})t^1_2t^2_1 \end{gathered} \tag{8.2.16}$$

define the 2×2 quantum matrix algebra $\mathbf{M}_q(2 \times 2; k)$. To define the quantum group $\mathbf{GL}_q(2;k)$ add an additional indeterminate z commuting with all the t^i_j and the relation

$$z\det_q \begin{pmatrix} t^1_1 & t^1_2 \\ t^2_1 & t^2_2 \end{pmatrix} = z(t^1_1t^2_2 - qt^1_2t^2_1) = 1. \tag{8.2.17}$$

When $q^L \neq q_R$ there results a two parameter deformation of the quantum matrix algebra and a two parameter two by two quantum group. (There is also a suitable definition of a quantum determinant in this case; see [**90**], p.16, formula (3).)

8.2.18. The case $n > 2$. Basically everything also works in general for $n > 2$. As an algebra the one parameter quantum group $\mathbf{GL}_q(n;k)$ is the algebra $k\langle t^i_j, z\rangle$ subject to the relations

$$t^s_r t^s_{\bar{r}} = q t^s_{\bar{r}} t^s_r \text{ if } \bar{r} > r; \quad t^s_r t^{\bar{s}}_r = q t^{\bar{s}}_r t^s_r \text{ if } \bar{s} > s; \quad t^r_{\bar{s}} t^{\bar{r}}_s = t^{\bar{r}}_s t^r_{\bar{s}} \text{ if } \bar{r} > r, \ \bar{s} > s$$
$$t^r_s t^{\bar{r}}_{\bar{s}} - t^{\bar{r}}_{\bar{s}} t^r_s = (q - q^{-1}) t^r_{\bar{s}} t^{\bar{r}}_s \text{ if } \bar{r} > r; \ t^r_s z = z t^r_s \tag{8.2.19}$$
$$z\det_q(t^i_j) = 1, \quad \text{where } \det_q(t^i_j) = \sum_\sigma (-1)^{\lg(\sigma)} t^1_{\sigma(1)} t^2_{\sigma(2)} \cdots t^n_{\sigma(n)}$$

where the sum is over all permutations σ and $\lg(\sigma)$, the length of σ, is the minimal length of an expression for σ as a product of adjacent transpositions $(i, i+1)$. The Hopf algebra structure is given by

$$\mu(t^i_j) = \sum_r t^i_r \otimes t^r_j, \quad \varepsilon(t^i_j) = \delta_{i,j}, \quad \mu(z) = z \otimes z, \quad \varepsilon(z) = 1.$$

A good way to remember the first four of the relations (8.2.19) is as follows. For all $\bar{r} > r$, $\bar{s} > s$ the commutation relations for the four indeterminates

$$\begin{matrix} t^r_s & t^r_{\bar{s}} \\ t^{\bar{r}}_s & t^{\bar{r}}_{\bar{s}} \end{matrix}$$

are the same as for the indeterminates

$$\begin{matrix} t^1_1 & t^1_2 \\ t^2_1 & t^2_2 \end{matrix}$$

in the case $n = 2$.

8.2.20. Definition (PBW algebras). An algebra over k is called a **PBW algebra** if there is a set of generators $x_1, x_2, \ldots$ such that the monomials

$$x^{r_1}_{i_1} x^{r_2}_{i_2} \cdots x^{r_m}_{i_m}, \quad i_1 \le i_2 \le \cdots \le i_m, \quad r_j \in \mathbf{N}$$

form a basis of the algebra. The acronym stands for Poincaré-Birkhoff-Witt and the PBW theorem says that the universal enveloping algebras over a field of characteristic zero are examples. There are many more PBW type theorems (including PBW theorems for quantum groups).

8.2.21. Multiparameter deformations (quantizations). Now consider the case that a suitable quotient of $k\langle t^i_j\rangle$ is to left coact on the quantum affine space algebra $k\langle X^1, X^2, \ldots, X^n\rangle/(X^i X^j - q^{i,j} X^j X^i)$ and to right coact on the quantum affine space algebra $k\langle Y_1, Y_2, \ldots, Y_n\rangle/(Y_r Y_s - q_{r,s} Y_s Y_r)$ where the $q_{r,s}$ are a second set of parameters or indeterminates commuting with the t's such that $q_{r,s} q_{s,r} = 1$ and $q_{r,r} = 1$.

This gives two sets of relations and unless there are good relations between the $q^{a,b}$ and $q_{r,s}$ the resulting algebra can be quite ugly. As a matter of fact there is a theorem to the effect that the relations from the left action together with the relations from the right action give a PBW algebra if and only if $q^{a,b}/q_{r,s} = \rho$ for all a, b, r, s, where ρ is one single fixed constant (or indeterminante) unequal to -1, [**315**].

This gives an $\binom{n}{2} + 1$ parameter family of quantum (?) deformations of the $n \times n$ matrix function algebra. There is also a suitable definition of a determinant in this multiparameter setting, see [**90**], and so there is an $\binom{n}{2} + 1$ parameter family, $\mathbf{GL}_Q(n;k)$, of quantum deformations of the general linear group of which the $\mathbf{GL}_q(n;k)$ are a one parameter subfamily. Such $\binom{n}{2}+1$ parameter families were constructed by several authors, independently of each other, all apparently in the winter of 1990/1991 or slightly earlier.

The families exhibited in [**35**], [**314**], [**315**], [**653**], [**659**] are all the same and are likely to be the same as the ones obtained by twisting as in [**587**].

The basic picture appears to be that the important quantum group is the single parameter quantum group $\mathbf{GL}_q(n;k)$, the central and most important one, and that the other ones are obtained from them by suitable twisting by two-cocycles, [**35**]. That is not to say that these twisted versions are not of interest; simply that they have not *yet* had significant applications.

8.2.22. Fundamental commutation relations. The first 4 relations of (8.2.19), which define the quantum matrix algebra $\mathbf{M}_q(n\times n,k)$, can be conveniently encoded as

$$RT_1T_2 = T_2T_1R \tag{8.2.23}$$

where T is the matrix of indeterminates

$$T = \begin{pmatrix} t_1^1 & \cdots & t_n^1 \\ \vdots & \ddots & \vdots \\ t_1^n & \cdots & t_n^n \end{pmatrix}, \quad T_1 = T\otimes I_n, \quad T_2 = I_n\otimes T \tag{8.2.24}$$

where the tensor product in (8.2.24) is the Kronecker product of matrices

$$A\otimes B = \begin{pmatrix} a_1^1B & \cdots & a_n^1B \\ \vdots & \ddots & \vdots \\ a_1^mB & \cdots & a_n^mB \end{pmatrix}.$$

Above and in the following there are frequently matrices of size $n^2\times n^2$. These will have there rows and columns labeled by pairs of indices (r,s), $r,s\in\{1,2,\dots,n\}$ in lexicographic order (see also example (8.2.30) below). Thus

$$(T_1T_2)_{r,s}^{a,b} = t_r^at_s^b, \quad (T_2T_1)_{r,s}^{a,b} = t_s^bt_r^a.$$

In (8.2.23) the $n^2\times n^2$ matrix R is given by

$$r_{c,d}^{a,b} = 0 \quad \text{unless } \{a,b\}=\{c,d\} \quad \text{as sets;}$$

$$r_{b,a}^{a,b} = \begin{cases} q^{-1}-q & \text{if } a<b \\ 0 & \text{if } a>b \end{cases}; \qquad r_{a,b}^{a,b} = \begin{cases} q^{-1} & \text{if } a=b \\ 1 & \text{if } a\neq b. \end{cases} \tag{8.2.25}$$

This is easily checked. The matrix equation (8.2.23) amounts to the n^4 equations

$$\sum_{c,d} r_{c,d}^{a,b}t_u^ct_v^d = \sum_{i,j} r_{u,v}^{i,j}t_j^bt_i^a. \tag{8.2.26}$$

There are 9 cases to consider.

Take first $a=b$, $u<v$.

Then the left hand side of (8.2.26) is equal to $r_{a,a}^{a,a}t_u^at_v^a = q^{-1}t_u^at_v^a$ and the right hand side is equal to $r_{u,v}^{u,v}t_v^at_u^a + r_{u,v}^{v,u}t_u^at_v^a = t_v^at_u^a$ and so $t_u^at_v^a = qt_v^at_u^a$ if $u<v$.

Now take $a<b$, $u=v$.

Then the left hand side of (8.2.26) is equal to $r_{a,b}^{a,b}t_u^at_u^b + r_{b,a}^{a,b}t_u^bt_u^a = t_u^at_u^b + (q^{-1}-q)t_u^bt_u^a$ and the right hand side is equal to $r_{u,u}^{u,u}t_u^bt_u^a = q^{-1}t_u^bt_u^a$. So that $t_u^at_u^b = qt_u^bt_u^a$ if $a<b$.

Now take $a<b$, $u<v$.

Then the left hand side of (8.2.26) is equal to $r^{a,b}_{a,b}t^a_u t^b_v + r^{a,b}_{b,a}t^b_u t^a_v = t^a_u t^b_v + (q^{-1}-q)t^b_u t^a_v$ and the right hand side is equal to $r^{u,v}_{u,v}t^b_v t^a_u + r^{v,u}_{u,v}t^b_u t^a_v = t^b_v t^a_u$. So that $[t^a_u, t^b_v] = (q-q^{-1})t^b_u t^a_v$ if $a < b$, $u < v$.

Next take $a > b$, $u < v$.
Then the left hand side of (8.2.26) is equal to $r^{a,b}_{a,b}t^a_u t^b_v + r^{a,b}_{b,a}t^b_u t^a_v = t^a_u t^b_v$ and the right hand side is equal to $r^{u,v}_{u,v}t^b_v t^a_u + r^{v,u}_{u,v}t^b_u t^a_v = t^b_v t^a_u$. And thus $t^a_u t^b_v = t^b_v t^a_u$ if $a > b$, $u < v$.

The other five cases bring nothing new. For instance, take $a = b$, $u > v$. Then the left hand side of (8.2.26) is $r^{a,a}_{a,a}t^a_u t^a_v = q^{-1}t^a_u t^a_v$ and the right hand side is equal to $r^{u,v}_{u,v}t^a_v t^a_u + r^{v,u}_{u,v}t^a_u t^a_v = t^a_v t^a_u + (q-q^{-1})t^a_u t^a_v$, so that in this case $qt^a_u t^a_v = t^a_v t^a_u + (q-q^{-1})t^a_u t^a_v$ and so $t^a_v t^a_u = qt^a_u t^a_v$ if $u > v$ which was obtained before.

Relations like (8.2.23) are called FCR (**Fundamental Commutation Relations**) and they form the starting point of the quantum inverse scattering method (Bethe Anzatz) developed by the Leningrad school, [**240**], [**241**].

8.2.27. Quantum Yang-Baxter equation. Let $R = (r^{a,b}_{u,v})$ be an $n^2 \times n^2$ matrix of unknowns. The **Yang-Baxter equation** is

$$R_{12}R_{13}R_{23} = R_{23}R_{13}R_{12}. \tag{8.2.28}$$

Here R is seen as determining an endomorphisms of the tensor square of an n-dimensional vector space V. I.e. if $\{e^1, e^2, \dots, e^n\}$ is a basis for V, then

$$R(e^a \otimes e^b) = \sum_{u,v} r^{a,b}_{u,v} e^u \otimes e^v.$$

In (8.2.28) R_{12} and R_{23} are the $n^3 \times n^3$ matrices $R_{12} = R \otimes I_n$, $R_{23} = I_n \otimes R$ and R_{13} is obtained by letting R act on the outer factors of $V \otimes V \otimes V$ and leaving the middle factor alone, i.e.

$$R_{13}(e^a \otimes e^b \otimes e^c) = \sum_{u,v} r^{a,c}_{u,v} e^u \otimes e^b \otimes e^v.$$

An interpretation of the quantum Yang-Baxter equation in terms of quantum scattering of particles on the line is as follows. Imagine three particles of the same kind that have n pure states. Then a state of one single particle is a vector in an n-dimensional space V and a state of a two particle system is a vector in $V \otimes V$ (as there are n^2 pure states for the two particles together). Now let the interaction (scattering) of two particles when they hit each other be given by the matrix R. Now consider three particles. The possible interaction patterns are

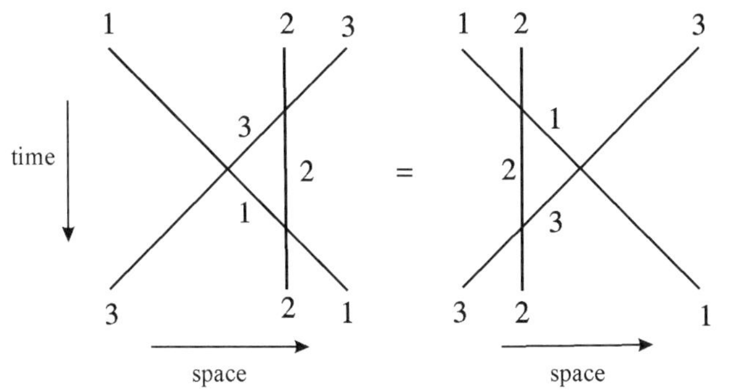

The so-called factorization requirement for the interaction says that it does not matter which of the two interaction patterns actually takes place. In terms of the interaction matrix the left hand pattern gives a total interaction $R_{12}R_{13}R_{23}$, and the right hand pattern gives a total interaction $R_{23}R_{13}R_{12}$, and the quantum Yang-Baxter equation results.

Which equation one actually gets depends on how the particles are numbered. If instead of as above they are numbered 1,2,3 from left to right throughout the interaction process (which is also natural) the picture becomes

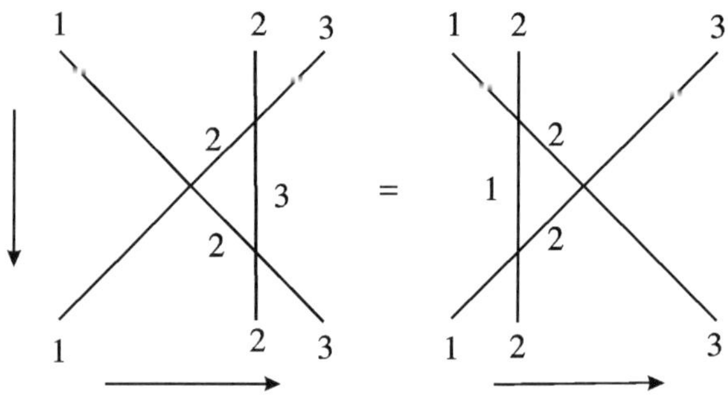

and the equation becomes the so-called braid version of the Yang-Baxter equation[5]

$$S_{23}S_{12}S_{23} = S_{12}S_{23}S_{12}. \tag{8.2.29}$$

A solution R of (8.2.28) gives a solution $S = \tau R$ of (8.2.29) and vice versa. Here τ is the switch (flip) isomorphism $x \otimes y \mapsto y \otimes x$.

The quantum Yang-Baxter is named after Chen-Ning Franklin Yang ([**699**], quantum scattering) and Rodney James Baxter ([**52**], [**53**], eight vertex model, hard hexagon model in statistical mechanics). In statistical mechanics the Yang-Baxter equation also goes under its original name in that specialism: star-triangle relation.

If the FCR (8.2.23) holds for a sufficiently general matrix T the matrix R satisfies the quantum Yang-Baxter equation.

Thus the quantum group matrix given by (8.2.25) is a solution of the QYBE (which can also be simply verified directly). This solution has the property that $r_{u,v}^{a,b} = 0$ if the sets $\{a, b\}$ and $\{u, v\}$ are not equal.

In [**315**] all solutions that satisfy this property are determined. They consist of blocks, which in turn are made up of components which are fitted together in not entirely trivial ways. Here is an example (one block of size three consisting of three components of size 1).

---

[5]Often also simply called the **braid equation**. A solution of the braid equation gives rise to representation of the Artin braid group, see section 8.5 below.

	1,1	1,2	1,3	2,1	2,2	2,3	3,1	3,2	3,3
1,1	ρ_1								
1,2		$x_{2,1}^{-1}$		y					
1,3			$x_{3,1}^{-1}$				y		
2,1				$x_{2,1}$					
2,2					ρ_2				
2,3						$x_{3,2}^{-1}$		y	
3,1							$x_{3,1}$		
3,2								$x_{3,2}$	
3,3									ρ_3

(8.2.30)

Here the $\rho \in \{-q, q^{-1}\}$ and $y = q^{-1} - q$. If all the ρ_i are chosen to be q^{-1} one finds the matrix defining (via the FCR) the $\binom{3}{2}+1$ parameter quantum matrix algebra $\mathbf{M}_q(3 \times 3; k)$ and the $\binom{3}{2}+1$ parameter quantum group $\mathbf{GL}_Q(3;k)$. It is obvious how to generalize this to arbitrary $n > 3$, giving an R-matrix for the general $\binom{n}{2}+1$ parameter quantum group $\mathbf{GL}_Q(n;k)$. Having such an explicit R-matrix can be useful on occasion.

Another example of a solution satisfying "$r^{a,b}_{u,v} = 0$ if the sets $\{a,b\}$ and $\{u,v\}$ are not equal" is

	1,1	1,2	1,3	1,4	2,1	2,2	2,3	2,4	3,1	3,2	3,3	3,4	4,1	4,2	4,3	4,4
1,1	ρ_1															
1,2		0			ρ_1											
1,3			$x_{3,1}^{-1}$						y							
1,4				$x_{4,1}^{-1}$									y			
2,1		ρ_1			0											
2,2						ρ_1										
2,3							$x_{3,2}^{-1}$			y						
2,4								$x_{4,2}^{-1}$						y		
3,1									$x_{3,1}$							
3,2										$x_{3,2}$						
3,3											ρ_2					
3,4												$x_{4,2}^{-1}$			y	
4,1													$x_{4,1}$			
4,2														$x_{4,2}$		
4,3															$x_{4,3}$	
4,4																ρ_3

In this matrix the two off-diagonal ρ_1 are at the positions (1,2)(2,1) and (2,1)(1,2) and the five y's are at positions (1,3)(3,1); (2,3)(3,2); (1,4)(4,1); (2,4)(4,2); (3,4)(4,3).

This one has one block of size 4 with three components: one of size 2 and two of size 1. What these solutions and also solutions with not all the ρ's equal, mean in terms of quantum groups remains to be explored.

The R-matrices defining the quantum versions of algebraic groups of types B^1, C^1, D^1, A^2 do not satisfy the condition: "$r^{a,b}_{u,v} = 0$ if the sets $\{a,b\}$ and $\{u,v\}$ are not equal" but satisfy the weaker condition: "$r^{a,b}_{u,v} = 0$ unless the sets $\{a,b\}$ and $\{u,v\}$ are equal or $(a,b) = (a, n+1-a)$ and $(u,v) = (u, n+1-u)$". It should be possible to find all solutions that satisfy this weaker condition. A good start has been made by Nico van den Hijligenberg, [**673**], but the job is by no means finished.

8.3. The quantum groups $U_q(\mathfrak{sl}(n;k))$

Dually there are the quantum deformations of universal enveloping algebras of Lie algebras. These exists quite generally, for instance for all finite dimensional semisimple Lie algebras and for Kac-Moody algebras with symmetrizable Cartan matrix.

8.3.1. Definition of the quantum enveloping algebra $U_q(\mathfrak{sl}(2;k))$. As an associative algebra $U_q(\mathfrak{sl}(2;k))$ is generated by four elements E, F, K, K^{-1} subject to the relations

$$KEK^{-1} = q^2E, \quad KFK^{-1} = q^{-2}F, \quad KK^{-1} = K^{-1}K = 1$$
$$EF - FE = \frac{K - K^{-1}}{q - q^{-1}}. \tag{8.3.2}$$

This is the description as in [**90**], p.46. There, there is also a nice algebraic description of the $U_q(\mathfrak{sl}(n;k))$ for larger n which avoids the quantum Serre relations (which tend to be difficult to handle); see loc. cit.

Let H be the element of $\mathfrak{sl}(2;k)$ that has the commutation relations $[H, E] = 2E$, $[H, F] = -2F$ as in the classical Lie algebra $\mathfrak{sl}(2;k)$. Then over fields like the real of complex numbers one can see K as

$$K = e^{hH}, \quad q = e^h.$$

Indeed $[H, E] = 2E$ is the same as $(H - 1)E = E(H + 1)$ and so

$$KE = e^h e^{h(H-1)} E = e^h E e^{h(H+1)} = q^2 E e^h e^{h(H-1)} = q^2 EK$$

giving the definition used in [**481**], p.83, with the small difference that in loc.cit. the element $e^{hH/2}$ is also supposed to be in $U_q(\mathfrak{sl}(2;k))$.

The comultiplication, counit, and antipode are given by (still following [**90**])

$$\begin{aligned} &\mu(K) = K \otimes K, \ \mu(E) = E \otimes 1 + K \otimes E, \ \mu(F) = F \otimes K^{-1} + 1 \otimes F \\ &\varepsilon(K) = 1, \ \varepsilon(E) = 0, \ \varepsilon(F) = 0 \\ &\iota(K) = K^{-1}, \ \iota(E) = -K^{-1}E, \ \iota(F) = -FK. \end{aligned} \tag{8.3.3}$$

Actually there are five possibilities that can be taken and that work; see [**1**].

8.3.4. Duality pairings. Given two bialgebras (Hopf algebras) U and H a **duality pairing** is a k-bilinear map

$$\langle\, ,\, \rangle : H \times U \longrightarrow k \tag{8.3.5}$$

such that

$$\begin{aligned} &\langle ab,\, u\rangle = \langle a \otimes b,\, \mu_U(u)\rangle, &&\langle a,\, uv\rangle = \langle \mu_H(a),\, u \otimes v\rangle \\ &\langle \varepsilon_H(a),\, \lambda\rangle = \langle a,\, e_U(\lambda)\rangle, &&\langle e_H(\lambda),\, u\rangle = \langle \lambda,\, \varepsilon_U(u)\rangle \end{aligned} \tag{8.3.6}$$

and if both H and U are Hopf algebras in addition

$$\langle \iota_H(a),\, u\rangle = \langle a,\, \iota_U(u)\rangle. \tag{8.3.7}$$

A pairing (8.3.5) is nondegenerate if

$\langle a,\, u\rangle = 0$ for all u implies $a = 0$ and $\langle a,\, u\rangle = 0$ for all a implies $u = 0$.

It is often more convenient and more elegant to work with nondegenerate pairings that with actual dual algebras.

Example 8.3.8. The duality pairing between the universal enveloping algebra of the Lie algebra of a connected Lie group and the Hopf algebra of differentiable functions on that Lie group. Let G be a connected Lie group. Let $\mathfrak{g}$ be its Lie algebra. The elements of this Lie algebra can be seen as left invariant vector fields on G, i.e. as left invariant differential operators of degree 1 on the differential functions on G. Then the elements of $U\mathfrak{g}$ become the left invariant differential operators (any order) on DiffFunc(G), the Hopf algebra of differential functions on G. The duality pairing is now

$$\langle\, , \rangle : U\mathfrak{g} \times \mathrm{DiffFunc}(G) \longrightarrow \mathbf{R}, \quad \langle D,\, f\rangle = (Df)(e)$$

where e is the unit element of the Lie group G.

8.3.9. Duality of $U_q(\mathfrak{sl}(2;k))$ and $\mathbf{SL}_q(2;k)$. Take as a candidate duality pairing

$$\langle K,\, a\rangle = q, \quad \langle K,\, d\rangle = q^{-1}, \quad \langle E,\, b\rangle = 1, \quad \langle F,\, c\rangle = 1$$

and all other pairings between generators K, E, F and a, b, c, d equal to zero, while the pairings with 1 on either side are given by the counit maps (as should be) so that

$$\langle 1,\, a\rangle = \langle 1,\, d\rangle = 1. \quad \langle K,\, 1\rangle = 1$$

and all other pairings with 1 on either side are zero. If there is a duality pairing that satisfies these data it is unique because of the requirements (8.3.6). It remains to check that the relations between the generators are respected. Here are a few

$$\langle K,\, ad - qbc\rangle = \langle K \otimes K,\, a \otimes d - qb \otimes c\rangle = q \cdot q^{-1} - 0 \cdot 0 = 1$$

$$\langle K,\, 1\rangle = 1$$

$$\langle E,\, ad - qbc\rangle = \langle E \otimes K + 1 \otimes E,\, a \otimes d - qb \otimes c\rangle = 0 \cdot 0 - q \cdot 0 + 1 \cdot 0 - 0 \cdot 0 = 0$$

$$\langle E,\, 1\rangle = 0$$

and similarly for F instead of E. So the determinant relation is OK.

Further, as $\mu(K) = K \otimes K$, $\langle K,\ \rangle = 0$ on all products of generators of $\mathbf{SL}_q(2;k)$ except ad and da. And $\langle K,\, ad - da\rangle = qq^{-1} - q^{-1}q = 0$ while $\langle K,\, bc\rangle = 0$. So $\langle K,\ \rangle$ respects all the relations of $U_q(\mathfrak{sl}(2;k))$.

Now $\mu(E) = E \otimes 1 + K \otimes E$, $\langle E,\ \rangle = 0$ on 1 and all products of generators of $\mathbf{SL}_q(2;k)$ except ba, bd, ab, db. Also

$$\langle E,\, ab\rangle = \langle E \otimes 1 + K \otimes E,\, a \otimes b\rangle = 0 \cdot 0 + q \cdot 1 = q$$
$$\langle E,\, qba\rangle = \langle E \otimes 1 + K \otimes E,\, qb \otimes a\rangle = q \cdot 1 + 0 \cdot 0 = q$$
$$\langle E,\, qdb\rangle = q\langle E \otimes 1 + K \otimes E,\, d \otimes b\rangle = q(0 \cdot 0 + q^{-1} \cdot 1) = 1$$
$$\langle E,\, bd\rangle = \langle E \otimes 1 + K \otimes E,\, b \otimes d\rangle = 1 \cdot 1 + 0 \cdot 0 = 1$$

and so $\langle E,\ \rangle$ also respects the relations. Similarly for $\langle F,\ \rangle$. For relations on the other side it is easy to see that $\langle EF - FE,\ \rangle$ is zero on $b, c, 1$, which is also the case for $\langle K - K^{-1},\ \rangle$. Further

$$\langle EF - FE,\, a\rangle = \langle E \otimes F - F \otimes E,\, a \otimes a + b \otimes c\rangle = 0 \cdot 0 + 1 \cdot 1 - 0 \cdot 0 - 0 \cdot 0 = 1$$

$$\langle \frac{K - K^{-1}}{q - q^{-1}},\, a\rangle = \frac{q - q^{-1}}{q - q^{-1}} = 1$$

$$\langle EF - FE,\, d\rangle = \langle E\otimes F - F\otimes E,\, c\otimes b + d\otimes d\rangle = 0\cdot 0 + 0\cdot 0 - 1\cdot 1 - 0\cdot 0 = -1$$

$$\langle \frac{K - K^{-1}}{q - q^{-1}},\, d\rangle = \frac{q^{-1} - q}{q - q^{-1}} = -1$$

and so the relation $EF - FE = (q - q^{-1})(K - K^{-1})$ is respected by the pairing.

Further $\langle KE,\ \rangle$ and $\langle q^2EK,\ \rangle$ are both zero on $1, a, c, d$ and

$$\langle KE,\, b\rangle = \langle K\otimes E,\, a\otimes b + b\otimes d\rangle = q\cdot 1 + 0\cdot 0 = q$$

$$\langle q^2EK,\, b\rangle = q^2\langle E\otimes K,\, a\otimes b + b\otimes d\rangle = q^2(0\cdot 0 + 1\cdot q^{-1}) = q$$

and so the relation $KE = q^2EK$ is also respected by pairing. Similarly for F. This proves the duality pairing statement.

8.4. R-matrices and QIST: Bethe Ansatz, FCR construction and the FRT theorem

Le superflu - chose tres nécessaire.

François-Marie Arouet (Voltaire)

In the previous section the quantum spaces $k\langle X^1, X^2, \ldots, X^n\rangle/(X^aX^b - q^{a,b}X^bX^a)$ and their duals $k\langle Y_1, Y_2, \ldots, Y_n\rangle/(Y_cY_d - q_{c,d}Y_dY_c)$ constituted the starting point and from them quantum groups were constructed by requiring that they coact on the left and on the right on these spaces. This section goes sort of the other way in that one starts with an $n^2\times n^2$ matrix R (which at least in the beginning need not satisfy any condition) and from it constructs a bialgebra $H_n(R)$ and algebras $A_{n,f}(\hat{R})$, $A^*_{n,f}(\hat{R})$ which are left and right comodule algebras over $H_n(R)$. Here $\hat{R} = \tau R$ where τ is the switch (flip) matrix, $\tau^{a,b}_{c,d} = \delta^a_d\delta^b_c$, so that $\hat{r}^{a,b}_{c,d} = r^{b,a}_{c,d}$ and f is any one variable polynomial over the ground ring k.

8.4.1. Quadratic algebras. Let L be a vector space with basis $Y_1, Y_2, \ldots, Y_n$. Then the tensor algebra $TM = k\oplus M\oplus M^{\otimes 2}\oplus\cdots$ is (can be seen as) the free associative algebra $k\langle Y_1, Y_2, \ldots, Y_n\rangle$ over k in the indeterminates (symbols) $Y_1, Y_2, \ldots, Y_n$. A **quadratic algebra** is one that is obtained from $k\langle Y_1, Y_2, \ldots, Y_n\rangle$ by quotienting out an ideal generated by a set of elements from $M^{\otimes 2} = \bigoplus_{i,j} kY_iY_j$. The algebras $k\langle X^1, X^2, \ldots, X^n\rangle/(X^aX^b - q^{a,b}X^bX^a)$, $k\langle Y_1, Y_2, \ldots, Y_n\rangle/(Y_cY_d - q_{c,d}Y_dY_c)$, and the quantum group $\mathbf{GL}_Q(n;k)$ of section 8.2 are examples of quadratic algebras (except for the determinant relation).

8.4.2. The FRT construction and theorem. Consider the free associative algebra $H_n = k\langle t^i_j\ :\ i,j\in\{1,2,\ldots,n\}\rangle$ over k generated by n^2 indeterminates t^i_j. With the comultiplication

$$\mu(t^i_j) = \sum_s t^i_s\otimes t^s_j. \tag{8.4.3}$$

Together with the counit $\varepsilon(t^i_j) = \delta^i_j$ (Kronecker delta) this becomes a bialgebra.

The comultiplication recipe can be succinctly and suggestively written

$$\mu(T) = T\tilde{\otimes}T, \text{which then means of course } \mu(T)^i_j = \sum_r t^i_r\otimes t^r_j.$$

There is a tilde over the tensor sign in order to avoid confusion with the Kronecker tensor product of matrices.

Lemma 8.4.4. *The two operations* $\tilde{\otimes}$ *and* $\otimes$ *are distributive over each other. This means that for all* $n \times n$ *matrices* A, B, C, D

$$(A\tilde{\otimes}B) \otimes (C\tilde{\otimes}D) = (A \otimes C)\tilde{\otimes}(B \otimes D). \tag{8.4.5}$$

Proof. Let W, X, Y, Z be not necessarily commuting variables and let $E^{(n)}_{i,j}$ be the $n \times n$ matrix with a 1 at spot (i, j) and zero everywhere else. Then by definition

$$E^{(n)}_{i,j} \otimes E^{(n)}_{r,s} = E^{(n^2)}_{(i-1)n+r,(j-1)n+s}.$$

Now calculate

$$\begin{aligned}
&(WE^{(n)}_{a,b}\tilde{\otimes}XE^{(n)}_{c,d}) \otimes (YE^{(n)}_{r,s}\tilde{\otimes}ZE^{(n)}_{u,v}) \\
&= ((W \otimes X)\delta_{b,c}E^{(n)}_{a,d}) \otimes ((Y \otimes Z)\delta_{s,u}E^{(n)}_{r,v} \\
&= (WY \otimes XZ)\delta_{b,c}\delta_{s,u}E^{(n^2)}_{(a-1)n+r,(d-1)n+v}
\end{aligned}$$

and

$$\begin{aligned}
&(WE^{(n)}_{a,b} \otimes YE^{(n)}_{r,s})\tilde{\otimes}(XE^{(n)}_{c,d} \otimes ZE^{(n)}_{u,v}) \\
&= WYE^{(n^2)}_{(a-1)n+r,(b-1)n+s}\tilde{\otimes}XZE^{(n^2)}_{(c-1)n+u,(d-1)n+v} \\
&= (WY \otimes XZ)\delta_{(b-1)n+s,(c-1)n+u}E^{(n^2)}_{(a-1)n+r,(d-1)n+v}.
\end{aligned}$$

But, looking at the Kronecker delta factors, $(b-1)n + s = (c-1)n + u$ if and only if $b = c$ and $s = u$. Indeed if $(b-1)n + s = (c-1)n + u$, $s \equiv u \mod n$ and as $s, u \in \{1, 2, \dots, n\}$, $s = u$ and hence also $b = c$. This proves the lemma. □

Now consider the n^4 quadratic relations[6]

$$RT_1T_2 = T_2T_1R, \quad T_1 = T \otimes I_n, \quad T_2 = I_n \otimes T.$$

The corresponding ideal is generated by n^4 elements

$$\sum_{c,d} r^{a,b}_{c,d}t^c_ut^d_v - \sum_{i,j} r^{i,j}_{u,v}t^b_jt^a_i. \tag{8.4.6}$$

Denote this ideal by $J_n(R)$. Then the claim is that $J_n(R)$ is a bialgebra ideal in H_n.

To see that $\varepsilon(J_n(R)) = 0$ is an easy observation. Further, one calculates

$$\begin{aligned}
\mu(RT_1T_2) &= R\mu(T \otimes I_n)\mu(I_n \otimes T) \\
&= R((T\tilde{\otimes}T) \otimes (I_n\tilde{\otimes}I_n))(I_n\tilde{\otimes}I_n) \otimes (T\tilde{\otimes}T)) \\
&= R((T \otimes I_n)\tilde{\otimes}(T \otimes I_n))((I_n \otimes T)\tilde{\otimes}(I_n \otimes T)) \\
&= R(T_1\tilde{\otimes}T_1)(T_2\tilde{\otimes}T_2) \\
&= RT_1T_2\tilde{\otimes}T_1T_2
\end{aligned}$$

and similarly

$$\mu(T_2T_1R) = RT_2T_1\tilde{\otimes}T_2T_1.$$

---

[6]There is danger in this notation. It is indeed the case that $T_1T_2 = (T \otimes I)(I \otimes T) = T \otimes T$ so that $(T_1T_2)^{a,b}_{u,v} = t^a_ut^b_v$; it is not true that $T_2T_1 = (I \otimes T)(T \otimes I)$ is equal to $T \otimes T$; instead $(T_2T_1)^{a,b}_{u,v} = t^b_vt^a_u$.

Now using the obvious remark that $AR\tilde{\otimes}B = A\tilde{\otimes}RB$, there results

$$\begin{aligned}\mu(RT_1T_2 - T_2T_1R) &= RT_1T_2\tilde{\otimes}T_1T_2 - T_2T_1\tilde{\otimes}T_2T_1R \\ &= RT_1T_2\tilde{\otimes}T_1T_2 - T_2T_1R\tilde{\otimes}T_1T_2 + T_2T_1\tilde{\otimes}RT_1T_2 - T_2T_1\tilde{\otimes}T_2T_1R \\ &= (RT_1T_2 - T_2T_1R)\tilde{\otimes}T_1T_2 + T_2T_1\tilde{\otimes}(RT_1T_2 - T_2T_1R)\end{aligned}$$

showing that indeed $\mu(J_n(R)) \subset J_n(R) \otimes H_n + H_n \otimes J_n(R)$.

8.4.7. Notation for the FRT construction and theorem. Fix a natural number n and an $n^2 \times n^2$ matrix R with entries $r^{a,b}_{u,v}$. Let $\hat{R} = \tau R$ so that the entries of $\hat{R}$ are $\hat{r}^{a,b}_{u,v} = r^{b,a}_{u,v}$. For a one variable polynomial $f(t) = c_0 + c_1t + \cdots + c_mt^m$, $f(\hat{R})$ denotes the $n^2 \times n^2$ matrix $f(\hat{R}) = c_0I_{n^2} + c_1\hat{R} + \cdots + c_m\hat{R}^m$ with entries denoted $\hat{r}(f)^{a,b}_{u,v}$.

V is the free module over k with basis $X^1, X^2, \ldots, X^n$ and

$$A_n = k\langle X^1, \ldots, X^n\rangle = TV$$

is the tensor algebra over V, i.e. the free associative algebra in the indeterminates X^i. Further $I_{n,f}(\hat{R})$ is the ideal in A_n generated by the n^2 elements $\sum\limits_{u,v} \hat{r}(f)^{a,b}_{u,v}X^uX^v$.

Dually V^* is the free module with biorthogonal basis $Y_1, Y_2, \ldots, Y_n$ (with respect to V and its given basis) and A^*_n is the tensor algebra over V^*. The ideal $I^*_{n,f}(\hat{R})$ is the one generated by the elements $\sum\limits_{a,b} \hat{r}(f)^{a,b}_{u,v}Y_aY_b$.

Write $H_n(R) = H_n/J_n(R)$ as before and $A_{n,f}(\hat{R}) = A_n/I_{n,f}(\hat{R})$, $A^*_{n,f}(\hat{R}) = A^*_n/I^*_{n,f}(\hat{R})$. Finally $\gamma_L : V \longrightarrow H_n \otimes V$, $\gamma^{\text{alg}}_L : A_n \longrightarrow H_n \otimes A_n$ are the left comodule and left comodule algebra structures defined by $X^a \mapsto \sum t^a_u \otimes X^u$, while $\gamma_R : V^* \longrightarrow V^* \otimes H_n$, $\gamma^{\text{alg}}_R : A^*_n \longrightarrow A^*_n \otimes H_n$ are the right comodule and comodule algebra structures defined by $Y_u \mapsto \sum\limits_a Y_a \otimes t^a_u$.

8.4.8. FRT theorem.[7] *Using the notations above the following statements hold*

(a) *$J_n(R)$ is a bialgebra ideal in H_n so that there is an induced bialgebra structure on $H_n(R)$.*

(b) *The left coaction γ_L induces a left coaction of $H_n(R)$ on the module V such that $\hat{R} : V \otimes V \longrightarrow V \otimes V$, $X^i \otimes X^j \mapsto \sum\limits_{c,d} \hat{r}^{i,j}_{c,d}X^c \otimes X^d$ is a morphism of left comodules.*

(c) *The following universality property holds. If B is a bialgebra with coacts on V on the left by a morphism $\gamma_B : V \longrightarrow B \otimes V$ such that $\hat{R} : V \otimes V \longrightarrow V \otimes V$ is a morphism of B-comodules there is a unique morphism of bialgebras $\alpha_B : H_n(R) \longrightarrow B$ such that $\gamma_B = (\alpha_B \otimes \mathrm{id})\gamma_L$.*

---

[7] Most of this theorem is already in the paper [**588**]. In fact all, except the universality statements. Going along with the order in which the authors are named in [**588**] the theorem should be called the RTF theorem.

There are more things in mathematics that go under the name "FRT theorem". For example the Frostman-Rogers-Taylor theorem on Hausdorff dimension (see e.g. [**67**]) and the Fast Rule theorem in cellular automata theory [**557**].

(d) *The comodule algebra structure* γ_L^{alg} *induces a comodule algebra structure* $A_{n,f}(\hat{R}) \longrightarrow H_n(R) \otimes A_{n,f}(\hat{R})$ *for every polynomial* f.

(e) *The right coaction* γ_R *induces a right* $H_n(R)$*-comodule structure on* V^* *such that* $\hat{R} : V^* \otimes V^* \longrightarrow V^* \otimes V^*$, $Y_u \otimes Y_v \mapsto \sum\limits_{i,j} \hat{r}^{i,j}_{u,v} Y_i \otimes Y_j$ *is a right comodule morphism.*

(f) Dual universality property. *If* V^* *has a right comodule structure* $\gamma^B : V^* \longrightarrow V^* \otimes B$ *over a bialgebra* B *such that* $\hat{R}$ *is a right* B*-comodule morphism there is a unique morphism of bialgebras* $\alpha^B : H_n(R) \longrightarrow B$ *such that* $(\mathrm{id} \otimes \alpha^B)\gamma_R = \gamma^B$.

(g) *The right comodule algebra structure given by* γ_R^{alg} *induces a right comodule algebra structure of* $H_n(R)$ *on* $A^*_{n,f}(\hat{R})$ *for every polynomial* f.

Proof. Statement (a) has already been proved above. For the remainder it is easier to work with the $\hat{r}$ coefficients than with the r coefficients. In these terms the elements defining the ideal $J_n(R)$ are

$$\sum_{c,d} \hat{r}^{a,b}_{c,d} t^c_u t^d_v - \sum \hat{r}^{i,j}_{u,v} t^a_i t^b_j \tag{8.4.9}$$

which can be seen either directly from (8.4.6) or by noting that $\tau T_1 \tau = T_2$, $\tau T_2 \tau = T_1$ so that

$$RT_1T_2 = T_2T_1R$$

is equivalent to

$$\hat{R}T_1T_2 = \tau RT_1T_2 = \tau T_2 \tau\tau T_1 \tau\tau R = T_1T_2\hat{R}$$

(and noting that $(T_1T_2)^{a,b}_{u,v} = t^a_u t^b_v$). Now let B be any bialgebra and let a left B-comodule structure on V be specified by

$$V \xrightarrow{\gamma_B} B \otimes V, \qquad X^i \mapsto \sum_c x^i_c \otimes X^c.$$

The induced B-module structure on $V \otimes V$ is given by

$$X^i \otimes X^j \mapsto \sum_{c,d} x^i_c x^j_d \otimes X^c \otimes X^d.$$

If the k-linear map $\hat{R} : V \otimes V \longrightarrow V \otimes V$ is to be a morphism of left B-comodules the following diagram must commute

$$\begin{array}{ccc} V \otimes V & \longrightarrow & B \otimes V \otimes V \\ \Big\downarrow{\scriptstyle \hat{R}} & & \Big\downarrow{\scriptstyle \mathrm{id}\otimes\hat{R}} \\ V \otimes V & \longrightarrow & B \otimes V \otimes V \end{array}$$

Tracing things out this means that

$$\sum_{c,d,u,v} \hat{r}^{i,j}_{u,v} x^u_c x^v_d \otimes X^c \otimes X^d = \sum_{c,d,u,v} \hat{r}^{u,v}_{c,d} x^i_u x^j_v \otimes X^c \otimes X^d$$

i.e.

$$\sum_{u,v} \hat{r}^{i,j}_{u,v} x^u_c x^v_d = \sum_{u,v} \hat{r}^{u,v}_{c,d} x^i_u x^j_v \quad \text{for all } i,j,c,d \in \{1,2,\ldots,n\}. \tag{8.4.10}$$

Replacing the x's with t's this turns precisely into the defining relations (8.4.9) showing (apart from the counit property (which is trivial to handle) that indeed

V is a comodule over $H_n(R)$ for which $\hat{R} : V \otimes V \longrightarrow V \otimes V$ is a morphism of comodules. This proves (b)[8]

Relation (8.4.10) also shows that the algebra morphism $H_n(R) \longrightarrow B$, induced by $t^i_j \mapsto x^i_j$ is well-defined. Moreover, the fact that V is a B-comodule means that the following diagram commutes

$$\begin{array}{ccc} V & \xrightarrow{\gamma_B} & B \otimes V \\ \big\downarrow{\scriptstyle \gamma_B} & & \big\downarrow{\scriptstyle \mu_B \otimes \mathrm{id}} \\ B \otimes V & \xrightarrow{\mathrm{id} \otimes \gamma_B} & B \otimes B \otimes V \end{array}$$

Tracing things this means that

$$\sum_c \mu_B(x^i_c) \otimes X^c = \sum_{c,d} x^i_d \otimes x^d_c \otimes X^c$$

and as the X's are a basis this implies $\mu_B(x^i_c) = x^i_d \otimes x^d_c$ showing that the algebra morphism induced by $t^i_j \mapsto x^i_j$ is a morphism of bialgebras. This proves statement (c).

Next to prove statement (d) it is necessary and sufficient to show that

$$\gamma_B^{\mathrm{alg}}(I_{n,f}(\hat{R})) \subset J_n(R) \otimes A_n + H_n \otimes I_{n,f}(\hat{R}).$$

Now the relations $\hat{R}(T_1T_2) = (T_1T_2)\hat{R}$ imply $f(\hat{R})(T_1T_2) = (T_1T_2)f(\hat{R})$ and so

$$\sum_{c,d} \hat{r}(f)^{a,b}_{c,d} t^c_u t^d_v - \sum_{i,j} \hat{r}(f)^{i,j}_{u,v} t^a_i t^b_j \in J_n(R). \tag{8.4.11}$$

Further $I_{n,f}(\hat{R})$ is generated by the elements $\sum\limits_{c,d} \hat{r}(f)^{a,b}_{c,d} X^c X^d$ and

$$\begin{aligned} \gamma_L^{\mathrm{alg}}(\sum_{c,d} \hat{r}(f)^{a,b}_{c,d} X^c X^d) &= \sum_{c,d,u,v} \hat{r}(f)^{a,b}_{c,d} t^c_u t^d_v \otimes X^u X^v \\ &= \sum_{c,d,u,v} \hat{r}(f)^{a,b}_{c,d} t^c_u t^d_v \otimes X^u X^v - \sum_{c,d,u,v} \hat{r}(f)^{i,j}_{u,v} t^a_i t^b_j \otimes X^u X^v \\ &+ \sum_{c,d,u,v} t^a_i t^b_j \otimes \hat{r}(f)^{i,j}_{u,v} X^u X^v \\ &\in J_n(R) \otimes A_n + H_n \otimes I_{n,f}(\hat{R}). \end{aligned}$$

This proves statement (d). The statements (e), (f), (g) are dealt with in the same way.

Remark 8.4.12. The calculation just done for the proof of statement (d) can be repeated almost manipulation for manipulation to give a proof that

$$\mu_{H_n(R)}(J_n(R)) \subset J_n(R) \otimes H_n + H_n \otimes J_n(R)$$

thus giving another proof of statement (a). This is meaning of the quote from Voltaire above.

---

[8]As $H_n(R)$ is a quotient of H_n and V is an H_n-comodule, it is automatically a $H_n(R)$-comodule, just like a module over a ring is automatically a module over any subring.

Example 8.4.13. For the quantum group $\mathbf{GL}_q(2;k)$ the R-matrix is

$$R = \begin{pmatrix} q^{-1} & & & \\ & 1 & q^{-1}-q & \\ & & 1 & \\ & & & q^{-1} \end{pmatrix}, \qquad \hat{R} = \begin{pmatrix} q^{-1} & & & \\ & 0 & 1 & \\ & 1 & q^{-1}-q & \\ & & & q^{-1} \end{pmatrix}.$$

Take $f(t) = t - q^{-1}$ so that the relations defining $A_{r,f}(\hat{R})$ are given by

$$\begin{pmatrix} 0 & & & \\ & -q^{-1} & 1 & \\ & 1 & -q & \\ & & & 0 \end{pmatrix} \begin{pmatrix} X^1X^1 \\ X^1X^2 \\ X^2X^1 \\ X^2X^2 \end{pmatrix} = 0$$

so that the quantum plane $k\langle X^1, X^2\rangle/(X^1X^2 - qX^2X^1)$ reappears.

8.4.14. Comment. From the purely algebraic point of view it would be more elegant to work exclusively with the matrix $\hat{R}$, the braid equation and the commutation relation $\hat{R}T_1T_2 = T_1T_2\hat{R}$. But that makes the relation with the original inspiring examples from quantum field theory and lattice statistical mechanics less elegant.

8.4.15. Addenda. If the R-matrix used to define $H_n(R)$ and the $A_{n,f}(\hat{R})$ satisfy the Yang Baxter equation there are kinds of extra nice properties: for instance $H_n(R)$ becomes a co-quasi-triangular bialgebra and $A_{n,f}(\hat{R})$ has the same Poincaré series as the commutative algebra $k[t^i_j] = H_n(I_n)$, see [**111**], section 5.4; [**588**], theorem 3, p.201.

The FRT theorem can be generalized considerably for instance to settings of complete monoidal categories, see e.g. [**64**].

8.4.16. Two dimensional lattice statistical mechanics models. In such models one considers a lattice of N sites which can be in several states, each state σ having a certain (interaction) energy $H(\sigma)$, attached to it with corresponding Boltzmann weight $e^{-H(\sigma)/kT}$ where k is the Boltzmann constant and T is the absolute temperature. One is now interested in the so-called partition function

$$Z_N = \sum_{\sigma} e^{-H(\sigma)/kT}$$

and a number of associated quantities such as the bulk free energy per site in the thermodynamic limit

$$\frac{-\psi}{kT} = \lim_{N\to\infty} N^{-1} \log Z_N$$

and how their behaviours changes as a function of T, especially discontinuities (phase transitions).

For instance there is the Ising model of a square lattice

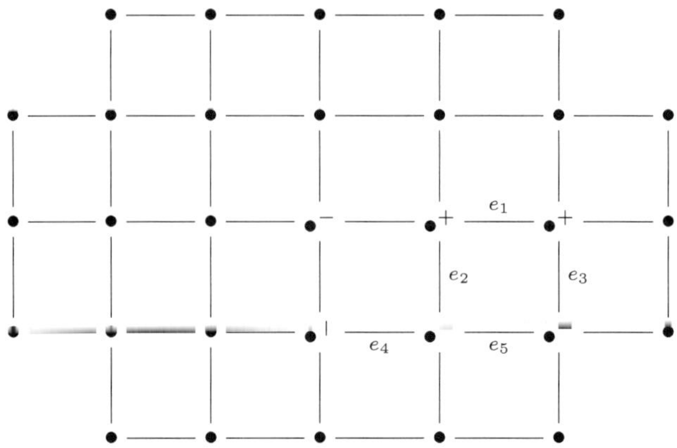

A state of the lattice is given by specifying for each site a spin $+1$ or -1 and each edge between two neighboring vertices contributes a term $-J\sigma_i\sigma_j$ to the Hamiltonian if they are horizontal nearest neighbors and a term $-J'\sigma_i\sigma_j$ if they are vertical neighbors. Thus the five labeled edges in the figure above contribute respectively terms $-J$, $+J'$, $+J'$, $+J$, $-J$. The partition function is therefore of the form

$$Z = \sum_{\sigma} \exp(K \sum_{(i,j)} \sigma_i\sigma_j + L \sum_{(k,l)} \sigma_k\sigma_l)$$

where the first inner sum is over all horizontal edges and the second inner sum is over all vertical edges and $K = -J/k_BT$, $L = -J'/k_BT$.

In the models to be considered sketchily below there are various labels attached to the edges that come from a fixed set $\{1, 2, \ldots, d\}$ (usually $d = 2$) and an energy $H(i, j, k, l)$ is assigned to each site depending on the four edges incident with it with corresponding Boltzmann weight denoted (for the current purposes) by $r^{i,k}_{j,l}$. Thus, imposing periodic boundary conditions, the partition function for the three by five lattice depicted below is

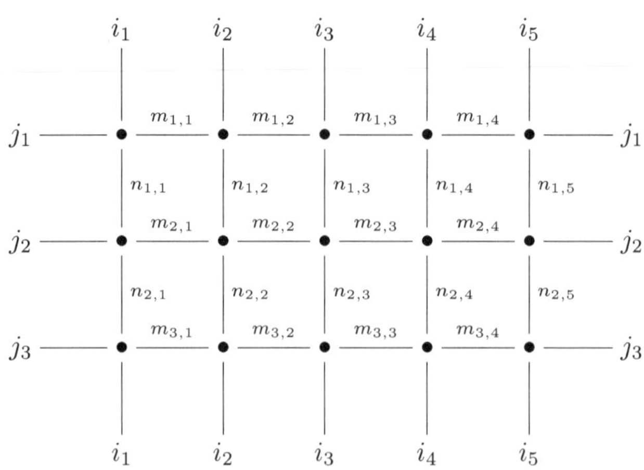

$$Z=\sum \prod_{(m,n)} r_{n_{s,t},m_{u,v}}^{n_{a,b},m_{c,d}} \tag{8.4.17}$$

where the sum is over all $d^{3\cdot 5+5\cdot 4}$ labelings of the edges, the product under the sum sign is over all fifteen sites of the lattice, and $m_{k,0}=j_k$, $n_{l,0}=i_l$ (in order not to burden the notation still more).

It looks like, and indeed is, a formidable task to say anything useful about expressions like (8.4.17) (and to take the thermodynamic limit in letting both the horizontal and vertical size of the lattice go to infinity).

It turns out that the quantum matrix algebras $M_q(n\times n;\mathbf{C})$ (or, practically equivalently, the quantum linear groups, $\mathbf{GL}_q(n;\mathbf{C})$, or quantum special linear groups $\mathbf{SL}_q(n;\mathbf{C})$ can help in this a good deal for a number of interesting special cases such as the six vertex model, the eight vertex model and the hard hexagon model. This goes via (tensor) products of representations of these quantum algebras (quantum groups).

More precisely what one could call parametrized representations.

8.4.18. Parametrized quantum Yang-Baxter equation. As a first step consider the 4×4 matrices

$$R(\lambda)=\begin{pmatrix}\lambda+1 & 0 & 0 & 0\\ 0 & \lambda & 1 & 0\\ 0 & 1 & \lambda & 0\\ 0 & 0 & 0 & \lambda+1\end{pmatrix}=\lambda I_4+P \tag{8.4.19}$$

where P is the matrix of the switch morphism $x\otimes y\mapsto y\otimes x$, and examine the question of when the equation

$$R_{1,2}(\nu)R_{1,3}(\lambda)R_{2,3}(\mu)=R_{2,3}(\mu)R_{1,3}(\lambda)R_{1,2}(\nu) \tag{8.4.20}$$

holds. Now

$$\begin{aligned} &R_{1,2}(\nu)R_{1,3}(\lambda)R_{2,3}(\mu)(x\otimes y\otimes z)\\ &=R_{1,2}(\nu)R_{1,3}(\lambda)(\mu(x\otimes y\otimes z)+(x\otimes z\otimes y))\\ &=R_{1,2}(\nu)(\lambda\mu(x\otimes y\otimes z)+\lambda(x\otimes z\otimes y)+\mu(z\otimes y\otimes x)+(y\otimes z\otimes x))\\ &=\nu\lambda\mu(x\otimes y\otimes z)+\nu\lambda(x\otimes z\otimes y)+\nu\mu(z\otimes y\otimes x)+\nu(y\otimes z\otimes x)\\ &\quad+\lambda\mu(y\otimes x\otimes z)+\lambda(z\otimes x\otimes y)+\mu(y\otimes z\otimes x)+(z\otimes y\otimes x)\end{aligned} \tag{8.4.21}$$

and just so

$$\begin{aligned} &R_{2,3}(\mu)R_{1,3}(\lambda)R_{1,2}(\nu)(x\otimes y\otimes z)\\ &=\mu\lambda\nu(x\otimes y\otimes z)+\mu\lambda(y\otimes x\otimes z)+\mu\nu(z\otimes y\otimes x)+\mu(z\otimes x\otimes y)\\ &\quad+\lambda\nu(x\otimes z\otimes y)+\lambda(y\otimes z\otimes x)+\nu(z\otimes x\otimes y)+(z\otimes y\otimes x).\end{aligned} \tag{8.4.22}$$

Most of the terms of (8.4.21) and (8.4.22) agree (because of commutativity of scalar multiplies of the identity with themselves and with P and because of the permutation identity $(12)(13)(23)=(23)(13)(12)$, but the terms with just one parameter do not quite match (because e.g. $(12)(13)\neq(13)(12)$ and for (8.4.21) and (8.4.22) to be equal it is necessary and sufficient that $\nu=\lambda-\mu$, so that for the operators (8.4.19)

$$R_{1,2}(\lambda-\mu)R_{1,3}(\lambda)R_{2,3}(\mu)=R_{2,3}(\mu)R_{1,3}(\lambda)R_{1,2}(\lambda-\mu) \tag{8.4.23}$$

is an example of parametrized quantum Yang-Baxter equation or QYBE with spectral parameter. This equation fits with the illustration given in 8.2.27 if the parameters λ and μ are interpreted as (non-relativistic) velocities.

8.4.24. Parametrized FCR. Similarly one considers parametrized fundamental commutation relations

$$R_{1,2}(\lambda,\mu)T_1(\lambda)T_2(\mu) = T_2(\mu)T_1(\lambda)R_{1,2}(\lambda,\mu) \tag{8.4.25}$$

where $R_{1,2}(\lambda,\mu)$ is a family of endomorphisms of $V_0 \otimes V_0$ parametrized by two parameters λ, μ, where V_0 is a so-called auxiliary vector space, $T(\lambda)$ is a family of endomorphisms of $V_0 \otimes V$ parametrized by λ, $T_1(\lambda)$ is a family of endomorphisms of $V_0 \otimes V_0 \otimes V$ obtained by letting $T(\lambda)$ act on the outer two factors and leaving the middle vector alone, i.e. if $T(v_0 \otimes v) = \sum v_{0,i} \otimes v_i$, then $T_1(v_0 \otimes v'_0 \otimes v) = \sum v_{0,i} \otimes v'_0 \otimes v_i$ and $T_2(\mu)$ is a family of endomorphisms of $V_0 \otimes V_0 \otimes V$ obtained by letting T act on the last two factors and leaving the first one alone, i.e. $T_2(\mu) = \mathrm{id}_{V_0} \otimes T(\mu)$.

If (8.4.25) holds then so does the same equation with $R(\lambda,\mu)$ replaced by $R(\lambda,\mu) + \nu P$ which is useful to make the matrices $R(\lambda,\mu)$ invertible in (8.4.25).

For (8.4.25) to work in what follows it is not necessary that the dependence of $R(\lambda,\mu)$ on the two parameters be linear in the parameters; any parametrized family will do, and e.g. nonlinear dependence does occur in e.g. relativistic factorised scattering.

It now makes eminent sense to take any family of $d \times d$ matrices $R(\lambda,\mu)$ and to repeat the FRT construction at the beginning of this section to obtain a Hopf algebra $H_n(R(\lambda,\mu))$ which is the algebra generated by (noncommuting) indeterminates $t^i_j(\lambda)$, one for each i, j and λ, subject to the relations

$$R_{1,2}(\lambda,\mu)T_1(\lambda)T_2(\mu) = T_2(\mu)T_1(\lambda)R_{1,2}(\lambda,\mu)$$

where

$$T = \begin{pmatrix} t^1_1(\lambda) & \cdots & t^1_n(\lambda) \\ \vdots & \ddots & \vdots \\ t^n_1(\lambda) & \cdots & t^n_n(\lambda) \end{pmatrix}.$$

The comultiplication and counit are as before:

$$t^i_j(\lambda) \mapsto \sum_{u=1}^{d} t^i_u(\lambda) \otimes t^u_j(\lambda), \qquad \varepsilon : t^i_j(\lambda) \mapsto \delta^i_j.$$

Once a basis of V_0 has been chosen an endomorphism T of $V_0 \otimes V$ is a $d \times d$ matrix (T^i_j) with entries that are endomorphisms of V.

Then if $T(\lambda)$ satisfies (8.4.25), $t^i_j(\lambda) \mapsto T^i_j(\lambda)$ defines a representation of $H_n(R(\lambda,\mu))$ on V. In particular if $R(\lambda)$ satisfies a parametrized QYBE

$$R_{1,2}(\lambda-\mu)R_{1,3}(\lambda)R_{2,3}(\mu) = R_{2,3}(\mu)R_{1,3}(\lambda)R_{1,2}(\lambda-\mu)$$

$t^i_j(\lambda) \mapsto R^i_j(\lambda)$ is a representation of $H_n(R(\lambda-\mu))$. Here $R^i_j(\lambda)^k_l = r^{i,k}_{j,l}(\lambda)$.

Now, because $H_n(R(\lambda,\mu))$ is a Hopf algebra products of representations can be constructed (one of the original motivations for studying Hopf algebras, see section 3.1. This will be very important in the considerations below.

8.4.26. Lattice statistical mechanics models (continued). Consider again a two dimensional lattice statistical model as described by (8.4.17) and the picture just above that formula.

Take a family of $d\times d$ matrices $R(\lambda)$ depending on one or more parameters that satisfies a QYBE, so that there are d possible 'spins' for each edge. For instance the two parameter family of 4×4 matrices

$$R(a,b)=\begin{pmatrix} a+b & 0 & 0 & 0\\ 0 & a & b & 0\\ 0 & b & a & 0\\ 0 & 0 & 0 & a+b\end{pmatrix}=aI_4+bP$$

which satisfies the QYBE

$$\begin{aligned}&R_{1,2}(a_1b_2-a_2b_1,a_2b_2)R_{1,3}(a_1,b_1)R_{2,3}(a_2,b_2)\\ &=R_{2,3}(a_2,b_2)R_{1,3}(a_1,b_1)R_{1,2}(a_1b_2-a_2b_1,a_2b_2).\end{aligned}$$

Then the $d^2\times d^2$ matrices $R^i_j(a,b)=(R^i_j(a,b))^u_v$ define a representation of a corresponding quantum matrix algebra. Let $T(\lambda)=(T^i_j(\lambda))$ be the N-fold product of this representation (where N is the number of columns of the lattice involved). Here the $T^i_j(\lambda)$ are $d^N\times d^N$ matrices so that T is a $d\times d$ block matrix whose entries are $d^N\times d^N$ matrices. It is called the row to row **transfer matrix**. In the setting of spin chains, see below, it is called the **monodromy matrix**.

It is simple matter to write down the entries of the $T^i_j(\lambda)$.

$$(T^i_j(\lambda))^{k_1,k_2,\dots,k_N}_{l_1,l_2,\dots,l_N}=\sum_{m_1,m_2,\dots,m_{N-1}} r^{i,k_1}_{m_1,l_1}r^{m_1,k_2}_{m_2,l_2}\cdots r^{m_{N-1},k_N}_{j,l_N}.$$

Pictorially this is the situation

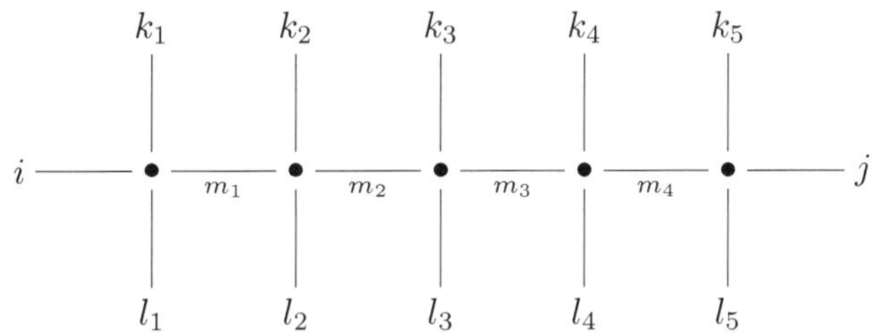

Now let

$$F(\lambda)=\sum_{i=1}^{d}T^i_i(\lambda) \tag{8.4.27}$$

be the operator trace of this row transfer matrix. It follows from the formula (8.4.17) that the partition function is now given by the trace of the M-th power of $F(\lambda)$

$$Z_{M\times N}(\lambda)=\mathrm{Tr}(F(\lambda)^M). \tag{8.4.28}$$

Now as the matrix of the N-fold product representation of the starting representation the transfer matrix T satisfies the same QYBE as the starting R-matrix R[9])

$$RT_1(\lambda)T_2(\mu)=T_2(\mu)T_1(\lambda)R \tag{8.4.29}$$

---

[9])This can also be seen directly of course. That is the way it is done in [**481**], section 4.4 and [**239**], just above formula (45).

where the matrices $T_i(\lambda)$ act on the space $V \otimes V \otimes V^{\otimes N}$. By their definition they are $d^2 \times d^2$ matrices whose entries are endomorphisms of $V^{\otimes N}$ and also by definition the entry

$$(T_1(\lambda)T_2(\mu))_{j,l}^{i,k} = T(\lambda)_j^i T(\mu)_l^k$$

so that

$$\mathrm{Tr}_{V\otimes V}(T_1(\lambda)T_2(\mu)) = \mathrm{Tr}_V(T(\lambda))\mathrm{Tr}_V(T(\mu)) = F(\lambda)F(\mu).$$

Further if τ is the switch on $V \otimes V$ and leaves $V^{\otimes N}$ alone

$$\tau T_1(\lambda)\tau = T_2(\lambda).$$

The QYBE (8.4.29) implies

$$(\tau R)T_1(\lambda)T_2(\mu) = \tau T_2(\mu)\tau\tau T_1(\lambda)\tau(\tau R) = T_1(\mu)T_2(\lambda)(\tau R)$$

and so, taking operator traces,

$$F(\lambda)F(\mu) = F(\mu)F(\lambda)$$

which implies all kinds of good and useful properties concerning the eigenvalues and eigenvectors of the family of matrices $F(\lambda)$ which in turn has implications of the partition function Z of the model via formula (8.4.28).

Actually the QYBE for the row to row transfer matrix implies a good deal more than can be used. See the discussion on the Bethe Ansatz below.

Example 8.4.30. It is an entertaining and illuminating (not entirely trivial) exercise to calculate the transfer matrix and partition function for the family of QYBE solutions

$$\begin{pmatrix} x & 0 & 0 & 0 \\ 0 & 0 & x & 0 \\ 0 & x & 0 & 0 \\ 0 & 0 & 0 & x \end{pmatrix} = xP.$$

The result is that the operator trace, $F_N(x)$, of the transfer matrix is of the form

$$x^{2^N} F_N(1)$$

where $F_N(1)$ is a permutation matrix on 2^N letters. Labeling these letters

$$0, 1, \ldots, 2^N - 1$$

and writing them in binary notation as a string of 0's and 1's, the permutation in question can be described combinatorically as follows

$$\varepsilon_1\varepsilon_2 \cdots \varepsilon_{N-2}\varepsilon_{N-1} \mapsto \varepsilon_{N-1}\varepsilon_0\varepsilon_1 \cdots \varepsilon_{N-2}, \quad \varepsilon_j \in \{0, 1\}. \tag{8.4.31}$$

Arithmetically this means for a natural number z in the range $0, 1, \ldots, 2^N - 1$

$$z \mapsto \begin{cases} z/2 & \text{if } z \text{ is even} \\ 2^{-1}(z + 2^N - 1) & \text{if } z \text{ is odd.} \end{cases} \tag{8.4.32}$$

It is an amusing fact that from the combinatorical description (8.4.31) it is immediate that all cycles have length a divisor of N while this is not at all clear from arithmetic description (8.4.32).

Thus, for example, the permutation $F_N(1)$ for $N = 5$ has two cycles of length 1 and six cycles of length 5. For $N = 6$ there are two cycles of length 1, one cycle of length 2, 2 cycles of length 3 and 9 cycles of length 6.

For the partition function $Z_{M,N}(x) = \mathrm{Tr}(F_N(x)^M)$ it follows that

$$\lim_{N,M\to\infty} Z_{M,N}(x) = \begin{cases} \infty & \text{if } x > 1 \\ 0 & \text{if } x < 1 \end{cases}$$

while the situation at $x = 1$ is undecided in that

$$\liminf_{N,M\to\infty} Z_{M,N}(1) = 2, \quad \text{and} \quad \limsup_{N,M\to\infty} Z_{M,N}(1) = \infty$$

a sort of phase transition.

A model that is truly interesting from the point of view of physics is the eight vertex model which works via solutions of parametrized QYBE of the form

$$\begin{pmatrix} w_0(\lambda)+w_3(\lambda) & 0 & 0 & w_1(\lambda)-w_2(\lambda) \\ 0 & w_0(\lambda)-w_3(\lambda) & w_1(\lambda)+w_2(\lambda) & 0 \\ 0 & w_1(\lambda)+w_2(\lambda) & w_0(\lambda)-w_3(\lambda) & 0 \\ w_1(\lambda)-w_2(\lambda) & 0 & 0 & w_0(\lambda)+w_3(\lambda) \end{pmatrix}$$

Such a family of matrices satisfies a parametrized QYBE if

$$w_{\sigma(0)}(\lambda-\mu)w_{\sigma(3)}(\lambda)w_{\sigma(2)}(\mu) - w_{\sigma(1)}(\lambda-\mu)w_{\sigma(2)}(\lambda)w_{\sigma(3)}(\mu)$$

$$+w_{\sigma(2)}(\lambda-\mu)w_{\sigma(1)}(\lambda)w_{\sigma(0)}(\mu) - w_{\sigma(3)}(\lambda-\mu)w_{\sigma(0)}(\lambda)w_{\sigma(1)}(\mu) = 0$$

and there are actually solutions depending on two more parameters in terms of the Jacobi elliptic function sn. See [**53**], section 10.4; and also [**564**], section 3.3; [**614**], pp.22-23.

8.4.33. Magnetic spin chains (Heisenberg models). To start with, here is the model itself. Consider a Hilbert space that is the tensor product of N copies of $V = \mathbf{C}^m$. Consider $\mathfrak{sl}(2;\mathbf{C})$, the Lie algebra of 2×2 complex matrices with trace zero. This three dimensional Lie algebra has a basis consisting of the three so-called Pauli matrices

$$\sigma^1 = \sigma^x = \begin{pmatrix} 0 & 1 \\ 1 & 0 \end{pmatrix}, \quad \sigma^2 = \sigma^y = \begin{pmatrix} 0 & -i \\ i & 0 \end{pmatrix}, \quad \sigma^3 = \sigma^z = \begin{pmatrix} 1 & 0 \\ 0 & -1 \end{pmatrix}.$$

On each V of dimension $m \geq 1$ there is a unique irreducible representation of $\mathfrak{sl}(2;\mathbf{C})$, sometimes called (in the physics literature) the spin $(1/2)(m-1)$ representation, see section 1.6. The first nontrivial one is the spin $1/2$ representation which is the identity representation. Take anyone of these representations. Denote the image of the $\frac{1}{2}\sigma^i$ by S^i for this representation. Let S^i_n be the operator (matrix) that acts on $\mathcal{H} = V^{\otimes N}$ as S^i in the n-th factor and as the identity everywhere else. This gives a set of operators satisfying the commutation relations

$$[S^i_n, S^j_m] = i\varepsilon_{i,j,k} S^k_n \delta_{n,m}$$

where $\varepsilon_{i,j,k}$ is the Levi-Civita symbol that is zero if two of the indices are equal and equal to the sign of the corresponding permutation (written as words) if they are all three different. The Hamiltonian of the XXX_s (spin s XXX) model is now

$$H = \sum_{i,n} S_n^i S_{n+1}^i - \frac{1}{4} \quad \text{where } S_{N+1}^i = S_1^i.$$

The corresponding anisotropic model (XYZ_s model) has Hamiltonian[10]

$$H = \sum_{i,n} J_X S_n^1 S_{n+1}^1 + J_Y S_n^2 S_{n+1}^2 + J_Z S_n^3 S_{n+1}^3 - \frac{1}{4}.$$

These models can be sort of seen as N atoms in a circle with nearest neighbour interaction. Solving these models (exactly) means calculating the eigenvectors and eigenvalues of these Hamiltonians (matrices).

The first two for the spin 1/2 XXX-model are

$$H_2 = \frac{1}{2}\begin{pmatrix} 0 & 0 & 0 & 0 \\ 0 & -1 & 1 & 0 \\ 0 & 1 & -1 & 0 \\ 0 & 0 & 0 & 0 \end{pmatrix}, \quad H_3 = \frac{1}{2}\begin{pmatrix} 0 & 0 & 0 & 0 & 0 & 0 & 0 & 0 \\ 0 & -1 & 1 & 0 & 0 & 0 & 0 & 0 \\ 0 & 1 & -2 & 0 & 1 & 0 & 0 & 0 \\ 0 & 0 & 0 & -1 & 0 & 1 & 0 & 0 \\ 0 & 0 & 1 & 0 & -1 & 0 & 0 & 0 \\ 0 & 0 & 0 & 1 & 0 & -2 & 1 & 0 \\ 0 & 0 & 0 & 0 & 0 & 1 & -1 & 0 \\ 0 & 0 & 0 & 0 & 0 & 0 & 0 & 0 \end{pmatrix}.$$

These two can still be handled by hand: the first one has an eigenvalue 0 of multiplicity 3 and an eigenvalue $-1/2$ of multiplicity 1; the second one has an eigenvalue 0 of multiplicity 4, and eigenvalues $-1/2$ and $-3/2$ both of multiplicity 2. The corresponding eigenvectors can also be readily written down. It seems clear, however, that things probably get rapidly out of hand as the dimensions grow (as 2^N), even for computer calculations. Also one is really interested mostly in what happens in the limit as $N \to \infty$.

Here again the quantum QYBE equations and FCR relations can help. Consider the **Lax operator**[11]

$$L_a(\lambda) = \begin{pmatrix} \lambda I_d + iS^3 & tS^- \\ iS^+ & \lambda I_d - iS^3 \end{pmatrix}$$

where $S^- = S^1 - iS^2$. For the spin 1/2 case this works out as $\sigma^- = \begin{pmatrix} 0 & 1 \\ 0 & 0 \end{pmatrix} = E$, $\sigma^+ = \begin{pmatrix} 0 & 0 \\ 1 & 0 \end{pmatrix} = F$, the matrices traditionally used by mathematicians when

[10] Here the Plank constant $\hbar$ has been set normalized to 1.

[11] These are (often) called Lax operators because they play a rather similar role as the Lax operators in the classical inverse scattering method.

dealing with $\mathfrak{sl}(2)$. Written out in the spin 1/2 case the Lax matrix becomes

$$\begin{pmatrix} \lambda+\frac{i}{2} & & & \\ & \lambda-\frac{i}{2} & i & \\ & i & \lambda-\frac{i}{2} & \\ & & & \lambda+\frac{i}{2} \end{pmatrix}$$

a matrix from the QYBE solution family that has occurred before. In the higher spin case the Lax operators satisfy the corresponding FCR relations. Now again as in the case of the 2D lattice statistical mechanics models considered above take the N-fold tensor product of these representations of the appropriate R-matrix defined quantum matrix algebra

$$T_{a,N}(\lambda) = L_{a,1}(\lambda)L_{a,2}(\lambda)\cdots L_{a,N}(\lambda)$$

on $\mathcal{H} = V^{\otimes N}$, i.e. $T_{a,N}(\lambda) = L_{a,1}(\lambda)L_{a,2}(\lambda)\cdots L_{a,N}(\lambda)$ acts on $\mathbf{C}^2 \otimes \mathcal{H}$, the tensor product of the auxiliary space $\mathbf{C}^2$ with $\mathcal{H}$, and is to be viewed as a 2×2 matrix $\begin{pmatrix} A(\lambda) & B(\lambda) \\ C(\lambda) & D(\lambda) \end{pmatrix}$ with entries that are operators on $\mathcal{H} = V^{\otimes N}$ and where $L_{a,i}$ acts on $\mathbf{C}^2 \otimes V$ as L_a, where V is the i-th factor of $\mathcal{H}$ and as the identity on the other $N-1$ factors V. All this practically exactly as in the case of the transfer matrix in the 2D lattice models discussed above[12].

Again take the trace with respect to the auxiliary space

$$F_N(\lambda) = \mathrm{Tr}_{V_0}(T_N(\lambda)) = A(\lambda) + D(\lambda).$$

Then, again as in the 2D lattice model case, the $F_N(\lambda)$ are a family of commuting operators (with as entries matrices of size $d^N \times d^N$). However, the FCR in this case also imply a good deal more as will be discussed in the next subsubsection on the so-called Bethe Ansatz.

The matrix $F(\lambda)$ is a matrix valued polynomial of degree N in λ

$$F(\lambda) = 2\lambda^N + \sum_{i=0}^{N-2} Q_i \lambda^i$$

yielding $N-1$ commuting operators. The Hamiltonian is (in certain sense, see below) among them. So these Q_i are conserved quantities and what is here is in some sense a completely integrable system.

When the parameter λ is equal to $i/2$ the matrix governing the FCR is equal to the switch matrix on $V_0 \otimes V_0$ and so could be of special importance. It turns out that the Hamiltonian is equal to

$$H = \frac{i}{2}\frac{d}{d\lambda} \ln F(\lambda)|_{\lambda=i/2} - \frac{N}{2}.$$

8.4.34. Algebraic Bethe Ansatz. The situation that turned up in both the spin chain case and the 2D lattice model case is as follows.

---

[12]The only difference is that in the present case the auxiliary space and the starting representation space V are not necessarily equal.

(i) There is a family of 2×2 matrices $T(\lambda)=\begin{pmatrix}A(\lambda) & B(\lambda)\\ C(\lambda) & D(\lambda)\end{pmatrix}$ with entries that are endomorphisms of $V^{\otimes N}$, i.e. $d^N \times d^N$ matrices, of which it is desired to calculate the eigenvalues and eigenvectors of its operator trace $F(\lambda)=A(\lambda)+D(\lambda)$.

(ii) The matrices $T(\lambda)$ satisfy fundamental commutation relations

$$R(\lambda,\mu)T_1(\lambda)T_2(\mu)=T_2(\mu)T_1(\lambda)R(\lambda,\mu) \tag{8.4.35}$$

where

$$R(\lambda,\mu)=\begin{pmatrix}1 & 0 & 0 & 0\\ 0 & b(\lambda-\mu) & c(\lambda-\mu) & 0\\ 0 & c(\lambda-\mu) & b(\lambda-\mu) & 0\\ 0 & 0 & 0 & 1\end{pmatrix}.$$

There is a third ingredient present in both cases. Take again a look at the Lax operator on which the whole situation is based:

$$L_a(\lambda)=\begin{pmatrix}\lambda I_d+iS^3 & iS^-\\ iS^+ & \lambda I_d-iS^3\end{pmatrix}.$$

There is a vector $\omega\in V$ such that with respect to the auxiliary space V_0 it acts triangularly

$$L_a(\lambda)\omega=\begin{pmatrix}a(\lambda) & *\\ 0 & d(\lambda)\end{pmatrix}\omega$$

meaning that

$$(\lambda I_d+iS^3)\omega=a(\lambda)\omega,\quad (\lambda I_d-iS^3)\omega=d(\lambda)\omega,\quad (iS^+)\omega=0.$$

In case of the spin 1/2 chain model $\omega=\begin{pmatrix}1\\0\end{pmatrix}$; in the higher spin models ω is the highest weight vector of the irreducible spin m representation of $\mathfrak{sl}(2;\mathbf{C})$, $d=2m+1$, see section 1.6.

It follows that for the vector $\Omega=\omega^{\otimes N}\in V^{\otimes N}$, sometimes called the vacuum vector,

(iii) $A(\lambda)\Omega=a(\lambda)^N\Omega,\ D(\lambda)\Omega=d(\lambda)^N\Omega,\ C(\lambda)\Omega=0.$

Of course this makes Ω a (rather special) eigenvector of $F(\lambda)$. The **Bethe Ansatz**[13] is now the guess or conjecture that there are other eigenvectors of the form

$$B(\lambda_1)B(\lambda_2)\cdots B(\lambda_N)\Omega \tag{8.4.36}$$

possibly enough of them, in which case it is said that the Bethe Ansatz is complete.

This is perhaps a quite understandable guess in that for $N=1$ the operator $B=iS^-=F$ is a creation operator in that when applied repeatedly to the highest weight vector ω it produces a basis of the representation space V consisting of eigenvectors for S^3 and $\lambda I+iS^3$ and $\lambda I-iS^3$.

The FCR of (8.4.35) when written out explicitly give

$$A(\lambda)A(\mu)=A(\mu)A(\lambda)$$

---

[13]'Ansatz' is a German noun meaning something like 'guess', 'conjecture', 'starting point'. As a German noun it should always be written with a capital initial letter.

$$D(\lambda)D(\mu) = D(\mu)D(\lambda)$$
$$B(\lambda)B(\mu) = B(\mu)B(\lambda) \tag{8.4.37}$$
$$A(\lambda)B(\mu) = c(\mu-\lambda)^{-1}B(\mu)A(\lambda) - \frac{b(\mu-\lambda)}{c(\mu-\lambda)}B(\lambda)A(\mu) \tag{8.4.38}$$
$$D(\lambda)B(\mu) = c(\lambda-\mu)^{-1}B(\mu)D(\lambda) - \frac{b(\lambda-\mu)}{c(\lambda-\mu)}B(\lambda)D(\mu). \tag{8.4.39}$$

Repeated use of these last three equations leads to the following theorem

Theorem 8.4.40. *If*

$$\frac{a(\lambda_j)^N}{d(\lambda_j)^N} = \prod_{k\neq j} \frac{c(\lambda_k - \lambda_j)}{c(\lambda_j - \lambda_k)} \tag{8.4.41}$$

then $B(\lambda_1)B(\lambda_2)\cdots B(\lambda_N)\Omega$ *is an eigenvector of* $F(\lambda)$.

The equations (8.4.41) are called the **Bethe Ansatz equations**.

The calculations leading to theorem 8.4.40, starting from (8.4.37)-(8.4.39) have been written down many times and it hardly seems worth the expenditure of ink and paper to do that again here; especially in a book on algebras.

There is quite a good deal more that can be (and perhaps needs to be) said about QISM and Bethe Ansatz and exact solvability of lattice statistical mechanics models. For this the reader is directed to the classics [**53**], [**240**], [**239**], [**241**], [**646**], and for some idea of how far the Bethe Ansatz reaches and other things it is related to, [**8**], [**102**], [**366**], [**367**], [**422**], [**448**], [**477**], [**525**], [**564**], [**601**], [**614**].

Nor is this all concerning algebraically interesting (even fascinating) aspects of QYBE, QISM, FRT construction, and Bethe Ansatz, see, as a small selection, [**64**], [**91**], [**112**], [**111**], [**281**], [**368**], [**512**], [**513**], [**514**], [**526**], [**575**], [**685**].

8.5. Knot invariants from quantum groups. Yang-Baxter operators

A **knot** is (the image of) an embedding of the circle in three space or, equivalently in the three sphere. Two knots are equivalent if there is a diffeomorphism of the ambient space, i.e. three space, that takes the one knot into the other. Technically this called an ambient isotopy. It amounts exactly to the intuitive pictorial idea that two knows are the same if one can be obtained from the other by moving, streching, bending without tearing or passing a knot through itself.

A link is the image of an embedding of any finite number of circles into three space.

Knots and links are usually depicted by their diagrams, that is a suitable projection on a plane such as in the following examples shown in figure 8.5.1.

A nice complicated example is represented in figure 8.5.2 on the left side. A famous example of a link is formed by the so-called Borromean rings depicted in figure 8.5.2 on the right side.

Another well known example is that of the Olympic rings shown in figure 8.5.3. All four pictures 8.5.1 - 8.5.3 come from the article 'Knots' from the Wikipedia; we are grateful to be allowed to use them.

The mathematical problem is to decide whether two links are the same or not.

Even to prove rigorously that the unknot and the trefoil knot labelled 3_1 above are not the same is a nontrivial task.

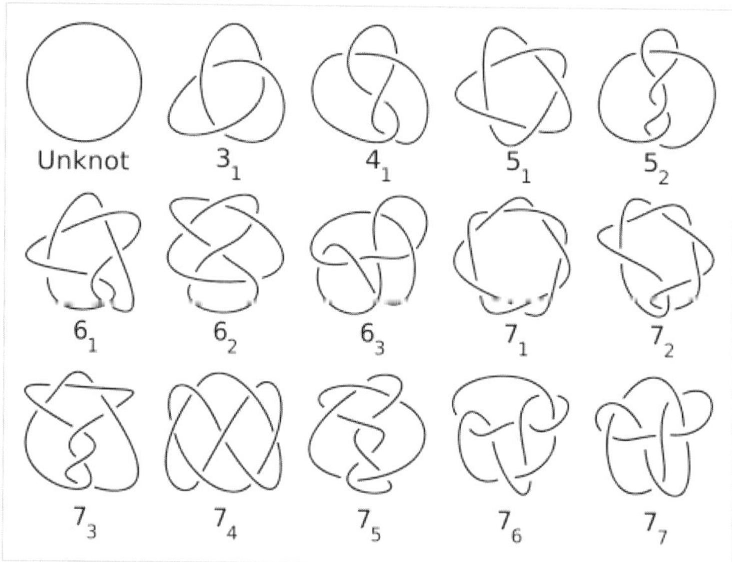

Figure 8.5.1.

But it can de done using an ingenious idea called 'tricoloring', see [**375**], p. 21ff; [**451**], p. 32ff. It is quite a good deal harder to show that the trefoil knot and it's mirror image are not the same. The Jones polynomial was the first invariant that could do that.

Figure 8.5.2.

An invariant of a knot or link is some mathematical object that can be attached to it and that is the same for equivalent knots and links. Many invariants are polynomials. One of the first was the Alexander polynomial (Alexander-Conway polynomial). That one distinguishes between the unknot and the trefoil, but it does not distinguish the trefoil from its mirror image.

Figure 8.5.3.

In the last 30 years or so the theory of knows and links has undergone something like a revolution. It is an amusing and instructive exercise to compare the contents of a 'classical' book like [**602**], with that of modern treatises like [**500**]; [**570**]; [**667**], or the valuable collection of reprints [**394**].

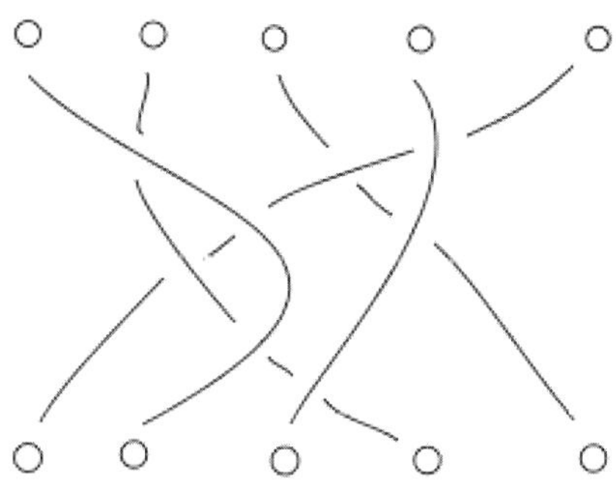

Figure 8.5.4. Braid on five strands.

8.5.1. Braids. A **braid** on n strands is an object consisting of two parallel (horizontal) planes P_0, P_1 in three space with on one plane a line $L_0 \subset P_0$, with n ordered points $a_1, \dots, a_n$ on it (starting points), and a parallel line $L_1 \subset P_1$ with n ordered points $b_1, \dots, b_n$ directly below it (end points) and n nonintersecting strands (arcs) connecting each a_i to a b_j such that each strand intersects each parallel plane between P_0 and P_1 exactly once.

It is customary to depict such an object by means of a projection to a vertical plane orthogonal to the P_0, P_1 with for each crossing in the projection indicated which strand overcrosses the other one (i.e. is at the near side at that projection point).

For instance figure 8.5.4 is one of a braid on five strands.

Two braids are considered equivalent (the same) if they can be obtained from each other by moving the strands about, by moving the planes up or down (as if the strands were elastic) by moving the lines (in parallel), by moving the points on the lines, all without crossings.

As an example the following two braids are equivalent, figure 8.5.5. Just what is the meaning of the equation

$$\sigma_1\sigma_2\sigma_1 = \sigma_2\sigma_1\sigma_2$$

in the caption of that picture will be explained in a moment.

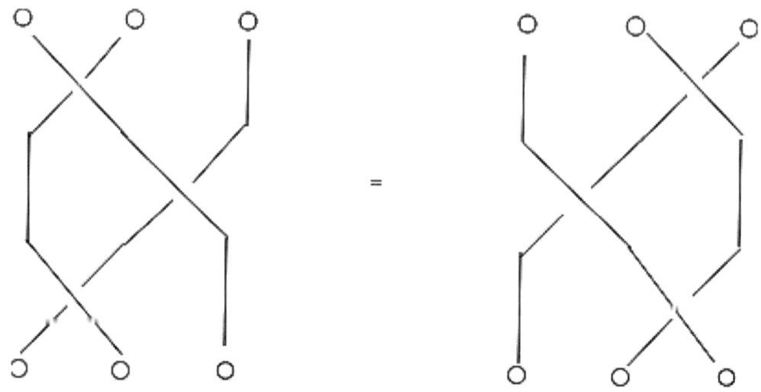

Figure 8.5.5. $\sigma_1\sigma_2\sigma_1 = \sigma_2\sigma_1\sigma_2$ in B_3.

8.5.2. Artin braid group. Two (equivalence classes of) braids on the same number of strands can be multiplied by putting them on top of each other.

This is illustrated by figure 8.5.6 in case of braids on two strands. There is a neutral element for this operation, viz the braid with no crossings. There are also inverses: the inverse of a braid (equivalence class) is obtained by taking its horizontal reflection. Thus braids on n strands form a group, called the Artin braid group and denoted B_n. The group B_1 is the trivial group on one element; the group B_2 is the infinite cyclic group $\mathbf{Z}$; and for $n > 2$, B_n is an infinite non-Abelian group that contains a copy of the free group on two generators.

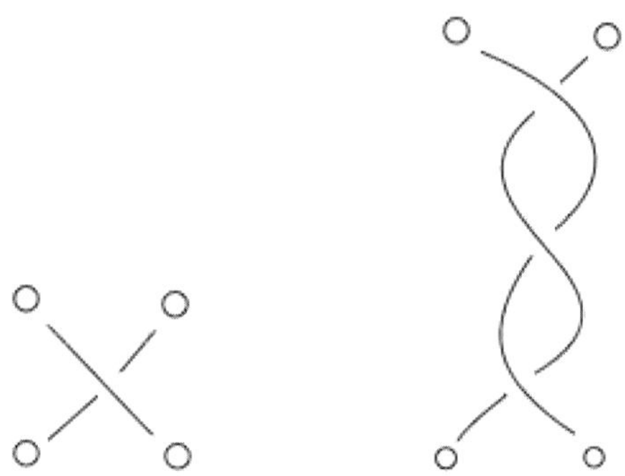

Figure 8.5.6. σ_1 on two strands $\quad$ σ_1^3 on two strands.

In B_n consider the braid with one crossing σ_i defined as follows. The first $i-1$ strands go straight down (and involve no crossings); the i-th strand goes to the $(i+1)$-th endpoint and overcrosses the next strand which goes from $i+1$ to i; the last $n-i-1$ strands again go straight down and involve no crossings. This is illustrated in figure 8.5.7 just below (and also in figure 8.5.5 in subsubsection 8.5.1.

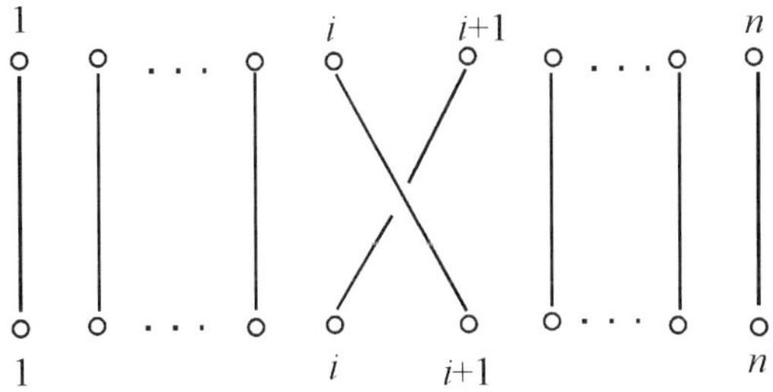

Figure 8.5.7.

It is rather obvious that the $\sigma_1, \ldots, \sigma_{n-1}$ form a set of generators for the group B_n.

8.5.3. Presentation of the Artin braid group. Consider again the braid group B_n and its set of generators $\sigma_1, \ldots, \sigma_{n-1}$. It is rather obvious that the following relations hold.

$$\begin{array}{ll} \sigma_i\sigma_{i+1}\sigma_i = \sigma_{i+1}\sigma_i\sigma_{i+1}, & i = 1, \ldots, n-2 \\ \sigma_i\sigma_j = \sigma_j\sigma_i, & \text{if } |i-j| \geq 2; i, j \in \{1, \ldots, n-1\}. \end{array} \tag{8.5.4}$$

Far less obvious is that these are all relations (needed). I.e.

Theorem 8.5.5. [**34**]. *The generators $\sigma_1, \ldots, \sigma_{n-1}$ and the relations* (8.5.4) *form a presentation of the braid group B_n.*

8.5.6. Closing braids. Take a braid. Now add arcs from each endpoint to the starting point above it without introducing any crossings (neither of the new arcs among themselves, nor of any of the new arcs with the original strands (arcs). The resulting 'closed braid' is a link. This is illustrated by figure 8.5.8 in the case of the braid $\sigma_1^3 \in B_2$. The result, as can also be seen easily from figure 8.5.8 is, in this case, one of the trefoil knots.

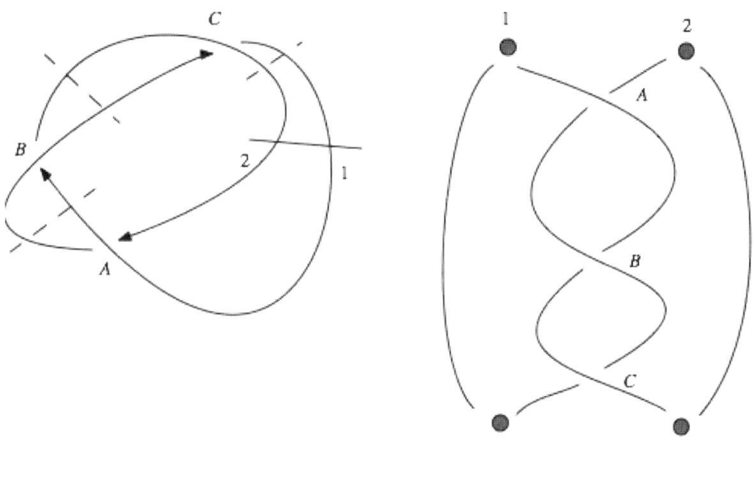

Figure 8.5.8. Braid closure of the trefoil knot.

The other, mirror image, trefoil knot results from closing the braid $\sigma_1^{-3} \in B_3$.

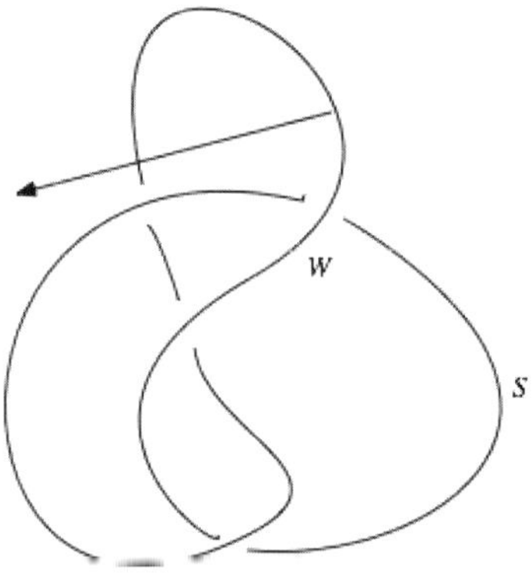

Figure 8.5.9.

It does not matter how the closing arcs are taken provided the non-crossing condition is observed. This is not at all difficult to see.

The question now immediately arises whether all links can be obtained in this way. They can and that is the content of the Alexander theorem, [**10**].

8.5.7. Notes on the proof of the Alexander theorem. Consider again the trefoil knot as traditionally drawn in the left part of figure 8.5.8. Observe that there is an axis vertical to the drawing plane, somewhere in the center of the knot such that the knot circles always in the same direction around this axis. Now snip all the strands along some radius emanating from this axis, as indicated in figure 8.5.9 (solid line), to find a braid whose closure is the given trefoil.

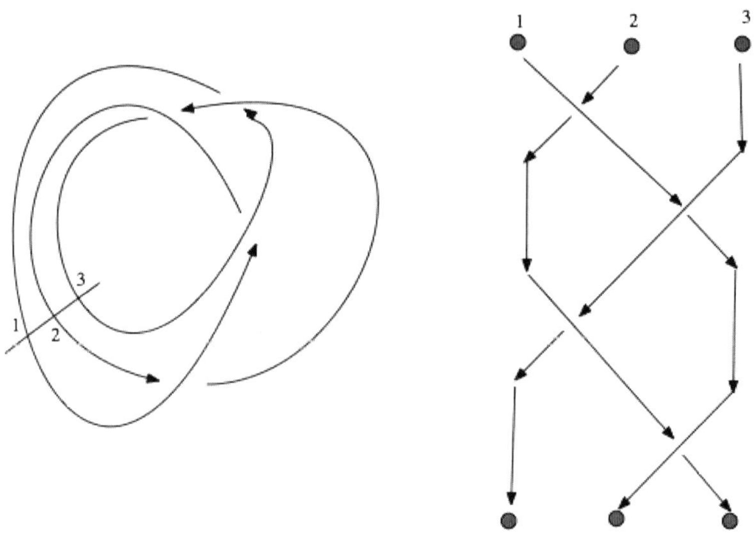

Figure 8.5.10. Braid closure form of figure 8 knot

Not all knots, as traditionally drawn, have this 'axis property'. Consider for instance the standard picture of the figure eight knot as drawn in figure 8.5.9. If one starts at the point marked S and moves clockwise things start to go wrong around the point W. However, this can be remedied by flipping over part of the knot as indicated. The result, after some more minor adjustments is something like figure 8.5.10.

The idea of the proof is now to do this repeatedly until a version of the diagram of the knot is reached that does have the 'axis property'. This works, but making the proof idea really rigorous is not all that easy [14]. See, however, [**69**], p. 42ff.

8.5.8. Markov moves. It is dead easy to construct examples of braids that are definitely different that upon closure give the same knot or link. For instance braids with different numbers of strands such as $\sigma_1 \in B_2$ and the trivial braid in B_1 which both give the unknot upon closure. There are two ways of changing a braid in a way for which one can see rather immediately that they do not change the closure link. They are known as **Markov moves** and are as follows.

Markov move 1. 'Adding a generator'. Consider the braid group B_n and consider it as a subgroup of the braid group B_{n+1} on one more strand by sending the generator σ_i on n strands to the generator σ_i on $n+1$ strands.

Now take a braid b in B_n and consider the braids $b\sigma_n^{\pm 1}$ in B_{n+1} (which are not in B_n). The claim is that upon closure these three braids yield the same link. This proved by figure 8.5.11.

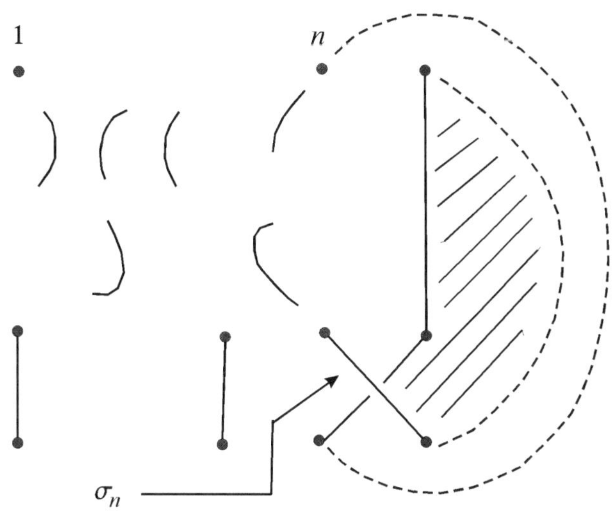

Figure 8.5.11. Flipping the arched region with its dashed boundary to the left shows that the addition of σ_n makes no difference for the link obtained from it by closure.

Markov move 2. 'Conjugation'. Consider two braids $b, g \in B_n$. The claim is that the three braids b and $g^{-1}bg$ and gbg^{-1} have the same link as their closures. Obviously it suffices to prove this in the case that g is one of the generators, and then figure 8.5.12 (and a similar one for an inverse generator) proves the claim.

Clearly, an indicated by the arrows, the outer solid-broken-drawn closing arc of the conjugated braid together with the overcrossing strands of the conjugating

[14]For instance the proof sketch on pages 91-93 in [**375**] is rather unsatisfactory.

generator can be pushed over to the dashed-broken-drawn closing arc of the original braid.

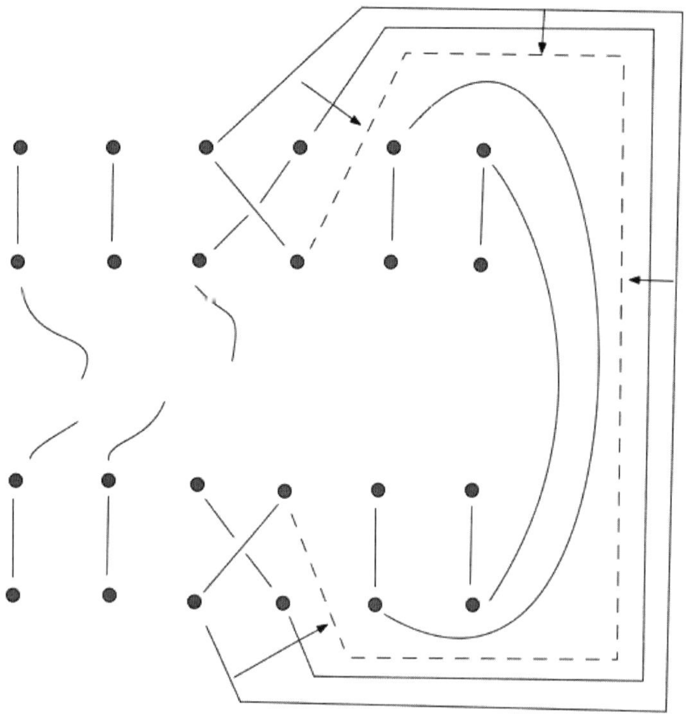

Figure 8.5.12.

8.5.9. Markov theorem. The Markov theorem, [**504**][15], now says that two braids give the same link upon closure if and only if they can be obtained from each other by a series of Markov moves (and inverse Markov moves).

See [**69**], section 2.2, for a proof and discussion of the proof of the Markov theorem.

8.5.10. Strategy for constructing knot and link invariants. The Alexander and Markov theorems suggest the following kind of strategy for the construction of link invariants.

Take a family of mappings of the braid groups

$$J_n : B_n \longrightarrow A \tag{8.5.11}$$

where A is a unital commutative ring, such that

$$J_n(g^{-1}bg) = J_n(b), \quad b, g \in B_n$$

$$\exists \alpha \in R \text{ such that } J_{n+1}(b\sigma_n) = \alpha J_n(b), \quad J_{n+1}(b\sigma_n^{-1}) = \alpha^{-1} J_n(b). \tag{8.5.12}$$

Then an invariant is defined by the formula

[15]This Andrej Andreevich Markov (1903-1979) is the son of Andrej Andreevich Markov (1856-1922) of Markov chains (and father and grandfather of more mathematicians called A. A. Markov).

$$J(L) = \alpha^{-w(b)} J_n(b) \tag{8.5.13}$$

where $b = \sigma_{i_1}^{a_1}\sigma_{i_2}^{a_2}\cdots\sigma_{i_k}^{a_k} \in B_n$ is a braid with as closure the link L, and where[16]

$$w(b) = a_1 + a_2 + \cdots + a_k. \tag{8.5.14}$$

This number is sometimes called the Tait number, and it is not an invariant of links but is an invariant of oriented links.

Suitable families of mappings J_n can arise from taking a family of representations ρ_n of the braid groups and taking $J_n = \mathrm{Tr}(\rho_n)$.

It remains to construct suitable sequences of representations complete with suitable trace-like functionals.

For the original construction of the Jones polynomial certain towers of algebras of projections in von Neumann algebras were used (Temperley-Lieb algebras), [**353**], see also [**355**], p. 11ff.

For the HOMFLYPT[17] polynomial, [**258**], [**571**], (Iwahori-)Hecke algebras are appropriate.

The Kauffman polynomial was originally constructed via statistical mechanics model ideas (so that it arises as a partition function), [**374**]. However, it can also be obtained with the approach outlined above. Here the algebras involved are the BMW algebras (Joan S. Birman, Jun Murakami, Hans Wenzl), [**70**].

The remainder of this section is mainly devoted to the implementation of this general "braid-trace" strategy using quantum groups as done first by Vladimir G. Turaev, [**668**].

8.5.15. Historical context. In 1984 there appeared the first of the 'new' knot and link polynomials invariants, the Jones polynomial. This was soon followed, in 1985, by the HOMFLY polynomial and the Kauffman polynomial. The central inspiration of [**668**] is that both polynomials can be obtained from the R-matrices of quantum group theory.

8.5.16. Extended Yang-Baxter operators. Let V be a finite rank free module over a commutative base ring k. Given an endomorphism of the iterated tensor product (over k) $V^{\otimes n} = V \otimes_k V \otimes_k \cdots \otimes_k V$ one can take the 'operator trace' with respect to the last factor (or any other factor for that matter). Such operator traces have occurred before in section 8.4 above where, for example, operator traces were taken with respect to an auxiliary space (in the quantum field setting).

Explicitly, let $v_1, \ldots, v_m$ be a basis of V, then an endomorphism F of $V^{\otimes n}$ determines and is determined by a multi-index matrix via the formula

$$F(v_{i_1} \otimes v_{i_2} \otimes \cdots \otimes v_{i_n}) = \sum_{j_1, j_2, \ldots, j_n} f_{i_1, i_2, \ldots, i_n}^{j_1, j_2, \ldots, j_n} v_{j_1} \otimes v_{j_2} \otimes \cdots \otimes v_{j_n}.$$

[16]There is an invariant of oriented links called the writhe or Tait number, see e.g. [**375**],p.19 and p.33. If a link L is obtained as a closed braid and oriented by letting the strands go down (and the closing arcs go up), the writhe of L is precisely this number $w(b)$.

[17]This polynomial is usually called the HOMFLY polynomial (after the initial letters of the names of the six authors; HOMFLYPT is better (two more authors) and that would have been the generally accepted appellation if the post in Poland had not been so slow.

In terms of this multi-index matrix the operator trace with respect to the last factor is given by

$$\mathrm{Tr}_n(F)(v_{i_1} \otimes v_{i_2} \otimes \cdots \otimes v_{i_{n-1}}) =$$
$$\sum_{j_1,j_2,\ldots,j_{n-1},j} f^{j_1,j_2,\ldots,j_{n-1},j}_{i_1,i_2,\ldots,i_{n-1},j} v_{j_1} \otimes v_{j_2} \otimes \cdots \otimes v_{j_{n-1}}.$$

The endomorphism $\mathrm{Tr}_n(F)$ of $V^{\otimes(n-1)}$ thus defined does not depend on the choice of basis.

An **extended Yang-Baxter operator** over V is a quadruple

$$S = (R, \mu, \alpha, \beta) \tag{8.5.17}$$

where R is an invertible endomorphism of the k-module $V^{\otimes 2} = V \otimes_k V$, μ is an endomorphism of V and α, β are invertible elements of k, such that

R satisfies the braid version of the QYBE (see 8.2.29) :

$$R_{12}R_{23}R_{12} = R_{23}R_{12}R_{23} \tag{8.5.18}$$

$$\mu^{\otimes 2} : V^{\otimes 2} \longrightarrow V^{\otimes 2} \text{ commutes with } R \tag{8.5.19}$$

$$\mathrm{Tr}_2(R^{\pm}(\mu \otimes \mu)) = \alpha^{\pm}\beta\mu. \tag{8.5.20}$$

8.5.21. Representations of the braid group. Let R be an invertible endomorphism that satisfies (8.5.18). Then for any n it defines a representation of the braid group B_n on the n-fold tensor product $V^{\otimes n}$ as follows. For $i \in \{1, 2, \ldots, n-1\}$ take

$$R_i = \mathrm{id}_V^{\otimes(i-1)} \otimes R \otimes \mathrm{id}_V^{\otimes(n-i-1)}.$$

So R_i acts as the identity on the first $(i-1)$ factors and on the last $(n-i-1)$ factors and acts as R on the remaining two factors. If $n = 3$, in the notation used in 8.2, $R_1 = R_{12}$ and $R_2 = R_{23}$. It is an immediate consequence of (8.5.18) that

$$R_iR_{i+1}R_i = R_{i+1}R_iR_{i+1}, \quad i \in \{1, 2, \ldots, n-2.\} \tag{8.5.22}$$

Further

$$R_iR_j = R_jR_i, \text{ if } |i-j| \geq 2, \quad i, j \in \{1, 2, \ldots, n-1\} \tag{8.5.23}$$

because, under these circumstances, the factors where R_i and R_j do something non-identity have nothing to do with one another (are disjoint). Thus, by the Artin braid group theorem,

$$\sigma_i \mapsto R_i, \quad i \in \{1, 2, \ldots, n-1\}$$

defines a representation of the braid group on n strands. Denote this representation, i.e. homomorphism $B_n \longrightarrow \mathrm{Aut}(V^{\otimes n})$, by b_R.

8.5.24. Invariant of oriented links defined by an extended Yang-Baxter operator. For an endomorphism $\mu : V \longrightarrow V$ let $\mu^{\otimes n} : V^{\otimes n} \longrightarrow V^{\otimes n}$ denote its n-th tensor power. Let $S = (R, \mu, \alpha, \beta)$ be an extended Yang-Baxter operator. Now for a braid $\xi \in B_n$ put

$$T_S(\xi) = \alpha^{-w(\xi)}\beta^{-n}\mathrm{Tr}(b_R \circ \mu^{\otimes n} : V^{\otimes n} \longrightarrow V^{\otimes n}). \tag{8.5.24a}$$

Theorem 8.5.25. *The expression* (8.5.24a), *i.e.* $T_S(\xi)$, *is invariant under the Markov moves and hence defines a link invariant.*

The proof is not difficult, see [**668**], page 531, and mainly depends on some properties of "Trace' with respect to certain special compositions and tensor products of morphisms.

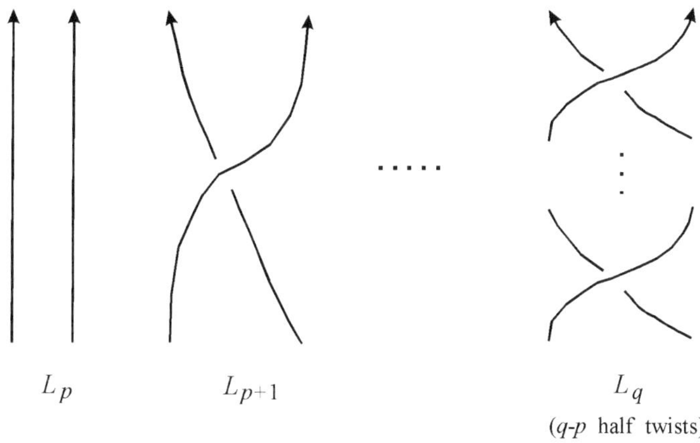

Figure 8.5.13.

To make a link with Conway-type relations (skein relations). The following notion is needed. Let τ be a mapping from the set of oriented isotopy classes of links into A and let $f(t) = \sum_{i=p}^{q} a_i t^i \in A[t, t^{-1}]$ be a Laurent polynomial. Then $f(t)$ is said to annihilate τ, notation $f(t) * \tau = 0$, if for any oriented links $L_p, L_{p+1}, \ldots, L_q$ which have the same diagrams outsize some disk and which look like the ones in figure 8.5.13 inside that disk one has

$$\sum_{i=p}^{q} a_i \tau(L_i) = 0. \tag{8.5.26}$$

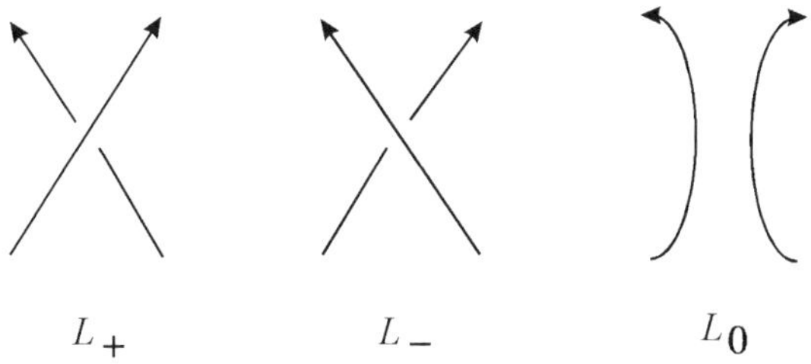

Figure 8.5.14.

In particular if the Laurent polynomial is of the form $a_{-1}t^{-1} + a_0 + a_1 t$ the resulting annihilation relation is a skein type relation between links that are the same outside some disk and inside that disk look like in figure 8.5.14.

Theorem 8.5.27. *Let $S = (R, \mu, \alpha, \beta)$ be an extended Yang-Baxter operator. If the automorphism R of $V^{\otimes 2}$ satisfies an equation $\sum_{i=p}^{q} a_i R^i = 0$ then the polynomial $\sum_{i=p}^{q} a_i \alpha^i t^i$ annihilates T_S.*

8.5.28. The A^1 series invariants. The fundamental vector representation of the simple Lie algebra A^1_{m-1} gives rise to the following Yang-Baxter operator

$$R = -q\sum_i E_i^i \otimes E_i^i + \sum_{i \neq j} E_j^i \otimes E_i^j + (q^{-1} - q)\sum_{i<j} E_i^i \otimes E_j^j. \tag{8.5.29}$$

see [**668**]; [**351**]. The inverse is given by a very similar formula: change q to q^{-1} and switch i and j.

Theorem 8.5.30. *Set $\mu_i = q^{2i-m-1}$ for $i = 1, \ldots, m$; $\alpha = -q^m$; $\beta = 1$. Then $S = (R, \mu = \mathrm{diag}(\mu_1, \ldots, \mu_m), \alpha, \beta)$ is an extended Yang-Baxter operator such that for each triple of links as in figure* 8.5.14

$$\begin{aligned} q^m T_S(L_+) - q^{-m} T_S(L_-) &= (q - q^{-1}) T_S(L_0) \\ T_S(\text{unknot}) &= (q^m - q^{-m})/(q - q^{-1}). \end{aligned} \tag{8.5.31}$$

Thus by the considerations above (see theorem 8.5.25) there is a two parameter family of invariants attached to the A^1 series of simple Lie algebras. A certain amount of ingenious reparametrization now allows to combine these into a Laurent polynomial N in two variables t and q that is an isotopy invariant of links and satisfies

$$tN(L_+) - t^{-1}N(L_-) = (q - q^{-1})N(L_0). \tag{8.5.32}$$

The Jones polynomial can be obtained from this one; just replace t by q^{-2}, see [**372**].

8.5.33. Open problem. The R-matrices of (8.5.29) satisfy the support condition

$$R^{k,l}_{i,j} = 0 \quad \text{unless the sets } \{i,j\} \text{ and } \{k,l\} \text{ are equal.}$$

In [**315**] all solutions of the Yang-Baxter equation that satisfy this support condition are determined. It looks like all these extend to extended Yang-Baxter equations and thus to (oriented) link invariants. It is uninvestigated whether these invariants yield anything new beyond the ones discussed above.

8.5.34. The invariants coming from the B^1, C^1, D^1, A^2 series of simple Lie algebras. Fix a $\nu \in \{1, -1\}$. In the following if m is odd ν is taken to be -1. Now for $i = 1, \ldots, m$ define

$$\begin{aligned} i' &= m + 1 - i \\ \bar{i} &= \begin{cases} i - \nu/2 & \text{if } 1 \leq i \leq (m+1)/2 \\ i & \text{if } i = (m+1)/2 \text{ where } m \text{ is supposed to be odd} \\ i + \nu/2 & \text{if } (m+1)/2 < i \leq m \end{cases} \\ \varepsilon(i) &= \begin{cases} 1 & \text{if } 1 \leq i \leq (m+1)/2 \\ -\nu & \text{if } (m+1)/2 \leq i \leq m \end{cases} \end{aligned} \tag{8.5.35}$$

The fundamental representations of these series of simple Lie algebras now give rise to the following Yang-Baxter operators (R-matrices):

$$\begin{aligned} R_\nu &= q \sum_{i,i\neq i'} E_i^i \otimes E_i^i + \sum_{i,i=i'} E_i^i \otimes E_i^i + \sum_{\substack{i,j \\ i\neq j,j'}} E_j^i \otimes E_i^j \\ &+q^{-1} \sum_{i,i\neq i'} E_{i'}^i \otimes E_i^{i'} + (q-q^{-1}) \sum_{i<j} E_i^i \otimes E_j^j \\ &+(q^{-1}-q) \sum_{i<j} \varepsilon(i)\varepsilon(j) q^{\bar{i}-\bar{j}} E_{j'}^i \otimes E_j^{i'}. \end{aligned} \tag{8.5.36}$$

Here for the Lie algebras $B_n^1, C_n^1, D_n^1, A_n^2$ the pairs (m,ν) are respectively $(2n+1,-1)$, $(2n,1)$, $(2n,-1)$, $(n+1,-1)$. In the case of an odd m the ring of Laurent series $A = \mathbf{Z}[q,q^{-1}]$ is replaced by $A = \mathbf{Z}[q^{1/2}, q^{-1/2}]$. See [**668**]; [**351**].

All these R-matrices can be extended to extended Yang-Baxter operators with μ the diagonal matrix with diagonal entries $\mu_i = q^{2i-m-1}$, $\mu_i = q^{2i-m-1}$, $\beta = 1$ and hence give rise to invariants.

As in the A^1 case these can be combined and reparametrized to yield an invariant Q_ν for oriented links and an invariant $\tilde{Q}_\nu(L) = x^{w(L)} Q_\nu(L)$ for links (i.e. which does not depend on the orientation) with values in $\mathbf{Z}[x,x^{-1},y,y^{-1}]$. The invariant Q_1 was introduced by Kauffman.

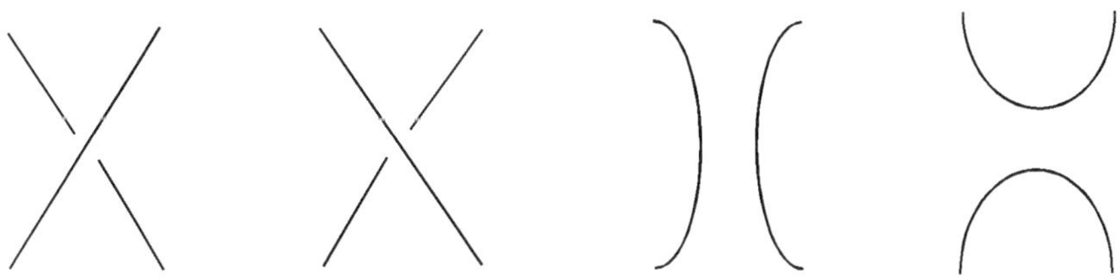

D_+ D_- D_0 D_∞

Figure 8.5.15.

If four links are the same outside some disk and inside that disk look like shown in figure 8.5.15 then there is the following skein-type relation

$$\begin{aligned} &\tilde{Q}_\nu(D_+) + \nu \tilde{Q}_\nu(D_-) = \tilde{Q}_\nu(D_0) + \tilde{Q}_\nu(D_\infty) \\ &Q_\nu(\text{unknot}) = 1. \end{aligned} \tag{8.5.37}$$

8.5.38. Open problem. The extended Yang-Baxter operators of type A^1 and (for instance) D^1 can be combined in nontrivial ways to yield invariants that are possibly more powerful than the separate A^1 and D^1 invariants together. There are positive indications to this effect in [**315**].

8.5.39. Notes. What has been presented above is the situation with quantum knot and link invariants roughly up to 1988. Since then much much more has happened. In particular the following four topics have not been mentioned above.

- Vassiliev invariants and their relation to quantum knot and link invariants. These work via singular links, i.e. links where (transversal) self intersections are permitted. Recommended further reading: [**372**]; [**570**], [**13**].

- Quantum invariants of three manifolds. This rests essentially on integer (ribbon) surgery presentations of three manifolds) (surgery along a framed link in S^3).

Recommended reading: [**570**] and for the relations to modular categories: [**667**]. Also [**376**], [**377**].

- State space approach (partition functions as in statistical mechanics) to link invariants. Recommended reading: [**375**], [**378**], [**668**], §5.

- The relations with topological quantum field theory, see [**12**]; [**667**]; [**687**]. The last paper is considered not rigorous by mathematical standards, but on the other hand it just uses techniques that have been successfully used in theoretical physics for many many years. For topological quantum theory itself see [**256**]; [**41**].

8.6. Quiver Hopf algebras

8.6.1. Path algebra and path coalgebra. In this section everything is over a base field k. Much can also be done over the integers, but not all. For instance at certain points one needs n-th roots of unity or even 'algebraically closed'. Unadorned tensor products are tensor products over k.

Let Q be a quiver with set of vertices Q_0, set of arrows Q_1, source (or starting vertex) map $s : Q_1 \to Q_0$ and target (or ending vertex) map $t : Q_1 \to Q_0$. So if $\alpha \in Q_1$ it is an arrow from $s(\alpha)$ to $t(\alpha)$.

A **path** in Q is a sequence of arrows $p = [\alpha_1, \alpha_2, \dots, \alpha_m]$ such that $t(\alpha_i) = s(\alpha_{i+1})$ for $i = 1, 2, \dots, m$. Such a path is said to be a path from $i = s(\alpha_1)$ to $j = t(\alpha_m)$ of length m, denoted $\lg(p)$. The vertices are seen as the paths of length 0. See [**317**], page 274. Sometimes it is very convenient to indicate explicitly the source and target of a path and to use a notation like $[i|\alpha_1, \alpha_2, \dots, \alpha_m|j]$[18].

Two paths can be multiplied as follows

$$\begin{aligned}&[i|\alpha_1, \alpha_2, \dots, \alpha_m|j][u|\beta_1, \beta_2, \dots, \beta_n|v]\\ &= \delta_{j,u}[i|\alpha_1, \alpha_2, \dots, \alpha_m, \beta_1, \beta_2, \dots, \beta_n|v].\end{aligned} \tag{8.6.2}$$

The vector space over k with paths as a basis and this multiplication is the **path algebra** $Path_k(Q)$ of Q also denoted kQ or kQ^a; see [**317**], page 275. The paths of length zero, i.e. the vertices, are a set of orthogonal idempotents. Their sum, if Q_0 is finite, is the unit element of $Path_k(Q)$.

There is also a comultiplication on the vector space spanned by the paths. It is 'cut'. I.e.

$$\begin{aligned}&\mu([\alpha_1, \alpha_2, \dots, \alpha_m]) = s(\alpha_1) \otimes [\alpha_1, \alpha_2, \dots, \alpha_m]\\ &+ \sum_{i=1}^{m-1} [\alpha_1, \alpha_2, \dots, \alpha_i] \otimes [\alpha_{i+1}, \alpha_2, \dots, \alpha_m]\\ &+[\alpha_1, \dots, \alpha_m] \otimes t(\alpha_m).\end{aligned} \tag{8.6.3}$$

This is coassociative. There is also a counit

$$\varepsilon(p) = \begin{cases} 1 & \text{if } \lg(p) = 0 \\ 0 & \text{if } \lg(p) \geq 1. \end{cases} \tag{8.6.4}$$

This gives the **path coalgebra** $CoPath_k(Q)$, also written kQ^c, first introduced by Chin and Montgomery, [**142**]. The multiplication (8.6.2) and the comultiplication (8.6.4) do NOT combine to define a Hopf algebra.

The central question in this section is when $Path_k(Q)$ and $CoPath_k(Q)$ admit (graded) Hopf algebra structures (and to describe these when they exist). The two

[18] In many publications paths are written to be read from right to left. Like compositions of morphisms. In that notation a path from i to j would be a sequence $[\alpha_m, \alpha_{m-1}, \dots, \alpha_1]$ with (again) $s(\alpha_1) = i$, $t(\alpha_m) = j$, $t(\alpha_i) = s(\alpha_{i+1})$

questions are dual to each other. Overall the path coalgebra question comes out just a bit nicer than the path algebra one and we will concentrate on the path coalgebra case.

Some special cases have already occurred in this volume. Take a quiver with one vertex and as arrows a set of loops indexed by an index set I. The the path algebra is the free algebra $k\langle X_i : i \in I\rangle$ and the path coalgebra is the vector space with basis all words over the alphabet I, including the empty word, under 'cut' This is the **graded-cofree coalgebra** over the vector space with basis I.

Both this free algebra and this graded-cofree coalgebra carry (several) Hopf algebra structures.

8.6.5. The graded-cofree coalgebra of a bicomodule over a coalgebra. Let C be a coalgebra over k and M a bicomodule over C. This means that there is both a left comodule structure and a right comodule structure

$$\gamma^M_{\text{left}} : M \longrightarrow C \otimes M, \quad \gamma^M_{\text{right}} : M \longrightarrow M \otimes C \tag{8.6.6}$$

that are compatible in the sense that

$$\begin{aligned} &(\gamma^M_{\text{left}} \otimes \text{id}_C) \circ \gamma^M_{\text{right}} : M \longrightarrow M \otimes C \longrightarrow C \otimes M \otimes C \\ &= (\text{id}_C \otimes \gamma^M_{\text{right}}) \circ \gamma^M_{\text{left}} : M \longrightarrow C \otimes M \longrightarrow C \otimes M \otimes C \end{aligned} \tag{8.6.7}$$

which is precisely the commutativity of the diagram obtained by reversing all arrows in the diagram that describes the compatibility condition for a bimodule over an algebra. The category of bicomodules over a coalgebra C is denoted ${}^C\text{Mod}^C$. (The category of left comodules over C is ${}^C\text{Mod}$; that of right comodules is Mod^C.)

Given two bicomodules M and N over C their cotensor product is defined by

$$M \odot_C N = \text{Ker}(M \otimes_k N \xrightarrow{\gamma^M_{\text{right}} \otimes \text{id}_N - \text{id}_M \otimes \gamma^N_{\text{left}}} M \otimes_k C \otimes_k N) . \tag{8.6.8}$$

See also [**102**], p.93ff. Taking the kernel in (8.6.8) is dual to quotienting out the relations $xa \otimes_k y - x \otimes_k ay$ when forming the tensor product of two bimodules over an algebra.

The vector space $M \otimes_C N$ is again a bicomodule: for the left comodule structure use the left comodule structure on M and for the right comodule structure use the right comodule structure on N. The cotensor product is associative.

The **cotensor coalgebra** of M over C is now

$$CoTen^C(M) = M^{\odot 0} \oplus M^{\odot 1} \oplus M^{\odot 2} \oplus \cdots \oplus M^{\odot i} \oplus \cdots \tag{8.6.9}$$

where

$$M^{\odot 0} = C, \quad M^{\odot 1} = M, \quad M^{\odot (i+1)} = M^{\odot i} \odot M^{\odot 1}$$

as a k-vectorspace where the coalgebra structure is induced by 'cut'. That is

$$\begin{aligned} &\mu|_C = \mu_C, \quad \mu|_M = \gamma^M_{\text{left}} \oplus \gamma^M_{\text{right}}, \\ &\mu(m_1 \otimes m_2 \otimes \cdots \otimes m_n) = \gamma^M_{\text{left}}(m_1) \otimes (m_1 \otimes m_2 \otimes \cdots \otimes m_n) \\ &+ m_1 \otimes (m_2 \otimes \cdots \otimes m_n) + \cdots + (m_1 \otimes m_2 \otimes \cdots \otimes m_{n-1}) \otimes m_n \\ &+ (m_1 \otimes m_2 \otimes \cdots \otimes m_n) \otimes \gamma^M_{\text{right}}(m_n) \end{aligned}$$

which is in

$$\begin{aligned} &(C \otimes_k M^{\odot n}) \oplus (M \otimes_k M^{\odot (n-1)}) \oplus \cdots \oplus (M^{\odot (n-1)} \otimes_k M) \oplus (M^{\odot n} \otimes_k C) \\ &\subset CoTen^C(M) \otimes_k CoTen^C(M). \end{aligned}$$

The counit is given by $\varepsilon|_C = \varepsilon_C$, $\varepsilon(M^{\odot n}) = 0$ for $n \geq 1$.

As is to be expected the cotensor coalgebra construction has a universal property.

Theorem 8.6.10. [**677**], p. 579. *Let $\psi : D \longrightarrow CoTen^C(M)$ be a coalgebra morphism. Denote by ψ_n the composition of ψ with the canonical projection $CoTen^C(M) \longrightarrow M^{\odot n}$. Then $\psi_0 : D \longrightarrow C$ is a morphism of coalgebras and $\psi_1 : D \longrightarrow M$ is a morphism of C-bicomodules where the left and right comodule structures on D are induced by ψ_0; that is, the left comodule structure on D is given by the composition of μ_D and $\psi_0 \otimes \mathrm{id}$ and the right comodule structure on D is the composition of μ_D and $\mathrm{id} \otimes \psi_0$. For $n \geq 2$ ψ_n is exactly the C-bicomodule morphism*

$$D \xrightarrow{\mu_D} D \otimes D \xrightarrow{\mu_D \otimes \mathrm{id}} D \otimes D \otimes D \longrightarrow \cdots \longrightarrow D^{\otimes n} \xrightarrow{\psi_1^{\otimes n}} M^{\otimes n}$$

so that ψ_n is uniquely determined by ψ_0 and ψ_1.

Conversely let $\psi_0 : D \longrightarrow C$ be a coalgebra morphism and $\psi_1 : D \longrightarrow M$ a C-bicomodule morphism. Let ψ_n be the composition above. Then the image of ψ_n is in $M^{\odot n}$ and $\psi_n : D \longrightarrow M^{\odot n}$ is a morphism of C-bicomodules. If for each $x \in D$ there are only finitely many i such that $\psi_i(x) \neq 0$ the direct sum of the ψ_n is a coalgebra morphism $\psi : D \longrightarrow CoTen^C(M)$.

The last condition is fulfilled if D is a graded coalgebra, so that $CoTen^C(M)$ has a universal property for morphisms from a graded coalgebra.

8.6.11. Path coalgebra vs cotensor coalgebra. Consider a quiver Q. Give kQ_0, the vector space with basis Q_0, the set coalgebra structure, see 2.4.24. Give kQ_1 the bicomodule structure over kQ_0 defined by

$$\gamma_{\text{left}}(\alpha) = s(\alpha) \otimes \alpha, \quad \gamma_{\text{right}}(\alpha) = \alpha \otimes t(\alpha). \tag{8.6.12}$$

Now apply the cotensor coalgebra construction to the pair (kQ_0, kQ_1). The result is the path coalgebra $CoPath_k(Q)$. Indeed, for instance, $\alpha_1 \otimes \alpha_2$ is in $kQ_1 \odot kQ_1$ if and only if $t(\alpha_1) = s(\alpha_2)$.

8.6.13. The tensor algebra of a bimodule over an algebra. Let A be an associative (but not necessarily commutative) algebra over k and let M be a bimodule over A. Then one can form the tensor algebra of M over A. As a k-vector space

$$Ten_A(M) = M^{\otimes_A 0} \oplus M^{\otimes_A 1} \oplus \cdots \oplus M^{\otimes_A n} \oplus \cdots \tag{8.6.14}$$

where

$$M^{\otimes_A 0} = A, \quad M^{\otimes_A 1} = M, \quad M^{\otimes_A (n+1)} = M \otimes_A M^{\otimes_A n}.$$

There is a natural (and obvious) algebra structure on $Ten_A(M)$ given by

$$a(m_1 \otimes_A m_2 \otimes_A \cdots \otimes_A m_i) = (am_1) \otimes_A m_2 \otimes_A \cdots \otimes_A m_i$$
$$(m_1 \otimes_A m_2 \otimes_A \cdots \otimes_A m_i)a = m_1 \otimes_A m_2 \otimes_A \cdots \otimes_A (m_i a)$$
$$(m_1 \otimes_A m_2 \otimes_A \cdots \otimes_A m_i)(m_1' \otimes_A m_2' \otimes_A \cdots \otimes_A m_j') =$$
$$(m_1 \otimes_A m_2 \otimes_A \cdots \otimes_A m_i \otimes_A m_1' \otimes_A m_2' \otimes_A \cdots \otimes_A m_j')$$

and the unit is the unit of $A = M^{\otimes_A 0}$. The tensor algebra $Ten_A(M)$ has a universal property dual to the universal property of the cotensor algebra of theorem 8.6.10.

Theorem 8.6.15. *Let* $Ten_A(M) \xrightarrow{\varphi} B$ *be a morphism of associative algebras. Then this morphism is uniquely determined by its restrictions to* $A = M^{\otimes_A 0}$ *and* $M = M^{\otimes_A 1}$. *Conversely if* $\varphi_0 : A \to B$ *is a morphism of unital associative algebras (which also makes* B *an* A*-bimodule) and* $\varphi_1 : M \to B$ *is a morphism of* A*-bimodules, then there is a unique morphism of algebras* $Ten_A(M) \xrightarrow{\varphi} B$ *that restricts to the given* φ_0 *on* $A = M^{\otimes_A 0}$ *and the given* φ_1 *on* $M = M^{\otimes_A 1}$.

This is a freeness-like property, just like theorem 8.6.10, together with the sentence following it, is a cofreeness-like property.

8.6.16. Tensor algebra vs path algebra. Let $Q = (Q_0, Q_1)$ be a quiver. This time make kQ_0 an algebra by declaring the elements of Q_0 to be a full set of orthogonal idempotents and define a kQ_0-bimodule structure on kQ_1 by

$$v\alpha = \begin{cases} \alpha & \text{if } v = s(\alpha) \\ 0 & \text{if } v \neq s(\alpha) \end{cases}, \quad \alpha v = \begin{cases} \alpha & \text{if } v = t(\alpha) \\ 0 & \text{if } v \neq t(\alpha) \end{cases} \tag{8.6.17}$$

where $v \in Q_0, \alpha \in Q_1$.

Now apply the free algebra construction of 7.6.13 to this algebra-bimodule pair (Q_0, Q_1). The result is the path algebra of the quiver Q. Indeed view the set of paths $Path(Q)$ as a subset of the set of all words $Word(Q_1)$ on the alphabet Q_1. This gives an inclusion of length graded algebras

$$Path_k(Q) \longrightarrow kWord(Q_1) = \bigoplus_{n=0}^{\infty} kQ_1^{\otimes_k n}.$$

Compose this with the projection

$$\bigoplus_{n=0}^{\infty} kQ_1^{\otimes_k n} \longrightarrow \bigoplus_{n=0}^{\infty} kQ_1^{\otimes_{kQ_0} n} = Ten_{kQ_0}(kQ_1)$$

to obtain an isomorphism of graded algebras $Path_k(Q) \xrightarrow{\simeq} Ten_{kQ_0}(kQ_1)$. Indeed, consider for instance an arbitrary length two word $\alpha_1 \otimes \alpha_2$. Then

$$\begin{aligned} \alpha_1\alpha_2 \mapsto \alpha_1 \otimes_k \alpha_2 \mapsto \alpha_1 \otimes_{kQ_0} \alpha_2 &= \alpha_1 t(\alpha_1) \otimes_{kQ_0} \alpha_2 = \alpha_1 \otimes_{kQ_0} t(\alpha_1)\alpha_2 \\ &= \begin{cases} 0 & \text{if } t(\alpha_1) \neq s(\alpha_2) \\ \alpha_1 \otimes_{kQ_0} \alpha_2 & \text{if } t(\alpha_1) = s(\alpha_2). \end{cases} \end{aligned}$$

8.6.18. Bi-Hopf-modules. Let H be a Hopf algebra over k. A module M over k is a **bi-Hopf-module** over H (also called H-Hopf-bimodule) if it is both a bimodule over the algebra H and a bicomodule over the coalgebra H such that the following compatibility conditions (Hopf module conditions) are satisfied.

$$\begin{aligned} \gamma_{\text{left}}(hmh') &= \sum h_1 m_{-1} h'_1 \otimes h_2 m_0 h'_2, \\ \gamma_{\text{right}}(hmh') &= \sum h_1 m_0 h'_1 \otimes h_2 m_1 h'_2. \end{aligned} \tag{8.6.19}$$

Here the γ's denote the left and right comodule coactions and the module actions are denoted by simple juxtaposition and Sweedler notation is used

$$\gamma_{\text{left}}(m) = \sum m_{-1} \otimes m_0, \quad \gamma_{\text{right}}(m) = \sum m_0 \otimes m_1,$$

$$\mu_H(h) = \sum h_1 \otimes h_2.$$

The category of bi-Hopf-modules over H is denoted ${}^H_H\mathrm{Mod}^H_H$ where the two upper H-s stand for the left and right comodule structures and the two lower H-s for the left and right module structures.

Proposition 8.6.20. *Let H be a graded Hopf algebra (as always in this text, graded by $\mathbf{N} \cup \{0\}$). Then H_0 is a sub-Hopf-algebra and H_1 is a bi-Hopf-module over it.*

8.6.21. Hopf algebra structures on tensor algebras, [**539**], see also [**609**]; [**135**]. Now let H be a Hopf algebra and let M be a bi-Hopf-module over it. Then there is a (natural) graded Hopf algebra structure on $Ten_H(M)$ with a freeness (universality) property and $Ten_H(M)_0 = H$, $Ten_H(M)_1 = M$. This graded Hopf algebra is constructed as follows. Consider the morphisms

$$\mu : H \longrightarrow H \otimes_k H \stackrel{\subset}{\longrightarrow} Ten_H(M) \otimes_k Ten_H(M)$$

$$\gamma_{\text{left}} \oplus \gamma_{\text{right}} : M \longrightarrow (H \otimes_k M) \oplus (M \otimes_k H) \stackrel{\subset}{\longrightarrow} Ten_H(M) \otimes_k Ten_H(M).$$

It is immediate that $Ten_H(M) \otimes_k Ten_H(M)$ is an H-bi-Hopf-module via the first of these two maps, which is an algebra morphism, and that the second one is a morphism of H-bi-Hopf-modules. It follows, by the universality property of $Ten_H(M)$, that there is a morphism of algebras

$$\mu : Ten_H(M) \longrightarrow Ten_H(M) \otimes_k Ten_H(M).$$

Similarly (but easier) one finds an algebra morphism

$$\varepsilon : k \longrightarrow Ten_H(M).$$

These two morphisms make the algebra $Ten_H(M)$ into a bialgebra. It remains to construct an antipode. This can be done in several ways. One is to use the following lemma from [**658**], see [**520**], lemma 5.2.10, p. 64.

8.6.22. Let C be a coalgebra and A an algebra (over k). Then $f \in \mathrm{Hom}(C, A)$ is convolution invertible if and only if its restriction to C_0 is convolution invertible. Here C_0 is the coradical. The proof of the lemma, which is neither long nor difficult, uses the fact that things are over a field and uses the coradical filtration (see 2.14). To see that the lemma applies note that the coradical of $Ten_H(M)$ (as a coalgebra) is contained in $H = Ten_H(M)_0$ which can be seen by noting that the intersection of two sub coalgebras of a coalgebra is a subcoalgebra (which again uses that one is working over a field).

8.6.23. Hopf algebra structures on cotensor algebras, [**539**], see also [**609**]; [**135**]. Again let H be a Hopf algebra and let M be a bi-Hopf-module over it. Then there is a (natural) graded Hopf algebra structure on $CoTen_H(M)$ with a cofreeness (universality) property and $CoTen_H(M)_0 = H$, $CoTen_H(M)_1 = M$. This graded Hopf algebra is constructed as follows.

Consider the coalgebra morphism

$$m_0 : CoTen_H(M) \otimes_k CoTen_H(M) \longrightarrow H$$

obtained by first projecting onto the zero components and then composing with

$$m_H : H \otimes_k H \longrightarrow H.$$

Further, consider the morphism of bi-Hopf-modules

$$m_1 : CoTen_H(M) \otimes_k CoTen_H(M) \longrightarrow M$$

determined by

$$m_1 : (CoTen_H(M) \otimes_k CoTcn_H(M))_1 = (H \otimes M) \oplus (M \otimes H) \longrightarrow M$$
$$(h \otimes n) \oplus (n' \otimes h') \mapsto hn + n'h'.$$

By the universality (cofreeness) property of $CoTen_H(M)$ these two morphisms determine a unique coalgebra morphism

$$m : CoTen_H(M) \otimes_k CoTen_H(M) \longrightarrow CoTen_H(M).$$

Further $k \longrightarrow CoTen_H(M)$, $1 \mapsto 1_H$ defines a unit morphism and it is not difficult to check that the result is a bialgebra structure on $CoTen_H(M)$. An antipode is obtained in the same way as in the case of $Ten_H(M)$ above in 8.6.18, so that all in all this gives a Hopf algebra structure on $CoTen_H(M)$.

8.6.24. Symmetric function Hopf algebra determined by a pair (H, M) consisting of a Hopf algebra H and a bi-Hopf-module M over it.

Define $\mathbf{Symm}_H(M)$ as the subalgebra of the Hopf algebra $CoTen_H(M)$ generated by H and M. It is clear that this is a sub-Hopf-algebra.

8.6.25. Duality, see [**135**]. Given appropriate compatible nondegenerate pairings

$$H \times H' \longrightarrow k, \quad M \times M' \longrightarrow k$$

there is a unique nondegenerate graded pairing

$$CoTen_H(M) \times CoTen_{H'}(M') \longrightarrow k$$

that extends the given ones. If, moreover, $H = H'$, $M = M'$, the graded Hopf algebra $\mathbf{Symm}_H(M)$ is graded selfdual.

Examples 8.6.26. Let I be an index set. Then there is the important Hopf algebra $LieHopf_k(I)$ which as an algebra is the free associative algebra $k\langle U_i : i \in I\rangle$ and whose comultiplication is induced by

$$U_i \mapsto 1 \otimes U_i + U_i \otimes 1.$$

There is also the equally important Hopf algebra $\mathit{Shuffle}_k(I)$ with as basis all words in the alphabeth I, cut as comultiplication, and shuffle as multiplication.

These two are graded dual to each other where the grading is in both cases is given by length. There are also other gradings possible; for instance by assigning to each i and corresponding U_i a weight and grade things by giving each word or monomial the sum of the weights of their components.

Now consider the quiver $Q = (Q_0, Q_1)$ where Q_0 is a single vertex and is a set of loops indexed by the index set I.

In this particular case $k = kQ_0$ is the trivial Hopf algebra and $k = kQ_1$ is a bi-Hopf module over it. Thus the constructions of 8.6.21 and 8.6.23 apply. One finds for this quiver

$$Path_k(Q) = LieHopf_k(I), \quad CoPath_k(Q) = \mathit{Shuffle}_k(I).$$

8.6.27. Hopf quiver. In general it is not true that the pair (kQ_0, kQ_1) coming from a quiver Q by the constructions of 8.6.13 and 8.6.16 is a pair consisting of a

Hopf algebra and a bi-Hopf-module over it. Indeed, kQ_0 is in fact a bialgebra, but it has an antipode if and only if there is only a single vertex (see 3.3.8). Worse, kQ_1 is a bi-Hopf-module over kQ_0 if and only if every arrow is a loop. So this gives back the examples in 8.6.26 above.

Thus other constructions are called for. These can only be done for very special quivers: Hopf quivers. These are formed as follows. Let G be a group. A ramification datum[19] on G is simply a map from the conjugacy classes $Conj(G)$ of G to $\mathbf{N} \cup \{0\}$.

$$\chi : G \longrightarrow \mathbf{N} \cup \{0\}.$$

Actually more general ramification data can be (and have been) considered, leading to quivers with possibly infinitely many arrows between two vertices, see [**712**].)

Now let (G, χ) be a group with ramification datum. Associated to this pair is the following quiver. The vertices are the elements of the group G, and for every vertex g and element $u \in C \in Conj(G)$ there are $\chi(C)$ arrows from g to ug.

The quivers obtained in this way are called **Hopf quivers**. A Hopf quiver is called **Schurian** if it comes from a pair (G, χ) with $\chi(C) \in \{0, 1\}$ for all $C \in Conj(G)$. The notion of a Hopf quiver is due to Cibils and Rosso, [**146**].

Example 8.6.28. Consider the cyclic group of order n and choose a generator g, so that the group is $\{g, g^2, \ldots, g^{n-1}, g^n = e\}$. Take the Schurian ramification datum

$$\chi(g) = 1, \chi(g^m) = 0 \quad \text{for all other elements of the group.}$$

The associated Hopf quiver is called the fundamental cycle of size n. It consists of n vertices numbered $0, 1, 2, \ldots, n-1$ with for each i precisely one arrow from i to $i+1$, where these numbers are to be taken modulo n. This example is denoted Z_n. Pictorially one has a regular oriented n-gon. The example for $n = 6$ is shown in figure 8.6.1.

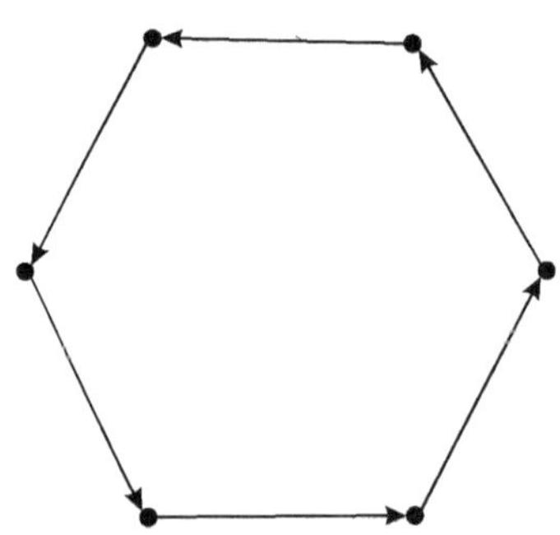

Figure 8.6.1. Fundamental cycle of size 6.

This is actually the example that started the whole business of Hopf algebra structures on path algebras and path coalgebras and related matters, [**145**]. In loc. cit. Cibils determines all Hopf algebra structures on the path algebra $Path_k(Z_n)$.

[19]In the published literature one mostly finds 'a ramification data'. This is wrong and grates upon all who know a little bit of Latin. 'Data' is a plural and so one cannot speak of 'a data'.

The central theorem is the following.

Theorem 8.6.29. [**147**], see also [**677**]. *Let Q be a quiver. Then the following are equivalent*

(i) *Q is a Hopf quiver*

(ii) *Q_0 is a group and kQ_1 is a bi-Hopf module over kQ_0 with the left and right comodule structures given by $\gamma_{\text{left}}(\alpha) = s(\alpha) \otimes \alpha$, $\gamma_{\text{right}}(\alpha) = \alpha \otimes t(\alpha)$ (which are the same comodule structures as used in* (8.6.12)).

(iii) *$CoPath_k(Q)$ admits a graded Hopf algebra structure with length grading.*

The fact that the grading involved is the length grading is quite important and it is mostly an open problem what happens if other gradings are allowed. Note that nothing is said about uniqueness of a Hopf algebra structure on $CoPath_k(Q)$. There can in fact be several different ones; see below.

A partial dual of this result, for path algebras rather then path coalgebras, was obtained earlier, [**146**].

Example 8.6.30. The path Hopf algebra $CoPathHopf_k(Z_n)(q)$. Consider again the fundamental cycle Z_n of length n with vertex set $\{g, g^2, \ldots, g^{n-1}, g^n = e\}$. Let p_i^l denote the (unique) path that starts at i and is of length l. Thus $p_i^0 = g^i$ and $p_i^1 = \alpha_i$, the arrow from g^i to g^{i+1}. Then for each root of unity q Cibils and Rosso, [**147**], define a Hopf algebra structure on $CoPath_k(Z_n)$ by the multiplication

$$p_i^l p_j^m = q^{jl} \binom{l+m}{l}_q p_{i+j}^{l+m}. \tag{8.6.31}$$

Here

$$\binom{l+m}{l}_q = \frac{(l+m)!_q}{l!_q m!_q} \quad \text{the Gaussian binomial coefficient} \tag{8.6.32}$$

$$m!_q = 1_q 2_q \cdots m_q, \quad i_q = 1 + q + \cdots + q^{i-1}.$$

(If $l = 0$ or $m = 0$ (or both), the Gaussian coefficient (8.6.32) is taken to be 1.)

There is also an explicit formula for the antipode, viz

$$\iota(p_i^l) = (-1)^l q^{-l(l+1)/2-il} p_{n-l-i}^l.$$

Example 8.6.33. The path Hopf algebra $CoPathHopf_k^d(Z_n)(q)$. Now consider the finite dimensional subcoalgebra $CoPath_k(Z_n)$ of the path coalgebra consisting of all paths of length strictly less than $d \geq 2$. If q is an n-th root of unity in k of (multiplicative) order d (same d), the binomial coefficient (8.6.32) is zero for $l, m \leq d-1$ and $l + m \geq d$. It now follows from the defining formula (8.6.31) that these paths of length smaller than d form a sub Hopf algebra of $CoPathHopf_k(Z_n)(q)$. This sub Hopf algebra is (here) denoted $CoPathHopf_k^d(Z_n)(q)$.

Example 8.6.34. Hopf algebra structures on the path coalgebra $CoPath_k(Z_\infty)$. It seems logical to use the notation Z_∞ for the Schurian Hopf quiver whose underlying group is the group of integers and which has precisely one arrow from i to $i+1$ for all $i \in \mathbf{Z}$. In [**147**]; [**677**] the notation A_∞^∞ is used.

For any $q \neq 0$ in k (root of unity or not) formula (H7.6.29) defines a Hopf algebra structure on the path coalgebra $CoPath_k(Z_\infty)$. This Hopf algebra is denoted $CoPathHopf_k(Z_\infty)(q)$.

If q is a root of unity of order d the same remark as above shows that the paths of lengths strictly less than d form a sub Hopf algebra of $CoPathHopf_k(Z_\infty)(q)$ which will be denoted $CoPathHopf_k^d(Z_\infty)(q)$. This is of course an infinite dimensional Hopf algebra.

8.6.35. Simple-pointed Hopf algebra. A Hopf algebra H is simple pointed if it is pointed, not cocommutative, and all proper sub Hopf algebras are contained in the coradical H_0. The notion was introduced by Radford, [**576**] and he classified all finite dimensional simple-pointed Hopf algebras over an algebraically closed field, loc. cit.

Theorem 8.6.36. [**677**]. *Let Q be a Schurian quiver and let H be a sub Hopf algebra of a length graded Hopf algebra structure on the path coalgebra $CoPath_k(Q)$, and let* $\mathrm{char}(k) = 0$. *Then H is simple-pointed if and only if as a Hopf algebra it is isomorphic to one of the following*

(i) $CoPathHopf_k^d(Z_n)(q)$ *for q a root of unity of multiplicative order $d \geq 2$.*
(ii) $CoPathHopf_k(Z_n)(1)$ *for some $n \geq 1$*
(iii) $CoPathHopf_k(Z_\infty)(q)$ *for q not a root of unity (and unequal to zero)*
(iv) $CoPathHopf_k^d(Z_\infty)(q)$ *for q a primitive d-th root of unity and $d \geq 2$*
(v) $CoPathHopf_k(Z_\infty)(1)$

8.6.37. More classification and related results. There are quite a few more classification and other results linked to what one may loosely call the path coalgebra approach to Hopf algebras.

For instance in [**133**] there are the following results. The path coalgebra $CoPath_k^d(Z_n)$ admits a Hopf algebra structure if and only if d divides n; the graded ones are the generalized Taft algebras $A_{n,d}(q)$, while the not necessarily length graded ones are the $A(n, d. \mu, q)$ studied (using other techniques) by Radford, [**573**] and Andruskiewitch-Schneider, [**18**].

An algebra is monomial if it is the quotient of a path algebra by a so-called admissible ideal. These are ideals I generated by a set of paths such that $J^2 \subseteq I \subseteq J^m$ where J is the ideal of all paths of length ≥ 1. This notion generalizes the older idea of a monomial quotient of a free algebra $k\langle X_i : i \in I\rangle$.

Dually a subcoalgebra C of a path coalgebra $CoPath_k(Q)$ is monomial if $Q_0, Q_1 \subset C$, $C \subseteq CoPath_k^m(Q)$ for some m and C has a basis consisting of paths. It turns out, also [**133**], that there is a one to one Galois type correspondence between isomorphism classes of nonsemisimple monomial Hopf algebras and isomorphism classes of ramification data on groups.

The techniques and results on 'path Hopf algebras' relate with such things as the Gabriel theorem on elementary algebras. An algebra A is basic if its quotient A/R is a product of division algebras where R is the Jacobson radical of A; it is elementary if this quotient is a product of copies of k. This notion is dual to that of a pointed coalgebra. The Gabriel theorem alluded to says that an elementary finite dimensional algebra is the quotient of the path algebra of a unique quiver $Q(A)$ by an admissible ideal. The quiver $Q(A)$ is called the Gabriel quiver[20] of A. For definitions, see [**318**], p.262ff; [**40**], p.59ff; [**43**], section III.1.

[20] This quiver is called the 'ordinary quiver of A' in [**40**].

There is a dual notion due to Chin and Montgomery, [**142**], and a corresponding dual Gabriel theorem.

For more on all this see among others [**450**]; [**134**].

There are also various relations with quantum groups starting with the result of Rosso to the effect that the upper triangular part, U_q^+, of the quantized enveloping algebra associated to a symmetrizable Cartan matrix is isomorphic (as a Hopf algebra) to the subalgebra generated by the elements of degree 0 and 1 of the cotensor Hopf algebra (path Hopf algebra) associated with a suitable bi-Hopf-module over the group algebra of $\mathbf{Z}^n$, [**609**]. That is, it is a symmetric Hopf algebra in the sense of 8.6.24 above. See e.g. [**289**]; [**320**] for more material.

The 'paths in quivers' techniques outlined in this section supplement other techniques, such as the Andruskiewitch-Schneider lifting approach, for pointed Hopf algebras and immediate generalizations. And sometimes they intermingle. A selection of references not already mentioned is: [**18**]; [**19**]; [**20**]; [**21**]; [**22**]; [**23**]; [**24**]; [**25**]; [**26**]; [**27**]; [**28**]; [**289**]; [**319**]; [**320**]; [**321**]; [**333**]; [**334**]; [**574**]; [**576**]; [**621**]; [**642**]; [**715**]; [**716**]; [**713**]; [**714**]; [**708**]; [**712**].

8.7. Ringel-Hall Hopf algebras

The **classical Hall-Steinitz algebras of partitions**, [**649**]; [**307**], are concerned with finite length modules over complete discrete valuation rings like $\mathbf{Z}_p$, the p-adic integers, the rings of integers of finite extensions of $\mathbf{Q}_p$ and $k[[T]]$, the ring of formal power series over a finite field k. Such a module is a direct sum of finite monogenic (also called cyclic) modules. For instance in the case of $\mathbf{Z}_p$ such a module is (isomorphic to) a direct sum

$$M \simeq \mathbf{Z}/(p^{\lambda_1}) \oplus \mathbf{Z}/(p^{\lambda_2}) \oplus \cdots \oplus \mathbf{Z}/(p^{\lambda_r}) \tag{8.7.1}$$

and it may as well be assumed that $\lambda_1 \geq \lambda_2 \geq \cdots \geq \lambda_r > 0$. Thus a module of length n (which is also the number of terms $\neq 0$ in a composition series of M) is classified by a partition of n and all partitions occur. The partition $\lambda_1 \geq \lambda_2 \geq \cdots \geq \lambda_r > 0$ is said to be the type of M.

There is not really much more to be said about these modules themselves.

Things change when one looks at extensions and starts counting subgroups, N, of a given type μ with prescribed type of the quotient M/N. The type of this quotient is called the cotype of N in M.

Note that if

$$0 \longrightarrow N \longrightarrow M \longrightarrow M/N \longrightarrow 0 \tag{8.7.2}$$

is an exact sequence

$$\mathit{length}(M) = \mathit{length}(N) + \mathit{length}(M/N). \tag{8.7.3}$$

Let A be a complete discrete valuation ring. For given types λ, μ, ν take the corresponding module M of type λ over the complete discrete valuation ring and define

$$G^{\lambda}_{\mu,\nu}(A) = \text{number of submodules } N \text{ of } M \text{ of type } \mu \text{ and cotype } \nu. \tag{8.7.4}$$

To get a first feeling for the phenomena here are some very simple examples.

Example 8.7.5. Take $A = \mathbf{Z}_p$, $\lambda = (1^2)$, $\mu = \nu = (1)$. So M can be taken to be equal to

$$M = \mathbf{Z}/(p) \oplus \mathbf{Z}/(p)$$

and the simple task is to count the number of cyclic subgroups of order p of M. (It is automatic in this case that the cotype of such a submodule is (1).) Every element $\neq (0,0)$ generates such a subgroup and two such subgroups either coincide or intersect only in $(0,0)$. Thus these subgroups partition[21] the $p^2 - 1$ nonzero elements of M into blocks of size $p-1$ and so there are $p+1$ subgroups N; i.e.

$$G^{(1^2)}_{(1),(1)}(\mathbf{Z}_p) = p+1.$$

Example 8.7.6. Take $A = \mathbf{Z}_p$, $\lambda = (2)$, $\mu = \nu = (1)$. So M can be taken to be equal to $\mathbf{Z}/(p^2)$. This time there is just one subgroup of order p, viz $p\mathbf{Z}/(p^2)$ and so

$$G^{(2)}_{(1),(1)}(\mathbf{Z}_p) = 1. \tag{8.7.7}$$

Example 8.7.8. Take $A = \mathbf{Z}_p$, $\lambda = (2,1)$, $\mu = (1)$, $\nu = (1^2)$. So M can be taken to be equal to

$$M = \mathbf{Z}/(p^2) \oplus \mathbf{Z}/(p). \tag{8.7.9}$$

Each subgroup of order p is generated by a nonzero element of the form (pa, b) and each such element generates a subgroup of order p. There are $p^2 - 1$ such elements, and hence as in example 8.7.5 there are $(p+1)$ subgroups of order p. It is easy to check that for each subgroup generated by a (pa, b) with $b \neq 0 \pmod p$ the cotype is (2). Further the cotype of the subgroup generated by a $(pa, 0)$ with $a \neq 0 \pmod p$ is obviously $(1,1)$. Thus

$$G^{(2,1)}_{(1),(1^2)}(\mathbf{Z}_p) = 1, \quad G^{(2,1)}_{(1),(2)}(\mathbf{Z}_p) = p. \tag{8.7.10}$$

Example 8.7.11. Take $A = \mathbf{Z}_p$, $\lambda = (3)$, $\mu = (1)$. So M can be taken to be equal to $M = \mathbf{Z}/(p^3)$. There is just one subgroup of order p viz $p^2\mathbf{Z}/(p^3)$and it has cotype (2). Thus

$$G^{(3)}_{(1),(2)}(\mathbf{Z}_p) = 1, \quad G^{(3)}_{(1),(1^2)}(\mathbf{Z}_p) = 0. \tag{8.7.12}$$

Example 8.7.13. Take $A = \mathbf{Z}_p$, $\lambda = (1^3)$, $\mu = (1)$. So M can be taken to be equal to $M = \mathbf{Z}/(p) \oplus \mathbf{Z}/(p) \oplus \mathbf{Z}/(p)$. By the same arguments as before there are $p^2 + p + 1$ different subgroups of order p. All of them have cotype (1^2). Thus

$$G^{(1^3)}_{(1),(1^2)}(\mathbf{Z}_p) = p^2 + p + 1, \quad G^{(1^3)}_{(1),(2)}(\mathbf{Z}_p) = 0. \tag{8.7.14}$$

Remark 8.7.15. The same formulas result, but with q replacing p, when one does these simple calculations for finite length modules over a complete discrete valuation ring with residue field k of q elements.

8.7.16. Duality. Let A be a complete discrete valuation ring with residue field k of q elements and let π be a uniformizing element, i.e. a generator of the maximal ideal of A. Then multiplication with π^r gives an imbedding $A/(\pi^s) \longrightarrow A/(\pi^{r+s})$

[21]Different meaning of the word partition.

which identifies $A/(\pi^s)$ with $\pi^r A/(\pi^{r+s})$. Take the injective limit E_A. This is the injective hull of k in the category of A-modules.

The dual of a finite length A-module is the module $M^* = \mathrm{Mod}_A(M, E_A)$. It is a fairly simple matter, using that the functor $\mathrm{Mod}_A(-, E_A)$ commutes with direct sums, to show that as A-modules M and M^* are isomorphic[22]. So, as E_A is injective an exact sequence

$$0 \longrightarrow N \longrightarrow M \longrightarrow M/N \longrightarrow 0 \tag{8.7.17}$$

gives rise to a dual exact sequence

$$0 \longrightarrow (M/N)^* \longrightarrow M^* \longrightarrow N^* \longrightarrow 0. \tag{8.7.18}$$

Also the canonical double dual morphism

$$M \longrightarrow M^{**}, \quad m \mapsto (\varphi_m : f \mapsto f(m)) \tag{8.7.19}$$

is an isomorphism.

8.7.20. Multiplication. Now define a graded multiplication on the free graded Abelian group with all partitions as basis as follows. For two partitions μ, ν define the multiplication (depending on A) by

$$\mu \cdot_A \nu = \sum_\lambda G^\lambda_{\mu,\nu}(A)\lambda. \tag{8.7.21}$$

Examples 8.7.22. From examples 8.7.5, 8.7.6, 8.7.8, 8.7.11, 8.7.13 one sees that

$$\begin{aligned} &(1) \cdot_{\mathbf{Z}_p} (1) = (1+p)(1^2) + (2), \\ &(1) \cdot_{\mathbf{Z}_p} (1^2) = (2,1) + (1+p+p^2)(1^3) \\ &(1) \cdot_{\mathbf{Z}_p} (2) = (3) + p(2,1). \end{aligned} \tag{8.7.23}$$

These formulas could already suggest (if one is of a sufficiently optimistic nature) that

(a) These partition algebras are freely generated (over the integers) by the partitions (1^r), $r = 1, 2, 3, \ldots$.

(b) The coefficients involved are polynomials in q, the number of elements in the residue field of A.

Both are true.

8.7.24. Commutativity. The multiplication (8.7.21) is commutative. This follows from duality (see 8.7.16 above): each subgroup of type μ and cotype ν of M gives rise to a subgroup $(M/N)^*$ of $M^* \simeq M$ and vice versa (by double duality).

8.7.25. Associativity. The multiplication (8.7.21) is also associative. This is seen by observing that a length 2 filtration of a module M

$$0 \subset N_1 \subset N_2 \subset M$$

can be obtained in two ways.

First take a proper subgroup N_2 of M and then take a proper subgroup of N_2.

Second, take a proper subgroup N_1 of M, then take a proper subgroup $\overline{N}_2$ of M/N_1 and take N_2 to be the inverse image of $\overline{N}_2$ under the projection $M \longrightarrow M/N_1$.

---

[22] This is rather similar to the fact that for finite Abelian groups $\mathbf{Ab}(M, \mathbf{Q}/\mathbf{Z}) \simeq M$

The first way gives a contribution from $(\rho \cdot \sigma) \cdot \tau$ where ρ is the type of N_1, σ is the type of N_2/N_1 and τ is the type of M/N_2.

The second way corresponds to a contribution from $\rho \cdot (\sigma \cdot \tau)$.

Quite generally for a given M of type λ each filtration of M with prescribed types of the filtration quotients as indicated in (8.7.26) below.

$$0 \underset{\rho_1}{\subset} N_1 \underset{\rho_2}{\subset} N_2 \subset \cdots \subset N_{r-1} \underset{\rho_r}{\subset} N_r = M \tag{8.7.26}$$

contributes 1 to the coefficient of λ in the product $\rho_1 \cdot \rho_2 \cdot \quad \cdots \quad \cdot \rho_r$.

8.7.27. Hall-Steinitz algebra. The free Abelian group with as basis all partitions and the multiplication (8.7.21) is called the Hall-Steinitz algebra (or Hall algebra[23]) and denoted $\mathrm{HS}(A)$ (or $\mathrm{H}(A)$). The empty partition, which corresponds to the zero module serves as the unit element. They are all isomorphic (over the integers) to the free commutative (unital associative) algebra in a countable infinity of indeterminates.

So they are all isomorphic to **Symm** and the Hall-Steinitz algebras thus also belong in the list of incarnations of the symmetric functions in section 4.12 above.

It is also true that there exist polynomials $g^{\lambda}_{\mu,\nu}(t)$, the Hall polynomials, such that

$$G^{\lambda}_{\mu,\nu}(A) = g^{\lambda}_{\mu,\nu}(q) \tag{8.7.28}$$

where q is the number of elements of the residue field k of A.

For proofs of these statements see e.g. [**472**], section II.1, or [**610**].

8.7.29. Hopf algebra structure. The algebra **Symm** carries for instance the graded Hopf algebra structure

$$e_n \mapsto \sum_{i=0}^{n} e_i \otimes e_{n-i} \tag{8.7.30}$$

and the question arises whether this coproduct has a good interpretation in terms of finite A-modules. This is almost the case.

8.7.31. Counting exact sequences. For each partition λ let $M(\lambda)$ be the standard module of this type. I.e.

$$M(\lambda) = A/(\pi^{\lambda_1}) \oplus A/(\pi^{\lambda_2}) \oplus \cdots \oplus A/(\pi^{\lambda_r}). \tag{8.7.32}$$

Let $P^{\lambda}_{\mu,\nu}$ be the number of exact sequences

$$0 \longrightarrow M(\mu) \xrightarrow{\alpha} M(\lambda) \xrightarrow{\beta} M(\nu) \longrightarrow 0 \tag{8.7.33}$$

i.e. the number of pairs (α, β) consisting of an injection α and a surjection β such that $\mathrm{Im}(\alpha) = \mathrm{Ker}(\beta)$. Counting such exact sequences is different from counting subgroups of given type and cotype. Indeed two different injections

$$M(\mu) \xrightarrow{\alpha,\alpha'} M(\lambda)$$

may well have the same image N; the difference is an automorphism of $M(\mu)$, viz the automorphism

$$M(\mu) \xrightarrow{\alpha'} N \xrightarrow{\alpha^{-1}} M(\mu).$$

[23]The term 'Hall algebra' is also used to refer to free nilpotent Lie algebras.

Similarly two possibly different surjections $M(\lambda) \xrightarrow{\beta,\beta'} M(\nu)$ may well have the same kernel and the difference is an automorphism of $M(\nu)$. (One can also use duality to see this.) It follows that

$$P^{\lambda}_{\mu,\nu}(A) = G^{\lambda}_{\mu,\nu}(A)a_{\mu}a_{\nu} \tag{8.7.34}$$

where a_{μ} is the number of automorphisms of $M(\mu)$.

It is a nice simple exercise to verify this for the simplest case

$$0 \longrightarrow \mathbf{Z}/(p) \xrightarrow{\alpha} \mathbf{Z}/(p) \oplus \mathbf{Z}/(p) \xrightarrow{\beta} \mathbf{Z}/(p) \longrightarrow 0. \tag{8.7.35}$$

An injection in (8.7.35) is given by $\alpha(1) = (a,b) \neq (0,0)$; a surjection β by $\beta(1,0) = c$, $\beta(0,1) = d$, not both zero. The condition for exactness is $ac + bd = 0$. All calculations are in $\mathbf{Z}/(p)$. If $a \neq 0$, $c = a^{-1}bd$. Then if $c = 0$, $d \neq 0$ and $b = 0$. There are thus $(p-1)^2$ of these exact sequences. If $a \neq 0$ and $c \neq 0$ and both b and d are unequal to zero one determines the other. Thus there are $(p-1)^3$ of such exact sequences. Finally, if $a = 0$, $b \neq 0$, and so $d = 0$, giving an additional $(p-1)^2$ exact sequences. So in total there are $(p-1)^2 + (p-1)^3 + (p-1)^2 = (p-1)^2(p+1)$ exact sequences (8.7.35). As $a_{(1)} = p-1$ and $G^{\lambda}_{\mu,\nu}(\mathbf{Z}_p) = p+1$ this verifies formula (8.7.34) in this simplest case.

8.7.36. Automorphisms. There is a general polynomial formula for the number a_{λ}, of automorphisms of $M(\lambda)$. It is

$$a_{\lambda} = q^{\mathrm{wt}(\lambda)+2n(\lambda)} \prod_{i \geq 1} \varphi_{m_i(\lambda)}(q^{-1}) \tag{8.7.37}$$

where $\mathrm{wt}(\lambda) = \lambda_1 + \lambda_2 + \cdots = \mathrm{length}(M(\lambda))$, $n(\lambda) = \sum_i (i-1)\lambda_i$, $m_i(\lambda)$ is the number of indices j such that $\lambda_j = i$ and $\varphi_m(t) = (1-t)(1-t^2)\cdots(1-t^m)$.

Again it is a nice little exercise to verify this in one of the simplest cases. Take $\lambda = (1^2)$, so that over $\mathbf{Z}_p$ $M(\lambda) = \mathbf{Z}/(p) \oplus \mathbf{Z}/(p)$. Then

$$\mathrm{wt}(\lambda) = 2,\ n(\lambda) = 1,\ m_1(\lambda) = 2,\ m_i(\lambda) = 0 \text{ for } i \geq 2$$

and the formula gives $a_{(1^2)} = p^4(1-p^{-1})(1-p^{-2})$. On the other hand an automorphism of this $M(\lambda)$ is given by $(1,0) \mapsto (a,b) \neq (0,0)$ and $(0,1) \mapsto (c,d)$ where (c,d) is not a multiple of (a,b). So there are $(p^2-1)(p^2-p)$ automorphisms. This fits.

8.7.38. Hopf algebra structure on the Hall-Steinitz algebra. Now define a (possible) comultiplication on $\mathrm{HS}(A)$ over the rational numbers as follows.

$$\mu_{\mathrm{HS}(A)_{\mathbf{Q}}}(\lambda) = \sum_{\mu,\nu} \frac{P^{\lambda}_{\mu,\nu}}{a_{\lambda}} \mu \otimes \nu. \tag{8.7.39}$$

For example, using some of the calculations above,

$$\mu_{\mathrm{HS}(\mathbf{Z}_p)_{\mathbf{Q}}}(1^2) = (\ ) \otimes (1^2) + \frac{(p-1)^2(p+1)}{(p^2-1)(p^2-p)}(1) \otimes (1) + (1^2) \otimes (\ )$$
$$= (\ ) \otimes (1^2) + p^{-1}(1) \otimes (1) + (1^2) \otimes (\ )$$

showing that denominators really occur. For the partitions (1^n) this works out as

$$(1^n) \mapsto \sum_{i=0}^{n} q^{-i(n-i)}(1^i) \otimes (1^{n-i}) \tag{8.7.40}$$

(which fits with the example above).

Theorem 8.7.41. *The algebra isomorphism*

$$(1^n) \mapsto q^{-n(n-1)/2} e_n : \mathrm{HS}(A)_{\mathbf{Q}} \longrightarrow \mathbf{Symm}_{\mathbf{Q}} \tag{8.7.42}$$

where e_n *is the* n*-th elementary symmetric function, takes the comultiplication* (8.7.39) *into the comultiplication* $e_n \mapsto \sum_{i=0}^{n} e_i \otimes e_{n-i}$ *of* $\mathbf{Symm}_Q$ *and hence* $\mathrm{HS}(A)_Q$ *is a Hopf algebra isomorphic to* $\mathbf{Symm}_Q$.

The proof that (8.7.42) takes the one comultiplication into the other is simply the observation that

$$i(n-i) + i(i-1)/2 + (n-i)(n-i-1)/2 = n(n-1)/2.$$

Of course the game of counting subobjects and filtrations can be played in much more general contexts: any k-linear (Abelian) category with a sufficiency of finiteness conditions will do, see e.g. [**624**], lecture 1. This gives rise to what are called Ringel-Hall algebras and that is the theme that will be taken up next.

The outline below will concentrate on the case of modules over an hereditary algebra. This includes the case of representations of finite connected acyclic quivers, or, equivalently, modules over their path algebras; see below. Another source of hereditary algebras is provided by coherent sheafs over an algebraic curve. Those will not be discussed here, but see [**50**]; [**75**]; [**106**]; [**364**]; [**589**]; [**623**]; [**624**]; [**625**].

Automorphic forms and motives form still another source of Ringel-Hall algebras, see [**364**] and [**359**] respectively.

8.7.43. Hereditary algebras. Recall that an algebra over a field k is right (resp. left) hereditary if every right (resp. left) ideal is projective; it is **hereditary** if it is both left and right hereditary, see e.g. [**317**], section 5.5.

Also recall that an algebra over a field k is **basic** iff the quotient $R/J(R)$ is a product of division algebras; it is **split** if this quotient is a product of full matrix rings, $R/J(R) = M_{n_1}(k) \times \cdots \times M_{n_s}(k)$ and it is split basic if $R/J(R) = k \times \cdots \times k$. Here $J(R)$ is the Jacobson radical of R, see e.g. [**317**], p. 262.

Theorem 8.7.44. [**40**], section VII.1, theorem 1.7, p. 248.

(a) *If* Q *is a finite, connected, and acyclic quiver, then the path algebra* kQ *is hereditary and the Gabriel quiver of it is equal to* Q.

(b) *If* A *is a split basic, connected hereditary algebra, then*

(i) *the Gabriel quiver,* $Q(A)$*, of it is finite connected and acyclic, and*

(ii) A *and* $kQ(A)$ *are isomorphic.*

Recall that an algebra is connected if it is not the direct product of two proper subalgebras, i.e. if 0 and 1 are the only central idempotents. For the definition of the **Gabriel quiver** (also called 'ordinary quiver') of an algebra see e.g. [**40**], section II.3. and also [**317**], chapter 11.

8.7.45. Ringel-Hall algebra of an hereditary algebra over a finite field. Let R be an hereditary algebra over a finite field k. Let ${}_R\mathrm{FinMod}$ denote the category of left R-modules that are finite as sets. The cardinality of such a module, X, is denoted $|X|$. It is also assumed that R is finitary in the sense that $|\mathrm{Ext}^1_R(S,S')|$ is finite for all simple modules from ${}_R\mathrm{FinMod}$. This is automatic if R is finitely generated over k but for instance does not hold for the polynomial algebra in countable many generators $k[X_1, X_2, \ldots]$.

$\mathcal{P}$ denotes the set of isomorphism classes of modules in ${}_R\mathrm{FinMod}$. The class of a module M is denoted $[M]$ and if $\lambda \in \mathcal{P}$ is an isomorphism class, u_λ denotes a (fixed) choice of an element in that isomorphism class.

For $\alpha, \beta, \lambda \in \mathcal{P}$ let $G^\lambda_{\alpha,\beta}$ be the number of submodules X of u_λ such that $[X] = \alpha$, $[u_\lambda/X] = \beta$. The Hall algebra of R is now defined like before by the multiplication[24]

$$\alpha \bullet \beta = \sum_\lambda G^\lambda_{\alpha,\beta}\lambda. \tag{8.7.46}$$

More precisely, the Hall algebra $H(R)$ is the free Abelian group with basis $\mathcal{P}$, and multiplication defined by (8.7.46). This is associative as is again seen by looking at filtrations and there is a unit element viz the class of the zero module.

In contrast with the case of the classical Hall-Steinitz algebra $H(R)$ is not necessarily commutative. Noncommutativity happens for instance for $R = \begin{pmatrix} k & k \\ 0 & k \end{pmatrix}$, the path algebra of the quiver of two vertices and one arrow between them.

Let Δ be one of the Dynkin diagrams (graphs) of type A, D, or E, let $\vec{\Delta}$ be any quiver with underlying graph Δ, and let $\Lambda = k\vec{\Delta}$ be the path algebra of $\vec{\Delta}$. Let u_i be the isomorphism class of the simple (irreducible) representation that has k at the vertex i and zero everywhere else. Then for a suitable numbering of the vertices (which always exists) there are the following relations. Let $i < j$, then

If there is no arrow from i to j, $u_i \bullet u_j = u_j \bullet u_i$

If there is an arrow from i to j

$$\begin{aligned} u_i^{\bullet 2} \bullet u_j - (q+1)u_i \bullet u_j \bullet u_i + qu_j \bullet u_i^{\bullet 2} &= 0 \\ u_i \bullet u_j^{\bullet 2} - (q+1)u_i \bullet u_j \bullet u_i + qu_j^{\bullet 2} \bullet u_i &= 0. \end{aligned} \tag{8.7.47}$$

Here q is the number of elements of the finite field k. These relations are rather similar to the ones between the 'E-generators' of the quantum universal enveloping algebra of the corresponding Dynkin diagram, but not quite spot on. See e.g. [**236**], p. 78, or [**597**], p.5 for these quantum enveloping algebra relations.

8.7.48. The twisted Ringel-Hall algebra of a hereditary algebra over a finite field. Let $\mathcal{I} \subset \mathcal{P}$ be the set of isomorphism classes of simple modules. Thus the u_i, $i \in \mathcal{I}$ form a complete set of simple modules (simple elements of ${}_R\mathrm{FinMod}$).

The Grothendieck group[25] $K_0({}_R\mathrm{FinMod})$ can be (and will be) identified with the free Abelian group $\mathbf{Z}\mathcal{I} = \{\sum_{i\in\mathcal{I}} \nu_i i : \nu_i \in \mathbf{Z}\}$. For $M \in {}_R\mathrm{FinMod}$, $\mathbf{dim}(M)$,

[24] A fat dot is used here, instead of just juxtapositon, because a twisted version of this multiplication, to be described below, is more important.

[25] with respect to short exact sequences of modules

denotes the corresponding element of this Grothendieck group. So the coefficient of $i \in \mathcal{I}$ in $\mathbf{dim}(M)$ is the number of times an isomorphic copy of u_i shows up in a composition series for M and $\mathbf{dim}(u_i) = i \in \mathcal{I}$.

In [**595**] Ringel introduced a slightly modified, but important, multiplication of isomorphism classes of modules. This 'twist' fixes two things: it turns the relations between the isomorphism classes of simple modules over $\Lambda = k\vec{\Delta}$ into exactly the 'E-relations' of the corresponding quantum enveloping algebra and it removes the dependence of $H(\Lambda)$ on the orientation of $\vec{\Delta}$.

To this end, for a general hereditary k-algebra R, define for $M, N \in {}_R\mathrm{FinMod}$

$$e(M,N) = \dim_k(\mathrm{Hom}_R(M,N)) - \dim_k(\mathrm{Ext}^1_R(M,N)). \tag{8.7.49}$$

The integer $e(M,N)$ depends only on $\mathbf{dim}(M)$ and $\mathbf{dim}(N)$, and hence defines a bilinear form on $\mathbf{Z}\mathcal{I}$

$$\langle\ ,\ \rangle_R : \mathbf{Z}\mathcal{I} \times \mathbf{Z}\mathcal{I} \longrightarrow \mathbf{Z}. \tag{8.7.50}$$

As $\mathrm{Ext}^n_R(M,N) = 0$ for $n \geq 2$ because R is hereditary, see e.g. [**318**], theorem 5.4.1, p. 231, the expression (8.7.49) is a kind of Euler characteristic. Correspondingly the form (8.7.50) is called the Euler form (of ${}_R\mathrm{FinMod}$). In case R is the path algebra of a quiver it coincides with the Euler form of that quiver, see e.g. [**188**], p. 50.

Now let $A = \mathbf{Z}[v, v^{-1}]$ where $v^2 = q = |k|$ and define the twisted Ringel-Hall algebra, $H_{tw}(R)$ as the free module over A with basis $\mathcal{P}$ and multiplication

$$u_\alpha u_\beta = \sum_{\lambda \in \mathcal{P}} v^{\langle \alpha, \beta \rangle} G^\lambda_{\alpha,\beta} u_\lambda. \tag{8.7.51}$$

Is is not difficult to check that this defines an $(\mathbf{N} \cup \{0\})$-graded associative (but not necessarily commutative) A-algebra with unit element given by the class of zero modules.

For the case of the classical Hall-Steinitz algebras of complete discrete valuation rings, nothing changes (the twist does nothing) because in that case the Euler form is identically zero.

8.7.52. Hopf algebra structure on $H_{tw}(R)$, [288]. For each $\alpha \in \mathcal{P}$ let a_α be the number of automorphism of u_α. Now define an A-linear morphism $\mu : H_{tw}(R) \longrightarrow H_{tw}(R) \otimes H_{tw}(R)$ by setting

$$\mu(u_\lambda) = \sum_{\alpha,\beta \in \mathcal{P}} v^{\langle \alpha, \beta \rangle} G^\lambda_{\alpha,\beta} \frac{a_\alpha a_\beta}{a_\lambda} u_\alpha \otimes u_\beta.$$

In loc. cit. Green proves, among other things, that this defines a structure of a Hopf algebra on $H_{tw}(R)$.

The proof is mostly straightforward with the exception of the Hopf property, which is the statement that the comultiplication is an algebra morphism (or, equivalently, that the multiplication is a coalgebra morphism). This is handled in [**288**] by a formula dealing with pairs of submodules, nowadays known as the Green formula. There are several special papers on it, see e.g. [**441**]; [**598**]; [**706**]; [**701**].

8.7.53. Relation of $H_{tw}(\Lambda)$ with the quantum algebra $U_q(\mathfrak{n}^+(\Delta))$. Let Δ be an $n \times n$ symmetric matrix with all diagonal entries equal to 2 and all off-diagonal entries equal to 0 or -1 (a simply laced generalized Cartan matrix). Such a matrix gives rise to a graph, also denoted Δ, with n vertices and an edge from i

to j if and only if the (i, j) entry is equal to -1. The algebra $U'_q(\mathfrak{n}^+(\Delta))$ over the field of rational functions $\mathbf{Q}(v)$ is the algebra with generators $E_1, E_2, \ldots, E_n$ and relations

$$\begin{aligned} &E_iE_j - E_jE_i = 0 \text{ if } a_{i,j} = 0 \\ &E_i^2E_j - (v + v^{-1})E_iE_jE_i + E_j^2E_i = 0 \text{ if } a_{i,j} = -1. \end{aligned} \tag{8.7.54}$$

Now define a kind of divided powers of the E_i by

$$E_i^{(m)} = \frac{1}{[m]!}E_i^m, \quad [m]! = \prod_{j=1}^{m}[j], \quad [j] = \frac{v^j - v^{-j}}{v - v^{-1}} \tag{8.7.55}$$

and define the upper triangular part of the universal quantum enveloping algebra corresponding to Δ as the $A = \mathbf{Z}[v, v^{-1}]$ subalgebra of $U'_q(\mathfrak{n}^+(\Delta))$ generated by the $E_i^{(m)}$.

There is now the somewhat startling and unexpected theorem:

Theorem 8.7.56. *There is an isomorphism of A-algebras*

$$\eta : U_q(\mathfrak{n}^+(\Delta)) \longrightarrow H_{tw}(\vec{\Delta}) \tag{8.7.57}$$

given by $\eta(E_i) = u_i$.

For details of the proof [**597**] is especially recommended. The whole business of Hall algebras vs quantum groups started with [**591**] (and [**595**]) and since then there have been many developments, some of which will be discussed below.

The theorem above immediately raises some questions such as:

1. The full quantum universal enveloping Hopf algebra (Drinfel'd-Jimbo algebra) has a triangular decomposition

$$U_q(\Delta) = U_q(\mathfrak{n}^+(\Delta)) \otimes U^0 \otimes U_q(\mathfrak{n}^-(\Delta)). \tag{8.7.58}$$

The plus part $U_q(\mathfrak{n}^+(\Delta))$ is not a sub Hopf algebra, but the Borel part

$$U_q(\mathfrak{b}^+(\Delta)) = U_q(\mathfrak{n}^+(\Delta)) \otimes U^0$$

is a sub Hopf algebra of $U_q(\Delta)$. Is there a way to fix up the Hall algebra part of theorem 8.7.56 so that the isomorphism becomes part of a Hopf algebra morphism?

2. Is there a way of obtaining the full quantum universal enveloping algebra via Hall algebras?

The answer to both[26] questions is positive. An outline will be given below.

8.7.59. Valued graphs, valued quivers, and species. A **valued graph** is a pair $(\Gamma, \mathbf{d})$ consisting of a finite set (of vertices) and for each pair (i, j) a nonnegative integer $d_{i,j}$ such that all $d_{i,i}$ are zero and there exist positive integers ε_i such that

$$d_{i,j}\varepsilon_j = d_{j,i}\varepsilon_i \text{ for all } i, j \in \Gamma.$$

Note that this implies that $d_{i,j} = 0$ if and only if $d_{j,i} = 0$. A pair of of vertices $\{i, j\}$ is said to form an edge if $d_{i,j} \neq 0$.

[26]Of course there is a third question. Using the isomorphism of theorem 8.7.56 the Green Hopf algebra structure on the twisted Hall algebra can be transferred to the plus part $U_q(\mathfrak{n}^+(\Delta))$ of the quantum group $U_q(\Delta)$. Does this Hopf algebra structure have an interpretation in the quantum group setting? Nothing seems to have been done about this (so far).

There is an associated symmetrizable generalized Cartan matrix given by $a_{i,j} = -d_{j,i}$ if $i \neq j$, $a_{i,i} = 2$. The symmetrizing diagonal matrix is $\text{diag}(\varepsilon_1, \ldots, \varepsilon_n)$.

An **orientation** Ω on a valued graph is specified by giving for each edge an order indicated by an arrow from one of the vertices of that edge to the other. This is denoted $i \to j$. A valued graph together with an orientation is called a **valued quiver**[27].

From now on in this section a valued quiver is connected (as a graph) and without loops or oriented cycles.

Given a valued quiver $(\Gamma, \mathbf{d}, \Omega)$ a **species**[28] $\mathbf{S} = (K_i, M_{i,j})$, $i, j \in \Gamma$ over a field k of this type is given by a collection of division algebras K_i over k and a family $M_{i,j}$ of K_i-K_j-bimodules such that $\dim_k(K_i) = \varepsilon_i$, the left and right actions of k induced by respectively the left action of K_i on $M_{i,j}$ and the right action of K_j are the same and, finally, $\dim(M_{i,j})_{K_j} = d_{i,j}$, where $\dim(M_{i,j})_{K_j}$ denotes the dimension of $M_{i,j}$ as a (right) vector space over K_j.

For an arbitrary field k a species[29] of a given type may not exist because there may be no division algebras of the required dimensions. Over a finite field a species of a prescribed type always exists and the division algebras involved are commutative[30] (and unique).

The tensor algebra $\Lambda = \text{Ten}(\mathbf{S})$ of a species $\mathbf{S}$ is defined as the graded algebra

$$A = \text{Ten}^0(\mathbf{S}) = \bigoplus_{i \in \Gamma_0} K_i, \quad \text{Ten}^1(\mathbf{S}) = \bigoplus_{i \to j} M_{i,j}, \quad \text{Ten}^n(\mathbf{S}) = \text{Ten}^1(\mathbf{S}) \otimes \text{Ten}^{n-1}(\mathbf{S}).$$

From the assumptions on $(\Gamma, \mathbf{d}, \Omega)$ that it be without loops or oriented cycles and is connected (as a graph) it follows that Λ is a finite dimensional, connected, basic, hereditary algebra over k. A representation of a species as defined e.g. in [**188**]; [**201**] is the same as a left Λ-module (just as in the case of path algebras and quivers). The category of finite left Λ-modules is denoted ${}_\Lambda\text{FinMod}$ and $H_{tw}(\mathbf{S}) = H_{tw}(\Lambda)$ is the twisted Hall algebra of this category. Like the path algebras of quivers.

8.7.60. Extended twisted Hall algebra. Like before $\mathcal{P}$ denotes the isomorphism classes of modules in ${}_\Lambda\text{FinMod}$; $\mathcal{I}$ is the set of isomorphism classes of

---

27 In [**188**] a seemingly slightly more general notion of a valued quiver is used. It is simply a quiver with to each vertex attached a positive integer ε_i and to each arrow $i \to j$ a nonnegative integer $m_{i,j}$ that is divisible by both ε_i and ε_j; see loc. cit. definition 3.1, p.128. Taking $m_{i,j} = d_{i,j}\varepsilon_j$ gives a valued quiver in this sense starting from a valued quiver in the sense of 8.7.59. One can go also the other way.

28 See also [**318**], section 2.9 (where a slightly different notion is specified). It does not matter much. More precisely, if $i \to j$ it does not matter much what are the modules $M_{j,i}$. As long as there are no cycles of length two, they do not enter in the definition of the tensor algebra Λ and that is the algebra that governs everything. But the numbers $d_{j,i} = d_{i,j}\varepsilon_j/\varepsilon_i$ are useful to have. Note also that in [**201**], see p.2, for a modulation (species) it is required that $M_{j,i} = \text{Hom}_{K_i}(M_{i,j}, K_i)$ so that $\dim(M_{j,i})_{K_i} = d_{j,i}$ follows from $\dim(M_{i,j})_{K_j} = d_{i,j}$ (given the symmetrization condition $d_{i,j}\varepsilon_j = d_{j,i}\varepsilon_i$). It is probably best to start from some notion of species and then derive the underlying valued quiver from that as is done in [**590**] and [**318**].

29 A species is called a modulated quiver in [**188**]. For the inconvenience of the reader the term 'species' is not mentioned in the index.

30 Because of the Wedderburn theorem on division rings to the effect that finite division rings are commutative, see [**611**], p.425ff.

simple (i.e. irreducible) modules; $\mathbf{Z}\mathcal{I}$ is identified with the Grothendieck group $K_0({}_\Lambda\text{FinMod})$; there is again the **dim** mapping from ${}_\Lambda$FinMod to $\mathbf{Z}\mathcal{I}$; there is an Euler bilinear form on $\mathbf{Z}\mathcal{I}$; and u_α denotes, as the case may be, the isomorphism class α or a fixed module from this isomorphism class; and u_i is an isomorphism class from $\mathcal{I}$ or a chosen representative from it. The Euler form is explicitly given by

$$\langle \mathbf{a}, \mathbf{b}\rangle = \sum_{i\in\mathcal{I}} \varepsilon_i a_i b_i - \sum_{i\to j} d_{i,j}\varepsilon_j a_i b_j. \tag{8.7.61}$$

The symmetric Euler form is given by

$$(\mathbf{a}, \mathbf{b}) = \langle \mathbf{a}, \mathbf{b}\rangle + \langle \mathbf{b}, \mathbf{a}\rangle.$$

In terms of the symmetric Euler form the associated generalized Cartan matrix of $(\Gamma, \mathbf{d}, \Omega)$ and the symmetrizing diagonal matrix are given by the familiar formulas

$$a_{i,i} = 2,\quad a_{i,j} = \frac{2(i,j)}{(i,i)},\quad \varepsilon_i = \frac{(i,i)}{2}.$$

This is independent of the orientation.

Let $R = \mathbf{Q}[v, v^{-1}]$ (or an integral domain extending it). The extended twisted Hall algebra $H_{ext,tw}(\Lambda)$ is now the free R-module with basis

$$\{K_v u_\alpha^+; v \in \mathbf{Z}\mathcal{I}, \alpha \in \mathcal{P}\} \tag{8.7.62}$$

multiplication

$$\begin{aligned} u_\alpha^+ u_\beta^+ &= v^{\langle\alpha,\beta\rangle} \sum_{\lambda\in\mathcal{P}} G_{\alpha,\beta}^\lambda u_\lambda^+ \\ K_\rho u_\alpha^+ &= v^{\langle\rho,\alpha\rangle+\langle\alpha,\rho\rangle} u_\alpha^+ K_\rho \\ K_\rho K_{\rho'} &= K_{\rho+\rho'} \end{aligned} \tag{8.7.63}$$

where $\alpha, \beta \in \mathcal{P}$, $\rho, \rho' \in \mathbf{Z}\mathcal{I}$ and the unit is $K_0 = u_0^+ = K_0 u_0^+ = 1$ (which has already been used in the notation above), and comultiplication

$$\begin{aligned} \mu(u_\lambda^+) &= \sum_{\alpha,\beta\in\mathcal{P}} v^{\langle\alpha,\beta\rangle} G_{\alpha,\beta}^\lambda \tfrac{a_\alpha a_\beta}{a_\lambda} u_\alpha^+ K_\beta \otimes u_\alpha^+ \\ \mu(K_\rho) &= K_\rho \otimes K_\rho \end{aligned} \tag{8.7.64}$$

for $\lambda \in \mathcal{P}$, $\rho \in \mathbf{Z}\mathcal{I}$ with counit $\varepsilon(u_\alpha^+) = 0$ for $\alpha \neq 0$, $\varepsilon(K_\rho) = 1$. Here K_β is short for $K_{\mathbf{dim}(u_\beta)}$. There is also an explicit formula for the antipode, see [**185**]; [**187**]; [**695**] on which this outline is mainly based. It is stimulating to note that

$$\mu(u_i^+) = u_i^+ \otimes 1 + K_i \otimes u_i^+,\ \iota(u_i^+) = -K_{-i}u_i^+. \tag{8.7.65}$$

These formulas have a rather familiar form! It is a theorem, see loc. cit. that these formulas define a Hopf algebra. This is a rather brute force positive answer to the first of the two questions just above 8.7.59. Brute force because there seems to be (so far) no good interpretation of the K_ρ in ${}_\Lambda$FinMod.

Note that the ideal generated by the $K_\rho - 1$ is a Hopf ideal and killing it off gives back the Hopf algebra $H_{tw}(\Lambda)$.

8.7.66. Skew pairing. Let A and B be two Hopf algebras over a field k. A **Hopf pairing** between them is a bilinear map

$$\varphi : A \times B \longrightarrow k$$

such that

$$\begin{aligned}&\varphi(1,b)=\varepsilon_B(b),\quad \varphi(a,1)=\varepsilon_A(a)\\&\varphi(a,bb')=(\varphi\otimes\varphi)(\mu_A(a),b\otimes b')\\&\varphi(aa',b)=(\varphi\otimes\varphi)(a\otimes a',\mu_B(b))\\&\varphi(\iota_A(a),b)=\varphi(a,\iota_B(b))\end{aligned}\tag{8.7.67}$$

where $\varphi\otimes\varphi$ on $(A\otimes A)\times(B\otimes B)$ is defined by $(\varphi\otimes\varphi)(a\otimes a',b\otimes b')=\varphi(a,b)\varphi(a',b')$.

A **skew Hopf pairing** between two Hopf algebras A,B is a Hopf pairing between A and B_{opp} where B_{opp} is the bialgebra with the same multiplication but the opposite comultiplication. It is a Hopf algebra if the antipode ι_B of the Hopf algebra B is invertible, and then the antipode of B_{opp} is $(\iota_B)^{-1}$.

Now assume that the antipode ι_B of the Hopf algebra B is invertible. Then given a skew Hopf pairing there is a Hopf algebra structure on $A\otimes B$ defined as follows, see [**357**], p. 71; [**695**], p. 104

$$\begin{aligned}&(a\otimes 1)(a'\otimes 1)=aa'\otimes 1\\&(1\otimes b)(1\otimes b')=1\otimes bb'\\&(a\otimes 1)(1\otimes b)=(a\otimes b)\\&(1\otimes b)(a\otimes 1)=\sum_{(a),(b)}\varphi(a_1,\iota_B(b_1))(a_2\otimes b_2)\varphi(a_3,b_3).\end{aligned}\tag{8.7.68}$$

This is the **Drinfel'd double** $\mathcal{D}(A,B,\varphi)$. Here

$$\sum_{(a)}a_1\otimes a_2\otimes a_3=\mu^{(3)}(a)=(\mu\otimes\mathrm{id})\mu(a)=(\mathrm{id}\otimes\mu)\mu(a)$$

and similarly for b.

8.7.69. Double Ringel-Hall algebra. In subsection 8.7.60 the extended twisted Ringel-Hall algebra $H_{ext,tw}(\Lambda)$ was defined. Denote this one by $H^+_{ext,tw}(\Lambda)$. There is a corresponding Hopf algebra $H^-_{ext,tw}(\Lambda)$ with basis

$$\{K_\rho u^-_\alpha:\rho\in\mathbf{Z},\alpha\in\mathcal{P}\}.$$

The formulas are the same except for a couple of minus signs; see [**695**], p. 132.

There is a (nondegenerate) skew Hopf pairing between these Hopf algebras defined by

$$\varphi(K_\rho u^+_\alpha,K_\sigma u^-_\beta)=v^{-(\rho,\sigma)-(\alpha,\sigma)+(\rho,\beta)}\frac{|u_\alpha|}{a_\alpha}\delta_{\alpha,\beta}.\tag{8.7.70}$$

Let $DH'(\Lambda)$ be the corresponding Drinfel'd double:

$$DH'(\Lambda)=\mathcal{D}(H^+_{ext,tw}(\Lambda),H^-_{ext,tw}(\Lambda),\varphi).\tag{8.7.71}$$

In the Hopf algebra $DH'(\Lambda)$ consider the ideal J generated by the $K_\rho\otimes K_{-\rho}-1\otimes 1$. This is a Hopf ideal. Quotienting out this ideal gives a Hopf algebra

$$DH(\Lambda)=DH'(\Lambda)/J\tag{8.7.72}$$

called the **double Ringel-Hall (Hopf) algebra** of Λ (or of the species $\mathbf{S}$).

There is visibly a triangular decomposition

$$DH(\Lambda)=H^+_{tw}(\Lambda)\otimes\mathcal{T}\otimes H^-_{tw}(\Lambda)\tag{8.7.73}$$

where $\mathcal{T}$ is the torus sub Hopf algebra with basis $\{K_\rho:\rho\in\mathbf{Z}\mathcal{I}\}$.

8.7.74. Composition algebra. Consider for the moment again the (original) Ringel-Hall algebra $H(\Lambda)$ (no twist, no extending). This one has $\{u_\alpha : \alpha \in \mathcal{P}\}$ as basis over the integers. The subalgebra generated by the simple modules u_i is called the composition algebra of Λ. If Λ is of finite representation type it coincides with $H(\Lambda)$.

The name derives from the observation that in a product

$$u_{i_1} u_{i_2} \cdots u_{i_m} = \sum_{\lambda \in \mathcal{P}} c^\lambda u_\lambda$$

the coefficient c^λ counts the number of composition series for a module of isomorphism class λ with successive quotients in the isomorphism classes u_{i_r} (in the given order).

Now return to the case of the double Ringel-Hall algebra. The subalgebra generated by the u_i^+, u_i^-, K_i is called the **double composition algebra** $\mathcal{C}(\Lambda)$. It is a sub Hopf algebra of $DH(\Lambda)$ and has an obvious triangular decomposition

$$\mathcal{C}(\Lambda) = \mathcal{C}^+(\Lambda) \otimes \mathcal{T} \otimes \mathcal{C}^-(\Lambda). \tag{8.7.75}$$

8.7.76. Generic composition algebra. As it is, because $v^2 = q$, ν is basically a number. What one wants (and needs) is a composition algebra (over $\mathbf{Z}[v, v^{-1}]$ or $\mathbf{Q}[v, v^{-1}]$ as the case may be) with v a variable. This can be imitated[31] by considering composition algebras over an infinity of different finite fields.

So let $\mathcal{K}$ be a set of finite fields such that the set of cardinalities

$$\{q_k = |k| : k \in \mathcal{K}\}$$

is infinite. Let R be an integral domain containing the rational numbers and containing for each $k \in \mathcal{K}$ an element v_k such that $v_k^2 = q_k$. For each $k \in \mathcal{K}$ there is the hereditary algebra Λ_k, with its isomorphism classes of modules $u_{i,k}^+$, its extended twisted Ringel-Hall algebra $H_{ext,tw}^+(\Lambda_k)$ and the (extended twisted) composition algebra $\mathcal{C}^+(\Lambda_k)$, the subalgebra of $H_{ext,tw}^+(\Lambda_k)$ generated by the $u_{i,k}^+$ and K_i, which is a sub Hopf algebra of $H_{ext,tw}^+(\Lambda_k)$.

Now consider the direct product

$$H_*^+(\Gamma) = \prod_{k \in \mathcal{K}} H_{ext,tw}^+(\Lambda_k)$$

and in it the elements $t, t^{-1}, u_{i,*}^+, K_{i,*}$, whose components in the $H_{ext,tw}^+(\Lambda_k)$ are respectively, $v_k, v_k^{-1}, u_{i,k}^+, K_i$. The elements $t, t^{-1}, u_{i,*}^+, K_{i,*}$ generate a sub algebra of $H^*(\Gamma)$ that is denoted $\tilde{\mathcal{C}}_*^+(\Gamma)$. The element t is central in $\tilde{\mathcal{C}}_*^+(\Gamma)$ and satisfies no nonzero polynomial, so $\tilde{\mathcal{C}}_*^+(\Gamma)$ can be regarded as an algebra over $\mathbf{Q}[t, t^{-1}]$ and tensoring it up with the rational function field gives $\mathcal{C}_*^+(\Gamma) = \mathbf{Q}(t) \otimes_{\mathbf{Q}[t,t^{-1}]} \mathcal{C}_*^+(\Gamma)$ which is called the **extended twisted generic composition algebra** of $(\Gamma, \mathbf{d}, \Omega)$.

Similarly one constructs $\mathcal{C}_*^-(\Gamma)$ and from the skew Hopf pairings on $H_{ext,tw}^+(\Lambda_k) \times H_{ext,tw}^-(\Lambda_k)$ one constructs a skew Hopf pairing

$$\mathcal{C}_*^+(\Gamma) \times \mathcal{C}_*^-(\Gamma) \longrightarrow \mathbf{Q}(t),$$

[31] This is really the adage (adagium) that one knows a rational function if one knows it at an infinity of points.

which in turn gives the Drinfel'd double $\mathcal{D}(\mathcal{C}_*^+(\Gamma), \mathcal{C}_*^-(\Gamma), \varphi)$ and finally, quotienting out the ideal generated by the $K_{i,*} \otimes K_{i,*} - 1$, there results the double generic composition algebra $\mathcal{C}_*(\Gamma)$.

Theorem 8.7.77. *Let $(\Gamma, \mathbf{d}, \Omega)$ be a finite connected oriented valued quiver without cycles (and without loops), let C be the associated symmetrizable generalized Cartan matrix, and let U be the Drinfel'd-Jimbo quantum group defined by C. Then $u_{i,*}^+ \mapsto E_i$, $u_{i,*}^- \mapsto F_i$, $K_{i,*} \mapsto K_i$, induces a Hopf algebra isomorphism $\mathcal{C}_*(\Gamma) \xrightarrow{\simeq} U$.*

There is a somewhat obvious triangular decomposition structure on $\mathcal{C}_*(\Gamma)$ and of course the standard triangular decomposition structure on U. The isomorphism above respects these triangular decompositions.

This gives a positive answer to the second question of 8.7.59 above.

8.7.78. More stuff. There is much much more to the theme of Hopf algebras, Hall algebras, and quantum groups (and their representations). In fact, more than enough to devote a full length monograph to the subject. For instance, the survey lectures [**624**] fill over a hundred pages and, as the author says, there is a substantial list of aspects that are not discussed there. However, as the subject is still developing energetically, it is probably to early for such a monograph.

In this last subsection there are a few references to further literature on the topic of Hall algebras.

There are a number of survey papers, [**187**]; [**340**]; [**610**]; [**624**]. See also part 4 of [**188**], especially chapter 11.

On the business of U_q^+ versus the twisted Ringel-Hall algebra, see also, besides the papers already quoted, [**136**]; [**257**]; [**441**]; [**544**]; [**546**]; [**507**]; [**592**]; [**594**]; [**599**]; [**598**]; [**663**]; [**701**].

For more on Hall polynomials, especially Hall polynomials in the general setting of hereditary algebras of finite representation type and cyclic quivers, see e.g. [**30**]; [**340**]; [**407**]; [**592**]; [**593**]; [**600**]; [**697**].

For more on composition algebras see e.g. [**133**]; [**295**]; [**335**]; [**545**]; [**596**]; [**703**].

Theorems 8.7.56 and 8.7.77 above should probably be seen as further developments in the topic of representations of quivers and (positive) roots of generalized Kac-Moody algebras, which started with the Gabriel theorem on the bijective correspondence between indecomposable representations of Dynkin quivers of ADE type and positive roots of the simple Lie algebras associated to those Dynkin diagrams, [**262**], later extended to representations of species' and positive roots of the other types of simple Lie algebras, [**201**]; [**263**]. There is more on the representation theoretic part, notably concerning a theorem and a conjecture of Victor G.Kac. To start with see e.g. [**134**]; [**186**]; [**257**]; [**636**]; [**637**]; [**638**].

In theorem 8.7.77 above the case that there are oriented cycles or loops present (so that the associated tensor algebra Λ becomes infinite dimensional) was explicitly excluded. This applies in particular to the fundamental cycle quiver, which is an affine quiver. A selection of papers dealing with cyclic and/or affine quivers in the context of Hall algebras and quantum groups is: [**30**]; [**257**]; [**294**]; [**295**]; [**336**]; [**335**]; [**340**]; [**544**]; [**545**]; [**596**]; [**622**]; [**697**]; [**703**]; [**702**]; [**705**]; [**717**].

For more on double Hall algebras and their relations to full Drinfel'd-Jimbo quantized universal enveloping algebras (quantum groups), see [**185**]; [**334**]; [**443**]; [**444**]; [**445**]; [**637**]; [**683**]; [**695**].

There are a good many topics that have strong relations with Hall algebras in their interactions with quantum groups. Some of these are canonical bases, crystal bases, and PBW bases; cluster algebras; BGP symmetries; Lusztig symmetries; derived categories; Boson algebras; .... A selection of references to some of these topics is: [**145**]; [**184**]; [**233**]; [**337**]; [**339**]; [**364**]; [**363**]; [**407**]; [**408**]; [**442**]; [**444**]; [**445**]; [**446**]; [**447**]; [**457**]; [**458**]; [**459**]; [**460**]; [**507**]; [**581**]; [**582**]; [**583**]; [**599**]; [**622**]; [**625**]; [**663**]; [**696**]; [**706**]; [**712**].

To conclude this section 8.7 let it be remarked that there is a second geometry/algebra/representation approach to the full quantized universal enveloping algebra based on q-Schur algebras. See [**56**].

8.8. More interactions of Hopf algebras with other parts of mathematics and theoretical physics

The subject of Hopf algebras and their applications in and relations with other parts of mathematics and theoretical physics, especially quantum physics, is a large one. A search in 2004 (using ZMATH) gave 10589 hits and showed that there is virtually no part of mathematics and quantum physics that does not interact with Hopf algebras and/or quantum groups, [**316**].

In this final section brief indications, heavily bibliographical, are given to some more topics in which Hopf algebras play a significant role and which have not been discussed above.

8.8.1. Capelli identities and other formulas for matrices and determinants with mildly noncommuting entries. There are quite a few famous (and important) formulas for determinants and matrices with commuting entries that have noncommutative (and quantum) analogues as long as the entries of the matrices involved are only mildly noncommuting. These include the Capelli identity, Cayley-Hamilton-Newton identity, Turnbull identity, MacMahon master theorem, Sylvester theorem, Cauchy-Binet formulas, Plücker relations, ... Some of these will be briefly discussed in this subsection.

8.8.1.1. Manin matrices. One good setting for many of these generalizations is that of Manin matrices (also called right quantum matrices[32]) and q-Manin matrices.

A matrix $M = (M_{i,j})$ is called a **Manin matrix** if

(i) elements in the same column commute

(ii) commutators of cross entries are equal; i.e. for all i, j, r, s

$$[M_{i,r}, M_{j,s}] = [M_{j,r}, M_{i,s}] \qquad (8.8.1.2)$$

[32]Unfortunately, the word 'quantum' is overworked nowadays. Also this terminology causes confusion with q-Manin matrices.

as illustrated in the following picture

$$\begin{array}{c} \\ i \\ \\ j \end{array} \begin{pmatrix} r & s \\ & \\ \bullet & \bullet \\ & \\ \bullet & \bullet \end{pmatrix}.$$

The terminology derives from the fact that such matrices play a significant role in quantum groups and noncommutative geometry, see [**496**], section 6.1, see also [**494**]; [**498**]; [**499**].

In terms of the Kronecker (tensor) product (such as used in section 8.4) the condition of being Manin for a matrix with entries in an associative ring A over a field k is

$$[M \otimes I_n, I_n \otimes M] = P[I_n \otimes M, M \otimes I_n] \tag{8.8.1.3}$$

where P is the switch matrix giving the endomorphism $x \otimes y \mapsto y \otimes x$ of the tensor product $A^n \otimes_k A^n$; i.e $P_{i,j}^{j,i} = 1$ for all i, j and all other entries are zero. So for $n = 2$ for example

$$P = \begin{pmatrix} 1 & 0 & 0 & 0 \\ 0 & 0 & 1 & 0 \\ 0 & 1 & 0 & 0 \\ 0 & 0 & 0 & 1 \end{pmatrix}.$$

For a thorough preliminary[33] account of Manin matrices and their applications, see [**138**].

A matrix T is q-**Manin** if for every 2×2 submatrix as illustrated below

$$\begin{array}{c} \\ i \\ \\ j \end{array} \begin{pmatrix} r & s \\ & \\ a & b \\ & \\ c & d \end{pmatrix} \tag{8.8.1.4}$$

$ca = qac, db = qdb$ (q-commutation of elements in the same column)

$ad - da = q^{-1}cb - qbc$ (cross q-commutation relation).

Here $i < j$ and $r < s$, i.e. no coinciding of columns or rows (as is permitted in the definition of a Manin matrix).

A matrix is a **quantum group matrix** if it and its transpose are both q-Manin. This amounts to the relations (still referring to (8.8.1.4))

$ca = qac, db = qdb$ (q-commutation of elements in the same column)

$ba = qab, dc = qcd$ (q-commutation of elements in the same row)

$ad - da = q^{-1}cb - qbc$ (first cross q-commutation relation)

$cb = bc$ (second cross q-commutation relation).

[33]'Preliminary' in the sense that there is almost certainly much more to come.

(Note that the last relation, given the third, is equivalent to the perhaps more elegant cross commutation relation $ad - da = q^{-1}bc - qcb$ as long as $q^2 \neq -1$).

In terms of tensor products this can be formulated in terms of the fundamental commutation relations as

$$R(T \otimes \mathrm{id})(\mathrm{id} \otimes T) = (\mathrm{id} \otimes T)(T \otimes \mathrm{id})R$$

where the R-matrix is given by

$$R = q^{-1} \sum_i E_{i,i} \otimes E_{i,i} + \sum_{i \neq j} E_{i,i} \otimes E_{j,j} + (q^{-1} - q) \sum_{i>j} E_{i,j} \otimes E_{j,i}$$

(see also section 8.4 on R-matrices and the FCR above).

Given the nature and form of the matrix equations defining Manin and q-Manin matrices it should come as no surprise that there might be applications to (quantum) integrable system theory. And so there are, notably to the Gaudin model and quantum spin chains as treated in section 8.4 above. See [**138**] and [**661**].

8.8.1.5. Column determinants and row determinants. When forming a determinant like expression in the entries of a matrix with (possibly) noncommuting entries it matters how the monomials making up the determinant are ordered.

The **column determinant** of an $n \times n$ matrix $M = (m_{ij})$ is, by definition, equal to

$$\det{}^{\mathrm{column}}(M) = \sum_{\sigma \in S_n} \mathrm{sign}(\sigma) m_{\sigma(1),1} m_{\sigma(2),2} \cdots m_{\sigma(n),n}. \tag{8.8.1.6}$$

That is entries of the monomials making up the determinant are ordered according to their column order.

The **row determinant** is similarly defined by

$$\det{}^{\mathrm{row}}(M) = \sum_{\sigma \in S_n} \mathrm{sign}(\sigma) m_{1,\sigma(1)} m_{2,\sigma(2)} \cdots m_{n,\sigma(n)}. \tag{8.8.1.7}$$

I.e. this time the ordering of the terms making up the determinant are in the order of the rows in which the entries occur.

Finally the **quantum determinant** (q-determinant) of a quantum matrix, i.e. a matrix M such that both itself and its transpose are q-Manin, is equal to

$$\begin{aligned} \det_q(M) &= \sum_{\sigma \in S_n} (-q)^{-\mathrm{lg}(\sigma)} m_{1,\sigma(1)} m_{2,\sigma(2)} \cdots m_{n,\sigma(n)} \\ &= \sum_{\sigma \in S_n} (-q)^{-\mathrm{lg}(\sigma)} m_{\sigma(1),1} m_{\sigma(2),2} \cdots m_{\sigma(n),n}. \end{aligned} \tag{8.8.1.8}$$

Here $\mathrm{lg}(\sigma)$, the length of σ, is the minimal number of terms in a product of adjacent transpositions, $(i, i+1)$, that is equal to σ.

8.8.1.9. Capelli identities. The **classical Capelli identity**[34], [**114**], is

$$\det\begin{pmatrix} E_{11}+n-1 & E_{12} & \cdots & E_{1,n-1} & E_{1n} \\ E_{21} & E_{22}+n-2 & \cdots & E_{2,n-1} & E_{2n} \\ \vdots & \vdots & \ddots & \vdots & \vdots \\ E_{n-1,n} & E_{n-1,2} & \cdots & E_{n-1,n-1}+1 & E_{n-1,n} \\ E_{n1} & E_{n2} & \cdots & E_{n,n-1} & E_{n,n}+0 \end{pmatrix}$$

$$= \det\begin{pmatrix} x_{11} & \cdots & x_{1n} \\ \vdots & \ddots & \vdots \\ x_{n1} & \cdots & x_{nn} \end{pmatrix} \det\begin{pmatrix} \frac{\partial}{\partial x_{11}} & \cdots & \frac{\partial}{\partial x_{1n}} \\ \vdots & \ddots & \vdots \\ \frac{\partial}{\partial x_{n1}} & \cdots & \frac{\partial}{\partial x_{nn}} \end{pmatrix}. \qquad (8.8.1.10)$$

Here the E_{ij} are the so-called polarization operators

$$E_{ij} = \sum_{r=1}^{n} x_{ir}\frac{\partial}{\partial x_{jr}}$$

and on the right of the Capelli identity, the " ∂-determinant", one finds the Cayley process used for obtaining new invariants from old ones. See e.g. [**261**], p. 507ff; [**330**]; [**409**], §9 for some information on the role the Capelli identity in classical invariant theory.

Consider the left and right action of square invertible $m\times m$ and $n\times n$ matrices on the space of $m\times n$ matrices. Then there is a corresponding infinitesimal action of the Lie algebra $\mathfrak{gl}(m)\oplus\mathfrak{gl}(n)$ given by the polarization operators, which gives a homomorphism from the universal enveloping algebra of $\mathfrak{gl}(m)\oplus\mathfrak{gl}(n)$ into the Weyl algebra of polynomial coefficient differential operators on the space of $m\times n$ matrices. This morphism is important in invariant theory, [**331**].

The entries on the left of the Capelli identity do not commute so some care must be taken in writing down its determinant. The relevant one is the column determinant defined by (8.8.1.6).

There is also a version of the Capelli identity for symmetric matrices, due to Turnbull, [**670**].

There are a number of generalizations of the Capelli identity in various contexts, see [**116**]; [**137**]; [**138**]; [**330**]; [**331**]; [**348**]; [**519**]; [**524**]; [**533**]; [**536**]; [**537**]; [**551**]; [**552**]. Here [**116**]; [**138**][35] are in the context of Manin matrices or something like

[34]There is a second, at first sight different (but not unrelated), notion called "Capelli identity". The m-th Capelli polynomial (Capelli polynomial of height m) is

$$d_m(x;y) = d_m(x_1,\ldots,x_m;y_1,\ldots,y_{m-1})$$
$$= \sum_{\sigma\in S_m}\operatorname{sign}(\sigma)x_{\sigma(1)}y_1x_{\sigma(2)}y_2\cdots x_{\sigma(m-1)}y_{m-1}x_{\sigma(m)}$$

where the sum is over all permutations in m letters. An algebra (with unit element) is said to satisfy a Capelli identity if this polynomial is identically zero on it. In [**580**] this notion is linked to the height of the cocharater sequence (c.c.s) of a PI algebra (polynomial identity algebra). See [**57**]; [**59**] for the notion of a cocharacter sequences for PI algebras, and their uses. In [**58**] the cocharacter sequence of the algebra of 3×3 matrices over a field of characteristic zero is worked out in detail (by way of example). There are also quantum versions of the Capelli polynomials.

[35]These two papers also contain generalized Turnbull identities in the context of Manin matrices.

it, while [**331**]; [**348**] place Capelli identities in the context of dual pairs (Howe duality), which is especially interesting.

There is also a quantum version of the Capelli identity:

$$z(q^{n\ 1}) = q^{\binom{1}{2}}\det_q(x_{ij})\det_{q^{-1}}(\partial_{ji})$$

where $z(t)$ denotes the central element of $U_q(\mathfrak{gl}(n))$ defined by Jimbo in [**350**] and $\det_q$ is the quantum determinant; see (8.8.1.8) above. For these quantum Capelli identities there is also a dual pair setting, [**542**]; [**543**].

8.8.1.11. Cauchy-Binet formulas. The classical Cauchy-Binet theorem says the following. Let A and B be $m \times n$ matrices with entries from a commutative ring R. Let I and J be subsets of cardinality r, then the following formula holds

$$\det((A^TB)_{I,J}) = \sum_{\substack{L\subset\{1,\dots,m\}\\ |L|=r}} \det((A^T)_{I,L})\det(B_{L,J}).$$

Here $M_{I,J}$ denotes the submatrix of a matrix M obtained by removing all rows not in I and all columns not in J.

There are a variety of generalizations to the mildly noncommutative case. Here is one ([**116**], p. 4).

Let R be a not necessarily commutative ring and let A and B be $m \times n$ matrices with entries from R. Suppose that $[a_{ij}, b_{kl}] = -h_{ik}\delta_{jl}$ for certain elements $h_{ik} \in R$ and suppose that A is Manin[36]. Then for any $I, J \subset \{1, 2, \dots, m\}$ of cardinality r

$$\begin{aligned} &\det{}^{\text{column}}((AB^T)_{I,J} + Q_{\text{column}}) \\ &= \textstyle\sum_{\substack{L\subset\{1,\dots,m\}\\ |L|=r}} \det{}^{\text{column}}(A_{I,L})\det{}^{\text{column}}((B^T)_{L,J}) \end{aligned} \qquad (8.8.1.12)$$

here the "quantum" correction Q_{column} is the matrix $(Q_{\text{column}})_{\alpha,\beta} = (r-\beta)h_{i_\alpha,j_\beta}$.

The reader who has seen such things before will recognize this as a Capelli identity for suitable matrices A, B (and $m = n = r$).

Thus (8.8.1.12) is something like an abstract Capelli identity for minors of matrices.

See [**116**]; [**138**] for much more on this theme.

8.8.1.13. Cayley-Hamilton-Newton formulae. The Newton formula relates the power sum symmetric functions (recursively) to the elementary symmetric functions or to the complete symmetric functions, and the latter two are linked through the Wronski relations.

The Cayley-Hamilton theorem says the following. Let X be a square matrix with commuting entries and $\chi(t) = \det(t - X)$ its characteristic polynomial, then $\chi(X) = 0$.

In the course of research to generalize these to the setting of quantum matrices and quantum groups it turned out that there is a kind of common basis to the two, now called (quantum) **Cayley-Hamilton-Newton relations**. Here is a brief sketch of some of this, based on [**343**].

---

[36] In [**116**] a Manin matrix is called a row pseudo-commutative matrix.

The setting is that of an R-matrix, i.e. an $n^2 \times n^2$ matrix that satisfies the Yang-Baxter equation (braid form):

$$R_1R_2R_1 = R_2R_1R_2, \quad \text{where } R_1 = R \otimes I_n, R_2 = I_n \otimes R \tag{8.8.1.14}$$

and also the Hecke algebra type relation

$$R^2 = I_n + (q - q^{-1})R. \tag{8.8.1.15}$$

Such a matrix is called a Hecke R-matrix. The classical case arises when $q = 1, R = P$, the switch matrix. When notationally or conceptually convenient R is viewed as an endomorphism of the second tensor power $V^{\otimes 2}$ of a standard vector space.

Let $T = (t_{ij})$ be a matrix of (noncommuting) indeterminates. As before in this chapter consider the associative unital algebra generated by them subject to the relations

$$RT_1T_2 = T_2T_1R.$$

Such an algebra is called an RTT algebra.

Now define the following sequence of elements of this RTT algebra.

$$\begin{aligned} p_k(T) &= \mathrm{Tr}(R_1R_2 \dots R_{k-1}T_1T_2 \dots T_k) \\ e_k(T) &= q^k \mathrm{Tr}(A^{(k)}T_1T_2 \dots T_k) \\ h_k(T) &= q^{-k} \mathrm{Tr}(S^{(k)}T_1T_2 \dots T_k) \end{aligned} \tag{8.8.1.16}$$

where $A^{(k)} \in \mathrm{End}(V^{\otimes k})$ and $S^{(k)} \in \mathrm{End}(V^{\otimes k})$ are q-antisymmetrizing and q-symmetrizing operators recursively defined by

$$\begin{aligned} A^{(1)} = I_n, \quad A^{(k)} &= \tfrac{1}{k_q} A^{(k-1)}(q^{k-1} - (k-1)_q R_{k-1})A^{(k-1)} \\ S^{(1)} = I_n, \quad S^{(k)} &= \tfrac{1}{k_q} S^{(k-1)}(q^{k-1} + (k-1)_q R_{k-1})A^{(k-1)} \end{aligned} \tag{8.8.1.17}$$

Here

$$k_q = \frac{q^k - q^{-k}}{q - q^{-1}} = q^{k-1} + q^{k-3} + \dots + q^{-k+1}$$

where q is not a root of unity (except that $q = 1$ or -1 is allowed), so that $k_q \neq 0$.

The notation in (8.8.1.16) and (8.8.1.17) (from loc.cit.) is slightly ambiguous in that e.g. in (8.8.1.16) the meaning of R_i and T_i depends on k. Given k

$$R_i = \underbrace{I \otimes \dots \otimes I}_{i-1} \otimes R \otimes \underbrace{I \otimes \dots \otimes I}_{k-i-1}$$

$$T_i = \underbrace{I \otimes \dots \otimes I}_{i-1} \otimes T \otimes \underbrace{I \otimes \dots \otimes I}_{k-i}.$$

Finally, introduce two versions each of the powers, exterior powers, and symmetric powers of the q-matrix T:

$$\begin{aligned} T^{\underline{k}} &= \mathrm{Tr}_{(1,\dots,k-1)}(R_1R_2 \dots R_{k-1}T_1T_2 \dots T_k) \\ T^{\overline{k}} &= \mathrm{Tr}_{(2,\dots,k)}(R_1R_2 \dots R_{k-1}T_1T_2 \dots T_k) \\ T^{\wedge\underline{k}} &= \mathrm{Tr}_{(1,\dots,k-1)}(A^{(k)}R_1R_2 \dots R_{k-1}T_1T_2 \dots T_k) \\ T^{\wedge\overline{k}} &= \mathrm{Tr}_{(2,\dots,k)}(A^{(k)}R_1R_2 \dots R_{k-1}T_1T_2 \dots T_k) \\ T^{S\underline{k}} &= \mathrm{Tr}_{(1,\dots,k-1)}(S^{(k)}R_1R_2 \dots R_{k-1}T_1T_2 \dots T_k) \\ T^{S\overline{k}} &= \mathrm{Tr}_{(2,\dots,k)}(S^{(k)}R_1R_2 \dots R_{k-1}T_1T_2 \dots T_k). \end{aligned} \tag{8.8.1.18}$$

Here, quite generally, $\mathrm{Tr}_{(i_1,\ldots,i_r)}$ applied to an endomorphism of $V^{\otimes n}$ means taking the partial trace with respect to the tensor factors $i_1, \ldots, i_r$. Such partial traces were also used in section 8.5 above. Thus the entities defined in (8.8.1.18) are all $n \times n$ matrices.

Then there are the following **Cayley-Hamilton-Newton identities**

$$\begin{aligned} m_q T^{\wedge \underline{m}} &= \sum_{i=1}^{m} (-1)^{i+1} e_{m-i}(T) T^{\underline{i}} \\ m_q T^{\wedge \overline{m}} &= \sum_{i=1}^{m} (-1)^{i+1} T^{\overline{i}} e_{m-i}(T) \\ m_q T^{S\underline{m}} &= \sum_{i=1}^{m} h_{m-i}(T) T^{\underline{i}} \\ m_q T^{S\overline{m}} &= \sum_{i=1}^{m} T^{\overline{i}} h_{m-i}(T). \end{aligned} \tag{8.8.1.19}$$

As a corollary, by taking one more final trace of the first and third of these formulas, (i.e taking the trace of these $n \times n$ matrices), one obtains two Newton type relations and one Wronski relation

$$q^{-m} m_q e_m(T) = \sum_{i=1}^{m-1} (-1)^{m-i-1} e_i(T) p_{m-i}(T) + (-1)^{m-1} p_m(T)$$

$$q^{m} m_q h_m(T) = \sum_{i=1}^{m-1} h_i(T) p_{m-i}(T) + p_m(T)$$

$$\sum_{i=1}^{m} (-1)^{m-i} q^{2i} h_i(T) e_{m-i}(T) = 0$$

and under the additional assumption that R is an even Hecke R-matrix of rank n, which means by definition that $\mathrm{rank}(A^{(n)}) = 1$, $A^{(n+1)} = 0$ there are two Cayley Hamilton type formulas:

$$e_n(T)\mathfrak{D}_l + \sum_{i=1}^{n} e_{n-i}(T)(-T)^{\underline{i}} = 0$$

$$e_n(T)\mathfrak{D}_r + \sum_{i=1}^{n} (-T)^{\overline{i}} e_{n-i}(T) = 0$$

where $\mathfrak{D}_r = q^{-n} n_q \mathrm{Tr}_{(2,\ldots,n)} A^{(n)}$, $\mathfrak{D}_l = q^{-n} n_q \mathrm{Tr}_{(1,\ldots,n-1)} A^{(n)}$. This is obtained by taking $m = n$ in (8.8.1.19) and observing that under the extra conditions stated $e_n(T) = q^n \det_q(T)$ and $A^{(n)} T_1 \ldots T_n = A^{(n)} \det_q(T)$.

There is much more. For instance there are also Cayley-Hamilton-Newton relations for reflection equation algebras, which are algebras generated by the indeterminates in $S = (s_{ij})$ subject to the relations $RS_1RS_1 = S_1RS_1R$ and there is common generalization of these two cases. A selection of references on these matters is [**302**]; [**303**]; [**343**]; [**344**]; [**345**]; [**346**]; [**572**].

In [**702**] there is a rather different (looking) Cayley-Hamilton theorem for quantum matrices. It is, however, much related to the above, [**550**]. Quite another Cayley-Hamilton theorem is in [**347**].

A very general Cayley-Hamilton theorem for any square matrix over an arbitrary unital associative ring is in [**655**]. This one is of the form

$$(\lambda_0 I + C_0) + A(\lambda_1 I + C_1) + \cdots + A^{n-1}(\lambda_{n-1} I + C_{n-1}) + A^n(\lambda_n I + C_n) = 0$$

where the 'noncommutativity correction terms' have their entries in the additive subgroup of commutator differences $[R, R]$.

Still another Cayley-Hamilton theorem, a two variable one, which is less known than it deserves[37], occurs in the theory of 'operator vessels', [**452**].

8.8.1.20. MacMahon master theorem. The original (classical) MacMahon master theorem, [**473**], Volume 1, pp 97-98, says the following.

Let A be a square matrix of size r with entries from some commutative ring. Let x_i, $i = 1, \ldots, r$ be a set of commuting indeterminates and form $X_i = \sum_{j=1}^{r} a_{ij} x_j$. For any sequence of nonnegative integers $(m_1, m_2, \ldots, m_r)$ let $G(m_1, m_2, \ldots, m_r)$ be the coefficient of $x_1^{m_1} x_2^{m_2} \cdots x_r^{m_r}$ in $X_1^{m_1} X_2^{m_2} \cdots X_r^{m_r}$. Then the MacMahon master formula says

$$\sum_{(m_1, m_2, \ldots, m_r)} G(m_1, m_2, \ldots, m_r) = \frac{1}{\det(I - A)}. \tag{8.8.1.21}$$

Slightly more generally, putting in counting indeterminates $t_1, \ldots, t_n$ (that commute with everything in sight), and letting $T = \operatorname{diag}(t_1, t_2, \ldots, t_r)$ be the diagonal matrix in these indeterminates, one has

$$\sum_{(m_1, m_2, \ldots, m_r)} G(m_1, m_2, \ldots, m_r) t_1^{m_1} t_2^{m_2} \cdots t_r^{m_r} = \frac{1}{\det(I - TA)}.$$

Now consider the symmetric and exterior powers of the matrix A and take their traces. Then

$$\operatorname{Tr}(S^n(A)) = \sum_{m_1 + m_2 + \cdots + m_r = n} G(m_1, m_2, \ldots, m_r)$$

$$\det(I - tA) = \sum_{n=0}^{\infty} (-1)^n \operatorname{Tr}(\Lambda^n(A)) t^n$$

and their results the following equivalent identity

$$\frac{1}{\sum_{n=0}^{\infty} (-1)^n \operatorname{Tr}(\Lambda^n(A)) t^n} = \sum_{n=0}^{\infty} \operatorname{Tr}(S^n(A)) t^n$$

called the **Boson-Fermion correspondence**. Here t is yet another counting indeterminate.

Now take noncommuting indeterminates x_i, $i = 1, \ldots, r$ subject to the commutation relations $x_j x_i = q x_i x_j$ for $i < j$; i.e. the algebra of functions of (the simplest) r-dimensional quantum space, and consider a q-Manin matrix $A = (a_{ij})$. It is supposed that the x's commute with the a's . Let X_i, $G(m_1, m_2, \ldots, m_r)$ be as before (but now the order of the indeterminates matters) and take

$$\det_q^{\text{column}}(B) = \sum (-q)^{\lg(\sigma)} b_{\sigma(1),1} b_{\sigma(2),2} \cdots b_{\sigma(r),r}.$$

[37] And which has nothing to do with the present considerations.

Define

$$\mathrm{Ferm}(A) = \sum_{J \subset \{1,2,\ldots,r\}} (-1)^{|J|} \det_q^{\mathrm{column}}(A_J)$$

where the sum is over all subsets J of $\{1, 2, \ldots, r\}$ and A_J is the submatrix of A obtained by removing all columns and rows whose index is not in J. Taking $J = \{1, 2, \ldots, r\}$ gives the empty submatrix, whose determinant is taken to be 1. Also define

$$\mathrm{Bos}(A) = \sum_{(m_1, m_2, \ldots, m_r)} G(m_1, m_2, \ldots, m_r)$$

then the quantum MacMahon master formula says, [**265**]

$$\mathrm{Bos}(A) = \frac{1}{\mathrm{Ferm}(A)}.$$

There is also a more quantum parameter version, $x_j x_i = q_{ij} x_i x_j$, see [**395**]. For more information on this topic, such as further generalizations, alternative proofs, ..., see for instance [**237**]; [**252**]; [**253**]; [**254**]; [**306**]; [**396**].

Both knot invariants, see section 8.5 above, and the quantum MacMahon master formula sort of run on R-matrices. So one could suspect links between the two. And, indeed, there are, see [**342**].

8.8.1.22. Other classical determinantal and matrix formulas and their generalizations. The above hardly exhausts the quantum generalizations of well known and important determinant and matrix formulas. In addition there are for instance quantum and (mildly) noncommutative versions of:

- Sylvester determinant identity, [**138**]; [**397**].
- Cramer formula for the inverse of a matrix, [**138**].
- Lagrange-Desnanot-Carroll formula, [**138**].
- Weinstein-Aronsjan formula, [**138**].
- Jacobi ratio theorem, [**138**].
- Plücker relations (q-Grassmann coordinates), [**138**]; [**428**]; [**430**].
- Schur complement, [**138**].
- Lagrange inversion formula, [**266**]; [**267**]; [**268**]; [**273**]; [**274**]; [**410**]; [**644**].

8.8.2. Quantum symmetry. Much of what is touched upon in this subsection has to do with quantum groups. Just what is a quantum group is not agreed upon. Some authors reserve the term for Hopf algebras that are deformations of algebraic groups depending on a parameter q (or a specific member of such a family; i.e. the parameter is given a value; others use the term 'quantum group' in the (wider) sense of a quasi-triangular Hopf algebra (defined below). Still others consider the term 'quantum group' synonymous with 'Hopf algebra'.

There are quite a few monographs and sets of lecture notes on the topic of quantum groups. Here is a selection: [**16**]; [**39**]; [**90**]; [**124**]; [**126**]; [**188**]; [**236**]; [**259**]; [**329**]; [**349**]; [**357**]; [**371**]; [**388**]; [**403**]; [**423**]; [**481**]; [**486**]; [**498**]; [**496**]; [**518**]; [**560**]; [**562**]; [**569**]; [**641**]; [**647**]; [**633**]; [**650**]; [**665**]. In addition there is an even larger collection of edited volumes more or less dedicated to quantum groups.

In contradistinction there are (comparatively speaking) but relatively few monographs and sets of lecture notes on Hopf algebras in more generality.

The classical idea of symmetry is, roughly, concerned with a group acting on a space, a manifold, an algebra, ..., and one is interested in entities that are invariant under the given action. That is the situation is like

$$G \times M \xrightarrow{\alpha} M \tag{8.8.2.1}$$

where G is a group and M is something on which G acts; that is $\alpha(e,m) = m$ where e is the identity element of the group and $\alpha(g, \alpha(g', m)) = \alpha(gg', m)$ for all $g, g' \in G$. Often one writes $\alpha(g,m) = gm$. For instance one is interested in functions or other derived quantities, f, of $m \in M$ that are invariant under G.

Obviously the automorphism group of the structure M is important here as it is the largest group that can act faithfully[38] on M.

For instance, there is the very well known example of the general linear group $\mathbf{GL}(n; \mathbf{C})$ acting on square complex entry matrices A of size n by $(S, A) \mapsto SAS^{-1}$. Here one invariant is the multiset of eigenvalues of A. (I.e. eigenvalues with their multiplicities.)

Another example is that of size $m \times n$ matrices over a commutative unital ring R under the action of $\mathbf{GL}(m; R) \times \mathbf{GL}(n; R)$ given by $A \mapsto SAT^{-1}$. Assume that R is a principal ideal ring. Then under this action every matrix can be brought into the form

$$\left(\begin{array}{cccc|c} d_1 & 0 & \cdots & 0 & \\ 0 & d_2 & \ddots & \vdots & \mathbf{0} \\ \vdots & \ddots & \ddots & 0 & \\ 0 & \cdots & 0 & d_r & \\ \hline & & \mathbf{0} & & \mathbf{0} \end{array}\right)$$

(Smith canonical form) for certain elements $d_1, \ldots, d_r$ from R such that d_i divides d_{i+1}. The sequence of ideals $(d_1) \supset \cdots \supset (d_r)$ is now an invariant for this action.

A less known example is that of completely reachable matrix pairs over a field k. Here the group consists of all block triangular matrices over k of the form

$$\begin{pmatrix} S & 0 \\ K & T \end{pmatrix} \tag{8.8.2.2}$$

where S and T are square invertible matrices of sizes n and m respectively. This group is called the feedback group. The space on which it acts is the vector space of matrix pairs (A, B) of sizes $n \times n$ and $n \times m$ such that the rank of the block matrix

$$(B \quad AB \quad A^2B \quad \cdots \quad A^{n-1}B)$$

is n (the completely reachable condition). The action of a group element (8.8.2.2) on such a pair is

$$(A, B) \mapsto (SAS^{-1} + BK, SBT^{-1})$$

which reflects 'base change in state space' given by S, 'base change in control space' given by T, and 'state feedback' given by the matrix K for linear control systems

[38] That means that the subgroup of all elements that leave every element of M fixed consists just of the identity element.

$\dot{x} = Ax + Bu$. The only invariants in this case are the so-called reachability indices (Kronecker indices), or, equivalently the nondecreasing sequence of positive integers

$$\operatorname{rank}(B),\ \operatorname{rank}(B\ AB),\ \ldots,\ \operatorname{rank}(B\ AB\ \ldots A^{n-1}B)$$

See for instance [**361**]; [**313**] for some information on such algebraic aspects of control theory.

Now what happens if the space, algebra, ..., M is deformed; for instance from a commutative algebra to a noncommutative one. This has a tendency to destroy a lot of the potential symmetry; i.e. to make the automorphism group (much) smaller.

Consider for instance the case of the standard action of the general linear group over a field k on the coordinates of affine space over that field

$$x \mapsto Sx, \quad x = (x_1, x_2, \ldots, x_n) \in k^n,$$
$$S = (s_{ij}) \in \mathbf{GL}(n; k), \quad (Sx)_i = \sum_j s_{ij} x_j.$$

This induces an action on the algebra of polynomials in the coordinates x_i

$$\mathbf{GL}(n; k) \times k[x_1, x_2, \ldots, x_n] \longrightarrow k[x_1, x_2, \ldots, x_n].$$

Now deform the algebra of commutative polynomials to an algebra of quantum polynomials. Slightly more generally let $Q = (q_{ij})$ be an $n \times n$ matrix of elements of k such that $q_{ij}q_{ji} - 1$ for all i, j. Now consider the algebra of quantum polynomials

$$k_Q[X_1, X_1^{-1}, X_2, X_2^{-1}, \ldots, X_r, X_r^{-1}, X_{r+1}, \ldots, X_n] \tag{8.8.2.3}$$

that is, the quotient of the free associative algebra

$$k\langle X_1, X_1^{-1}, X_2, X_2^{-1}, \ldots, X_r, X_r^{-1}, X_{r+1}, \ldots, X_n\rangle, \quad 0 \le r \le n$$

by the relations

$$X_i X_i^{-1} = 1, \quad i = 1, 2, \ldots, r \quad \text{and } X_i X_j = q_{ij} X_j X_i, \quad i, j = 1, 2, \ldots, n$$

The algebra of quantum polynomials is called a **generic algebra of polynomials** if the q_{ij}, $i < j$ in k^* are independent. Then there is the following result from [**32**].

Let α be an endomorphism of a generic algebra of quantum polynomials such that there are at least three variables that are not in the kernel of α. Then there are an $\varepsilon \in \{-1, 1\}$ and elements $a_w \in k$ such that for all w, $\alpha(X_w) = a_w X\varepsilon_w^{\varepsilon}$. If $r < n$, $\varepsilon = 1$.

Thus, in particular, quantization kills practically the whole infinite dimensional (if $n \ge 2$) Cremona group of algebra automorphisms of $k[X_1, \ldots, X_n]$.

Restricting to homogenous automorphisms and the one parameter deformation given by the relations

$$X_j X_i = q X_i X_j \quad \text{for } i < j \tag{8.8.2.4}$$

much the same thing happens if $q^2 \neq 1$. A homogeneous automorphism is given by an invertible matrix $A = (a_{ij})$, $X_i \mapsto \sum a_{ij} X_j$, and a quick calculation shows that this implies

$$\begin{aligned} &q a_{is} a_{js} = a_{js} a_{is} \quad \text{for } i < j \\ &[a_{is}, a_{jt}] = q^{-1} a_{js} a_{it} - q a_{ii} a_{js} \quad \text{for } i < j, s < t. \end{aligned} \tag{8.8.2.5}$$

If the entries of the matrix A commute this implies that the matrix is diagonal. If they are not required to commute one finds the conditions for an invertible q-Manin matrix, see 8.8.1 above.

The question is now: can the lost symmetry be recovered? That is, in the present setting, indeed the case by considering Hopf algebra symmetries. In fact the conditions (8.8.2.5) for a homogenous coaction on the left plus similar conditions for a homogeneous coaction on the right on another algebra of quantum polynomials serve to define an $n(n-1)/212$ parameter quantum group. See [**315**]; see also [**32**]; [**33**]; [**35**]; [**496**]; [**497**].

The phenomenon just described: loss of symmetry upon deformation (quantization) and regain of symmetry by considering the more general kind of symmetry made possible by considering quantum groups and Hopf algebras, is not an isolated phenomenon. It happens all over in quantum theory, in noncommutative geometry, ... Here, very briefly described, are some more examples.

Consider the two dimensional sphere. There is a large symmetry (automorphism) group, viz the orthogonal group $SO(3;\mathbf{R})$ of rotations in Euclidean 3-space. When deforming (the algebra of functions on) the sphere to obtain the Podleś quantum sphere this symmetry disappears, [**569**], p.3; [**568**]. It reappears when quantum group symmetries are allowed.

A classical completely integrable system such as the Toda lattice has a very large symmetry group given by the so-called dressing transformations, see [**44**]; [**635**]. However, in general, these transformations do not preserve the underlying Poisson structure. This was first remarked by Michael Semenov Tian-Shansky, [**635**]. The symmetry group needs to be quantized as well[39].

A fourth example comes from quantum field theory: symmetries that are broken by going to a noncommutative underlying geometry are recovered by considering quantum symmetries, [**453**].

A fifth example is in [**640**] where a hidden $U_q(\mathfrak{sl}(2))$ symmetry on a q-version of the first Weyl algebra of differential operators is described.

Returning to the general situation of (8.8.2.1), take an algebra of functions, $\mathrm{Fun}(M)$ on the structure M (but leave the group alone) with action

$$(g\varphi)(m) = \varphi(g^{-1}(m)), \quad g \in G, m \in M, \varphi \in \mathrm{Fun}(G)$$

and pointwise multiplication. This is the same as a kG-module algebra structure on the function algebra $\mathrm{Fun}(M)$:

$$kG \otimes \mathrm{Fun}(M) \longrightarrow \mathrm{Fun}(M).$$

Taking functions on the group as well one finds that (8.8.2.1) turns $\mathrm{Fun}(M)$ into a $\mathrm{Fun}(G)$ -comodule algebra

$$\mathrm{Fun}(M) \longrightarrow \mathrm{Fun}(G) \otimes \mathrm{Fun}(M).$$

[39]This makes one wonder how physicists have gotten away with using non-quantum groups like SU(3) for so long. The answer could be that the representation theory of such a group and its generic quantum analogue is virtually the same.

Thus, generalizing the idea of symmetry to admit symmetries embodied by Hopf algebras and quantum groups, both modules (representations) and comodules (corepresentations) are important. Often module algebras and comodule algebras.

8.8.2.6. Delimitation. In this subsection 8.8.2 the focus is on Hopf algebra symmetries. That is H-modules and module algebras, H-comodules and comodule algebras[40], and the associated theories of invariants and coinvariants.

That looks like "noncommutative invariant theory", "quantum invariant theory", and "quantum invariant" might be good terms to use for surfing the web or consulting bibliographic data bases, or preprint collections to find material. That is not the case.

The term 'noncommutative invariant theory" is mostly used to refer to the situation of a non-quantum group acting on a noncommutative algebra. See [**383**] for a survey on this topic; see also e.g. [**206**].

A search on the web using "quantum invariant theory" mostly yields pages that have to do with what is called "Lewis-Riesenfeld quantum invariant theory", [**440**], which is about time dependent systems such as the time dependent harmonic oscillator; principally the behavior of eigenvalues and eigenstates. This has nothing to do with quantum symmetry in the sense of a Hopf algebra (co)acting on something. Further, "quantum invariant" often refers to invariants of knots and 3-manifolds obtainable via quantum groups and braid groups as discussed in section 8.5 above.

It is quite amazing how many of the constructions, definitions, and results of the theory of algebraic groups and Lie groups and Lie algebras have natural good analogues for their quantum versions, especially the quantum versions of semisimple groups and Lie algebras. Below some of these will be briefly described, at least in the form of giving some, hopefully appropriate, references.

Before doing that, here is another bit of motivation for studying deformed (quantized) versions of things. It can be briefly summed up in the adagium: "to really understand a mathematical object includes understanding all its possible deformations". Or: "deformations are good".

This is a substantial part of the interest in resolution of singularities for differential mappings, for resolution of singular varieties, the study of multiplicities in, for instance, commutative and homological algebra, the deformation of complex manifolds à la Spencer, Kodaira-Kuranishi, and Nijenhuis, and the theory of deformations of algebras (and Hopf algebras) according to Gerstenhaber, Shack, Giaquinto[41].

In the present case what we have is a family of Hopf algebras depending on a parameter q, both in the function case and the universal enveloping case. This invites looking at what happens when q goes to zero (for an appropriate parametrization). This was initiated by Masaki Kashiwara and led to his notion of crystal bases[42] (and the closely related notion of Lusztig's canonical bases, [**460**]; [**459**]; [**458**]; [**461**]). These bases in turn lead to crystal bases in the original (non-quantized)

[40]And various dual notions.

[41]In [**272**] quantum symmetry is discussed from the point of view of deformation theory of algebras and Hochschild cohomology.

[42]The terminology 'crystal' derives from what happens in physics at (absolute) zero temperature.

context and there are applications of these to e.g. character formulas, [**370**]; [**613**] and Macdonald polynomials, [**234**]; [**235**]. For crystal bases see also [**1**]; [**329**]; [**362**].

Yangian's are a special kind of "RTT function type quantum groups" which have a variety of applications to non-quantized situations; see below.

In [**418**]; [**419**] there are other investigations of what happens for Hecke algebras and general linear quantum groups at zero in the context of representations and the quasi-symmetric and noncommutative symmetric functions of chapter 6.

The term "quantum symmetry" also encompasses several concepts which will not be touched upon here such as the Ocneanu algebras of quantum symmetries associated to ADE graphs in connection with rational conformal field theories of $\mathbf{SU}(2)$ type. See [**115**]; [**173**]; [**172**]; [**171**]; [**279**]; [**548**].

After these preliminaries, here is a partial alphabetical list of well known classical concepts and results that have good quantum group analogues.

$6j$-symbols. Clebsch-Gordan theory is concerned with the decomposition of a tensor product of irreducible representations into a direct sum of irreducible representations. The Clebsch-Gordon coefficients are coefficients that describe the decompostion as a sum of irreducible representations of the tensor product of two irreducible representations. In quantum physics such coefficients turn up when considering angular momentum coupling. Wigner's $6j$, or equivalently the Racah W symbols, arise when there are three sources of angular momentum in the coupling problem. The book [**119**] treats the representation theory of both classical $\mathfrak{sl}_2$ and quantum $U_q(\mathfrak{sl}_2)$ with particular emphasis on the $6j$-symbols and with a view towards applications to the Turaev-Viro quantum invariants of three manifolds. As is mostly the case, at roots of unity things are different from the generic case. (See also 'representation theory' below.)

The Clebsch-Gordan coefficients for quantum group representations relate to the structure constants of the underlying algebra of the quantum group involved, [**182**]; [**183**].

Automorphism groups of finite structures. The simplest finite structure is a finite set. The automorphism group of a finite set of cardinality n is the symmetric group on n letters. Switching to (complex valued) functions on a finite set one can wonder what the quantum group of automorphisms could be. Here there is a surprise: for cardinality ≥ 4 it is much larger than S_n, [**682**]. Other papers on quantum groups of finite structures, such as graphs, and of compact metric spaces, are [**46**]; [**47**]; [**285**]; [**63**].

Borel-Weyl construction. The Borel-Weyl theorem describes a powerful technique for constructing irreducible representations. A quantum version is discussed in [**68**]; [**286**].

Boson-Fermion statistics and quantum symmetry. Quantum symmetry is ill adjusted to the split up into Boson particle statistics and Fermion statistics, i.e. the split up into anti-symmetric and symmetric parts. More general (symmetric group) statistics are called for, [**248**]; [**656**].

Character formulas. As the representation theory (in the generic case) of quantum group deformations of semisimple Lie algebras, (Kac-Moody algebras) is virtually the same as in the non-deformed case, see the section on *representation theory* below, things like the Kac-Weyl and Demazure character and denominator formulas for highest weight representations remain valid. Crystal bases, mentioned above, are an additional strong tool in character formula theory; see e.g. [**370**]; [**508**]; [**562**], chapter 10; [**613**].

Clifford algebra. The Clifford algebras are, among many other things, important for the construction of spin (spinor) representations, [**389**] p. 440ff. For a quantum analogue, both of the algebras and the representations, see [**310**].

Conservation laws, Noether theorem. In theoretical physics, especially mechanics, and the calculus of variations, the presence of continuous symmetries (Lie group symmetries, infinitesimal symmetries) is practically synonomous with the presence of conserved quantities. This is a very rough statement of the Noether theorem, [**541**]. For instance conservation of energy comes from the time invariance of the laws of physics. For quantum versions, see for instance [**6**]; [**341**]; [**387**].

Double commutant theorem, (Brauer-)Schur Weyl duality. The double commutant theorem for the group $\mathbf{GL}(n;\mathbf{C})$ says the following. Let V be the fundamental representation of $\mathbf{GL}(n;\mathbf{C})$, i.e. $V=\mathbf{C}^n$ and the action is the standard one of $n\times n$ matrices on n-vectors. Take the m-fold tensor product of this fundamental representation; i.e g acts on $V^{\otimes m}$ diagonally: $g(v_1\otimes v_1\otimes\cdots\otimes v_m)=g(v_1)\otimes g(v_2)\otimes\cdots\otimes g(v_m)$. The symmetric group S_m also acts on $V^{\otimes m}$ by permuting the tensor factors. These two actions commute. Even better the images of the group algebras $\mathbf{GL}(n;\mathbf{C})$ and S_m in $\mathrm{End}(V^{\otimes m})$ are exactly each other's commutant.

For the quantum version one needs the correct (for this purpose) quantum analogues of the symmetric groups, which are the Hecke and BMW algebras, see the paragraphs on the symmetric groups below.

Then the quantum double commutant theorem is as follows, [**608**], p.465.

(i) Let U_q be a quantum group of type A defined via the standard R-matrix R. Let V be the fundamental representation of U_q. Take the image of the Hecke algebra (at q^2) in $\mathrm{End}(V^{\otimes m})$ and the image of U_q in $\mathrm{End}(V^{\otimes m})$ under the m-fold tensor product representation. Then these two subalgebras are the commutants of each other.

(ii) Let U_q be a quantum group of type B, C, D defined via the standard R-matrix R (of the appropriate type). Let V be the fundamental representation of U_q. Take the image of the BMW algebra (at (q^2,q^{2n}) for type B, at $(q^2,-q^{(2n-1)/2})$ for type D, and at $(q^2,-q^{-(2n+1)})$ for type C, in $\mathrm{End}(V^{\otimes m})$ and the image of U_q in $\mathrm{End}(V^{\otimes m})$ under the m-fold tensor product representation. Then these two subalgebras are the commutants of each other. The BMW algebra is described below in the section on quantum analogues of the symmetric group algebra.

Part (i) is due to Jimbo [**352**], part (ii) to Reshetikhin, [**585**]; [**586**].

A more general double centralizer theorem is in [**251**].

Endomorphism algebra representations. Let H be a Hopf algebra and V an H-module. This notion only involves the algebra part of H. The comultiplication enters the picture in providing the tensor product of two representation with an H-module structure making the category of H-modules into a tensor category.

In the usual situation given a representation ρ of, say, a group G with carrier space V over a field k, one constructs a corresponding endomorphism representation on $\mathrm{End}_k(V)$ as follows. Let α be an endomorphism of V, then $g(\alpha) = \rho(g)\alpha\rho(g^{-1}) = g\alpha g^{-1}$ (where on the right and in the middle juxtaposition means composition in $\mathrm{End}_k(V)$). This construction is of some importance, [**330**]. It is a representation by *algebra* endomorphisms.

Let $V^* = \mathbf{Vect}_k(V, k)$ be the (dual) space of k-linear functionals on V. Then the representation ρ defines a contragredient (or dual) representation

$$\rho^*(g)(\varphi)(v) = \varphi(\rho(g^{-1})v), \quad g \in G, \varphi \in V, v^* \in V^*$$

Under the canonical isomorphism $V \otimes V^* \simeq \mathrm{End}_k(V)$, $(v \otimes \varphi)(w) = \varphi(w)v$, the endomorphism representation defined by ρ gets identified with the tensor product representation $\rho \otimes \rho^*$.

The analogues for Hopf algebras H and (left) H-modules V are as follows. The contragredient H-module (representation) is given by $h(\varphi)(v) = \varphi(\iota_H(h)v)$ and the endomorphism H-module is defined by the formula

$$h(\alpha) = \sum_h h_{(1)}\alpha\iota_H(h_{(2)}), \quad \text{where } \mu_H(h) = \sum_h h_{(1)} \otimes h_{(2)}$$

and where h_1, h_2 are seen as endomorphisms of V. It is a nice little exercise to show that this makes $\mathrm{End}_k(V)$ an H-module algebra. Indeed, by the coassociativity of μ_H

$$\mu_H^{(4)} = (\mu_H \otimes \mu_H)\mu_H = (\mathrm{id} \otimes \mu_H \otimes \mathrm{id})(\mu_H \otimes \mathrm{id})\mu_H$$

which written out means

$$\sum_h h_{(1)(1)} \otimes h_{(1)(2)} \otimes h_{(2)(1)} \otimes h_{(2)(2)} =$$

$$\sum_h h_{(1)(1)} \otimes h_{(1)(2)(1)} \otimes h_{(1)(2)(2)} \otimes h_{(2)}.$$

The H-module algebra requirement is

$$h(\alpha\beta) = \sum_h h_{(1)}(\alpha)h_{(2)}(\beta).$$

So, using the formula above

$$\begin{aligned}
&\sum_h h_{(1)}(\alpha)h_{(2)}(\beta) \\
&= \sum_h h_{(1)(1)}\alpha\iota_H(h_{(1)(2)})h_{(2)(1)}\beta\iota_H(h_{(2)(2)}) \\
&= \sum_h h_{(1)(1)}\alpha\iota_H(h_{(1)(2)(1)})h_{(1)(2)(2)}\beta\iota_H(h_{(2)}) \quad \text{(by the associativity formula above)} \\
&= \sum_h h_{(1)(1)}\alpha\varepsilon_H(h_{(1)(2)})\beta\iota_H(h_{(2)}) \quad \text{(by the antipode property)}
\end{aligned}$$

$$= \sum_h h_{(1)(1)} \varepsilon_H(h_{(1)(2)}) \alpha\beta\iota_H(h_{(2)}) \qquad \text{(because } \varepsilon(h_{(1)(2)}) \text{ is a scalar)}$$

$$= \sum_h h_{(1)} \alpha\beta\iota_H(h_{(2)}) = h(\alpha\beta) \qquad \text{(by the counit property)}$$

proving (apart from checking the counit property for End(V), not for V) that End(V) is indeed an H-module algebra.

Frobenius reciprocity, induction, restriction. There are (of course) analogues of these in the context of quantum groups; to start with, see [**286**]; [**555**].

Heisenberg algebra. The "classical" Heisenberg algebra is the Lie algebra on $2n+1$ dimensional vector space with basis $x_i, \partial/\partial x_i, 1$ and commutators $[x_i, \partial/\partial x_i] = -1$, $[\partial/\partial x_i, 1] = [x_i, 1] = 0$. For the quantum version and its representations see [**488**]; [**605**]; [**606**].

Homogenous spaces. In the classical case a homogeneous space X is the space of left (or right) cosets of a Lie group G with respect to a closed subgroup N, $X = G/N$ (resp. $X = N\backslash G$). There is of course an induced left (respectively right) action of G on X. In terms of functions this gives an injection of the algebras of functions $\mathcal{O}(X) \longrightarrow \mathcal{O}(G)$ with the extra property that $\mu(\mathcal{O}(X)) \subset \mathcal{O}(X) \otimes \mathcal{O}(G)$ (in the left coset space case).

Thus it is entirely natural for a quantum group with Hopf algebra of functions $\mathcal{O}_q(G)$ to consider a quantum subgroup with algebra of functions $\mathcal{O}_q(K)$. Here quantum subgroup means that $\mathcal{O}_q(K)$ comes together with a surjective morphism of Hopf algebras $\mathcal{O}_q(G) \xrightarrow{\pi} \mathcal{O}_q(K)$. Then there is a left and a right coaction of $\mathcal{O}_q(K)$ on $\mathcal{O}_q(G)$ given by

$$L_K = (\pi \otimes \mathrm{id})\mu : \mathcal{O}_q(G) \longrightarrow \mathcal{O}_q(K) \otimes \mathcal{O}_q(G)$$

$$R_K = (\mathrm{id} \otimes \pi)\mu : \mathcal{O}_q(G) \longrightarrow \mathcal{O}_q(G) \otimes \mathcal{O}_q(K)$$

and one defines

$$\begin{aligned} &\mathcal{O}_q(K)\backslash\mathcal{O}_q(G) = \{a \in \mathcal{O}_q(G) : L_K(a) = 1 \otimes a\} \\ &\mathcal{O}_q(G)/\mathcal{O}_q(K) = \{a \in \mathcal{O}_q(G) : R_K(a) = a \otimes 1\} \\ &\mathcal{O}_q(K)\backslash\mathcal{O}_q(G)/\mathcal{O}_q(K) = \mathcal{O}_q(K)\backslash\mathcal{O}_q(G) \cap \mathcal{O}_q(G)/\mathcal{O}_q(K) \end{aligned} \qquad (8.8.2.6a)$$

the left invariant, right invariant, and bi-invariant elements of $\mathcal{O}_q(G)$ with respect to the quantum subgroup $\mathcal{O}_q(K)$. These are the (coordinate algebras of the) left and right quantum homogenous spaces defined by this quantum subgroup.

One easily sees that these three submodules are subalgebras and one proves without difficulty that they satisfy

$$\mu_G(\mathcal{O}_q(K)\backslash\mathcal{O}_q(G)) \subset \mathcal{O}_q(K)\backslash\mathcal{O}_q(G) \otimes \mathcal{O}_q(G)$$

$$\mu_G(\mathcal{O}_q(G)/\mathcal{O}_q(K)) \subset \mathcal{O}_q(G) \otimes \mathcal{O}_q(G)/\mathcal{O}_q(K)$$

as should be. This is the approach taken in [**388**], 11.6.2.

In [**126**], section 13.2, there is a slightly different (but equivalent modulo the right ideas of duality) point of view. Consider a Lie group acting on a manifold X. Identify the left invariant vector fields on the Lie group with its Lie algebra $\mathfrak{g}$. Then the Lie algebra $\mathfrak{g}$ acts by derivations on the functions on X; that is

$$Z(f_1 f_2) = Z(f_1) f_2 + f_1 Z(f_2)$$

which can be written as

$$Z(f_1 f_2) = m_X \mu(Z)(f_1 \otimes f_2)$$

where m_X is the multiplication of functions on X and μ is (the restriction to $\mathfrak{g}$ of) the comultiplication on $U(\mathfrak{g}) \supset \mathfrak{g}$.

This action extends to an action of the universal enveloping algebra $U(\mathfrak{g})$ on the algebra of functions $C^\infty(X)$. Thus it is natural to define a quantum G space as an algebra A equipped with a $U_q(\mathfrak{g})$-module algebra structure.

All the classical homogeneous spaces seem to have their quantum analogues (often many different ones): spheres, projective spaces, Grassmann manifolds complete with Plücker coordinates and their relations, Schubert varieties, flag varieties, Stiefel manifolds, ... A selection of references is: [**18**]; [**99**]; [**191**]; [**249**]; [**250**]; [**286**], [**404**], [**480**], [**481**], section 6.1, [**500**], [**507**], [**508**], [**630**], [**657**].

Invariant theory and coinvariant theory. As the name indicates (classical) invariant theory is concerned with a group acting on a module or algebra and with the question of what things remain invariant. Special emphasis has been given to the study of the case of the general linear, special linear, orthogonal, symplectic groups acting on tensor products $V^{\otimes m} \otimes (V^*)^{\otimes n}$ where V is the fundamental (defining, basis, natural) representation of the group in question and V^* its contragredient (dual) representation, [**330**], and to the case of an algebraic group acting on (the algebra of functions of) algebraic varieties (geometric invariant theory, [**531**]).

In the noncommutative (Hopf algebra, quantum group) case, as indicated above, both modules and comodules are important.

Let H be a Hopf algebra and V a left H-module. Then the module of invariants is

$$V^H = \{v \in V : hv = v \text{ for all } h \in H\}$$

and if W is a left H-comodule, $\gamma : W \longrightarrow H \otimes W$

$$W^H = \{w \in W : \gamma_H(w) = 1 \otimes w\}.$$

If V (resp. W) is an H-module algebra (resp. H-comodule algebra), the module of invariants is an H-module algebra (resp. H-comodule algebra). The book [**521**] is a good starting point for learning about the action of Hopf algebras on rings.

A selection of papers on invariants and coinvariants for actions and co-actions of quantum groups is: [**203**]; [**204**]; [**205**]; [**246**]; [**245**]; [**250**]; [**282**]; [**283**]; [**284**]; [**305**]; [**406**]; [**435**]; [**436**].

According to Weyl, [**686**], no theory of invariants is complete without a first fundamental theorem (on finite generation of algebras of invariants) and a second fundamental theorem (finite generation of the relations (syzygies) between generators). And indeed these kinds of theorems are starting to appear in the quantum case, [**66**], [**204**], [**250**], [**283**].

Lattice statistical mechanics models. Consider again, as in section 8.4 above, a two-dimensional lattice statistical mechanics model. But now let the four edges impinging on a site be labelled by vectors from some space V and let the Boltmann weight of the site depend on these four vectors. Further suppose that the vector space V carries a (quantum) representation. The question arises, e.g., what statistical mechanical quantities remain invariant. Some results on this and related matters are in [**62**]; [**311**]; [**312**].

Lorentz and Poincaré groups. For quantum groups associated to a Cartan matrix (semisimple Lie algebra, Kac-Moody Lie algebra, the corresponding groups) there are the two constructions touched upon in section 8.4, the quantum universal enveloping algebra, $U_q(\mathfrak{g})$, and function quantum group, $\mathcal{O}_q(G)$, constructions, which are dual to each other.

In addition there is the C^*-based approach of Woronowicz, [**688**]; [**694**].

However, deforming (quantizing) non semisimple Lie groups (algebras) such as the Poincaré and Lorentz groups is rather a different matter and there seems to be no general recipe. Some references are: [**7**]; [**29**]; [**37**]; [**38**]; [**39**]; [**455**]; [**454**]; [**456**]; [**480**]; [**693**]; [**692**].

Minkowski space. A noncommutative version of Minkowski space has been introduced. One motivation being that it might be a carrier for quantum gravity. It is often referred to as κ-Minkowski space. Some papers on it are [**7**]; [**421**]; [**569**]. This of course also has to do with noncommutative geometry and symmetries of noncommutative geometries, a topic for which some references are given in subsection 8.8.3 below.

Peter-Weyl theory. Let G be a compact topological group and consider the space $L^2(G)$ of square integrable functions on it with respect to normalized Haar measure. This space carries a representation of G under left translation (the analogue of the left regular representation of a finite group). The original Peter-Weyl theorem, [**565**]; [**389**], §1.5, says that this representation breaks up into a topological direct sum of irreducible representations each occurring with a multiplicity equal to its degree.

The quantum version gives a similar decomposition of the algebra of functions for the quantum groups $\mathcal{O}_q(G)$, $G = \mathbf{SL}(n)$, $\mathbf{O}(n)$, $\mathbf{SO}(n)$, $\mathbf{Sp}(n)$, $\mathbf{GL}(n)$, and also for generic cosemisimple Hopf algebras, [**388**], §11.2, §11.5.

For quantum Haar measure and integrals for Hopf algebras see [**481**], section 1.7; [**388**], 4.1.6, 11.2.2, and chapter 3 in this volume.

Quantum field theory. There is a substantial amount of published work on (quantum) symmetries for quantum field theories (conformal, rational conformal, topological). Lack of space precludes giving enough background to explain how this works, but here is a selection of references: [**81**]; [**80**]; [**82**]; [**89**]; [**115**]; [**242**]; [**260**]; [**270**]; [**271**]; [**277**]; [**279**]; [**379**]; [**478**]; [**615**]; [**666**].

The paper [**89**] is especially recommended; the paper [**279**] pays special attention to the root of unity case.

Quasitriangular Hopf algebras. A pair (H, R) consisting of a Hopf algebra H and an invertible element $R = \sum_i a_i \otimes b_i \in H \otimes H$, is a quasi-triangular Hopf algebra if

$$\tau\mu_H(h) = R\mu_H(h)R^{-1},$$

$$\text{where, as usual } \tau \text{ is the switch map } \tau(x \otimes y) = y \otimes x \tag{8.8.2.7}$$

$$(\mu_H \otimes \mathrm{id})R = R_{13}R_{23}, \quad (\mathrm{id} \otimes \mu_H)R = R_{13}R_{12}.$$

The first condition is almost cocommutativity; i.e the Hopf algebra is cocommutative up to conjugation; the equations of the second condition are sometimes called hexagon equations. In those equations, of course,

$$R_{12} = \sum a_i \otimes b_i \otimes 1,\ R_{13} = \sum a_i \otimes 1 \otimes b_i,\ R_{23} = \sum 1 \otimes a_i \otimes b_i.$$

The pair (H, R) is a triangular Hopf algebra if in addition $R_{12}R_{21} = 1$.

The meaning and importance of the second axiom for quasi-triangularity is not very clear at first sight. One property that follows from these two equations is that if the element R is used to define a morphism $\tilde{R}_1 : H^* \longrightarrow H$ by $\tilde{R}_1(h^*) = (\mathrm{id}\otimes h^*)(R)$ and a morphism $\tilde{R}_2 : H^* \longrightarrow H$ by $\tilde{R}_2(h^*) = (h^* \otimes \mathrm{id})(R)$, then these equations imply that $\tilde{R}_2$ is an algebra and anticoalgebra morphism and that $\tilde{R}_1$ is an anti algebra and coalgebra morphism. This is exercise 2.1.5 on page 35 in [**481**].

The notation "R" is not an accident. The element $R \in H \otimes H$ satisfies

$$R_{12}R_{13}R_{23} = R_{23}R_{13}R_{12}$$

[**371**], theorem VIII.2.4, p.175. The proof of this uses the second axiom of (8.8.2.7). As a consequence, if $\rho : H \longrightarrow \mathrm{End}(V)$ is a representation of H the image $(\rho \otimes \rho)(R) \in \mathrm{End}(V \otimes V)$ is a solution of the Yang-Baxter equation[43].

An important consequence of almost commutativity (also called quasi-commutativity) is that if V and W are two H-modules then $V \otimes W \simeq W \otimes V$, [**521**], lemma 10.1.2, p. 178[44].

The function quantum groups $\mathcal{O}_q(G)$ of an affine algebraic group and the quantized universal enveloping algebras $U_q(\mathfrak{g})$ of a Kac Moody Lie algebra are examples of quasi-triangular Hopf algebras.

Representation theory. The representation theory of quantum groups like $U_q(\mathfrak{g})$ and their (dual) function algebra counterparts is very different depending on whether 'q is generic', meaning that q is not a root of unity or that q is treated as an additional variable (commuting with everything in sight), or whether it is a root of unity. See e.g. [**388**].

In the generic case things are practically the same as in the non-quantized case. In particular every representation of $\mathfrak{g}$ can be deformed to a representation of $U_q(\mathfrak{g})$, irreducible representations are still labelled by Young diagrams and apart from a possible tensor product with a one dimensional representation each one is obtained as such a deformation, [**607**]. See also e.g. [**129**].

There is also, for instance, a double commutant theorem, see above, and below in the paragraphs on the quantum analogues of the symmetric groups.

See e.g. [**202**]; [**388**], 3.3 and 7.5, for some information on representations of quantum groups at a root of unity (of sufficiently high order).

Symmetric group, Birman-Murakami-Wenzl (BMW) algebra. In the non-quantum representation theory (of, say, $\mathbf{GL}(n)$), the symmetric group plays an important role (**Schur-Weyl duality**).

Take a parameter $\varepsilon \in k$. Consider $m - 1$ indeterminates $g_1, g_2, \ldots, g_{m-1}$ and take the quotient of the free associative algebra in these indeterminates subject to

[43] in star-triangle form not in braid form.

[44] This is not always the case. For instance it does not hold when H is the dual of the group algebra of a finite noncommutative group.

the relations

$$g_i g_j = g_j g_i \quad \text{if } |i-j| \geq 2$$
$$g_i g_{i+1} g_i = g_{i+1} g_i g_{i+1}$$
$$g_i^2 = (\varepsilon - 1) g_i + \varepsilon.$$

This is the **Hecke algebra** $\mathcal{H}_\varepsilon(m)$. Note that for $\varepsilon = 1$ one finds the group algebra kS_m.

Let $U_q(\mathfrak{g})$ (or the function version $\mathcal{O}_q(G)$, borrowing a notation from algebraic geometry) be a quantized universal enveloping algebra (or quantum group) of type A based on its standard R-matrix R. As before, see sections 8.2 - 8.5, let

$$R_i = \underbrace{\mathrm{id} \otimes \mathrm{id} \otimes \cdots \otimes \mathrm{id}}_{i-1} \otimes R \otimes \underbrace{\mathrm{id} \otimes \mathrm{id} \otimes \cdots \otimes \mathrm{id}}_{m-i-1}$$

then the R_i satisfy the braid relations

$$R_i R_j = R_j R_i \quad \text{if } |i-j| \geq 2$$
$$R_i R_{i+1} R_i = R_{i+1} R_i R_{i+1}.$$

In addition there is the quadratic relation

$$R_i^2 = (q - q^{-1}) R_i + 1$$

for the R-matrix

$$q \sum_i E_{ii} \otimes E_{ii} + \sum_{i \neq j} E_{ij} \otimes E_{ji} + (q - q^{-1}) \sum_{i<j} E_{jj} \otimes E_{ii}$$

which is one of the standard R-matrices for defining $\mathbf{GL}_q(n)$, so that the qR_i define a representation of the Hecke algebra $\mathcal{H}_{q^2}(m)$ on $V = k^m$.

Now take two parameters $\varepsilon, r \neq 0$. Consider again $m-1$ indeterminates $g_1, g_2, \ldots, g_{m-1}$, this time together with their inverses and consider the quotient of the free algebra generated by them subject to the following relations

$$g_i g_j = g_j g_i \quad \text{if } |i-j| \geq 2$$
$$g_i g_{i+1} g_i = g_{i+1} g_i g_{i+1}$$
$$e_i g_i = r^{-1} g_i$$
$$e_i g_i^{\pm 1} e_i = r^{\pm 1} e_i$$

where by definition

$$e_i = 1 - (\varepsilon - \varepsilon^{-1})^{-1} (g_i - g_i^{-1}).$$

This is the **Birman-Murakami-Wenzl algebra** (BMW algebra), $BMW_{\varepsilon,r}(m)$. It came originally out of the theory of quantum link invariants. It plays the role of the symmetric group for the quantum deformations for types B, C, D. See [**381**]; [**291**]; [**431**]; [**275**]; [**433**]; [**150**]; [**612**]; [**332**]; [**432**]; [**434**] for some literature on BMW algebras.

The R-matrices for quantum deformations in the B, C, D cases define a representation of the BMW algebras in the tensor powers $V^{\otimes m}$.

Both Hecke algebras and BMW algebras are discussed also in [**388**], section 8.6.4, p. 293ff.

Schur algebra. Let R be a commutative ring. The classical **Schur algebra** can be described as the algebra $\mathrm{End}_{S_d}(V)$ of endomorphisms of the d-th tensor power of a free finite rank module over R that commute with the action of the symmetric group permuting the tensor factors, see e.g. [**209**] for type A and [**210**] for types B, C, D. Schur algebras are important in representation theory (Schur-Weyl duality), [**287**]; [**505**] (general linear case); [**200**] (symplectic case).

A quantum version was introduced in [**199**]. See [**207**]; [**208**]; [**651**]; [**660**]; [**678**] for some further information on **q-Schur algebras**. The book [**207**] is especially recommended. (The author describes it as a q-version of [**287**]).

Tannaka reconstruction, Tannaka-Krein duality. Consider a compact group G and its category of finite dimensional representations over $\mathbf{C}$. There is the tensor product of representations and there is also a conjugation operation. Tannaka reconstruction (Tannaka-Krein theory), [**664**]; [**417**] is a process for recovering the group from this category (with extra structure).

This is done by looking at the forgetful functor U that assigns to a representation its carrier space and its functorial endomorphisms. Such a morphism u is a collection (u_ρ) of endomorphisms of the carrier vector spaces V_ρ of the representations (ρ, V_ρ) of the compact group G such that for each morphism of representations (intertwining operator), $h : (\rho, V_\rho) \longrightarrow (\sigma, V_\sigma)$ the following diagram commutes

$$\begin{array}{ccc} V_\rho & \xrightarrow{u_\rho} & V_\rho \\ \downarrow h & & \downarrow h \\ V_\sigma & \xrightarrow{u_\sigma} & V_\sigma \end{array}$$

These form in any case a monoid. Now consider only those functorial endomorphisms that are tensor product preserving, self-conjugate, and such that $u_1 = \mathrm{id}$, where $\mathbf{1}$ is the trivial one dimensional representation. These form a group $\mathfrak{T}(G)$. (The invertibility follows by considering the contragredient representation (ρ^*, V^*) of a representation (ρ, V) and the morphism of representations $\rho^* \otimes \rho \longrightarrow \mathbf{1}$ given by the evaluation map $V^* \otimes V \longrightarrow \mathbf{C}$, $v^* \otimes w \mapsto v^*(w)$.)

Every element $g \in G$ defines an endomorphism of the functor U, viz. the functorial endomorphism $u(g)_\rho(v) = \rho(g)(v)$. This defines a morphism of groups $G \longrightarrow \mathfrak{T}(G)$ and one proves that this is in fact an isomorphism.

The Tannaka reconstruction procedure can be applied in various situations. There are reconstruction theorems for algebras, coalgebras, Hopf algebras, quasi-Hopf algebras, ... This is also a way of obtaining quantum groups starting with a suitable (forgetful) functor. So now the matter at hand is not reconstruction, but construction. This last bit was initiated by Lyubashenko, [**467**]; see also [**468**]; [**469**]; [**470**]; [**471**] and (for other ways) [**358**] section 12; and [**690**].

It may happen that Tannaka reconstruction gives more than the original object (hidden symmetries), cf [**561**]. To what extent various other 'hidden symmetries' such as those in [**9**]; [**14**]; [**72**]; [**277**]; [**640**] can be seen this way seems open.

A selection of other writings on the topic of Tannaka reconstruction is [**2**]; [**55**]; [**181**]; [**219**]; [**479**]; [**481**], section 9.4; [**561**]; [**527**]; [**538**]; [**616**].

Weyl algebra. The Weyl algebras (of differential operators with polynomial coefficients), which are the universal enveloping algebras of the Heisenberg Lie algebras (of quantum commutation relations) are important for the Segal-Shale-Weil representations (oscillator representations, metaplectic representations), [**255**], chapter 4; [**369**]. For quantum versions see [**310**].

Wigner-Eckart theorem. Let ρ be an irreducible representation of a compact group G with carrier space V_ρ of dimension m. Further, let the t^ρ_{ij} denote the matrix elements of this representation with respect to some basis. A set of operators $A^\rho = \{A^\rho_i : i = 1, \ldots, m\}$ acting on a Hilbert space $\mathcal{H}$ is said to be a tensor operator transforming under the representation ρ if there exists a continuous representation T of G on $\mathcal{H}$ such that

$$T(g)A^\rho_i T(g^{-1}) = \sum_{j=1}^{m} t^\rho_{ji}(g) A^\rho_j.$$

Thus, the space spanned by the operators from A^ρ is taken into itself by the endomorphism representation associated to T and it does so according to the irreducible representation ρ. The Wigner-Eckardt theorem gives a useful formula for the coefficients of the operators A^ρ_i in terms of the Clebsch-Gordan coefficients of the tensor products of the irreducible components of T with ρ.

A quantum group analogue is discussed in [**68**]; [**388**], section 3.6.

Yangian. The Yangians, introduced by Drinfel'd are a special kind of "RTT-type", see section 8.4 above, quantum group. They are obtained by taking the R-matrix equal to $I + u^{-1}P$ where P is the switch matrix with entries $a^{ji}_{ij} = 1$ and all other entries equal to 0. There are also twisted Yangians, [**554**].

There is a substantial literature on (twisted) Yangians and their applications: [**3**]; [**31**]; [**61**]; [**73**]; [**92**]; [**125**]; [**128**]; [**211**]; [**213**]; [**385**]; [**384**]; [**386**]; [**393**]; [**405**]; [**438**]; [**439**]; [**449**]; [**515**]; [**516**]; [**517**]; [**518**]; [**534**]; [**535**]; [**537**]; [**554**]; [**553**]; [**709**]; [**710**]; [**711**]. The book [**518**] is especially recommended.

8.8.3. Hopf algebra symmetries in noncommutative geometry. Noncommutative geometry operates under a similar philosophy as quantum group theory and generalized symmetry (Hopf algebra symmetry): replace the commutative algebra of functions by a noncommutative algebra with enough extra structure that something nontrivial can be done and that there are solid real problems that can be addressed via the new noncommutative structures. In the present case C^*-algebras of operators.

Sightly more precisely, a noncommutative geometry is a (socalled) spectral triple $(A, \mathcal{H}, D)$, where A is an algebra (thought of as an algebra of functions) with a representation as a C^*-algebra of operators on a Hilbert space $\mathcal{H}$ and D is an unbounded operator with bounded resolvent that is designed to give the differentable, metric, etc. structures. The standard (introductory) literature is represented by: [**4**]; [**149**]; [**154**]; [**158**]; [**159**]; [**167**]; [**168**]; [**220**]; [**221**]; [**222**]; [**276**]; [**382**]; [**425**]; [**429**]; [**474**]; [**475**]; [**476**]; [**502**]; [**503**].

The book [**154**] is still the Bible (and is freely available in electronic form, as should be) but it does not treat of Hopf symmetry which is what the present scribblings are about.

Symmetries in noncomutative geometry. The symmetry situation seems to be as expected. Replacing a classical commutative object with a noncommutative one (for instance a deformation) often results in loss of symmetry which is recovered by admitting generalized symmetries represented by Hopf algebra module and comodule structures. Lack of space prevents explaining enough of noncommutative geometry to see how this works more concretely. So here is simply a list of selected references on the matter: [**4**]; [**36**]; [**38**]; [**49**]; [**54**]; [**74**]; [**94**]; [**97**]; [**101**]; [**161**]; [**164**]; [**165**]; [**166**]; [**304**]; [**480**]; [**484**]; [**485**]; [**487**]; [**523**]; [**679**]; [**680**]; [**719**]. There is overlap with the matter of covariant differential calculi discussed below and with quantum homogeneous spaces discussed in the previous section.

Spheres galore. It is somewhat remarkable how many different noncommutative spheres there are. Some literature on these is [**101**]; [**175**]; [**176**]; [**177**]; [**178**]; [**309**]; [**567**]; [**568**]; [**679**]. The Podleś quantum spheres seem to be regarded as the standard ones.

Finite noncommutative geometries. A noncommutative geometry can be finite (or zero dimensional) and that by no means makes them trivial: [**74**]; [**487**]; [**563**].

Links with other structures. Some rather unexpected links of Hopf algebras in noncommutative geometry with other structures which seem to be particularly interesting are [**174**]; [**365**]. This is in addition to the relations with algebraic number theory as in [**502**]; [**503**].

Reconstruction. Suppose that one has a spectral triple $(A, \mathcal{H}, D)$ that satisfies all the axioms and that the algebra A happens to be commutative. Does that mean that it comes from a 'classical' differentiable manifold. Almost, but not quite. The matter is examined in [**584**].

Hopf algebras of rooted trees. The (first) Hopf algebra of rooted trees constructed by Kreimer to disentangle the combinatorics of renormalization in perturbative quantum field theory is much related to the Hopf algebra of transverse foliations in noncommutative geometry, [**155**]; [**680**]. Also, as remarked in these two papers, there are universality properties present.

Actually this Hopf algebra was discovered much earlier by Butcher, [**107**], as remarked and discussed in [**85**].

Still less known is a Hopf algebra of rooted trees that came out of control theory, [**292**]; [**293**]. This last one has been proved isomorphic to the Kreimer Hopf algebra, [**327**]; [**556**].

Next there is a construction of free noncommutative Rota-Baxter algebras in [**229**]; [**230**]; [**231**]. See below for some notes on Baxter algebras. This construction is also based on rooted trees and certainly much related to the Kreimer and Grossman-Larson Hopf algebras. There are some remarks on that in the papers [**229**]; [**230**]; [**231**] and in [**493**], section 7.

Finally in [**718**] there is a connection between certain structures over the Grossman-Larson Hopf algebra and the Hopf algebra of quasisymmetric functions of chapter 6 of the present volume. Further there there are the papers [**300**]; [**301**] linking (the Hopf algebra of) renormalization to multiple zeta functions and the algebra of multiple zetafunctions is nothing else than the Hopf algebra of quasisymmetric functions, [**325**]; [**326**]; [**328**].

8.8.4. Hopf algebras in Galois theory. Galois theory, as usually understood, deals with field extensions L/K that are both normal and separable and then there is the Galois correspondence between intermediate fields and subgroups of the Galois group $\mathrm{Gal}(L/K)$. There is also, and has been for quite some time. a theory for purely inseparable extensions in terms of Lie algebras (derivations and higher derivations), see [**189**]. Hopf algebras generalize both Lie algebras and groups. So it is not unreasonable to suspect that Hopf algebras could be generalized carriers of symmetry for more general extensions. Such a theory was started in [**131**].

Here is an example[45]. Consider the non-normal extension $\mathbf{Q}(\sqrt[3]{2})$ of the rational numbers $\mathbf{Q}$. There is no automorphism of $\mathbf{Q}(\sqrt[3]{2})$ over $\mathbf{Q}$ except the identity. Yet there is a Hopf algebra acting on $\mathbf{Q}(\sqrt[3]{2})$ (in a nice way) such that the algebra of invariants is precisely $\mathbf{Q}$.

There are formulations of Hopf-Galois theory both in terms of H-comodule algebras (and co-invariants) and H-module algebras (and invariants). There are indications that the "co" picture is the right one; more precisely, formulations in terms of corings as discussed in [**102**]; [**105**], see [**280**], [**232**].

One of the definitions of a Hopf-Galois extension that is often used is the following. Let H be a Hopf algebra over a field k and A a right H-comodule algebra $\alpha : A \longrightarrow A \otimes H$. Let $R = \{a \in A : \alpha(a) = a \otimes 1\}$ be the algebra of co-invariants. Then the extension $R \subset A$ is called a Hopf-Galois extension if the map

$$A \otimes_R A \longrightarrow A \otimes H, \quad x \otimes y \mapsto (x \otimes 1)\alpha(y)$$

is bijective[46] and A is faithfully flat as an R-module. This definition goes back to [**416**] (which in turn generalizes definitions from [**130**]; [**131**].

A weakness of the definition is that the Hopf algebra H does not intrinsically arise as some sort of symmetry algebra and (for the case of field extensions) that there is often more than one Hopf algebra that does the job, [**117**]; [**290**].

There are a good many other extension type (and, dually, covering type) situations in mathematics where one can dream about some sort of Galois theory and Galois correspondence.

And in addition there is the matter of 'forms' of objects. Given, say, a field extension L/K two algebraic objects over K are said to be L/K-forms of each other if they become isomorphic over L. This topic involves Galois cohomology.

All in all there is a very considerable amount of literature on Hopf-Galois theory. A selection of references is: [**11**]; [**15**]; [**42**]; [**65**]; [**100**]; [**103**]; [**104**]; [**108**]; [**109**]; [**110**]; [**117**]; [**139**]; [**140**]; [**141**]; [**148**]; [**180**]; [**232**]; [**245**]; [**280**]; [**290**]; [**373**]; [**522**]; [**520**]; [**559**]; [**617**]; [**618**]; [**619**]; [**620**]; [**630**]; [**631**]; [**632**]; [**654**]; [**672**]; [**684**]; [**698**]; [**700**].

The lecture notes [**104**] are particularly recommended.

---

[45] This is the starting example from [**290**].

[46] Assuming that the situation is such that dimension and rank counting make sense, this implies that $\dim_k(H) = \mathrm{rank}_R(A)$. So things make sense. Also this fits with the Galois group case for separable normal field extensions.

There are many papers on the Galois theory of differential and difference equations (Picard-Vessiot theory). These have found a most nice unification in [**15**].

A vexing problem has been a normal basis theorem for the rings of integers of local and global fields. Here Hopf algebras also help, see [**110**]; [**141**].

Special mention should be made of the Grothendieck-Teichmüller group, defined by Drinfel'd in 1991, [**214**], using quasi Hopf algebras. It is closely related to the absolute Galois group $\mathrm{Gal}(\bar{\mathbf{Q}}/\mathbf{Q})$ of the algebraic closure of the rational numbers. This object is showing a strong tendency to turn up near everywhere. Some papers are [**48**]; [**308**]; [**626**], of which [**308**] is recommended to start with[47].

Finally, of course, Hopf-Galois theory does not live in isolation but interacts with other things, including noncommutative geometry and quantum invariant theory; a few papers that are interesting but difficult to place are [**179**]; [**501**]; [**547**].

8.8.5. Hopf algebras and renormalization. Renormalization (in perturbative quantum field theory) is a tricky business, conceptually as well as algebraically and combinatorially. Certain Hopf algebras of trees (there are several) help to organize things. Even before the swath of papers having to do with the Hopf algebra of trees approach to renormalization this was a vast subject. There is not space enough in this volume to explain things anywhere well enough to prepare a reader new to the subject. So here are just some pointers to the literature. It is important, as was remarked in the last section of 8.8.3 above, that the Hopf algebras of renormalization also turn up in a variety of other contexts far from quantum field theory as things are presently understood.

Besides Hopf algebras an important role in this approach to renormalization, is played by Birkhoff factorization (decomposition) and the Riemann-Hilbert problem.

Three basically introductory and accessible papers on the topic of Hopf algebras in renormalization are [**143**]; [**492**]; [**493**].

A selection of further material is [**60**]; [**76**]; [**77**]; [**78**]; [**79**]; [**81**]; [**82**]; [**83**]; [**84**]; [**85**]; [**86**]; [**87**]; [**88**]; [**89**]; [**144**]; [**155**]; [**156**]; [**157**]; [**160**]; [**162**]; [**163**]; [**223**]; [**224**]; [**225**]; [**227**]; [**247**]; [**327**]; [**411**]; [**412**]; [**413**]; [**414**]; [**415**]; [**427**]; [**490**]; [**491**]; [**509**]; [**510**]; [**556**]; [**674**]; [**675**]; [**676**].

There is an analogue, if one may call it that, of renormalization in number theory, is discussed in [**244**]; [**300**]; [**301**]. And these may well help to understand renormalization better.

Finally there are two most interesting papers linking Lie algebras of loops with the Hopf algebra of renormalization: [**264**]; [**669**].

8.8.6. Quantum calculi. *q*-numbers. There are q-analogs of many things in mathematics. For instance q-orthogonal polynomials, q-differentiation, q-binomial theorem, q-Vandermonde identity, q-hypergeometric series, q-analogs of Ramsey

---

[47] The Grothendieck-Teichmüller group is already 'quantum' in view of its definition; so the paper [**626**] is sort of double quantum.

theory and the Sperner lemma, ... These usually involve q-numbers. Here, depending on context, q is a (complex) number or an indeterminate. The most important are[48]

$$[n]_q = \frac{1-q^n}{1-q}, \ [n]_q! = [n]_q[n-1]_q \cdots [2]_q[1]_q, \quad (q - \text{factorials})$$

$$\binom{n}{m}_q = \frac{[n]_q!}{[m]_q![n-m]_q!} \quad (q - \text{binomial coefficients})$$

$$(a;q)_n = (1-a)(1-qa)\cdots(1-q^{n-1}a)$$

so that the binomial coefficients can also be written

$$\binom{n}{m}_q = \frac{(q;q)_n}{(q;q)_m(q;q)_{n-m}}.$$

That they are really some sort of binomial coefficients is perhaps well illustrated by the q-binomial theorem which says that if v, w are two elements in an algebra that q-commute, i.e. $wv = qvw$

$$(v+w)^n = \sum_{m=0}^{n} \binom{n}{m}_q v^m w^{n-m}.$$

There is also e.g. a q-analog of the Pascal triangle and a q-Vandermonde formula.

The q-binomial coefficients (Gaussian binomial coefficients) give the number of subspaces of dimension m of an n-dimensional vector space over the finite field of q elements. As q goes to 1 the q-binomial coefficients go to the usual ones which count the number of subsets of size k of an n-element set. This encourages one to think of sets as vector spaces over the field of one element; a way of thinking that has been fruitful both ways (from linear algebra to combinatorics and vice-versa).

The **q-derivative** is defined by

$$D_q f(x) = \frac{f(qx) - f(x)}{qx - x}.$$

It will turn up again later in this section. Meanwhile observe that

$$D_q(x^n) = [n]_q x^{n-1}.$$

Differential calculus on an algebra. Let M be a differential manifold. The vector fields on M are the sections of the cotangent bundle. Switching to functions there is the commutative algebra, $C^\infty(M)$ of differentiable functions on M and the vector fields form a module $\Omega^1(M)$ over this algebra. The exterior derivative is a morphism of modules $d : C^\infty(M) \longrightarrow \Omega^1(M)$ that obeys the Leibniz rule.

---

[48] In quantum group theory the numbers $\frac{q^n - q^{-n}}{q - q^{-1}}$ tend to turn up rather more frequently than the numbers $\frac{q^n - 1}{q - 1}$ and so the notations $[n]_q = \frac{q^n - q^{-n}}{q - q^{-1}}$, $[[n]]_q = \frac{q^n - 1}{q - q^{-1}}$ are sometimes used in papers and books on quantum groups, instead of the notation here; this is the case for instance in [**388**].

Let A be a unital associative algebra (but not necessarily commutative). A first order differential calculus (FODC) on A consists simply of an A-bimodule Ω together with a morphism of modules such that the Leibniz rule holds

$$d : A \longrightarrow \Omega, \quad d(ab) = (da)b + a(db)$$

and such that Ω is spanned by the $a(db)c$ for $a, b, c \in A$.

It follows from the Leibniz rule that $(da)b = d(ab) - a(db)$ so that

$$\Omega = A(dA)A = A(dA) = (dA)A.$$

There are many such FODC's. Here is the universal one (for a given A). Take

$$\Omega_u = \mathrm{Ker}(A \otimes A \xrightarrow{m_A} A), \quad d_u(a) = 1 \otimes a - a \otimes 1.$$

This one is universal in that any FODC for A has as module of differentials a quotient module of Ω_u and taking derivatives is composition of d_u with the quotient morphism.

More generally in the manifold case there is a whole differential complex

$$C^\infty(M) \xrightarrow{d} \Omega^1(M) \longrightarrow \Omega^2(M) \longrightarrow \cdots$$

This also generalizes to the noncommutative algebra case and there still is a universal example.

Covariant differential calculi. The classical differential calculus on, say, $\mathbf{R}^n$ or Minkowski space, is covariant under translations (indeed the whole group of motions, resp. the Poincaré group). So, to obtain something reasonable in the way of of differential calculi on quantum spaces and quantum groups it seems sensible to impose suitable covariance conditions.

The setting is as follows. There is an algebra A (thought of as a (noncommutative) algebra of functions; there is a Hopf symmetry given by a left H-comodule algebra structure on A, and there is a FODC on A

$$\alpha : A \longrightarrow H \otimes A, \quad d : A \longrightarrow \Omega.$$

Such a FODC is called left covariant if there exist a Hopf symmetry (H-comodule structure)

$$\varphi : \Omega \longrightarrow H \otimes \Omega$$

such that

(i) $\varphi(a\omega b) = \alpha(a)\varphi(\omega)\alpha(b)$ for all $a, b \in A$, $\omega \in \Omega$

(ii) $\varphi(da) = (\mathrm{id} \otimes d)\alpha(a)$ for all $a \in A$

The first condition says that φ is a morphism of bimodules where the A-bimodule structure on $H \otimes \Omega$ is defined by α, and the second condition says that d is morphism of left H-comodules.

If such an H-comodule structure exists (which is not always the case) it is unique. Also there is a reformulation in terms of a condition that does not involve φ and permits writing down a formula for φ if it is fulfilled; see [**388**], chapter 12; [**118**], chapter 3.

The condition is

$$(\sum a_i(db_i) = 0) \Rightarrow (\sum \alpha(x_i)(\mathrm{id} \otimes d)\alpha(b_i) = 0) \qquad (8.8.6.1)$$

and then the formula for the coaction of H on Ω is

$$\varphi(\sum a_i(db_i)) = \sum \alpha(a_i)(\mathrm{id} \otimes d)\alpha(b_i). \tag{8.8.6.2}$$

Right covariance is defined analogously.

When $A = H$ is itself a Hopf algebra (bialgebra) the comultiplication $H \xrightarrow{\mu_H} H \otimes H$ defines both a left and a right H-comodule structure on H and one can talk about bicovariant calculi (meaning both left and right covariant).

Examples. Consider a ring of polynomials $k[X]$ in one indeterminate (affine line). Suppose that as a left module Ω is free of rank 1 with basis dX. Let $p(X)$ be a polynomial and define the right module structure on Ω by $(dX)X = p(X)(dX)$. This defines a FODC on $k[X]$ for which

$$d(X^n) = (p(X)^{n-1} + Xp(X)^{n-2} + \cdots + X^{n-2}p(X) + X^{n-1})dX. \tag{8.8.6.3}$$

The question now arises when this is left covariant under the additive group structure on the affine line (translation) given by

$$X \mapsto X \otimes 1 + 1 \otimes X. \tag{8.8.6.4}$$

It is a simple exercise using formula (8.8.6.2) to show that this is the case if and only if the polynomial involved is of the form $p(X) = X + h$ for some constant $h \in k$. And then it is also right covariant, hence bicovariant. This is in fact difference calculus. Let $h \neq 0$ and

$$D_h f(X) = \frac{f(X+h) - f(X)}{h} \tag{8.8.6.5}$$

then, as is seen from (8.8.6.3) for the polynomial $X + h$

$$df = (D_h f)dX.$$

If the polynomial is $p(X) = X$ the usual differential calculus appears, see again (8.8.6.3).

As a second example consider the q-calculus given by the 'differentiation'

$$D_q f(X) = \frac{f(qX) - f(X)}{qX - X}, \quad df = (D_q f)dX. \tag{8.8.6.6}$$

This is not left covariant on the affine line if $q \neq 1$. (As q goes to 1, the expression (8.8.6.6) also goes to the standard differentiation.)

Instead, consider the algebra of Laurent polynomials $k[X, X^{-1}]$ (the circle), with the Hopf algebra structure

$$\mu(X) = X \otimes X.$$

The the q-derivative is left covariant (and also right covariant, so bicovariant).

There is much to say about these two calculi, especially the q-calculus; the very well written book [**360**] is recommended.

Classification of bicovariant calculi on quantum groups. The matter of finding all bicovariant differential calculi on quantum groups was pioneered by Woronowicz, [**689**]; [**691**]. By now, in fact since 1995 or thereabouts, things are settled for the quantum groups $\mathcal{O}_q(G)$ with G a general linear group, a special

[15] Amano Katsutoshi, Akira Masuoka and Mitsuhiro Takeuchi, *Hopf algebraic approach to Picard-Vessiot theory*, in: M. Hazewinkel, ed., *Handbook of Algebra, Volume 6*, Elsevier, 2009.

[16] H. H. Andersen, J. C. Jantzen and W. Soergel, *Representations of quantum groups at p-th roots of unity and of semisimple groups in charactieristic p: independence of p*, SMF, 1994.

[17] George E. Andrews, Li Guo, William Keigher and Ken Ono, *Baxter algebras and Hopf algebras*, Trans. Amer. math. Soc. **355** (2003), 4639–4656; preprint version: `arXiv:math/0407181v1[math.RA.]`

[18] Nicolas Andruskiewitsch and Hans-Jurgen Schneider, *Lifting of quantum homogeneous spaces and pointed Hopf algebras of order* p^3, J. of Algebra **209** (1998), 658–691.

[19] Nicolas Andruskiewitsch, *On the quiver theoretical quantum Yang-Baxter equation*, 2000; `arXiv:math/0402269v3 [math.QA]`.

[20] Nicolas Andruskiewitsch and Hans-Jurgen Schneider, *Finite quamtum groups and Cartan matrices*, Adv. in Math. **154** (2000), 1–45.

[21] Nicolas Andruskiewitsch and Hans-Jurgen Schneider, *Pointed Hopf algebras*, in S. Montgomery and H.-J. Schneider, eds., *New directions in Hopf algebras*, Cambr. Univ. Press, 2002, 1–68.

[22] Nicolas Andruskiewitsch and Hans-Jurgen Schneider, *On the classification of finite-dimensional pointed Hopf algebras*, 2005; `arXiv:math/0502157`.

[23] Nicolas Andruskiewitsch and Fernando Fantino, *On pointed Hopf algebras associated to unmixed conjugacy classes in* S_n, 2006; `arXiv:math/0608701v1`.

[24] Nicolas Andruskiewitsch and Fernando Pantino, *On pointed Hopf algebras associated with alternating and dihedral groups*, 2007; `arXiv:math/0702559v3 [math.QA]`

[25] Nicolas Andruskiewitsch and Shouchuan Zhang, *On pointed Hopf algebras associated to some conjugacy classes in* S_n, Proc. Amer. math. Soc. **135** (2007), 2723–2731.

[26] Nicolas Andruskiewitsch, Fernando Fantino and Souchuan Zhang, *On pointed Hopf algebras associated with the symmetric group*, 2008; `arXiv:0807.2406v2 [math.QA]`.

[27] Nicolas Andruskiewitsch, Fernando Fantino, Matias Grana and Leandro Vendramin, *On pointed Hopf algebras associated to symmetric groups II*, 2009; `arXiv:0904.3978v1 [math.QA]`.

[28] Nicolas Andruskiewitsch, Fernando Pantino, Matias Grana and Leandro Vendramin, *Finite dimensional pointed Hopf algebras with alternating groups are trivial*; 2009, `arXiv:0812.4628v3[math.QA]`.

[29] Aneva Boyka, *A realization of the quantum Lorentz group*, 2008; `arXiv:0804.1529v1 [math.ph]`.

[30] Pham Ngoc Anh, *Hall polynomials for* A_n, Archiv der Math. **78** (2002), 263–267.

[31] D. Arnaudon, J. Avan, L. Frappat, É. Ragaochy and M. Rossi, *On the quasi-Hopf structure of deformed double Yangians*, Lett. in math. Physics **51** (2000), 193–204.

[32] Artamonov, Vyacheslav A., *Actions of Hopf algebras on quantum polynomials*, in: F.U. Coelho and a.o., eds., *Representations of algebras*, Marcel Dekker, 2002, 11–20.

[33] Vyacheslav A. Artamonov, *Pointed Hopf algebras acting on quantum polynomials*, J. of Algebra **259** (2003), 323–352.

[34] Emil Artin, *Theorie der Zöpfe*, Abh. math. Sem. Hamburg **4** (1925), 47–72.

[35] Michael Artin, William Schelter and John Tate, *Quantum deformations of* GL_n, Commun. pure appl. Math. **44** (1991), 879–895.

[36] Tsuguhiko Asakawa, Masashi Mori and Satoshi Watamura, *Hopf algebra symmetry and string theory*, (2008); `arXiv:0805.2203`

[37] Paolo Aschieri, Marija Dimitrijevic, Frank Meyer and Julius Weiss, *Noncommutative geometry and gravity*, Max Planck Inst. für Physik, München, 2005; `arXiv:0510059`

[38] Paolo Aschieri, *Noncommutative symmetries and gravity*, Erwin Schrödinger int. Inst. for mathematical Physics, 2006; `arXiv:0608172`

[39] Paolo Aschieri, *Lectures on Hopf algebras, quantum groups and twists*, Centro Studi e Ricerche Enrico Fermi, 2007; `arXiv:07033013`

[40] Ibrahim Assem, Daniel Simson and Andrzej Skowronski, *Elements of the representation theory of associative algebras. Volume 1: Techniques of representation theory*, Cambridge Univ. Press, 2006.

[41] Michael Frrancis Atiyah, *Topological quantum field theory*, Publ. math. IHES **68** (1989), 175–186.

[42] Thomas Aubriot, *Classification des objets Galoisiennes d'une algèbre de Hopf*, Univ. de Strassbourg, 2007.

[43] Maurice Auslander, Idun Reiten and Sverre O.Smalø, *Representation theory of Artin algebras*, Cambridge Univ. Press, 1995.

[44] Olivier Babelon and Denis Bernard, *Dressing symmetries*, Comm. in math. Physics **149** (1992), 279–306.

[45] H. C. Baehr, Aristophanes Dimakis and Folkert Müller-Hoissen, *Differential calculi on commutative algebras*, Max Planck Inst. Physik, Münich, 1994.

[46] Teodor Banica, J. Bichon and G. Chenevier, *Graphs having no quantum symmetry*, 2006; `arXiv:math/0605257v2 [math.CO]`.

[47] Teodor Banica, Julien Bichon and Benoit Collins, *The hyperoctahedral quantum group*, 2007; `arXiv:math/0701859v2 [math.RT]`.

[48] Dror Bar-Natan , *On associators and the Grothendieck-Teichmüller group I*, Selecta math. NS **4** (1998). Preprint version: `arXiv:q-alg/9606021v4`.

[49] Eliezer Batista and Shahn Majid, *Noncommutative geometry of angular momentum space* $U(su(2))$, 2002; `arXiv: hep-th/0205128v2`.

[50] Pierre Baumann and Christian Kassel, *The Hall algebra of the category of coherent sheaves on the projective line*, J. reine und angew. Math. **533** (2001), 207–233. Preprint version: `arXiv:math/9906037`.

[51] G. Baxter, *An analytic problem whose solution follows from a simple algebraic identity*, Pacific J. of Math. **10** (1960), 731–742.

[52] Rodney J. Baxter, *Partition function of the eight vertex model*, Ann. of Physics **70** (1972), 193–228.

[53] Rodney J. Baxter, *Exactly solved models in statistical mechanics*, Acad. Press, 1982. An electronic version of the book is available through the homepage of Paul Pearce; `http://www.ms.unimelb.edu.au/~pap`

[54] Yuri Bazlov, *Nichols-Woronowicz algebra model for chubert calculus on Coxeter groups*, 2005; `arXiv: math/0409206v2 [math.QA]`.

[55] E. J. Beggs, and Shahn Majid, *Bar categories and star operations*, Algebras and Representation Theory **12** (2009), 103–152.

[56] A. Beilinson, George Lusztig and Robert MacPherson, *A geometric setting for the quantum deformation of* GL_n, Duke math. J. **61** (1990), 655–677.

[57] A. Ya. Belov and Louis Halle Rowen, *Polynomial identities: a combinatorial appoach*, A. K. Peters, 2005.

[58] Francesca Benanti, *On the cocharater sequence of 3 by 3 matrices*, Dip. mat., Univ. di Palermo, 2000.

[59] Allan Berele, *Application of Belov's theorem to the cocharacter sequence of p.i. algebras*, J. of Algebra **298** (2006), 208–214.

[60] Christoph Bergbauer and Dirk Kreimer, *The Hopf algebra of rooted trees and Epstein-Glaser renormalization*, Inst. Math. Freie Univ. Berlin, 2004.

[61] D. Bernard, *An introduction to Yangian symmetries*, in L. Bonora a.o., ed., Integrable quantum field theories, Plenum, 1993, 39–52.

[62] Denis Bernard and Giovanni Felder, *Quantum group symmetries in two-dimensional lattice quantum field theory*, Nuclear Physics B **365** (1991), 98–120.

[63] Jyotishman Bhowmick and Debashish Goswami, *Quantum isometries of classical spaces*, 2009; arXiv:0903.1322v2 [math.QA]

[64] Julien Bichon and Ross Street, *Militaru's D-Equation in monoidal categories*, Applied categorical structures **11**:4 (2003), 337–357.

[65] Julien Bichon, *Galois and bi-Galois objects over monomial non semisimple Hopf algebras*, 2004; `arXiv:0404059 v1`.

[66] Julien Bichon, *Free quantum analogues of the first fundamental theorem of invariant theory*, Proc. Edinburgh math. Soc. **47** (2004), 297–303.

[67] Chr. Bishop, *Measures, martigales, manifolds and mappings*, Notes of the 1998 Barrett lectures in Knoxville, TN, June 1998.

[68] L. C. Biedenharn and M. A. Lohe, *Induced representations and tensor operators for quantum groups*, in P. P. Kulish, ed., *Quantum groups*, Springer, 1992, 197–209.

[69] Joan S. Birman, *Braids, links, and mapping class groups*, Princeton Univ. Press, 1975.

[128] Vyjayanthi Chari and Andrew Pressley, *Yangians, integrable quantum systems and Dorey's rule*, Comm. in math. Physics **181** (1996), 265–302.

[129] Vyjayanthi Chari and Andrew Pressley, *Weyl modules for classical and quantum affine algebras*, Representation Theory **5** (2001), 191–223.

[130] S. U. Chase, D. K. Harrison and A. Rosenberg, *Galois theory and cohomology of commutative rings*, AMS, 1965.

[131] Stephen U. Chase and Moss E. Sweedler, *Hopf algebras and Galois theory*, Springer, 1969.

[132] Xueqing Chen, *Root vectors of the composition algebra of the Kronecker quiver*, Algebra and discrete Mathematics **1** (2004), 37–56.

[133] Xiao-Wu Chen, Hua-Lin Huang, Yu Ye and Pu Zhang, *Monomial Hopf algebras*, J. of Algebra **275** (2004), 212–232.

[134] Xiao-Wu Chen, Hua-Lin Huang and Pu Zhang, *Dual Gabriel theorem with applications*, Sci. China Ser. A Math. **49** (2006), 9–26. Preprint version: `arXiv:math/0404121v3 [math.RT]`.

[135] Xiao-Wu Chen, *Duality between quantum symmetric algebras*, Lett. in math. Physics **79** (2007), 39–50.

[136] Hangrong Chen and Bangming Deng, *Fundamental relations in Ringel-Hall algebras*, J. of Algebra **320** (2008), 1135–1149.

[137] A. Chervov and G. Falqui, *Manin matrices and Talalaev's formula*, 2008; `arXiv:0711.2236v2 [math.QA]`.

[138] A. Chervov, G. Falqui and V. Rubtsov, *Algebraic properties of Manin matrices*, 2009; `arXiv:0901.0235v1 [math.QA]`.

[139] Lindsay N. Childs , *On Hopf-Galois extensions of fields and number rings*, in: F. v. Oystaeyen and L. Le Bruyn, eds., *Perspectives in ring theory*, Kluwer Acad. Publ., 1988, 117–128.

[140] Lindsay N. Childs and David J. Moss, *Hopf algebras and local Galois module theory*, in: J. Bergen and S. Montgomery, eds., *Advances in Hopf algebras*, Marcel Dekker, 1994, 1–24.

[141] Lindsay N. Childs, *Taming wild extensions: Hopf algebras and local Galois module theory*, Amer. math. Soc., 2000.

[142] William Chin and Susan Montgomery, *Basic coalgebras*, in V. Chari and I. B. Penkov, eds., *Modular interfaces*, Amer. math. Soc., Int. Press, 1995, 41–47.

[143] Chryssomalis Chryssomalakos, *Physics, combinatorics and Hopf algebras*, UNAM, Mexico City, 2004.

[144] Chryssomalis Chryssomalakos, and Micho Durdevich, *Higher order measures, generalized quantum mechanics and Hopf algebras*, Mod. Physics Lett. A **19** (2004), 197-212; `arXiv:0309092v1`.

[145] Claude Cibils, *A quiver quantum group*, Comm. in math. Physics **157** (1993), 459–477.

[146] Claude Cibils and Marc Rosso, *Algèbres des chemins quantiques*, Adv. in Math. **125** (1997), 171–199.

[147] Claude Cibils and Marc Rosso, *Hopf quivers*, J. of Algebra **254** (2002), 241–251.

[148] Claude Cibils and Andrea Solotar, *Galois coverings, Morita equivalence and smash extensions of categories over a field*, Documenta math. **11** (2006), 143–258.

[149] Nicola Ciccoli, *From Poisson to quantum geometry, Electronic notes of the TOK symposium "Noncommutative geometry and quantum groups"*. Warsaw 2004/2005, Univ. of Warsaw, 2008; `http://toknotes.mimuw.edu.pl`

[150] Arjeh M. Cohen, Die A. H. Gijsbers and David B. Wales, *BMW algebras of simply laced type*, 2003; `arXiv:math/0310011v1 [math.RA]`.

[151] Miriam Cohen and Susan Montgomery, *Group graded rings, smash products, and group actions*, Trans. Amer. math. Soc. (1984), 237–258.

[152] Miriam Cohen and Davida Fishman, *Hopf algebra actions*, J. of Algebra **100** (1986), 363–379.

[153] Miriam Cohen, *Hopf algebra actions-revisited*, in M. Gerstenhaber and J. Stasheff, eds., *Deformation theory and quantum groups with applications to mathematical physics*, Amer. math. Soc., 1992, 1–18.

[154] Alain Connes, *Noncommutative geometry*, Academic Press, 1994. Free electronic version: `http://www.alainconnes.org/docs/book94bigpdf.pdf`.

[155] Alain Connes and Dirk Kreimer, *Hopf algebras, renormalization and noncommutative geometry*, IHES, Bures sur Yvette, France, 1998; `arXiv:hep-th/9808042`.

[156] AlainConnes and Dirk Kreimer, *Renormalization in quantum field theory and the Riemann Hilbert problem*, IHES, Bures sur Yvette, France, 1999; `arXiv:hep-th/9909126`.

[157] Alain Connes and Dirk Kreimer, *Renormalization in quantum field theory and the Riemann-Hilbert problem I: the Hopf algebra structure of graphs and the main theorem*, IHES, Bures sur Yvette, France, 1999.

[158] Alain Connes, *A short survey of noncommutative geometry*, J. math. Physics **41** (2000), 3832–3866. Preprint version: arXiv:hep-th/0003006v1.

[159] Alain Connes, *Noncommutative geometry. Year 2000*, 2000; arXiv:math/0011193.

[160] Alain Connes and Dirk Kreimer, *Renormalization in quantum field theory and the Riemann-Hilbert problem II: the beta function, diffeomorphisms and the renormalization group*, IHES, Bures sur Yvette, France, 2000.

[161] Alain Connes, and Henri Moscovici, *Cyclic cohomology and Hopf algebra symmetry*, Lett. in math. Physics **52** (2000), 1–28.

[162] Alain Connes and Dirk Kreimer, *From local perturbation theory to Hopf and Lie algebras of Feynman graphs*, Lett. in math. Physics **56** (2001), 3–15.

[163] Alain Connes and Dirk Kreimer, *Insertion and eliminaton: the doubly infinite Lie algebra of Feynman graphs*, Ann. Henri Poincaré **3** (2002), 411–433; arXiv:hep-th/0201157.

[164] Alain Connes and Henri Moscovici, *Rankin-Cohen brackets and the Hopf algebra of transverse geometry*, College de France, 2003.

[165] Alain Connes, *Cyclic cohomology, noncommutative geometry and quantum group symmetries*, in S. Doplicher and R. Longo, eds., *Noncommutative geometry*, Springer, 2004, 1–71.

[166] Alain Connes and Henri Moscovici, *Modular Hecke algebras and their Hopf symmetry*, Moscow math. J. **4** (2004), 67–109. Preprint version: arXiv:math.QA/0301089.

[167] Alain Connes and Matilde Marcolli, *A walk in the noncommutative garden*, 2006; arXiv:math/0601054v1 [math.QA]

[168] Alain Connes, *Noncommutative geometry and physics, Electronic notes of the TOK symposium "Noncommutative geometry and quantum groups"*. Warsaw 2004/2005, Univ. of Warsaw, 2008. http://toknotes.mimuw.edu.pl.

[169] Robert Coquereaux, A. O. Garcia and R. Trinchero, *Finite dimensional quantum group covariant calculus on a complex matrix algebra*, 1998; arXiv:9804021v1.

[170] Robert Coquereaux and G. E. Schieber, *Action of a finite quantum group on the algebra of complex $N \times N$ matrices*, 1998; arXiv:9807016v1.

[171] Robert Coquereaux and Roberto Trinchero, *On quantum symmetries of ADE graphs*, 2004; arXiv: hep-th/0401140v1.

[172] Robert Coquereaux, *Classical and quantum polyhedra. A fusion graph algebra point of view*, 2007; arXiv:hep-th/0105239v2.

[173] Robert Coquereaux, Ariel Garcia and Roberto Trinchero, eds., *Quantum symmetries in theoretical physics and mathematics*, American math. Soc., 2002.

[174] Cornelissen, Gunther, Matilde Marcolli, Kamran Reihani and Alina Vdovina, *Noncommutative geometry on trees and buildings*, 2006; arXiv:math/0604114v1 [math.QA]

[175] L. Dabrowski, Giovanni Landi and Tetsuya Masuda, *Instantons on the quantum 4-spheres S_q^4*, 2001; arXiv:math/0012103v2.

[176] L. Dabrowski and Giovanni Landi, *Instanton algebras and quantum 4-spheres*, 2002; arXiv:math/0101177v2 [math.QA].

[177] L. Dabrowski, *The garden of quantum spheres*, in: P. M. Hajac and W. Pusz, eds., *Noncommutative geometry and quantum groups*, PWN, 2003, 37–48.

[178] L. Dabrowski and A. Sitarz, *Dirac operator on the standard Podleś quantum sphere*, in: P. M. Hajac and W. Pusz, eds., *Noncommutative geometry and quantum groups*, PWN, 2003, 49–58.

[179] Ludwig Dabrowski, Harald Grosse and Piotr M. Hajac, *Strong connections and Chern-Connes pairing in Hopf-Galois theory*, Comm. in math. Physics **220** (2001), 301–331.

[180] Kenny De Commer, *Galois objects for algebraic quantum groups*, J. of Algebra **321** (2009), 1746–1785.

[181] Pierre Deligne and J. S. Milne, *Tannakian categories*, In: *Hodge cycles, motives and Shimura varieties*, Springer, 1982, 101–228.

[182] Gustav W. Delius, A. Huffmann, Mark D. Gould and Y.-Z. Zhang, *The quantum Lie algebras associated to $U_h(gl_n)$ and $U_h(sl_n)$*, J. of Physics A **29** (1996), 5611–5617; arXiv:q-alg/9508013.

[183] Gustav W. Delius, Christopher Gardner and Mark D. Gould, *The structure of the quantum Lie algebras for the classical series B, C, D*, 1997; arXiv:q-alg/9706029v2.

[184] Bangming Deng, *Quasi-hereditary algebras and quadratic forms*, J. of Algebra **239** (2001), 438–459.

[185] Bangming Deng and Jie Xiao, *On double Ringel-Hall algebras*, J. of Algebra **251** (2002), 110–140.

[186] Bangming Deng and Jie Xiao, *A new approach to Kac's theorem on representations of valued quivers*, Math. Zeitschrift **245** (2003), 183–199.

[187] Bangming Deng and Jie Xiao, *On Ringel-Hall algebras*, in: V. Dlab and C.-M. Ringel, eds., *Representations of finite dimensional algebras and related topics in Lie theory and geometry*, Amer. math. Soc., 2004, 319–348.

[188] Bangming Deng, Jie Du, Brian Parshall and Jianpan Wang, *Finite dimensional algebras and quantum groups*, Amer. math. Soc., 2008.

[189] James K. Deveney and John N. Mordeson, *Higher derivation Galois theory of inseparable field extensions*, in: M. Hazewinkel, ed., *Handbook of algebra, volume 1*, Elsevier, 1996, 187–220.

[190] A. Di Bucchianico and D. Loeb, *A selected survey of umbral calculus*, Electronic J. of Combinatorics (2000), D83.

[191] M. S. Dijkhuizen and T. H. Koornwinder, *Quantum homogeneous spaces, duality and quantum 2-spheres*, CWI, Amsterdam, 1993.

[192] Aristophanes Dimakis and Folkert Müller-Hoissen, *From continuum to lattice theory via deformation of the differential calculus*, Phys. Lett. B **300** (1993), 141–144.

[193] Aristophanes Dimakis and Folkert Müller-Hoissen, *Stochastic differential calculus, the Moyal *-product, and noncommutative geometry*, Lett. in math. Physics **28** (1993), 123–137.

[194] Aristophanes Dimakis, Folkert Müller-Hoissen and T. Striker, *Noncommutative differential calculus and lattice gauge theory*, J. Phys. A: Math. Gen. **26** (1993), 1927–1949.

[195] Aristophanes Dimakis and Folkert Müller-Hoissen, *Differential calculus and gauge theory on finite sets*, J. Phys. A: Math. Gen. **27** (1994), 3159–3178.

[196] Aristophanos Dimakis and Folkert Müller-Hoissen, *Discrete differential calculus: graphs, topologies, and gauge theory*, J. of math. Physics **34** (1994), 6703–6735.

[197] Aristophanes Dimakis, Folkert Müller-Hoisen and François Vanderseypen, *Discrete differential manifolds and dynamics on networks*, J. of math. Physics **36** (1995), 3771–3791.

[198] Aristophanes Dimakis and Folkert Müller-Hoissen, *Noncommutative geometry and integrable models*, Lett. in math. Physics **39** (1997), 69–79.

[199] Richard Dipper and G. D. James, *The q-Schur algebra*, Proc. London math. Soc. **59** (1989), 23–50.

[200] Richard Dipper, Stephen Doty and Jun Hu, *Brauer algebras, symplectic Schur algebras and Schur-Weyl duality*, 2005; `arXiv:math/0503545v2 [math.RT]`.

[201] Vlastimil Dlab and Claus Michael Ringel, *Indecomposable representations of graphs and algebras*, Amer. math. Soc., 1976.

[202] V. K. Dobrev, *Classification and characters of $U_q(sl(3,C))$ representations*, in: H.-D. Doebner and J.-D. Hennig, eds., *Quantum groups*, Springer, 1990, 107–117.

[203] M. Domokos, R. Fioresi and T. H. Lenagan, *Orbits for the adjont coaction on quantum matrices*, 2003; `arXiv:math/0207207v2 [math.QA]`.

[204] M. Domokos and T. H. Lenagan, *Weakly multiplicative coactions of quantized function algebras*, J. of pure and appl. Algebra **183** (2003), 45–60. Preprint version: `arXiv:math/0208142v1 [math.QA]`.

[205] M. Domokos and T. H. Lenagan, *Conjugation coinvariants of quantum matrices*, Bull. London math. Soc. **35** (2003), 117–127.

[206] Matyas Domokos, *A Hilbert-Nagata theorem in noncommutative invariant theory*, Trans. Amer. math. Soc. **350** (1998), 2797–2811.

[207] Stephen Donkin, *The q-Schur algebra*, Cambridge Univ. Press, 1998.

[208] Stephen Doty and Anthony Giaquinto, *Presenting quantum Schur algebras as quotients of the quantum universal enveloping algebra of gl_2*, 2001; `arXiv:math/0011164v2 [math.RT]`.

[209] Stephen Doty and Anthony Giaquinto, *Presenting Schur algebras*, 2002; `arXiv:math/0109046v3 [math.RT]`.

[210] Stephen Doty, Anthony Giaquinto and John Sullivan, *Presenting generalized Schur algebras in types B, C, D*, 2005; `arXiv:math/0504075v1 [math.RT]`.

[211] V. G. Drinfel'd, *Degenerate affine Hecke algebras and Yangians*, Funkts. Analiz i ego Prilozhen. **20** (1985), 69–70.

[212] V. G. Drinfel'd, *Quantum groups*, Zap. Nauchn. Sem. LOMI **155** (1986), 18–49. extended version of the report in Int. Congress of Math. in Berkeley, 1986.

[213] V. G. D rinfel'd, *New realizations of Yangians and quantum affine algebras*, Dokl. Akad. Nauk SSSR **296** (1987), 13–17.

[214] V. G. Drinfel'd, *On quasitriangular quasi-Hopf algebras and a group closely connected with* $Gal(\mathbf{Q}^{ac}/\mathbf{Q})$, Leningrad Math. J. **2** (1991), 829–860.

[215] Michel Dubois-Violetteand Peter W. Michor, *Derivations et calcul differentiel non-commutatif II*, 1994; `arXiv:q-alg/9608026v1`.

[216] M. Dubois-Violette and P. W. Michor, *More on the Fröhlicher-Nijenhuis bracket in non-commutative differential geometry*, J. of pure and applied Algebra **121** (1997), 107–136.

[217] Michel Dubois-Violette and Richard Kerner, *Universal q-differential calculus and q-analog of homological algebra*, 1996; `arXiv:q-alg/9608026v1`.

[218] Michel Dubois-Violette, *Lectures on graded differential algebras and noncommutative geometry*, 2000; `arXiv:math/9912017v3[math.QA]`.

[219] Micho Durdevic, *Quantum principal bundles & Tannaka-Krein duality theory*, Reports in mathematical Physics **38** (1986), 313–324.

[220] Micho Durdevich, *A brief introduction to quantum geometry*, Inst. Mat. UNAM, Mexico City, 1996; `http://www.matem.unam.mx/~micho`.

[221] Micho Durdevich, *Quantum geometry and new concept of space*, Inst. Mat. UNAM, Mexico City, 1996; `http://www.matem.unam.mx/~micho`.

[222] Micho Durdevich, *Introduction to quantum principal bundles*, Inst. Mat. UNAM, Mexico City, 1996; `http://www.matem.unam.mx/~micho`.

[223] Kurusch Ebrahimi-Fard, Li Guo and Dirk Kreimer, *Spitzer's identity in the algebraic Birkhoff decomposition in pQFT*, Inst. Henri Poincaré, 2004; `arXiv:hep-th/0407082`.

[224] Kurusch Ebrahimi-Fard, Li Guo and Dirk Kreimer, *Integrable renormalization II: the general case*, Inst. Henri Poincaré, 2004; `arXiv:hep-th/0403118`.

[225] Kurusch Ebrahimi-Fard, Li Guo and Dirk Kreimer, *Integrable renormalization I: the ladder case*, Inst. Henri Poincaré, 2004.

[226] Kurusch Ebrahimi-Fard and Li Guo, *Rota-Baxter algebras, dendriform algebras and Poincar-Birkhoff-Witt theorem*, 2005; `arXiv:0503647 v2`.

[227] Kurusch Ebrahimi-Fard and Dirk Kreimer, *Hopf algebra approach to Feynman diagram calculations*, 2005; `arXiv:0510202 v2`.

[228] Kurusch Ebrahimi-Fard and Li Guo, *Mixable shuffles, quasi-shuffles and Hopf algebras*, J. Algebraic Combinatorics **24** (2006), 83–101.

[229] Kurusch Ebrahimi-Fard and Li Guo, *On free Rota-Baxter algebras*, IHES, 2006.

[230] Kurusch Ebrahimi-Fard and Li Guo, *Rota-Baxter algebras in renormalization of perturbatic quantum field theory*, 2006; `arXiv:hep-th/0604116v2`.

[231] Kurusch Ebrahimi-Fard and Li Guo, *Free Rota-Baxter algebras and rooted trees*, J. Algebra Appl. **7** (2008), 167–194.

[232] L. El Kaoutit and J. Gomez-Torrecillas, *Comatrix corings, Galois corings, descent theory and a structure theorem for cosemisimple corings*, Math. Zeitschrift **244** (2003), 887–906.

[233] Karen Erdmann, Edward L. Green, Nicole Snashall and Rachel Taillefer, *Representation theory of the Drinfeld doubles of a family of Hopf algebras*, J. of pure and applied Algebra **204** (2006), 413–454.

[234] Pavel I. Etingof and Alexander A. Kirillov Jr., *Macdonald polynomials and representations of quantum groups*, Math. Research Letters **1** (1994), 279–296.

[235] Pavel I. Etingof and Alexander A. Kirillov Jr., *On Cherednik-Macdonald-Mehta identities*, 1997; `arXiv:q-alg/9712051`.

[236] Pavel Etingof and Olivier Schiffman, *Lectures on quantum groups*, International Press, 1998.

[237] Pavel Etingof and Igor Pak, *An algebraic extension of the MacMahon master theorem*, 2006; `arXiv:math/0608005v1 [math.CO]`.

[238] L. D. Faddeev and P. N. Pyatov, *The differential calculus on quantum linear groups*, 1994; `arXiv: hep-th/9402071v2`.

[239] Ludwig D. Faddeev, *How the algebraic Bether Ansatz works for integrable models*, 1995; `arXiv:hep-th/9605187 v1`.

[240] Ludwig D. Faddeev and Leon A. Takhtajan, *The quantum inverse scattering method of the inverse problem and the Heisenberg XYZ model*, Russian math. Surveys **34** (1979), 11–68.

[241] Ludwig D. Faddeev,*Integrable models in 1+1 dimensional quantum field theory*, in J.-B. Zuber and R. Stora, eds., *Recent advances in field theory and statistical mechanics*, Les Houches, session 39, North Holland, 1984, 563–608.

[242] Fernando Falceto and Krzysztof Gawedzki, *On quantum group symmetries of conformal field theories*, IHES, Bures sur Yvette, 1991.

[243] Bertfried Fauser and P. D. Jarvis, *A Hopf laboratory for symmetric functions*, J. of Physics A **37** (2004), 1633–1663. Preprint version: `arXiv:math-ph/0308043v1`.

[244] Bertfried Fauser, *Renormalization: a number theoretical model*, Comm. in math. Physics **277** (2008), 627–641.

[245] Vitor O. Ferreira, Lucia S. I. Murakami and A. Paques, *A Hopf-Galois correspondence for free algebras*, J. of Algebra **276** (2004), 407–416.

[246] Vitor O. Ferreira and Lucia S. I. Murakami, *Finitely generated invariants of Hopf algebras on free associative algebras*, Linear Algebra and its Applications **420** (2006), 70–78.

[247] Hector Figueroa and Jose M. Gracia-Bondia, *Combinatorial Hopf algebras in quantum field theory I*, Dept. Mat. Univ. Costa Rica, 2004.

[248] Gaetano Fiore and Peter Schupp, *Identical particles and quantum symmetries*, J. of math. Physics, 1996, 3437–3452; `arXiv:hep-th/9508047v3`.

[249] R. Fioresi and C. Hacon, *On infinite dimensional Grassmannians and their quantum deformations*, 2001; `arXiv: math/0111112v1`.

[250] R. Fioresi and C. Hacon, *Quantum coinvariant theory for the special quantum linear group and quantum Schubert varieties*, J. of Algebra **242** (2001), 433–446.

[251] Davida Fischman and Susan Montgomery, *A Schur double centralizer theorem for cotriangular Hopf algebras and generalized Lie algebras*, Dept. Math., Univ. of Southern California, Los Angeles, 1992.

[252] Dominique Foata and Guo-Niu Han, *Specializations and extensions of the quantum MacMahon master theorem*, 2006; `arXiv:math/0603466v1 [math.CO]`.

[253] Dominique Foata and Guo-Niu Han, *A new proof of the Garoufilidis-Le-Zeilberger quantum MacMahon master theorem*, 2006; `arXiv:math/0603463v1 [math.CO]`.

[254] Dominique Foata and Guo-Niu Han, *A basis for the right quantum algebra and the '1 = q' principle*, 2006; `arXiv:math/0603463v1 [math.CO]`.

[255] G. B. Folland, *Harmonic analysis in phase space*, Princeton Univ. Press, 1989.

[256] D. S. Freed, *Lectures in topological quantum field theory*, in: L. A. Ibort and M. A. Rodriguez, eds., *Integrable sytems, quantum groups, and quantum field theories*, KAP, 1993, 95–156.

[257] Igor B. Frenkel, Anton Malkin and Maxim Vybornov, *Affine Lie algebras and tame quivers*, Selecta matematica. New Series **7** (2001), 1–56.

[258] Peter Freyd, David Yetter, J. Hoste, W. B. R. Lickorish, K. Millett and Adrian Ocneanu, *A new polynomial invariant of knots and links*, Bull. Amer. math. Soc. **12** (1985), 239–246.

[259] Jürgen Fuchs, *Affine Lie algebras and quantum groups*, Cambridge University Press, 1992.

[260] Jürgen Fuchs, *On the quantum symmetry of rational field theories*, 1994; `arXiv:hep-th/9407013v1`.

[261] William Fulton and Joe Harris, *Representation theory. A first course*, Springer, 2004.

[262] Peter Gabriel, *Unzerlegbare Darstellungen I*, Manuscripta mathematica **6** (1972), 71–103.

[263] P. Gabriel, *Indecomposable representations II*, Symp. Math. INDAM 11 (Rome. 1971), Acad. Press, 1973, 81–104.

[264] We Liang Gan and Travis Schedler, *The necklace Lie coalgebra and renormalization algebras*, 2007; `arXiv:math-ph/0702055v3`.

[265] Stavros Garoufalidis, Thang T. Q. Le and Doron Zeilberger, *The quantum MacMahon master theorem*, 2005; `arXiv:math/0303319v3 [math.QA]`.

[266] Antonio M. Garsia, *A q-analogue of the Lagrange inversion formula*, Houston J. of Math. **7** (1981), 205–237.

[267] Antonio M. Garsia, and J. Remmel, *A novel form of q-Lagrange inversion*, Houston J. of Math. **12** (1986), 503–523.

[268] A. M. Garsia and M. Haiman, *A remarkable q,t-Catalan sequence and q-Lagrange inversion*, J. algebraic Combinatorics **5** (1996), 191–244.

[269] George Gasper and Mizan Rahman, *Basic hypergeometric series*, Cambridge Univ. Press, 1990.

[270] Krzysztof Gawedzki, *Classical origin of quantum group symmetries in Wess-Zumino. Witten conformal field theory*, Commun. math. Physics **139** (1991), 201–213.
[271] K. Gawedzki, *Quantum group symmetry in conformal field theory*, in: H. Araki, K. R. Ito, A. Kishimoto and I. Ojima, eds., *Quantum and noncommutative analysis*, KAP, 1993, 239–252.
[272] Murray Gerstenhaber, A. Giaquinto and S. Don Shack, *Quantum symmetry*, in: P. P. Kulish, ed., *Quantum groups*, Springer, 1992, 9–46.
[273] Ira Gessel, *A noncommutative generalization and q-analog of the Lagrange inversion formula*, Trans. of the Amer. math. Soc. **257** (1980), 455–482.
[274] Ira Gessel and Dennis Stanton, *Applications of q-Lagrange inversion to basic hypergeometric series*, Trans. of the Amer. math. Soc. **277** (1983), 173–201.
[275] Dié Gijsbers, *BMW algebras of simply laced type*, Techn. Univ. of Eindhoven, 2005.
[276] Victor Ginzburg, *Lectures on noncommutative geometry*, 2005; `arXiv:math/0506603v1 [math.AG]`.
[277] C. Gomez, Marti Ruiz-Altaba and G. Sierra, *Quantum groups in two dimensional physics*, in: L. A. Ibort and M. A. Rodriguez, eds., Cambridge Univ. Press, 1993.
[278] C. Gomez and G. Sierra, *A brief history of hidden quantum symmetries in conformal field theories*, in: L. A. Ibort and M. A. Rodriguez, eds., *Integrable sytems, quantum groups, and quantum field theories*, KAP, 1993, 25–44.
[279] Cesar Gomez and German Sierra, *Graph quantum groups*, (1993); `arXiv:hep-th/9312177v1`.
[280] Torrecillas Gómez, José and Joost Vercruysse, *Galois theory in bicategories*, 2007; `arXiv:0711.3642v1 [math.RA]`.
[281] Kenneth R. Goodearl, *Prime spectra of quantized coordinate rings*, 1999; `arXiv:math/9903091 v1`.
[282] Kenneth R. Goodearl and Thomas H. Lenagan, *Quantum determinantal ideals*, Duke math. J. **103** (2000), 165–190. Preprint version: `arXiv:math/0302035v1 [math.QA]`.
[283] Kenneth R. Goodearl, Thomas H. Lenagan and Laurent Rigal, *The first fundamental theorem of coinvariant theory for the quantum general linear group*, Publ. RIMS, Kyoto (2000), 269–296. Preprint version: `arXiv:math/0302035v1 [math.QA]`.
[284] Kenneth R. Goodearl and T. H. Lenagan, *Quantized coinvariants at transcendental q*, in: S. Caenepeel and F. v. Oystaeyen, eds., *Hopf algebras in noncommutative geometry and physics*, Marcel Dekker, 2005. Preprint version: `arXiv:math/0302035v1 [math.QA]`.
[285] Debashish Goswami, *Quantum isometry group of compact metric space*, 2009; `arXiv:0811.0095v5 [math.QA]`.
[286] A. R. Gover and R. B. Zhang, *Geometry of quantum homogeneous vector bundles and representation theory of quantum groups I*, 1997; `arXiv:q-alg/9705016v1`.
[287] J. A. Green, *Polynomial representations of GL_n*, Springer, 2007. Preprint version (lecture notes): Univ. of Warwick, 1980.
[288] James A. Green, *Hall algebras, hereditary algebras and quantum groups*, Inventiones mathematicae **120** (1995), 361–377.
[289] Edward L. Green and Ø. Solberg, *Basic Hopf algebras and quantum groups*, Math. Zeitschrift **229** (1998), 45–76.
[290] Cornelius Greither and Bodo Pareigis, *Hopf Galos theory for separable field extensions*, J. of Algebra **106** (1987), 239–258.
[291] Uwe Grimm, *Trigonometric R matrices related to "dilute" Birman-Wenzl-Murakami algebra*, Lett. in math. Physics **32** (1994), 183–188.
[292] Robert Grossman and Richard G.Larson, *Hopf-algebraic structure of families of trees*, J. of Algebra **126** (1989), 184–210.
[293] Robert Grossman and Richar G. Larson, *Hopf -algebraic structure of combinatorial objects and differential operators*, Israel J. Math, **72** (1990), 109–117.
[294] Jin Yun Guo, *The centre of the Hall algebra of a cyclic quiver*, in: V. Dlab and H. Lenzing, eds., Proc. of the 6-th int. conf. on representations of algebras, 1992.
[295] Jin Yun Guo, *The Hall and composition algebras of a cyclic quiver*, Comm. Algebra **26** (1998), 2745–2766.
[296] Li Guo, *Properties of free Baxter algebras*, Adv. in Math. **151** (2000), 346–374.
[297] Li Guo, *Baxter algebras and the umbral calculus*, Adv. in appl. Math. **27** (2000), 405 - 426.
[298] Li Guo and William Keicher, *Baxter algebras and shuffle products*, Adv. in Math. **150** (2000), 117–149.

[299] Li Guo, *Baxter algebras and differential algebras*, in L. Guo and a.o., eds., *Proc. of the international workshop, Rutgers Univ. Nov. 2000*, World Scientific, 2002, 281–305.

[300] Li Guo and Bin Zhang, *Renormalization of multiple zeta values*, 2006; `arXiv: math/0606076v3 [math.NT]`.

[301] Li Guo and Bin Zhang, *Differential Birkhoff decomposition and the renormalization of multiple zeta values*, 2007; `arXiv:0719.0432v1 [math.NT]`.

[302] D. I. Gurevich, *Algebraic aspects of the quantum Yang-Baxter equation*, Algebra i analiz **2** (1990), 119–148. Translation: Leningrad math. J. **2** (1991), 801–828.

[303] D. I. Gurevich, P. N. Pyatov and P. A. Saponov, *Cayley-Hamilton theorem for quantum matrix algebras of GL(m — n) type*, 2005. Translation: Leningrad math. J. **2** (1991), 801–828; `arXiv:math/0412192v2 [math.QA]`.

[304] Tom Hadfield and Shahn Majid, *Bicrossproduct approach to the Connes-Moscovici Hopf algebra*, 2008; `arXiv:math/0601581v2 [math.QA]`.

[305] Phung Ho Hai, *Realizations of quantum Hom-spaces, invariant theory, and quantum determinental ideals*, J. of Algebra **248** (2002), 50–84.

[306] Phung Ho Hai and Martin Lorenz, *Koszul algebras and the quantum MacMahon master theorem*, 2006; `arXiv:math/0603169v3[math.QA]`.

[307] Philip Hall, *The algebra of partitions, Proc. 4-th Canadian math. Congress, Banff, 1957*, Univ. of Toronto Press, 1959, 147–159.

[308] David Harbater and Leila Schneps, *Fundamental groups of moduli and the Grothendieck-Teichmüller group*, Trans. Amer. math. Soc. **352** (2000), 3117–3148.

[309] Eli Hawkins and Giovanni Landi, *Fredholm modules for quantum Euclidean spheres*, 2002; `arXiv:math/0210139v2 [math.KT]`.

[310] Takahiro Hayashi, *Q-analogues of Clifford and Weyl algebras-spinor and oscillator representations of quantum enveloping algebras*, Dept. of Math., Nagoya Univ., 1989.

[311] Takahiro Hayashi, *Quantum group symmetry of partition functions or IRF models and its application to Jones' index theory*, Commun. in math. Physics **157** (1993), 331–345.

[312] T. Hayashi, *Face algebras I - a generalization of quantum group theory*, J. math. Soc. Japan **50** (1998), 293–316.

[313] Michiel Hazewinkel, *(Fine) moduli (spaces) for linear systems: what are they and what are they good for?*, in: C. I. Byrnes and C. F. Martin, eds., *Geometric methods for the theory of linear systems*, Reidel publ. Cy, 1980, 125–193.

[314] Michiel Hazewinkel, *On quadratic Hopf algebras*, 1990. Unpublished.

[315] Michiel Hazewinkel, *Multiparameter quantum groups and multiparameter R-matrices*, Acta applicandae mathematicae **41** (1995), 57–98.

[316] Michiel Hazewinkel, *Hopf algebras: their status and pervasiveness*, 2004; `arXiv:0411536`.

[317] Michiel Hazewinkel, Nadiya Gubareni and Volodymir V. Kirichenko, *Algebras, rings and modules. Volume 1*, Kluwer acad. Publ., 2004.

[318] Michiel Hazewinkel, Nadiya Gubareni and Volodymir V. Kirichenko, *Algebras, rings and modules. Volume 2*, Springer, 2007.

[319] Istvan Heckenberger, *The Weyl-Brandt groupoid of a Nichols algebra of diagonal type*, 2004; `arXiv:math/00411477v1 [math.QA]`.

[320] I. Heckenberger and H. Yamane, *Drinfel'd doubles and Shapovalov determinants*, 2008; `arXiv:0910.1621v1 [math.QA]`

[321] Istvan Heckenberger, *Classification of arithmetic root systems*, 2008; `arXiv:math/0605795v1 [math.QA]`.

[322] Istvan Heckenberger and Hans-Jurgen Schneider, *Root systems and Weyl groupoids for Nichols algebras*, 2008; `arXiv:0807.0691v1[math.QA]`.

[323] Heinrich Eduard Heine, *Über die Reihe*

$$1+\frac{(q^\alpha-1)(q^\beta-1)}{(q-1)(q^\gamma-1)}x+\frac{(q^\alpha-1)(q^{\alpha+1}-1)(q^\beta-1)(q^{\beta+1}-1)}{(q-1)(q^2-1)(q^\gamma-1)(q^{\gamma+1}-1)}x^2+\cdots,$$

J. reine und angew. Math. **32** (1846), 210–212.

[324] Heinrich Eduard Heine, *Untersuchungen über die Reihe*

$$1+\frac{(q^\alpha-1)(q^\beta-1)}{(q-1)(q^\gamma-1)}x+\frac{(q^\alpha-1)(q^{\alpha+1}-1)(q^\beta-1)(q^{\beta+1}-1)}{(q-1)(q^2-1)(q^\gamma-1)(q^{\gamma+1}-1)}x^2+\cdots,$$

J. reine und angew. Math. **34** (1847), 285–328.

[325] M. E. Hoffman, *The algebra of multiple harmonic series*, J. of Algebra **194** (1997), 477–495.
[326] Michael E. Hoffman, *Quasi-shuffle products*, J. Algebraic Combinatorics **11** (2000), 49–68.
[327] Michael E. Hoffman, *Combinatorics of rooted trees and Hopf algebras*, Trans. Amer. Math. Soc. **355** (2003), 3795–3811.
[328] M. E. Hoffman, *Algebraic aspects of multiple zeta values*, in T. Aoki and et al., eds., *Zeta functions, topology and quantum physics*, Springer, 2005, 51–73. Preprint version: `arXiv:math/0309425`.
[329] Jin Hong and Seok-Jin Kang, *Introduction to quantum groups and crystal bases*, Amer. math. Soc., 2002.
[330] Roger E. Howe, *Remarks on classical invariant theory*, Trans. American Math. Soc. **313** (1989), 539–570.
[331] Roger Howe and Tôru Umeda, *The Capelli identity, the double commutant theorem, and multiplicity - free actions*, Mathematische Annalen **290** (1991), 565–619.
[332] Jun Hu, *BMW algebra, quantized coordinate algebra and type C Schur-Weyl duality*, 2009; `arXiv:0908.3009v2 [math.QA]`.
[333] Hua-Lin Huang, Libin Li and Yu Ye, *Self dual Hopf quivers*, Commun. in Algebra **33** (2005), 4505–4514.
[334] Hua-Lin Huang and Shilin Yang, *Quantum groups and double quiver algebras*, Lett. in math. Physics **71** (2005), 49–61.
[335] Andrew Hubery, *The composition algebra of an affine quiver*, 2004; `arXiv:math/0403206v1 [math.RT]`.
[336] Andrew Hubery, *Symmetric functions and the centre of the Ringel-Hall algebra of a cyclic quiver*, Math. Zeitschrift **251** (2005), 705–719.
[337] Andrew Hubery, *Acyclic cluster algebras via Ringel-Hall algebras*, Dept. Math., Univ of Leeds, 2006. Notes of a lecture at the university of Paderborn, 8 Nov. 2007.
[338] Andrew Hubery, *Ringel-Hall algebras*, Dept. Math., Univ of Leeds, 2007. Notes of a course.
[339] Andrew Hubery, *Derived Hall algebras*, Dept. Math., Univ of Leeds, 2007. Notes of a lecture at the university of Paderborn, 8 Nov. 2007.
[340] Andrew Hubery, *Hall polynomials for affine quivers*, 2007; `arXiv:math/0703178v3 [math.RT]`.
[341] Tobias Hurth and Kostas Skenderis, *Quantum Noether method*, 1998; `arXiv:hep-th/9803030v2`.
[342] Vu. Huynh and Thang T.Q. Le, *On the colored Jones polynomial and the Kashaev invariant*, 2005; `arXiv:math/0502396v1 [math.GT]`.
[343] A. Isaev, O. Ogievetsky and P. Pyatov, *Generalized Cayley-Hamilton-Newton identities*, Czechoslovak J. Phys. **48** (1998), 1369–1374. Preprint version: `arXiv:math/9809047v1[math.QA]`.
[344] A. Isaev, O. Ogievetsky and P. Pyatov, *On quantum matrix algbras satisfying the Cayley-Hamilton-Newton identities*, 1998. Preprint version: `arXiv:math/9809047v1 [math.QA]`; `arXiv:math/980970v1[math.QA]`.
[345] A. Isaev, O. Ogievetsky and P. Pyatov, *Q-multilinear algebra*, 1999. Preprint version: `arXiv:math/9809047v1 [math.QA]`; `arXiv:math/9912231v1 [math.QA]`.
[346] A. Isaev, O. Ogievetsky and P. Pyatov, *Cayley-Hamilton-Newton identities and quasitriangular Hopf algebras*, 1999. Preprint version: `arXiv:math/9809047v1 [math.QA]`; `arXiv:math/9912197v1[math.QA]`.
[347] Minoru Itoh, *A Cayley-Hamilton theorem for the skew Capelli elements*, J. of Algebra **242** (2001), 740–761.
[348] Minoru Itoh, *Capelli identities for reductive dual pairs*, Adv. in Math. **194** (2005), 345–397.
[349] Jens Carsten Jantzen, *Lectures on quantum groups*, American mathematical Society, 1996.
[350] Michio Jimbo, *A q-difference analogue of U(g) and the Yang Baxter equation*, Lett. in math. Physics **10** (1985), 63–69.
[351] M. Jimbo, *Quantum R matrix for the generalized Toda system*, Comm. in math. Physics **102** (1986), 537–548.
[352] Michio Jimbo, *A q-analogue of U(gl(N+1)), Hecke algebra and the Yang Baxter equation*, Lett. in math. Physics **11** (1986), 247–252.
[353] Vaughan F. R. Jones, *A polynomial invariant for knots via von Neumann algebras*, Bull. Amer. math. Soc. **12** (1985), 103–112.

[354] Vaughan F. R. Jones, *In and around the origin of quantum groups*, Dept. Math. Univ. of California at Berkeley, 2004.

[355] Vaughan F. R. Jones, *The Jones polynomial*, Dept. of Math., Univ. of California at Berkeley, 2005.

[356] S. A. Joni and G.-C. Rota, *Coalgebras and bialgebras in combinatorics*, in: R. Morris, ed., *Umbral calculus and Hopf algebras*, Amer. math. Soc., 1982, 1–47.

[357] Anthony Joseph, *Quantum groups and their primitive ideals*, Springer, 1997.

[358] André Joyal and Ross Street, *An introduction to Tannaka duality and quantum groups*, in: A. Carboni, M.C. Perdiccio and G. Rossolini, eds., *Category theory*, Como 1990, Springer, 1991, 411–492.

[359] Dominic Joyce, *Configurations in Abelian categories-II: Ringel-Hall algebras*, Adv. in Math. **210** (2007), 635–706.

[360] Victor Kac and Pokman Cheung, *Quantum calculus*, Springer, 2002.

[361] R. E. Kalman, P.L. Falb and M.A. Arbib, *Topics in system theory*, McGraw-Hill, 1969.

[362] Seok-Jin Kang, Jeong-Ah Kim and Dong-Uy Shin, *Crystal bases for quantum classical algebras and Nakajima's polynomials*, Publ. RIMS **40** (2004), 757–792.

[363] M. Kapranov and E. Vasserot, *Kleinian singularities, derived categories and Hall algebras*, 1999; `arXiv:math/9812016v2 [math.AG]`.

[364] M.M. Kapranov, *Eisenstein series and quantum affine algebras*, J. math. Sciences New York **84** (1997), 1311–1360.

[365] M. Kapranov, *Noncommutative geometry and path integrals*, 2006; `arXiv:math/0612411v1 [math.QA]`.

[366] Michael Karbach and Gerhard Müller, *Introduction to the Bethe Ansatz I*, 1998; `arXiv:cond-mat/9809162 v1`.

[367] Michael Karbach and Gerhard Müller, *Introduction to the Bethe Ansatz II*, 1998; `arXiv:cond-mat/9809163 v1`.

[368] M. Kasatani, and Y. Takeyama, *The quantum Kniznik-Zamolodchikov equations and nonsymmetric Macdonald polynomials*, 2006.

[369] Masaki Kashiwara and Michele Vergne, *On the Segal-shale Weil representations and harmonic polynomials*, Inv. Math. **44** (1978), 1–47.

[370] Masaki Kashiwara, *The crystal base and Littelmann's refined Demazure character formula*, Duke math. J. **71** (1993), 839–858.

[371] Christian Kassel, *Quantum groups*, Springer, 1995.

[372] Christian Kassel, Marc Rosso and Vladimir Turaev, *Quantum groups and knot invariants*, SMF, 1997.

[373] Christian Kassel and Hans-Jürgen Schneider, *Homotopy theory of Hopf Galois extensions*, 2004; `arXiv:0402034 v1`.

[374] Louis H. Kauffman, *An invariant of regular isotopy*, Trans. Amer. Math. Soc. **318** (1990), 417–471.

[375] Louis H. Kauffman, *Knots and physics*, World Scientific, 1991.

[376] Louis H. Kauffman, *Knots, spin networks and three manifold invariants*, Knots 90, de Gruyter, 1992, 271–287.

[377] Louis H. Kauffman and Sóstenes L.Lins, *Temperley-Lieb recoupling theory and invariants of three manifolds*, Princeton Univ. Press, 1994.

[378] Louis H. Kauffman, ed., *The interface of knots and physics*, Amer. Math. Soc., 1995.

[379] Yasuyuki Kawahigashi, *Topological quantum field theories and operator algebras*, 2003; `arXiv:math.QA/0306112v2`.

[380] Keigher, William, *On the ring of Hurwitz series*, Comm. in Algebra **25** (1997), 1845–1859.

[381] Kerov, S.V., *Characters of Hecke and Birman-Wenzl algebras*, in P. P. Kulish, ed., *Quantum groups*, Springer, 1992, 335–340.

[382] Masoud Khalkhali, *Very basic noncommutative geometry*, 2004; `arXiv:math/0408446 [math.KT]`.

[383] V. K. Kharchenko, *Fixed rings and noncommutative invariant theory*, in: M. Hazewinkel, ed., *Handbook of algebra, volume 2*, Elsevier, 2000, 359–398.

[384] S. Khoroshkin, D. Lebedev and S. Pakuliak, *Traces of intertwining operators for the Yangian double*, Lett. in math. Physics **41** (1997), 31–47.

[385] S. M. Khoroshkin and V. N. Tolstoy, *Yangian double*, Lett. in math. Physics **36** (1996), 373–402.

[386] A. N. Kirillov and N. Yu. Reshetikhin, *The Yangians, Bethe Ansatz and combinatorics*, Lett. in math. Physics **12** (1986), 199–208.

[387] M. Klimek, *Conservation laws for linear equations on braided spaces*, J. of math. Physics **40** (1999), 4165–4176.

[388] Anatoli U. Klimyk and Konrad Schmüdgen, *Quantum groups and their representations*, Springer, 1997.

[389] Anthony Knapp, *Representation theory of semisimple groups*, Princeton Univ. Press, 1986.

[390] H. T. Koelink, *Askey-Wilson polynomials and the quantum group SU(2): survey and applications*, Acta appl. math. **44** (1996), 295–352.

[391] H. T. Koelink, *8 lectures on quantum groups and q-special functions*, Mat. Inst. Univ. of Amsterdam, 1996.

[392] Erik Koelink and Yvette van Norden, *Pairings and actions of dynamical quantum groups*, Adv. in Math. **208** (2007), 1–39.

[393] Hitsohi Kohno, *Free field representation of level-k Yangian double* $DY(sl_2)_k$ *and deformation of Wakimoto modules*, Lett. in math. Physics **40** (1997), 321–336.

[394] Toshitake Kohno, ed., *New developments in the theory of knots*, World scientific, 1990.

[395] Matjaz Konvalinka and Igor Pak, *Non-commutative extensions of the MacMahon master theorem*, 2006; `arXiv:math/0607737v1`.

[396] Matjaz Konvalinka *A generalization of Foata's fundamental transformation and its application to the right-quantum algebra*, 2007; `arXiv:math/0703203v1`.

[397] Matjaz Konvalinka, *Non-commutative Sylvester's determinantal identity*, 2007; `arXiv:math/0703213v1 [math.CO]`.

[398] Tom H. Koornwinder, *Representations of the twisted* $SU(2)$ *quantum group and some q-hypergeometric orthogonal polynomials*, Indagationes math. **51** (1989), 97–112.

[399] Tom H. Koornwinder and Erik T. Koelink, *The Clebsch-Gordan coefficients for the quantum group* $SU_\mu(2)$ *and q Hahn polynomials*, Indagationes math. **51** (1989), 443–456.

[400] T. H. Koornwinder, *q-special functions: a tutorial*, Dept. Math. Univ. of Amsterdam, 1994; `arXiv:math/9403216v1 [math.CA]`.

[401] Tom H. Koornwinder, *Quantum groups and q-special functions*, in: V. Baldoni and M. A. Picardello, eds., *Representations of Lie groups and quantum groups*, Longman, 1994, 46–128.

[402] Tom H. Koornwinder, *q-special functions: an overview*, 2005; `arXiv:math/0511148v1 [math.CA]`.

[403] Leonid I. Korogodski and Yan S. Soibelman, *Algebras of functions on quantum groups: Part I*, Amer. math. Soc., 1998.

[404] L. I. Korogodsky, and L. L. Vaksman, *Quantum G-spaces and Heisenberg algebra*, in: P. P. Kulish, ed., *Quantum groups*, Springer, 1992, 56–66.

[405] S. M. Koroshkin and V. N. Tolstoy, *Yangian double*, Lett. in math. Physics **36** (1996), 373–402.

[406] A. N. Koryukin, *Noncommutative invariants of bialgebras*, Algebra and Logic **33** (1995), 366–380.

[407] Justyna Kosakowska, *The existence of Hall polynomials for posets of finite prinjective type*, J. of Algebra **308** (2007), 654–665.

[408] Justyna Kosakowska, *A specialization of prinjective Ringel-Hall algebra and the associated Lie algebra*, Acta math. Sinica. English Series **24** (2008), 1687–1702.

[409] Hanspeter Kraft and Claudio Procesi, *Classical invariant theory. A primer*, 1996. Preliminary version.

[410] Christian Krattenthaler, *Operator methods and Lagrange inversion: a unified approach to Lagrange formulas*, Trans. of the Amer. math. Soc. **305** (1988), 431–465.

[411] Dirk Kreimer, *Renormalization and knot theory*, Dept. of Physics, Univ. of Mainz, 1996.

[412] Dirk Kreimer, *Knots and Feynman diagrams*, Cambridge Univ. Press, 2000.

[413] Dirk Kreimer, *Combinatorics of (perturbative) quantum field theory*, Lyman Lab., Harvard Univ., 2000; `arXiv:hep-th/0010059`.

[414] Dirk Kreimer, *The residues of quantum field theory. Numbers we should know*, IHES, 2004; `arXiv:hep-th/0404090`.

[415] Dirk Kreimer, *Structures in Feynman graphs - Hopf algebras and symmetries*, in: M. Lyubich and L. Takhtajan, eds., *Graphs and patterns in mathematics and theoretical physics*, Amer. Math. Soc., 2005, 43–78. Preprint version: `arXiv:hep-th/0202110`.

[416] H. F. Kreimer and M. Takeuchi, *Hopf algebras and Galois extensions of an algebra*, Indiana Univ. Math. J. **11** (1981), 675–692.

[417] M. G. Krein, *A principle of duality for a bicompact group and square block algebra*, Dokl. Akad. Nauk SSSR **69** (1949), 725–728.

[418] D. Krob and J.-Y. Thibon, *Noncommutative symmetric functions V: a degenerate version of $U_q(\mathfrak{gl}_N)$*, Int. J. Algebra and Computation **9** (1997), 405–430.

[419] Daniel Krob and Jean-Yves Thibon, *Noncommutative symmetric functions IV: quantum linear groups and Hecke algebras at $q = 0$*, J. of algebraic Combinatorics **6** (1997), 339–376.

[420] P. P. Kulish and N. Yu. Reshetikhin, *Quantum linear problem for the sine-Gordon equations and higher representations*, Zap. Nauchn. Sem. LOMI **101** (1981), 101–110. English translation: J. Soviet Math. **23** (1983), 2435–2441.

[421] Petr P. Kulish, V. D. Lyakhovsky and A. I. Mudrov, *Twist related geometries on q-Minkowski space*, Proc. Steklov Inst. of Math. **226** (1999), 86–99.

[422] Petr P. Kulish, *Models solvable by Bethe Ansatz*, J. of generalized Lie Theory and Applications **2** (2008), 190–200.

[423] Larry A. Lambe and David E. Radford, *Introduction to the quantum Yang-Baxter equation: an algebraic approach*, KAP, 1997.

[424] G. Landi, G. Marmo and G. Vitasi, *An algebraic approach to integrability*, ESI, Vienna, 1993.

[425] G. Landi, *An introduction to noncommutative spaces and their geometry*, 1997; `arXiv:hep-th/9701078v1`.

[426] M. B. Landstad, *Duality for dual covariance algebras*, Comm. in math. Physics **52** (1977), 191–202.

[427] Alexander Lange, *The Epstein-Glaser approach to pQFT: graphs and Hopf algebras*, Max Planck Inst. for Mathematics, 2004.

[428] Aaron Lauve, *Quantum and quasi-Plücker coordinates*, J. of Algebra **296** (2006), 440–461.

[429] Lieven Le Bruyn, *Noncommutative geometryn*, 2000. Unpublished book manuscript `http://www.mat.ua.ac.be/~lebruyn`.

[430] Bernard Leclerc and Andrei Zelevinsky, *Quasicommuting families of quantum Plücker coordinates*, in G. I. Olshanski, ed., *Kirillov's seminar on representation theory*, Amer. math. Soc., 1996, 85–108.

[431] Robert Leduc and Arun Ram, *A ribbon Hopf algebra approach to the irreducible representations of centralizer algebras: the Brauer, Birman-Wenzl, and type A Iwahori Hecke algebras*, Adv. Math. **125** (1997), 1–94.

[432] G. I. Lehrer and R. B. Zhang, *Strongly multiplicity free modules for Lie algebras and quantum groups*, J. of Algebra **306** (2006), 138–174.

[433] G. I. Lehrer and R. B. Zhang, *A Temperley-Lieb analogue for the BMW algebra*, 2008; `arXiv:0806.0687`.

[434] G. I. Lehrer and R. B. Zhang, *On endomorphisms of quantum tensor space*, 2008; `arXiv:0806.3807v1 [math.RT]`.

[435] Frank Leitenberger, *Quantum deformation of invariants of higher binary forms*, 2005; `arXiv:math/0509677v1 [math.QA]`.

[436] Thomas H. Lenagan and Laurent Rigal, *Quantum analogues of Schubert varieties in the Grassmannian*, 2006; `arXiv:math/0610704v1`

[437] Cristian Lenart and Nigel Ray, *Hopf algebras of set systems*, Discrete Math. **180** (1998), 255–280.

[438] Serge Z. Levendorskii, *On PBW bases for Yangians*, Lett. in math. Physics **27** (1993), 37–42.

[439] S. Z. Levendorskii and A. Sudbery, *Yangian construction of the Virasoro algebra*, Lett. in math. Physics **37** (1996), 243–247.

[440] H. R. Lewis Jr. and W. B. Riesenfeld, *An exact quantum theory of the time dependent harmonic oscillator and of a charged particle in a time dependent electromagnetic field*, J. of math. Physics **10** (1969), 1458–1473.

[441] Libin Li and Pu Zhang, *Twisted Hopf algebras, Ringel-Hall algebras, and Green's categories*, J. of Algebra **231** (2000), 713–743.

[442] Libin Li and Pu Zhang, *Colored Ringel-Hall algebras: applications to quantized Borcherds super algebras*, in D. Happel and et al., eds., *Representations of algebras. Volume I*, Beijing normal Univ. Press, 2002, 308–320.

[443] Libin Li, *The coradical filtration for Drinfeld double Ringel-Hall algebras*, Comm. in Algebra **31** (2003), 1545–1559.
[444] YixiangLi and Zongzhu Lin, *A realization of quantum groups via product valued quivers*, 2006; `arXiv:math/0610447v1 [math.RT]`.
[445] Yixiang Li and Zongzhu Lin, *Generalized Kac-Moody Lie algebras and product quivers*, 2006; `arXiv:math/0610448v1 [math.RT]`.
[446] Yixiang Li and Zongzhu Lin, *Canonical bases of Borcherds-Cartan type*, 2008; `arXiv:0812.1449v1 [math.RT]`.
[447] Zongzhu Lin, *Lusztig's geometric approach to Hall algebras*, in V. Dlab and C.-M. Ringel, eds., *Representations of finite dimensional algebras and related topics in Lie theory and geometry*, Amer. Math. Soc., 2004, 349–364.
[448] JonLinks and Katrina Hibberd, *Bethe Ansatz solutions of the Bose - Hubbard dimer*, Symmetry, Integrability and Geometry. Methods and Applications **2** (2006), paper 095. `http://emis.de/journals/SIGMS`.
[449] Z. Lipinsky, *Yangian construction of a centrally extended Lie algebra of a diffeomorphism group of a torus*, J. of math. Physics **42** (2001), 2285–2292.
[450] Gong-Xiang Liu and Fang Li, *Basic Hopf algebras of finite representation type*, Zheziang Univ., Hangzhou, 2004.
[451] Charles Livingston, *Knot theory*, Math. Ass. of America, 1993.
[452] M. S. Livsic, N. Kravitsky, A. S. Markus and V. Vinnikov, *Theory of commuting nonselfadjoint operators*, Kluwer Acad. Publ., 1995.
[453] Fedele Lizzi, S. Vaidya and P. Vitale, *Twisted conformal symmetry in noncommutative two-dimensional quantum field theory*, 2006. Copies of the overheads of a lecture given in Vietri.
[454] J. Lukierski and A. Nowicki, *Quantum deformations of D=4 Poincaré algebra*, in: R. Gielerak, J. Lukierski and Z. Popowicz, eds., *Quantum groups and related topics*, KAP, 1992, 33–44.
[455] J. Lukierski, A. Nowicki and H. Ruegg, *Quantum deformations of D=4 Poincaré and conformal algebras*, in: H.- D. Doebner and V. K. Dobrev, eds., *Quantum symmetries*, World Scientific, 1993, 304–326.
[456] Jerzy Lukierski and Henri Ruegg, *Quantum κ-Poincaré in any dimension*, 1993; `arXiv:hep-th9310117v1`.
[457] George Lusztig, *Canonical bases arising from quantized enveloping algebras*, J. Amer. math. Soc. **3** (1990), 447–498.
[458] George Lusztig, *Canonical bases arising from quantized enveloping algebras II*, Progres of theor. Phys. Suppl. **102** (1990), 175–201.
[459] G. Lusztig, *Canonical bases in tensor products*, Proc. nat. Acad. Sci. USA **89** (1992), 8177–8179.
[460] George Lusztig, *Introduction to quantum groups*, Birkhäuser, 1993.
[461] George Lusztig, *Canonical bases and Hall algebras*, in A. Broer, ed., *Representation theories and algebraic geometry*, KAP, 1998, 365–400.
[462] V. Lychagin, *Quantizations of braided differential operators*, Schrödinger Inst., Vienna, 1993.
[463] V. Lychagin, *Braidings and quantizations over bialgebras*, Sophus Lie Centre, Moscow, 1993.
[464] Valentin Lychagin, *Differential operators and quantizations* (1), Mat. Inst. Univ. Oslo, 1993.
[465] Valentin Lychagin, *Colour calculus and colour quantizations*, Mat. Inst. Univ. Oslo, 1994.
[466] V. Lychagin, *Calculus and quantizations over Hopf algebras*, Acta applicandae mathematicae **51** (1998), 303–352.
[467] V. V. Lyubashenko, *Hopf algebras and vector symmetries*, Usp. mat. Nauk **41** (1986), 185–186.
[468] V. V. Lyubashenko, *Vector symmetries*, Math. Inst., Univ. of Stockholm, 1987.
[469] V. V. Lyubashenko, *Tensor categories and RCFT. II: modular transformations*, Inst. theoretical Physics, Akad. Nauk Ukrainian SSR, 1990.
[470] V.V. Lyubashenko, *Tensor categories and RCFT. I: Hopf algebras in rigid categories*, Inst. theoretical Physics, Akad. Nauk Ukrainian SSR, 1990.
[471] V. V. Lyubashenko, *Tangles and Hopf algebras in braided categories*, J. of pure and applied Algebra **98** (1995), 245–278.
[472] I. G. Macdonald, *Symmetric functions and Hall polynomials*, Clarendon Press, 1995.

[473] Percy A. MacMahon, *Combinatory analysis*, Cambridge Univ. Press, 1915. Chelsea one volume reprint, 1960.

[474] J. Madore, *An introduction to noncommutative differential geometry and its physical applications*, Cambridge university press, 1999.

[475] John Madore, *Introduction to noncommutative geometry*, in: H. Gausterer et. al., eds., *Geometry and quantum physics*, Springer, 1999, 231–273.

[476] John Madore, *Noncommutative geometry for pedestrians*, 1999; `arXiv:gr-qc/9906059v2`.

[477] J. M. Maillet, *Drinfel'd twists and algebraic Bethe Ansatz*, 1996; `arXiv:q-alg/9612012`.

[478] Shahn Majid, *Quasi-quantum groups as internal symmetries of topological field theories*, Lett. in math. Physics **22** (1991), 83–90.

[479] Shahn Majid, *Tannaka-Krein theorem for quasi-Hopf algebras and other results*, in: M. Gerstenhaber and J. Stasheff, eds., *Deformation theory and quantum groups with applications to mathematical physics*, Amer. math. Soc., 1992, 219–232.

[480] Shahn Majid, *Bicrossproduct structure of the κ-Poincaré group and non-commutative geometry*, Physics Lett. B **334** (1994), 348–354; `arXiv:hep-th/9405107v2`.

[481] Shahn Majid, *Foundations of quantum group theory*, Cambridge University Press, 1995.

[482] Shahn Majid, *Classification of bicovariant differential calculi*, J. of Geometry and Physics **25** (1998), 119–140. preprint versions: Erwin Schrodinger Inst. for math. Phys., 1996; `arXiv:q-alg/9608016v2`.

[483] Shahn Majid, *Quantum geometry of field extensions*, J. math. Physics **40** (1999), 2311–2323. Preprint version: `arXiv:q-alg/0706026v1`.

[484] Shahn Majid, *Quantum groups and noncommutative geometry*, J. of math. Physics **41** (2000), 3892–3942.

[485] Shahn Majid, *Noncommutative differential geometry and twisting of quantum groups*, 2000; `arXiv:math/0006151v1[math.QA]`.

[486] Shahn Majid, *Quantum groups. A primer*, Cambridge Univ. Press, 2002.

[487] Shahn Majid, *Classification of differentials on quantum doubles and finite noncommutative geometry*, 2003; `arXiv:math/0205150v2`.

[488] Marie-Paule Malliavin, *Algèbre d'Heisenberg quantique*, Bull. Sci. Math. II Ser. **118** (1994), 511–537.

[489] G. Maltsinotis, *Le language des espaces et des groupes quantiques*, Commun. math. Physics **151** (1993), 275–302.

[490] D. V. Malyshev, *Hopf algebra of graphs and the RG equations*, Dept. Math. Princeton Univ., 2004; `arXiv:hep-th/0408230`.

[491] Dominique Manchon, *Hopf algebras, from basics to applications to renormalization*, Univ. Blaise Pascal, Clermont-Ferrand, 2004.

[492] Dominique Manchon, *Hopf algebras in renormalisation*, in: M. Hazewinkel, ed., *Handbook of algebra, Volume 5*, Elsevier, 2008, 365–427.

[493] Dominique Manchon, *Quelques aspects combinatoires de la renormalization*, Gazette des Math. Janvier 2009 (2009), 5–15.

[494] Yu. I. Manin, *Some remarks on Koszul algebras and quantum groups*, Ann. de l'Inst. Fourier **37** (1987), 191–205.

[495] Yu. I. Manin, *Quantum automorphism groups of quadratic algebras*, 1988.

[496] Yuri I. Manin, *Quantum groups and noncommutative geometry*, Centre de recherches Mathématiques, Univ. de Montréal, 1988.

[497] Yu. I. Manin, *Multiparametric quantum deformation of the general linear supergroup*, Comm. in math. Physics **123** (1989), 163–175.

[498] Yuri I. Manin, *Topics in noncommutative geometry*, Princeton Univ. Press, 1991.

[499] Yu. I. Manin, *Notes on quantum groups and quantum de Rham complexes*, Theoretical and mathematical Physics **92** (1992), 425–450.

[500] Vasily O. Manturov, *Knot theory*, CRC Press, 2003.

[501] W. Marcinek, *Quantum-classical interactions and Galois type extensions*, in: P. M. Hajac and W. Pusz, eds., *Noncommutative geometry and quantum groups*, PWN, 2003, 103-109. Preprint version: `arXiv:0210479v1 [math.QA]`.

[502] Matilde Marcolli, *Lectures on arithmetic noncommutative geometry*, 2004; `arXiv:math/0409520 [math.QA]`.

[503] Matilde Marcolli, *Noncommutative geometry and number theory*, NAW 5/9 (2008), 109–112.

[504] A. A. Markov, *Über die freie Äquivalenz geschlossener Zöpfe*, Recueil math. Moscou **1** (1935), 73–78.

[505] Stuart Martin, *Schur algebras and representation theory*, Cambridge Univ. Pr., 1993.

[506] Tetsuya Masuda, Katsuhisa Mimachi, Yoshiomi Nakagami, Masatoshi Noumi and Kimio Ueno, *Representations of quantum groups and a q-analogue of orthogonal polynomials*, C. R. Acad. de Sci. Paris Sér. I **307** (1988), 559–564.

[507] Akira Masuoka, *Generalized q-boson algebras and their integrable modules*, 2009; `arXiv:0812.4495v2 [math.QA]`.

[508] D. J. Melville, *An A-form technique of quantum deformations*, in N. Jing and K. Misra, eds., *Recent developments in quantum affine algebras and related topics*, Amer. math. Soc., 1999, 359–375.

[509] Igor Mencattini and Dirk Kreimer, *Insertion and elimination Lie algebra: the ladder case*, Centre Math. Phys., Boston Univ., 2003; `arXiv:math.QA/0308042`.

[510] Igor Mencattini and Dirk Kreimer, *The structure of the ladder insertion-elimination Lie algebra*, Centre Math. Phys., Boston Univ., 2004; `arXiv:math-ph/0408053`.

[511] Miguel A. Mendez, *The umbral calculus of symmetric functions*, Adv. in Math. **124** (1996), 207–271.

[512] G. Militaru, *The Hopf modules category and the Hopf equation*, (1998); `arXiv:math/9807003 v1`.

[513] G. Militaru, *A class of non-symmetric solutions for the integrability of the Kniznik-Zamolodchikov equation: a Hopf algebra approach*, 1998; `arXiv:math/9807028 v1`.

[514] G. Militaru, *The Long dimodules category and nonlinear equations*, 1998; `arXiv:math/9807017 v1`.

[515] A. I. Molev, *Representations of twisted Yangians*, Lett. in math. Physics **26** (1992), 211–218.

[516] A. I. Molev, *Gelfand - Tsetlin basis for representations of Yangians*, Lett. in math. Physics **30** (1994), 53–60.

[517] A. I. Molev, *Finite dimensional irreducible representations of twisted Yangians*, J. of math. Physics **39** (1998), 5559–5600.

[518] Alexander Molev, *Yangians and classical Lie algebras*, American math. Soc., 2007.

[519] Alexander I. Molev and Maxim Nazarov, *Capelli identities for classical Lie algebras*, School of Math. and Stat., Univ. of Sydney, 1999. `arXiv:9712021v3`.

[520] S. Montgomery, *Hopf algebras and their actions on rings*, Amer. math. Soc., 1993.

[521] S. Montgomery, ed., *Group actions on rings*, Amer. math. Soc., 1985.

[522] Susan Montgomery, *Hopf Galois extensions*, in D. Haile and J. Osterburg, eds., *Azumaya algebras, actions, and modules*, Amer. math. Soc., 1992, 129–140.

[523] Henri Moscovici, and Bahram Rangipour, *Cyclic cohomology of Hopf algebras of transverse symmetries in codimension one*, 2006; `arXiv:0602020v2`.

[524] E. Mukhin, V. Tarasov and A. Varchenko, *A generalization of the Capelli identity*, Dept. math. Sci., Purdue Univ., 2006; `arXiv:math/0610799v1 [math.Q]`.

[525] E. Mukhin, V. Tarasov and A. Varchenko, *On separation of variables and completeness of the Bethe Ansatz*, 2007; `arXiv:math.QA/0712,0981 v1`.

[526] E. Mukhin, V. Tarasov and A. Varchenko, *Schubert calculus and representations of the general linear group*, 2007; `arXiv:math/0711.4079 v1`.

[527] M. Müger, J. E. Roberts and L. Tuset, *Representations of algebraic quantum groups and reconstruction theorems for tensor categories, Algebras and representation theory* **7** (2004), 517–573. Preprint version: `arXiv:math/0203206v1`; #2204.

[528] Folkert Müller-Hoissen, *Differential calculi on the quantum group $GL_{p,q}(2)$*, J. Phys. A: Math Gen **25** (1992), 1703–1734.

[529] Folkert Müller-Hoissen and C. Reuten, *Bicovariant differential calculi on $GL_{p,q}(2)$ and quantum subgroups*, J. Phys. A: Math. Gen. **26** (1993), 2955–2975.

[530] Folkert Müller-Hoissen, *Physical aspects of differential calculi on commutative algebras*, Inst. Theor. Physik, Univ. Göttingen, 1994.

[531] David Mumford, *Geometric invariant theory*, Springer, 1965.

[532] Y. Nakagami, *Dual action on a von Neumann algebra and Takesaki's duality for a locally compact group*, Publ. Res. Inst. math. Sci., Kyoto **12** (1977), 727–775.

[533] M. L. Nazarov, *Quantum Berizinian and the classical Capelli identity*, Letters in math. Physics **21** (1991), 123–131.

[534] M. L. Nazarov, *Yangians of 'strange' Lie super algebras*, in P. P. Kulish, ed., *Quantum groups*, Springer, 1992, 90–97.

[535] Maxim Nazarov and Vitaly Tarasov, *Yangians and Gelfand-Zetlin bases*, Publ. RIMS Kyoto **30** (1994), 459–478.

[536] Maxim L. Nazarov, *Capelli identities for Lie super algebras*, 1996; `arXiv:9610032`.

[537] Maxim L. Nazarov, *Yangians and Capelli identities*, 1996; `arXiv:q-alg/9601102v2`.

[538] M. Neuchl and P. Schauenburg, *Reconstruction in braided categories and a notion of commutative bialgebra*, J. of pure and appl. Algebra **124** (1998), 241–260.

[539] Warren D. Nichols, *Bialgebras of type one*, Comm. Algebra **6** (1978), 1521–1552.

[540] Warren Nichols and Moss Sweedler, *Hopf algebras and combinatorics*, in R. Morris, ed., *Umbral calculus and Hopf algebras*, Amer. math. Soc., 1982, 49–84.

[541] Emmy Noether, *Invariante Variationsprobleme*, Nachr. König. Ges. Wiss. Göttingen (1918), 235–257.

[542] Masatoshi Noumi, Toru Umeda and Masato Wakayama, *A quantum dual pair* (sl_2, o_n) *and the associated Capelli identity*, Lett. in math. Physics **34** (1995), 1–7.

[543] Masatoshi Noumi, Toru Omeda and Masato Wakayama, *Dual pairs, spherical harmonics and a Capelli identity in quantum group theory*, Compositio mathematicae **104** (1996), 227–277.

[544] Abdukadir Obul, *Minimal generating system of Ringel-Hall algebras of affine valued quivers*, J. of Algebra **297** (2006), 311–332.

[545] Abdukadir Obul and Guanglian Zhang, *PBW-bases of the twisted generic composition algebras of affine valued quivers*, J. of Algebra **297** (2006), 333–360.

[546] Abdukadir Obul, *Generalized tilting functors and Ringel-Hall algebras*, J. of pure and appl. Algebra **208** (2007), 445–488.

[547] Adrian Ocneanu, *Quantized groups, string algebras and Galois theory for algebras*, in: D. E. Evans and M. Takesaki, eds., *Operator algebras and applications, Volume 2*, Cambridge Univ. Press, 1988, 119–172.

[548] Adrian Ocneanu, *Paths on Coxeter diagrams: from Platonic solids and singularities to minimal models and subfactors*, in: R. Bhata a.o., eds., *Lectures on operator theory*, Amer. math. Soc., 2000.

[549] O. Ogievetsky and B. Zumino, *Reality in the differential calculus on q-Euclidean spaces*, Lett. in math. Physics **25** (1992), 121–130.

[550] O. Ogievetsky and A. Vahlas, *The relation between two types of characteristic equations for quantum matrices*, Lett. in math. Physics **65** (2003), 49–57.

[551] Andrei Okounkov, *Young basis, Wick formula, and higher Capelli identities*, 1996; `arXiv:q-alg/9602027v2`.

[552] Andrei Okounkov, *Quantum immanents and higher Capelli identities*, Transformation Groups **1** (1996), 99–126.

[553] G. I. Olshanskij, *Yangians and universal enveloping algebras*, Zap. Nauchn. Sem. LOMI **164** (1987), 142–150.

[554] G. I. Olshanskii, *Twisted Yangians and infinite classical dimensional Lie algebras*, in: P. P. Kulish, ed., *Quantum groups*, Springer, 1992, 104–119.

[555] Arupkumar Pal, *Induced representations and Frobenius reciprocity for compact quantum groups*, Proc. Indian Acad. Sci. (math. Sci.) **102** (1995), 157–167.

[556] Florin Panaite, *Relating the Connes-Kreimer and Grossman-Larson Hopf algebras built on rooted trees*, Lett. in math. Physics **51** (2000), 211–219.

[557] T. S. Papatheodorou, M. J. Ablowitz, Y. G. Saridakis, *A rule for the fast computation and analysis in cellular automata*, Stud. in Appl. Math. **79** (1988), 173–184.

[558] Bodo Pareigis, *A non-commutative non-cocommutative Hopf algebra in 'nature'*, J. of Algebra **70** (1981), 356–374.

[559] Bodo Pareigis, *Forms of Hopf algebras and Galois theory, Topics in algebra, part 1: rings and representations*, PWN, 1990, 75–93.

[560] Bodo Pareigis, *Vorlesungen über Quantengruppen und nichtkommutative Geometrie*, University of München, 1993.

[561] Bodo Pareigis, *Reconstruction of hidden symmetries*, J. of Algebra **183** (1996), 90–154.

[562] Brian Parshall and Jian-pan Wang, *Quantum linear groups*, American math. Soc., 1991.

[563] Mario Paschke and Andrzej Sitarz, *Discrete spectral triples and their symmetries*, J. of math. Phys. **39** (1998), 6191–6205.

[564] Paul A. Pearce, *Eight vertex and hard hexagon models*, Dept. Math. and Stat., Univ. of Melbourne, Australia, 2004.

[565] F. Peter, and H. Weyl, *Die Vollständigkeit der primitiven Darstellungen einer geschlossenen kontinuierlichen Gruppe*, Math. Ann. **97** (1927), 737–755.

[566] G. B. Podkolzin and L. I. Vainerman, *Quantum Stiefel manifold and double cosets of quantum unitary group*, Pacific J. of Math. **188** (1999), 179–199.

[567] P. Podleś, *Quantum spheres*, Lett. in math. Physics **14** (1987), 193–202.

[568] Piotr Podleś, *Symmetries of quantum spaces. Subgroups and quotient spaces of quantum $SU(2)$ and $SO(3)$ groups*, Comm. in math. Physics **170** (1995), 1–20.

[569] P. Podleś and E. Müller, *Introduction to quantum groups*, 1997; `arXiv:9704002 v2`.

[570] V. V. Prasolov and A. B. Sossinsky, *Knots, links, braids and 3-manifolds*, Amer. math. Soc., 1997.

[571] Jozef H. Przytycki and Pawel Traczyk, *Invariants of links of Conway type*, Kobe math. J. **4** (1988), 115–139.

[572] P. N. Pyatov and P. A. Saponov, *Characteristic relations for quantum matrices*, 1995; `arXiv:q-alg/9502012v3`.

[573] David E. Radford, *On the coradical of a finite dimensional Hopf algebra*, Proc. Amer. math. Soc. **53** (1975), 9–15.

[574] David E. Radford, *On the structure of pointed coalgebras*, J. of Algebra **77** (1982), 1–14.

[575] David E. Radford, *Solutions to the Yang-Baxter equation and the Drinfel'd double*, J. of Algebra **161** (1993), 20–32.

[576] David E. Radford, *Finite-dimensional simple pointed Hopf algebras*, J. of Algebra **211** (1999), 686–710.

[577] O. V. Radko and A. A. Vladimirov, *On the algebraic structure of differential calculus on quantum groups*, J. of math. Physics **38** (1997), 5434 5446.

[578] Serban Raianu, *Coalgebras from formulas*, in J. Bergen, S. Catoiu and W. Chin, eds., *Hopf algebras*, Marcel Dekker, 2004, 215–224.

[579] Nigel Ray, *Extensions of umbral calculus: prenumbral coalgebras and generalized Bernouilli numbers*, Adv. in Math. **61** (1986), 49–109.

[580] Amitai Regev, *Algebras satisfying a Capelli identity*, Israel J. of Mathematics **33** (1979), 149–154.

[581] M. Reineke, *Monomials in canonical bases of quantum groups and quadratic forms*, J. of pure and apllied Algebra **157** (2001), 310–310.

[582] Markus Reineke, *Quantum groups and geometry of quiver representations - an overview*, Univ. of Wuppertal, 2002. Synopsis Habilitationsschrift.

[583] Markus Reineke, *Framed quiver moduli, cohomology, and quantum groups*, J. of Algebra **320** (2008), 94–115.

[584] Adam Rennie and Joseph C. Varilly, *Reconstruction of manifolds in noncommutative geometry*, 2008; `arXiv:math/0610418v4[math.QA]`.

[585] N. Yu. Reshetikhin, *Quantized universal enveloping algebras, the Yang Baxter equation and invariants of links, I*, LOMI, Leningrad, 1988.

[586] N. Yu. Reshetikhin, *Quantized universal enveloping algebras, the Yang Baxter equation and invariants of links, II*, LOMI, Leningrad, 1988.

[587] N. Reshetikhin, *Multiparameter quantum groups and twisted quasitriangular Hopf algebras*, Lett. in math. Physics **20** (1990), 331–336.

[588] N. Yu. Reshetikhin, L. A. Takhtajan and L. D. Faddeev, *Quantization of Lie groups and Lie algebras*, Leningrad Math. J. **1** (1990), 193–225. Translation from the Russian of "Quantum groups and Lie algebras", Algebra i Analyz **1**:1 (1989), 178–205.

[589] M. Rietsch, *An introduction to perverse sheaves*, in V. Dlab and C.- M. Ringel, eds., *Representations of finite dimensional algebras and related topics in Lie theory and geometry*, Amer. math. Soc., 2004, 391–420.

[590] Claus Michael Ringel, *Representations of K-species and bimodules*, J. of Algebra **41** (1976), 269–302.

[591] Claus Michael Ringel, *Hall algebras and quantum groups*, Inventiones Math. **101** (1990), 583–592.

[592] Claus Michael Ringel, *Hall algebras*, in: S. Balcerzyk, T. Jozefiak, J. Krempa, D. Simson and W. Vogel, eds., *Topics in algebra. Volume 1: Rings and representations of algebras*, PWN, 1990, 433–447.

[593] Claus Michael Ringel, *Hall polynomials for the representation of finite hereditary algebras*, Adv. in Math. **84** (1990), 137–178.

[594] Claus Michael Ringel, *From representations of quivers via Hall and Loewy algebras to quantum groups*, in: L. A. Bokut, Y. L. Ershov and A. I. Kostrikin, eds., *Proceedings of the international conference on algebra in honor of A.I. Mal'cev*, Novosibirsk 1989., Amer math. Soc., 1992, 381–402.

[595] Claus Michael Ringel, *Hall algebras revisited*, in A. Joseph and S. Shnider, eds., *Quantum deformations of algebras and their representations*, Bar-Ilan University, 1993, 171–176.

[596] Claus Michael Ringel, *The composition algebra of a cyclic quiver*, Proc. of the London math. Soc. **66** (1993), 507–537.

[597] Claus Michael Ringel, *The Hall algebra approach to quantum groups. Notas de cursos, Escuela Latin Americana de Matematicas*, Sociedad Matematica Mexicana, 1993.

[598] Claus Michael Ringel, *Green's theorem on Hall algebras*, in: R. Bautista and et al., eds., *Representation theory of algebras and related topics*, Amer. math. Soc., 1996, 185–245.

[599] Claus Michael Ringel, *PBW-bases of quantum groups*, J. reine und angew. Math. **479** (1996), 51–87.

[600] Claus Michael Ringel, *The theorem of Bo Chen and Hall polynomials*, Nagoya math. J. **183** (2006), 145–160.

[601] Izfhak Roditi, *Integrable systems and quantum groups*, Brazilian J. of Physics **30** (2000). `http://www.scielo.br/`.

[602] Dale Rolfsen, *Knots and links*, Publish or Perish, 1973.

[603] Steven Roman and Gian-Carlo Rota, *The umbral calculus*, Adv. in Math. **27** (1978), 95–188.

[604] Steven Roman, *The umbral calculus*, Acad. Press, 1984.

[605] Alexander L. Rosenberg, *The spectrum of the algebra of skew differential operators and the irreducible representations of the quantum Heisenberg algebra*, Comm. in math. Physics **142** (1991), 547–588.

[606] Alexander L. Rosenberg, *The unitary irreducible representations of the quantum Heisenberg algebra*, Comm. in math. Physics **144** (1992), 41–51.

[607] Marc Rosso, *Finite dimensional representations of the quantum analogue of the enveloping algebra of a complex simple Lie algebra*, Comm. in math. Physics **117** (1988), 581–593.

[608] Marc Rosso, *Représentations des groupes quantiques*, Séminaire Bourbaki 1990/1991, Soc. math. de France, 1991, 443–484.

[609] Marc Rosso, *Quantum groups and quantum shuffles*, Inventiones math. **133** (1998), 399–416.

[610] Rafael Rouquier, *Hall algebras*, Inst. of pure Math., Univ. of Oxford, 2007.

[611] Louis Halle Rowen, *Graduate algebra: noncommutative view*, Amer. math. Soc., 2008.

[612] Hebing Rui and Jie Xu, *The representations of cyclotomic BMW algebras*, 2008; `arXiv:0801.0465v2 [math.QA]`.

[613] Steen Ryom-Hansen, *Littleman's refined Demazure character formula revisited*, Sém. Lotharingien de Combinatoire **49** (2003), B49d.

[614] Ladislav Samaj, *Introduction to exactly solvable 1D quantum lattice modes of solid state physics*, Inst. of Physics, Slovak Acad. of Sciences, 2004.

[615] Yuya Sasai, *Braided quantum field theories and their symmetries*, Yukawa Inst. for theor. Physics, 2007; `arXiv:0704.0822`.

[616] Peter Schauenburg, *Tannaka duality for arbitrary Hopf algebras*, Verlag Reinhard Fischer, 1991.

[617] Peter Schauenburg, *Hopf-Galois extensions of graded algebras*, Math. Inst. Univ. of München, 1994.

[618] Peter Schauenburg, *Quantum torsors and Hopf-Galois objects*, 2002; `arXiv:math/0208047v1 [math.QA]`.

[619] Peter Schauenburg and Hans-Jürgen Schneider, *On generalized Hopf-Galois extensions*, J. of pure and appl. Algebra **202** (2003), 168–194.

[620] Peter Schauenburg, *Hopf-Galois and bi-Galois extensions*, in G. Janelidze, B. Pareigis and W. Tholen, eds., *Galois theory, Hopf algebras, and semi-Abelian categories*, Amer. math. Soc., 2004, 469–516.

[621] Travis Schedler, *A Hopf algebra quantizing a necklace Lie algebra canonically associated to a quiver*, 2004; `arXiv:math/0406200v2 [mathQA]`.

[622] Olivier Schiffmann, *The Hall algebra of a cyclic quiver and canonical bases of Fock pace*, Int. math. Research Notes (2000), 413–440. Preprint version: arXiv:math/9906069v2 [math.OA].

[623] Olivier Schiffmann, *The Hall algebra of an elliptic curve II*, 2005. Preprint version: arXiv:math/9906069v2 [math.OA]; arXiv:math/0508551v1 [math.RT].

[624] Olivier Schiffmann, *Lectures on Hall algebras*, 2006; arXiv:math/0611617v1.

[625] Olivier Schiffmann and E. Vasserot, *The elliptic Hall algebra, Cherednik Hecke algebras and Macdonald polynomials*, 2009. Preprint version: arXiv:math/9906069v2 [math.OA]; arXiv:0802.1001v[math.QA].

[626] Karl-Georg Schlesinger, *On a quantum analog of the Grothendieck-Teichmüller group*, 2002; arXiv:math/0104275v3.

[627] Konrad Schmüdgen, *Noncommutative covariant differential calculi on quantum spaces and quantum groups*, Séminaire Sophus Lie **1** (1991), 125–132.

[628] Konrad Schmüdgen and Alex Schüler, *Classification of bicovariant differential calculi on quantum groups*, Commun. in math. Physics **170** (1995), 315–335.

[629] Konrad Schmüdgen and Axel Schüler, *Left covariant differential calculi on $SL_q(2)$ and $SL_q(3)$*, 1996; arXiv:q-alg/9601020v1.

[630] Hans-Jürgen Schneider, *Principal homogeneous spaces for arbitrary Hopf algebras*, Israel J. Math. **72** (1990), 167–195.

[631] Hans-Jürgen Schneider, *Representation theory of Hopf Galois extensions*, Israel J. of Math. **72** (1990), 196–231.

[632] H.-J. Schneider, *Hopf Galois extensions, crossed products and Clifford theory*, in: J. Bergen and S. Montgomery, eds., *Advances in Hopf algebras*, Marcel Dekker, 1994, 267–299.

[633] Hans-Jürgen Schneider, *Lectures on Hopf algebras*, Math. Sem. Univ. München, 1995.

[634] Peter Schupp, Paul Watts and Bruno Zumino, *Differential geometry on linear quantum groups*, Lett. in math. Physics **25** (1992), 139–148.

[635] Michael A. Semenov Tian-Shansky, *Dressing transformations and Poisson group actions*, Publ. RIMS **21** (1985), 1237–1260.

[636] Bert Sevenhant and Michel Van den Bergh, *On the number of absolutely indecomposable representations of a quiver*, J. of Algebra **221** (1999), 29–49.

[637] Bert Sevenhant and Michel Van den Bergh, *On the double of the Hall algebra of a quiver*, J. of Algebra **221** (1999), 135–160.

[638] Bert Sevenhant and Michel Van den Bergh, *A relation between a conjecture of Kac and the structure of the Hall algebra*, J. pure and appl. Algebra **160** (2001), 319–332.

[639] Dmitry Shklyarov, Sergey Sinel'shchikov and Leonid Vaksman, *Quantum matrix ball differential and integral calculi*, 1999; arXiv:math/9905035v2 [math.QA].

[640] Dmitry Shklyarov, Sergey Sinel'shchikov and Leonid Vaksman, *Hidden symmetries of some algebras of q-differential operators*, in: S. Duplij and J. Wess, eds., *Noncommutative structures in mathematics and physics*, Kluwer Acad. Publ., 2001, 309–320.

[641] Steven Shnider and Shlomo Sternberg, *Quantum groups. From coalgebras to Drinfeld algebras*, International Press, 1994.

[642] Daniel Simson, *Path coalgebras of profinite bound quivers, cotensor coalgebras of bound species and locally nilpotent representations*, Colloquium math. **109** (2007), 307–343.

[643] S. Sinel'shchikov and L. Vaksman, *Hidden symmetry of the differential calculus on the quantum matrix space*, J. of Phys. A: math. gen. **30** (1997), L23–L26. Preprint version: arXiv:q-alg/9603012v1.

[644] D. W. Singer, *q-analogues of Lagrange inversion*, Adv. in Math. **115** (1995), 99–116. Erratum: ibid. **137** (1998), 396–397.

[645] Evgeny K. Sklyanin, *On an algebra generated by quadratic relations*, Uspekhi Mat. Nauk **40**:2 (1985).

[646] Evgeny K. Sklyanin, *Quantum inverse scattering method. Selected topics*, 1992; arXiv:hep-th/9211111 v1.

[647] S. P. Smith, *Quantum groups for ring theorists*, In: S. Montgomery, L. Small (eds.), *Noncommutative rings*, Amer. math. Soc., 1992, 131–178.

[648] F. Spitzer, *A combinatorial lemma and its application to probability theory*, Trans. Amer. math. Soc. **82** (1956), 323–339.

[649] E. Steinitz, *Zur Theorie der Abel'schen Gruppen*, Jahresbericht Deutscher Math. Verein **9** (1901), 80–85.

[650] Ross Street, *Quantum groups. A path to current algebra*, Cambridge Univ. Press, 2007.

[651] Xiuping Su, *A generic multiplication in quantised Schur algebras*, 2008; arXiv:0812.0688v1 [math.QA].

[652] Yu. N. Subbotin, *Bernoulli polynomials*, in M. Hazewinkel, ed., *Encyclopaedia of mathematics, Vol.1 (A-B)*, Reidel, 1988, 361–362.

[653] A. Sudbery, *Quantum matrix groups determined by quadratic coordinate algebras*, Dept. Math., Univ. of York, Heslington, UK, 1990.

[654] Moss E. Sweedler, *The Hopf algebra of an algebra aplied to field theory*, J. of Algebra **8** (1968), 262–276.

[655] Jeno Szegeti, *Cayley-Hamilton theorem for matrices over an arbitrary ring*, Serdica math. J. **32** (2006), 269–276.

[656] K. Szlachanyi and P. Vecsernyes, *Quantum symmetry and braid group statistics in G-spin models*, Commun. in math. Physics **156** (1993), 127–168.

[657] Earl J. Taft, *Quantum deformation of the flag variety*, in P.P. Kulish, ed., Quantum groups, Springer, 1992, 138–141.

[658] M. Takeuchi, *Free Hopf algebras generated by coalgebras*, J. math. Soc. Japan **23** (1971), 561–582.

[659] Mitsuhiro Takeuchi, *Matrix bialgebras and quantum groups*, Israel J. Math. **72** (1990), 232–251.

[660] M. Takeuchi, *The group ring of $GL_n(q)$ and the q-Schur algebra*, J. math. Soc. Japan **48** (1996), 259–274.

[661] D. Talalaev, *Quantization of the Gaudin system*, Inst. for theoretical and experimental physics, Moscow, 2004; arXiv:hep-th/0404153v1.

[662] Dmitri Tamarkin and Boris Tsygan, *The ring of differential operators on forms in noncommutative calculus*, in: M. Lyubich and L. Takhtajan, eds., *Graphs and patterns in mathematics and theoretical physics*, Amer. math. Soc., 2005, 105–131. Preprint version: arXiv:hep-th/0202110.

[663] Youjun Tan, *The q-analogue of Bosons and Hall algebras*, Comm. in Algebra **30** (2002), 4335–4347. Corrigenda et addenda: arXiv:0904.1789v1 [math.QA].

[664] T. Tannaka, *Über den Dualitätsatz der nichtkommutativen Gruppen*, Tohôku math. J. **45** (1939), 1–12.

[665] Thomas Timmermann, *An invitation to quantum groups and duality. From Hopf algebras to multiplicative unitaries and beyond*, AMS, 2008.

[666] I. T. Todorov, *Quantum groups as symmetries of chiral conformal algebras*, in: H.-D. Doebner and J.-D. Hennig, eds., *Quantum groups*, Springer, 1990, 231–277.

[667] V. G. Turaev, *Quantum invariants of knots and three manifolds*, de Gruyter, 1994.

[668] V. G. Turaev, *The Yang-Baxter equation and invariants of links*, Inv. math. **92** (1988), 527–553.

[669] Vladimir G. Turaev, *Loops on surfaces, Feynman diagrams and trees*, J. of Geom. and Phys. **53** (2005), 461–482. Preprint version: arXiv:hep-th/0403266.

[670] H. W. Turnbull, *Symmetric determinants and the Cayley and Capelli operators*, Proc. Edinburgh math. Soc. **8** (1948), 76–86.

[671] L. L. Vaksman and Ya. S. Soibelman, *The algebra of functions on the quantum group SU(2)*, Funk. Anal. Pril. **22** (1988), 1–14.

[672] Alphons Van Daele and Y. Zhang, *Multiplier Hopf algebras of discrete type*, J. of Algebra **214** (1999), 400–412.

[673] Nico W. van den Hijligenberg, *Special solutions of the quantum Yang-Baxter equation*, CWI, Amsterdam, 1996.

[674] Pepijn van der Laan, *Some Hopf algebras of trees*, Math. Inst., Univ. of Utrecht, 2002.

[675] Pepijn van der Laan, *Operads and the Hopf algebras of renormalization*, Math. Inst., Univ. of Utrecht, 2003.

[676] Pepijn van der Laan and Ieke Moerdijk, *Families of Hopf algebras of trees and pre-Lie algebras*, Math. Inst., Univ. of Utrecht, 2004.

[677] Fred van Oystaeyen and Pu Zhang, *Quiver Hopf algebras*, J. of Algebra **280** (2004), 277–289.

[678] Michela Varangolo and Eric Vasserot, *On the decomposition matrices of the quantum Schur algebra*, 1999; arXiv:math/9803023v3[math.QA].

[679] Joseph C. Varilly, *Quantum symmetry groups of noncommutative spheres*, (2001); arXiv: math/0102065v3 [math.QA].

[680] Joseph C. Varilly, *Hopf algebras in noncommutative geometry*, 2001; `arXiv:hep-th/0109077v2`.

[681] N. Ya. Vilenkin and A. U. Klimyk, *Representation of Lie groups and special functions. Vol.3: Classical and quantum groups and special functions*, KAP, 1992.

[682] S. Wang, *Quantum symmetry groups of finite spaces*, Comm. in math. Phys. **195** (1998), 195–211.

[683] Yanxin Wang and Jie Xiao, *The double Ringel-Hall algebras of valued quivers*, Chinese Annals of Math. **26B** (2006), 701–722.

[684] Sara Westreich and Tadashi Yanai, *More about a Galois type correspondence theory*, J. of Algebra **246** (2001), 629–640.

[685] Sara Westreich, *Hopf algebras of type A_n, twistings and the FRT construction*, Algebras and representation theory **11** (2008), 63–82.

[686] Hermann Weyl, *The classical groups*, Princeton Univ. Pr., 1946.

[687] Edward Witten, *Quantum field theory and the Jones polynomial*, in: C. N. Yang and M.-L. Ge, eds., *Braid group, knot theory and statistical mechanics*, World Scientific, Singapore, 1989, 239–329.

[688] S. L. Woronowicz, *Compact matrix pseudogroups*, Comm. in math. Physics **111** (1987), 613–665.

[689] S. L. Woronowicz, *Twisted SU(2) group. An example of noncommutative differential calculus*, Publ. Res. Inst. Math. Sci. **23** (1987), 117–181.

[690] S. L. Woronowicz, *Tannaka - Krein duality for compact matrix pseudogroupes*, Inv. math. **93** (1988).

[691] S. L. Woronowicz, *Differential calculus on compact matrix pseudogroups (quantum groups)*, Comm. in math. Physics **122** (1989), 125–170.

[692] S. L. Woronowicz, *Quantum E(2) group and its Pontryagin dual*, Lett. in math. Physics **23** (1991), 251–264.

[693] S.L. Woronowicz and S. Zakrezewski, *Quantum deformation of the Lorentz group. The Hopf *-algebra level*, Comp. math. **90** (1994), 211–243.

[694] S. L. Woronowicz, *Compact quantum groups*, in: A. Connes a.o., eds., Quantum symmetries. Proc. of the Les Houches summer school, session LXIV, 1995, North Holland, 1998, 845–884.

[695] Jie Xiao, *Drinfeld double and Ringel - Green theory of Hall algebras*, J. of Algebra **190** (1997), 100–144.

[696] Jie Xiao and Shilin Yang, *BGP reflection functors and Lusztig symmetries: a Ringel-Hall algebra approach to quantum groups*, J. of Algebra **341** (2001), 204–246.

[697] Tao Xie, *Hall polynomials for affine quivers of type $A_m(m \geq 1)$*, Acta math. Sinica **24** (2008), 387–396.

[698] Tadashi Yanai, *Galois correspondence theorem for Hopf algebra actions*, in J. de la Peña, ed., *Algebraic structures and their representations*, Amer. math. Soc., 2005, 393–411.

[699] C.-N. Yang, *Some exact results for the many body problem in one dimension with repulsive delta-function interaction*, Physics Res. Lett. **19** (1968), 1312–1314.

[700] Kenji Yogogawa, *Non-commutative Hopf Galois extensions*, Osaka J. of Math. **18** (1981), 63–73.

[701] Guanglian Zhang and Shuai Wang, *The Green formula and heredity of algebras*, Science in China Ser. A: Mathematics **48** (2005), 610–617.

[702] J. J. Zhang, *The quantum Cayley -Hamilton theorem*, J. pure and appl. Algebra **129** (2001), 101–109.

[703] Pu Zhang, *Composition algebras of affine type*, J. of Algebra **206** (1998), 505–540.

[704] Pu Zhang, *Representations as elements in affine composition algebras*, Trans Amer. math. Soc. **353** (2001), 1221–1249.

[705] Pu Zhang and Ying-Bo Zhang, *Minimal generators of Ringel-Hall algebras of affine quivers*, J. of Algebra **239** (2001), 675–704.

[706] Pu Zhang, *Generalized Green classes*, J. of Algebra **250** (2002), 709–730.

[707] Pu Zhang, *Skew differential operator algebras*, Adv. in Math. **183** (2004), 80–126.

[708] Pu Zhang, *Hopf algebras on Schurian quivers*, Comm. in Algebra **34** (2006), 4065–4082.

[709] R. B. Zhang, *The quantum super-Yangian and Casimir operators of $U_q(gl(M|N))$*, Lett. in math. Physics **33** (1995), 263–272.

[710] R. B. Zhang, *Representations of super Yangians*, J. of math. Physics **36** (1995), 3854–3865.

[711] R. B. Zhang, *The $gl(M|N)$ super Yangian and its finite dimensional representations*, Lett. in math. Physics **37** (1996), 419–434.

[712] Shouchuan Zhang and Xijing Guo, *Path Hopf algebras and co-path Hopf algebras*, Dept. Math., Hunan Univ, 2004; `arXiv:0410150v2`.

[713] Shouchuan Zhang and Yao-Zhong Zhang, *Structures and representations of generalized path algebras*, Algebras and representation theory **10** (2007), 117–134.

[714] Shouchuan Zhang, Yao-Zhong Zhang and Hui-Xiang Chen, *Classification of PM quiver Hopf algebras*, 2007; `arXiv:math/0410150 [math QA]`.

[715] Shouchuan Zhang and Hui-Xiang Chen, *Classification of quiver Hopf algebras and pointed Hopf algebras of Nichols type*, 2008; `arXiv:math/0802.3488v1 [math QA]`.

[716] Shouchuan Zhang, Hui-Xiang Chen and Yao-Zhong Zhang, *Classification of quiver Hopf algebras and pointed Hopf algebras of type one*, 2008; `arXiv:math/0802.3488v4 [math QA]`.

[717] Shunhua Zhang, *Lie algebras determined by cyclic quivers*, Southeast Asian Bull. Math. **27** (2003), 385–393.

[718] Wenhua Zhao, *A noncommutative symmetric system over the Grossman-Larson Hopf algebra of labeled rooted trees*, J. of algebraic Combinatorics **28** (2008), 235–260.

[719] Ionnis P. Zois, *Noncommutative versus gauge symmetry*, School of Math., Cardiff Univ., 2002.

[720] Bruno Zumino, *Differential calculus on quantum spaces and quantum groups*, Dept. Physics, Univ. of California at Berkeley, 1992.

[721] B. M. Zupnik, *Differential calculus on the quantum sphere and deformed self-duality equation*, Dubna, preprint E2-94-487, 1994.

Index